Animal Manure Recycling

Animal Manure Recycling

Treatment and Management

Edited by

SVEN G. SOMMER
Institute of Chemical Engineering, Biotechnology and Environmental Technology,
University of Southern Denmark, Denmark

MORTEN L. CHRISTENSEN
Department of Biotechnology, Chemistry and Environmental Engineering,
Aalborg University, Denmark

THOMAS SCHMIDT
Technology Transfer Office, Aarhus University, Denmark

LARS S. JENSEN
Department of Plant and Environmental Sciences, University of Copenhagen,
Denmark

Library of Congress Cataloging-in-Publication Data

Animal manure : recycling, treatment, and management / edited by Sven Gjedde Sommer, Lars Stoumann Jensen, Morten L. Christensen, Thomas Schmidt.
   pages cm
   Includes index.
   ISBN 978-1-118-48853-9 (cloth)
   1. Animal waste–Recycling. 2. Biomass energy. 3. Farm manure. 4. Manures. I. Sommer, Sven Gjedde, 1955– editor of compilation. II. Jensen, Lars Stoumann, editor of compilation. III. Christensen, Morten L., editor of compilation. IV. Schmidt, Thomas, editor of compilation.
   S655.A55 2013
   628′.7466–dc23

                                                                                                    2013015045

A catalogue record for this book is available from the British Library.

ISBN: 9781118488539

Typeset in 10/12pt Times by Aptara Inc., New Delhi, India

1   2013

Contents

7 Solid–Liquid Separation of Animal Slurry   **105**
Morten L. Christensen, Knud V. Christensen and Sven G. Sommer

8 Gaseous Emissions of Ammonia and Malodorous Gases   **131**
Sven G. Sommer and Anders Feilberg

List of Contributors

Morten Birkved, Department of Management Engineering, Technical University of Denmark, Denmark

Dionysis Bochtis, Department of Engineering, Aarhus University, Denmark

Sander Bruun, Department of Plant and Environmental Sciences, University of Copenhagen, Denmark

David Chadwick, School of Environment, Natural Resources & Geography, Bangor University, Environment Centre for Wales, UK

Knud V. Christensen, Institute of Chemical Engineering, Biotechnology and Environmental Technology, University of Southern Denmark, Denmark

Morten L. Christensen, Department of Biotechnology, Chemistry and Environmental Engineering, Aalborg University, Denmark

Tim J. Clough, Faculty of Agriculture and Life Sciences, Lincoln University, New Zealand

Anders Feilberg, Department of Engineering, Aarhus University, Denmark

Lars S. Jensen, Department of Plant and Environmental Sciences, University of Copenhagen, Denmark

James J. Leahy, Department of Chemical and Environmental Sciences, University of Limerick, Ireland

Teruo Matsunaka, Faculty of Dairy Science, Rakuno Gakuen University, Japan

Oene Oenema, Environmental Sciences, Wageningen University, Netherlands

Søren O. Petersen, Department of Agroecology, Aarhus University, Denmark

Alan Rotz, USDA-ARS Pasture Systems and Watershed Management Research Unit, USA

Thomas Schmidt, Technology Transfer Office, Aarhus University, Denmark

Sven G. Sommer, Institute of Chemical Engineering, Biotechnology and Environmental Technology, University of Southern Denmark, Denmark

Claus A.G. Sørensen, Department of Engineering, Aarhus University, Denmark

Peter Sørensen, Department of Agroecology, Aarhus University, Denmark

Marieke ten Hoeve, Department of Plant and Environmental Sciences, University of Copenhagen, Denmark

Björn Vinnerås, Department of Energy and Technology, Swedish University of Agricultural Sciences; National Veterinary Institute, Sweden

Alastair J. Ward, Department of Engineering, Aarhus University, Denmark

Preface

There are several reasons for studying environmentally friendly technologies for managing animal waste efficiently and in a sustainable manner.

For the engineer the challenges lie in the development of the technology and in helping to implement the technology on animal production farms. Since environmentally friendly technologies for managing biowaste are increasingly in demand on a rapidly growing global market, many engineers will find their future jobs in this field.

Expertise in animal production and its environmental impact is also greatly needed, because a rapidly changing and expanding livestock and poultry production sector is causing a range of environmental problems at the local (odour and ammonia emissions), regional (nitrate leaching to the aquatic environment) and global scale (greenhouse gas (GHG) emissions). Thus, for professionals, academics and students interested in contributing to the reduction of pollution caused by animal farming, this book will give an understanding of the mechanisms behind the pollution. This knowledge is required if one wishes to make a useful contribution to alleviating the problems caused by pollution from animal farming.

Thus, this book presents "state-of-the-art" knowledge and provides information necessary for graduate studies on animal biowaste engineering and management. The target groups for the book are professionals, consultants, academics, and MSc and PhD students, whose diverse backgrounds may be in engineering or natural sciences (agronomy, biology, microbiology, chemistry or the like).

Internationally, there is increasing focus on reducing pollution from animal production. The policy is to reduce pollution from animal waste management by regulations that support the use of environmental technologies to mitigate environmental problems in industrialised animal production.

Thus, there will be a need for specialists in environmental technologies and management of animal waste. Biowaste specialists will be needed in industry and consultancy work in the agricultural advisory service, in public service and in research and development.

The intention of this book is to introduce the engineers, consultants and academics to the biological, physical and chemical processes controlling pollution from animal production and the technologies needed to manage animal waste. A short introduction is given to the need for environmentally friendly technologies for treating and managing animal waste.

The book describes the physical and chemical characteristics of animal manure and microbial processes. Gaseous emissions of ammonia and GHGs are presented. Through the example of odour and ammonia emissions, interactions between meteorological physics, liquid chemistry, chemical processes, pH buffer systems and so on are introduced. Environmentally friendly technologies for reducing emissions of ammonia, odour and the GHGs nitrous oxide and methane are presented, and reduction of plant nutrient losses using separation technologies is introduced. Energy production in biogas plants, combustion of waste and the effect on GHG emissions are also covered.

The book introduces the reader to the sustainable use of animal manure for crop fertilisation and for soil amelioration. It presents management strategies for efficient recycling of manure as a means of reducing leaching and runoff loss of plant nutrients. Finally, and most importantly, it describes methods to commercialise and transfer knowledge about technologies to end-users.

A readers guide to the different elements in the chapters:

Text Box – Basic: These provide terminology definitions and fundamental knowledge, which is assumed as prerequisite to understand the main text. Engineers would probably need to read Text Box – Basics on agronomic topics and Agronomists the Text Box – Basics on technical topics.

Text Box – Advanced: These provide deeper insight and more advanced knowledge on specific topics, which may be read by those interested, but the information Text Box – Advanced will not be a prerequisite for understanding the main text, and can thus be skipped for those less interested.

Examples: These are typically provided to illustrate the principles of more advanced chemical reactions or calculation models. They may be utilized as models for solving problems.

Acknowledgements

The editing of this book was supported by grants from Stiftelsen Hofmansgave and The Danish Industry Foundation. For the editors, it was very encouraging that a farmers' foundation and an industrial fund both decided to support this project, because it indicated that users of this book will be found both within the agricultural sector, which will use the environmentally friendly technology, and in the industrial sector, which will manufacture the technology.

1

Animal Manure – From Waste to Raw Materials and Goods

Sven G. Sommer

Institute of Chemical Engineering, Biotechnology and Environmental Technology, University of Southern Denmark, Denmark

Societies will inevitably have to recognise that animal excreta are not just a waste material requiring disposal, but a crucial raw material needed to boost plant production to feed a growing world population. If used appropriately, animal excreta can replace significant amounts of mineral fertilisers in areas with livestock production. Manure comprises animal excreta dissolved in water or mixed with straw, a substance made up of organic matter and used as an organic fertiliser in agriculture, where it contributes to the fertility of the soil by adding plant nutrients and organic matter (Figure 1.1). In the management chain before it is applied to soil, manure can also be used for energy production.

The increasing focus on developing and using new technologies and management methods for manure handling is the consequence of both a huge increase in livestock production worldwide and specialisation in agriculture. Thus, in new production systems, traditional farms with a mixture of livestock and crop production are often replaced with landless livestock production units. These new livestock production systems may not have the capacity to recycle manure on-farm, which was a feature of many farming system in the past.

The plant nutrients in manure can, if used appropriately, replace significant amounts of mineral fertilisers, and the organic matter can boost soil fertility (Text Box – Basic 1.1) and can be used for energy production. On the other hand, improper management and utilisation of manure results in loss of plant nutrients (Bouwman *et al.*, 2012)(Figure 1.2), which are a limited resource, and this can be a risk to the global feed and food supply. For example, phosphorus (P) is a limited resource, with the mineable phosphate-rich rocks used for P fertiliser production projected to become exhausted within the next 60–130 years (Figure 1.3). In a 14-month period during 2007–2008, the global food crisis led to phosphate rock and fertiliser demand exceeding supply and prices increased by 700% (Cordell *et al.*, 2009). This increase in cost may be mitigated by reducing P

Animal Manure Recycling: Treatment and Management, First Edition. Edited by Sven G. Sommer, Morten L. Christensen, Thomas Schmidt and Lars S. Jensen.
© 2013 John Wiley & Sons, Ltd. Published 2013 by John Wiley & Sons, Ltd.

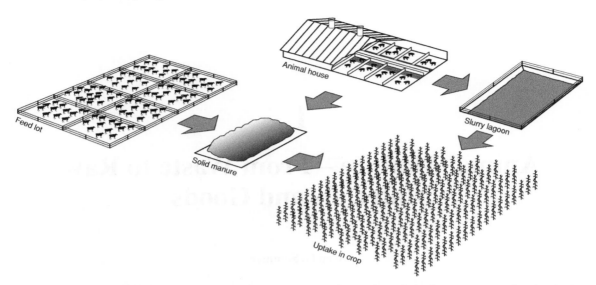

Figure 1.1   *Animal manure management (bold arrows) is a chain of interlinked operations and technologies, of which the major steps are collection of excreta in animal houses or beef feedlots, storage of manure in-house and/or outside, treatment of the manure (not shown), transport to fields and spreading in the fields. At each stage there is a risk of emission of components, which represents a loss to the farmer and a risk to the environment. (© University of Southern Denmark.)*

losses. It is estimated that close to 25% of the 1 billion tonnes of P mined since 1950 has ended up in water bodies or is buried in landfill (Rosmarin, 2004).

Text Box – Basic 1.1   Soil and environmental terminologies

Soil fertility: The ability of soil to provide plants with sufficient, balanced and non-toxic amounts of nutrients and water, and to act as a suitable medium for root development, in order to assure proper plant growth and maturity. Soil fertility is basically controlled by the inherent mineralogy and soil texture as determined by location and geology, and by the dynamic parameters of soil organic matter content, acidity, nutrient concentration, porosity and water availability, all of which can be influenced by human activity and management.

Soil organic matter (SOM): The total organic matter in soil, except for materials identifiable as undecomposed or partially decomposed biomass, is called humus and is the solid, dark-coloured component of soil. It plays a significant role in soil fertility and is formed by microbial decay of added organic matter (e.g. plant residues and manur) and polymerisation of the cycled organic compounds. Carbon content in soil organic matter ranges from 48% to 57%.

Eutrophication: An increase in the concentration of chemical nutrients in terrestrial and aquatic ecosystems to the extent that it increases the primary productivity of the ecosystem. Subsequent negative environmental effects in watercourses, such as anoxia and severe reductions in water quality, fish stocks and other animal populations, may occur. On land, the negative effect is seen as a change in the existing plant community composition, which becomes dominated by species that prefer a high plant nutrient level. As a consequence, the enrichment in plant nutrient content is associated with a decline in biodiversity.

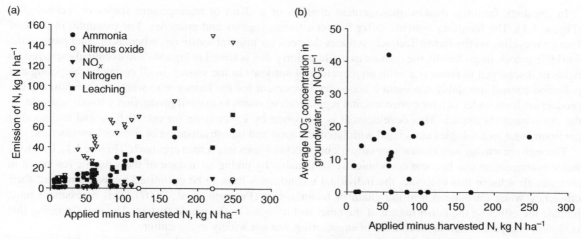

Figure 1.2   (a) Nitrogen emissions related to surplus N application to agricultural land, here calculated as N added to agricultural land in fertilisers and animal manure minus uptake by plants. (b) Nitrate concentration in water boreholes related to N surplus. (Data taken with permission from Oenema et al. (2007). © 2007 Elsevier.)

In the development of new technologies and management practices for improving the quality of the livestock product and for reducing production costs, the management of externalities, which in this case is manure, is often unchanged. This tendency is because the producers and experts who develop the new livestock production system often overlook the fact that the existing management of manure needs to be adapted to new livestock production systems. In livestock production this is reflected in a surplus of plant nutrients in regions where livestock production has increased. Thus, plant nutrient surpluses have been documented in regions in America, Europe and Asia. In Asia, such surpluses are commonly centred around cities (Gerber *et al.*, 2005), because of consumer demand for meat to be slaughtered immediately before sale, and in these countries living animals are not transported long distances. Increasing livestock densities (livestock units ha^{-1}) will lead to surplus plant nutrients as documented for nitrogen (N) on livestock farms in Europe (Olesen *et al.*, 2006) and these surpluses may end up polluting the environment (i.e. eutrophication of ecosystems) (Figure 1.2).

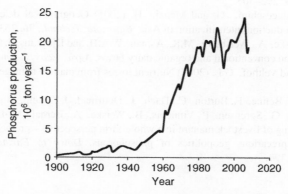

Figure 1.3   Global production of mined P. (Adapted with permission from Cordell et al. (2009). © 2009 Elsevier.)

In livestock farming, manure management consists of a chain of management stages or technologies (Figure 1.1). The handling systems differ between farms, regions and countries. For example, in parts of Europe recycling on the farm effectively reduces the need for mineral fertilisers, whereas in other parts of the world, livestock farms handle the manure as a dilute slurry that is stored in lagoons and eventually sprayed on fields or discharged to rivers (i.e. with no recycling of nutrients in the waste). In all countries, recycling and pollution control inevitably represent a necessary investment for the farmer who wants to maintain a given production level under stricter environmental regulations or wants to expand production without aggravating the environmental impact. This development is supported by lower costs for establishing and maintaining environmental technologies associated with intensification and industrialisation of livestock production.

Through optimising new environmentally friendly technologies in a "chain approach" (Figure 1.1), livestock waste management can become economically sustainable by taking advantage of the valuable resources in manure. To achieve this outcome, the individual technologies have to be optimised by assessment of their efficiency when introduced into the chain of technologies (Petersen *et al.*, 2007). This assessment must include the effect on the performance of the other technologies in the whole system. The tool for doing this is system analysis, which is much used in engineering, but not widely in agriculture.

This leads us back in time to the late nineteenth century, when researchers at experimental stations at Rothamsted in England and Askov in Denmark carried out field studies comparing manure and fertiliser efficiency to convince farmers that mineral fertiliser was useful and could increase plant production at a low cost. Today, mineral fertilisers are costly, because it takes much energy to produce them and because the sources are approaching exhaustion. As a consequence, there is a burgeoning need for technologies and management practices to use animal manure as a valuable nutrient source for the production of crops and food, as well as for energy production. Collaboration between different types of professionals (e.g. engineers, agronomists and natural scientists) on the development of manure management and utilisation technologies is therefore necessary and requires a mutual insight and understanding of processes, technologies and management. This book is written to facilitate such collaboration.

References

Bouwman, L., Goldewijk, K.K., Van Der Hoek, K.W., Beusen, A.H.W., Van Vuuren, D.P., Willems, J., Rufino, M.C. and Stehfest, E. (2012) Exploring global changes in nitrogen and phosphorus cycles in agriculture induced by livestock production over the 1900–2050 period. *Proc. Natl. Acad. Sci. USA*, Early Edition, doi: 10.1073/pnas.1012878108.

Cordell, D., Drangert, J.-O. and White, S. (2009) The story of phosphorus: global food security and food for thought. *Global Environ. Change*, **19**, 292–305.

Gerber, P., Chilonda, P., Franceschini, G. and Menzi, H. (2005) Geographical determinants and environmental implications of livestock production intensification in Asia. *Bioresour. Technol.*, **96**, 263–276.

Olesen, J.E., Schelde, K., Weiske, A., Weisbjerg, M.R., Asman, W.A.H. and Djurhuus, J. (2006) Modelling greenhouse gas emissions from European conventional and organic dairy farms. *Agric. Ecosyst. Environ.*, **112**, 207–220.

Oenema, O., Oudendag, D. and Velthof, G.L. (2007) Nutrient losses from manure management in the European Union. *Livest. Sci.*, **112**, 261–272.

Petersen, S.O., Sommer, S.G., Béline, F., Burton, C., Dach, J., Dourmad, J.Y., Leip, A., Misselbrook, T., Nicholson, F., Poulsen, H.D., Provolo, G., Sørensen, P., Vinnerås, B., Weiske, A., Bernal, M.-P., Böhm, R., Juhász, C. and Mihelic, R. (2007) Recycling of livestock manure in a whole-farm perspective – preface. *Livest. Sci.*, **112**, 180–191.

Rosmarin, A. (2004) The precarious geopolitics of phosphorous. *Down to Earth*, **2004**, 27–31; http://www.downtoearth.org.in/node/11390.

2

Animal Production and Animal Manure Management

Sven G. Sommer[1] and Morten L. Christensen[2]

[1]*Institute of Chemical Engineering, Biotechnology and Environmental Technology,*
University of Southern Denmark, Denmark
[2]*Department of Biotechnology, Chemistry and Environmental Engineering,*
Aalborg University, Denmark

2.1 Introduction

Globally, livestock production varies from systems where the animals move freely to find their feed to systems where the animals are housed for periods of up to an entire year (Figure 2.1). In some countries livestock waste or manure may be collected in the form of slurry and used to fertilise fields, while in others it is collected manually in the form of solid excreta, which is used for incineration.

Livestock production systems can broadly be classified into: (i) grazing systems, (ii) mixed systems and (iii) landless or industrial systems. Grazing systems are entirely land-based systems, with stocking rates less than one livestock unit per hectare (Text Box – Basic 2.1). In mixed systems a part of the value of production comes from activities other than animal production, while part of the animal feed is often imported. Industrial systems have stocking rates greater than 10 livestock units ha^{-1}, and they depend primarily on outside supplies of feed, energy and other inputs.

Manure is collected on farms where the animals are kept in confined environments, which may be fenced beef feedlots or livestock houses with and without outdoor exercise areas. Housing has been developed to give shelter and provide a comfortable and dry environment for animals, with the purpose of increasing production and facilitating feeding. In some dry climates, such as the North American prairies, there is less need for

Animal Manure Recycling: Treatment and Management, First Edition. Edited by Sven G. Sommer, Morten L. Christensen,
Thomas Schmidt and Lars S. Jensen.

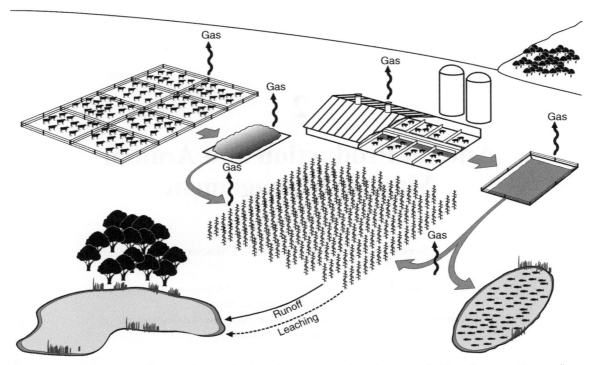

Figure 2.1 *Manure management from collection of manure in animal houses to field application. Manure flows are indicated by arrows. (© University of Southern Denmark.)*

shelter and both dairy cows and calves for beef production are raised in open feedlots, even at temperatures below −20 °C.

Text Box – Basic 2.1 Livestock production systems

The Food and Agriculture Organization of the United Nations (FAO) definition of a livestock system is that more than 90% of the feed to livestock originates from rangeland, pasture, annual forage and purchased feed, and less than 10% of the total value of production comes from non-livestock farming activities (Seré and Steinfeld, 1996). Non-livestock activities are households having a few animals fed on organic waste, cuttings from road verges and so on. The FAO defines three categories: grazing, mixed and industrial systems.

Grazing systems: The livestock are fed according to the definition of livestock production systems mentioned above. The important definition of the grazing system is that the livestock travel to find feed (mobile), depend on local communal pasture (sedentary) or have access to sufficient feed within the boundaries of the farm (ranching and grassland). Annual stocking rates are less than 10 livestock units ha⁻¹ agricultural land. Grazing production systems can be found in arid, semi-arid, sub-humid and humid regions, and in temperate and tropical highlands. In terms of total production, grazing systems supply only 9% of global meat production.

Mixed systems: Mixed systems are defined as farms where: (i) more than 10% of the dry matter fed to livestock comes from crop byproducts (e.g. feed from industrial processing of food) and/or stubble, or (ii) more than 10% of the value of production comes from non-livestock farming activities. Thus, the feed for the livestock in these systems comes from communal grazing, crop residues and crops, cut-and-carry processes, on-farm production, and external feed. Globally, mixed farming systems produce the largest share of total meat (54%) and milk (90%), and mixed farming is the main system for smallholder farmers in many developing countries. In Europe, farmers must have access to more than approximately 1.5-ha fields per livestock unit because this ensures that there is a sufficient cropped area that can utilise the plant nutrients produced in manure. Most of the feed for the livestock come from these fields and the production therefore falls under the mixed category of livestock production systems.

Industrial systems: Industrial systems have average stocking rates greater than 10 livestock units ha^{-1} of agricultural land and less than 10% of the dry matter fed to livestock is produced on the farm. The production systems in focus are poultry production (broilers and layers), pig production, ruminant feedlot meat production and large-scale dairy production. The industrial livestock production systems depend on outside supplies of feed, energy and other inputs, and the demand for these inputs can thus have effects on the environment in regions other than those where production occurs. Industrial systems provide more than 50% of global pork and poultry meat production, and 10% of beef and mutton production. Examples of these production systems are landless pig and poultry farms. In North America, pig farms with slurry spraying on fields of a few hectares, which receive manure from animal houses supporting production of several thousand pigs, fall into this category, as do beef and dairy cattle feedlots that accommodate up to 100 000 head of livestock.

The manure is spread on fields to fertilise crops, is used to fertilise algae and water plants in fish ponds (the plants being eaten by herbivorous fish) or is used for energy production. The system used for manure collection, storage and end-use depends on climate, tradition and production system.

2.2 Housing, Feedlots and Exercise Areas

In cold and wet climate zones, the animal houses or barns provide a warm and dry indoor environment, whereas in the tropics the objective is to provide a cool and dry environment. Housing design is related not only to the climate, but also to the animal category being housed and the objective of the production.

2.2.1 Cattle

Cattle are divided into the categories of calves, heifers, bulls and cows (Text Box – Basic 2.2). These categories relate to the age of the animal, the gender and their part in production on the farm.

Most cattle buildings are naturally ventilated with air flowing through openings in the walls or through open gates. In warm climates the ventilation can be forced with fans, creating an open air flow, or with closed tunnel ventilation systems where ventilator fans are positioned at the gable end of the house. These create an air flow through the length of the cattle house, with air coming in from openings in the other gable end.

Text Box – Basic 2.2   Cattle categories (Pain and Menzi, 2003)

Cow: Bovine female bearing her second calf, thus after giving birth to a first calf, a heifer becomes a cow.
Calf: The offspring of a cow.
Heifer calf: A female calf.
In-calf heifer: A pregnant heifer.
Dairy cow: Animal bred for producing milk and for rearing calves for a dairy herd. One should bear in
   mind that a cow has to rear calves in order to produce milk.
Bull: Bovine male.
Bull calf: A male calf.
Beef cattle: Cattle kept for slaughtering at 450–550 kg live weight, which may be at an age of 13–16
   months for intensive feeding or 17–30 months for grazed animals.
Steer or *bullock:* A castrated bull.

The most commonly used house design for cattle is the *loose housing system*, where the animals are free
to walk around in the house. In these houses the manure management may be based on slurry (i.e. mixed
excreta) collected below the slatted floor. A slatted floor is in most cases constructed of concrete surfaces
with 1- to 1.5-m long and a few-centimetre-wide openings or slots between the slats (beams). The excreta,
urine and spilt drinking water fall through the openings and are collected as *slurry* in the pit or channel below
the slatted floor. In houses where the entire floor is slatted, the floor is defined as *fully slatted*. A *partly slatted
floor* is found in houses where the slatted floor is restricted to the walking alleys (Figure 2.2). In these houses
the building is divided into rows of individual stalls or *cubicles* in which animals lie when at rest, but are
not restrained. The cubicles have a *solid floor*, which is constructed of a hard, impermeable material such as
concrete. The solid floor may be strewn with straw, sawdust, wood shavings, sand or peat. The walking alleys
may also be solid floors of concrete, asphalted concrete or concrete covered with rubber. Walking alleys with
a solid floor are cleaned at least once per day (e.g. by a tractor-mounted scraper or more frequently by an
automatic scraper) (Monteny and Erisman, 1998).

In *tied housing* systems, dairy cows are tied by the neck to a fence at the feeding trough with ropes, chains
or bars and have restricted freedom to move in their living area, which is typically a concrete floor covered
with bedding material that may be straw, sawdust or sand. At the rear of the animal there may be a channel
covered by a metal grid or a concrete slatted floor. In this channel the faeces and urine together with bedding
material are collected as *slurry*. In other systems the faeces, urine and litter are collected in a 5- to 10-cm
deep gutter. The faeces and litter are scraped out of the gutter as *solid manure or farmyard manure* and the
urine is drained by gravity to a *liquid manure* store.

Calves for beef production are often housed in animal buildings with a solid floor covered with bedding material, on which urine and excreta are deposited (Figure 2.3). Such systems are gaining increasing
importance in Europe for larger cattle (heifers, beef cattle, suckling cows) for animal welfare reasons. In the
Scandinavian countries specifically, housing systems with both solid manure and liquid manure storage are
disappearing for reasons of animal welfare. The solid floor houses are often constructed with the living area
below ground, because a thick layer of litter is placed on the floors on which the animals walk. In the start of
a production period the cattle walk below surface level on a litter layer of 20–40 cm, but more litter is added
weekly and with time the animals end up walking on a thick layer of deep litter, which is compacted by their
hooves. The deep litter provides comfort for the animals and absorbs moisture. The material used are straw,
chopped straw, sawdust, wood shavings and peat.

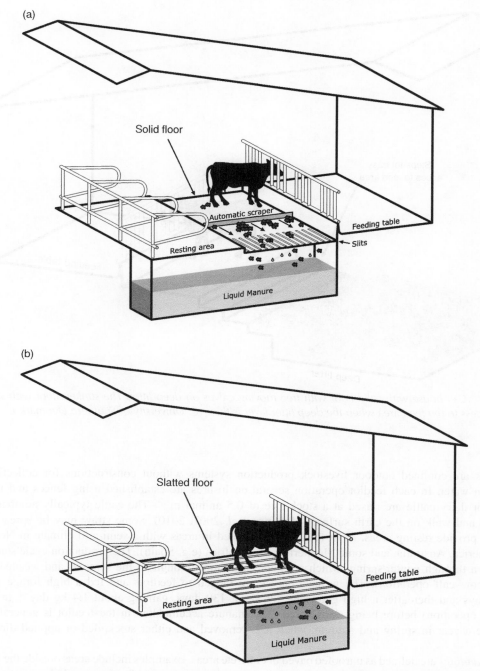

Figure 2.2 *(a) A dairy cow house with solid floors in resting areas, walking and excretion alleys and the feeding area, where the faeces/urine mixture is scraped to the channel at the gable to the right. (b) Heifer house with a solid floor in the resting area and a slatted floor in the exercise alley and feed area. (© University of Southern Denmark.)*

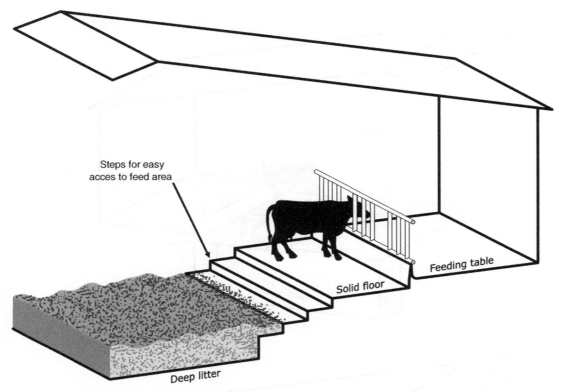

Steps for easy
acces to feed area

Feeding table

Solid floor

Deep litter

Figure 2.3   *Calf house with solid floor with free moving calves on deep litter. The sunken floor with steps for ease of access to the feed area when the deep litter layer is thin. (© University of Southern Denmark.)*

Feedlots are confined outdoor livestock production systems without constructions for collecting liquid or rain water. In each feedlot operation several enclosures are established using fences and in these the beef or dairy cattle are raised at a stock rate of 0.5 animal m^{-2}. The cattle typically number in the thousands and walk on the earth surface (McGinn *et al.* 2007, 2010). Some straw may be spread in the feedlot to provide resting areas. Most feedlots are situated in areas with a semi-arid climate in North and South America, Australia, and some Mediterranean countries (e.g. Spain). The production cycle starts with calves born through winter/spring, which are kept on pasture throughout the summer and weaned in late autumn (cow–calf operation). The weaned calves are moved to feedlots and fed a high forage diet for 80–100 days and thereafter a high grain diet for about 130 days, gaining about 1.4 kg day^{-1} in weight (finishing operation) before being slaughtered. The manure accumulated in the feedlot is generally handled twice a year in spring and autumn, when it is removed and either stockpiled or applied directly to the field.

Hard standings are defined as unroofed paved or concrete areas. Examples include areas outside the milking parlour, where dairy cows congregate prior to milking, or an exercise yard for dairy cattle kept in tied stalls, as is required in some countries for animal welfare reasons. Urine and faeces deposited on the hard standing are typically cleaned off by scraping (hand-held or tractor-mounted) and, less commonly, yards may be washed.

2.2.2   Pigs

The main pig categories on a pig farm are sows, weaners, piglets and fattening pigs (Text Box – Basic 2.3), each representing a phase of the production cycle.

Text Box – Basic 2.3   Pig categories (Pain and Menzi, 2003)

Sow: An adult female pig after having produced her first litter of piglets.
Gestating sow: A pregnant sow.
Farrowing sow: A sow giving birth (parturition) to piglets.
Litter of piglets: A group of piglets farrowed by a sow.
Suckling piglets: Young pigs still nursing the sow.
Weaners: Suckling piglets after being removed from the sow (weaned). The weaners are removed from the sow's milk at 3–6 weeks and are termed weaners until the age of 10 weeks (25–30 kg live weight).
Fattener: Definition of a pig after 10 weeks (25–30 kg), where they grow to become a fattener. The pig is a fattening pig from 25–30 to 100–120 kg weight or in some countries even higher end weight.
Grower pigs: A fattening pig category from 25–30 kg to about 60 kg.
Finishers: A fattening pig category between about 60 kg and slaughter.
Boar: An adult male pig.
Hog: A castrated male pig. Boar meat has an unpleasant flavour and boars are therefore castrated.

The basic unit of a pig house is the pen (Figure 2.4). A pen is a confined area of the house fenced with bars or walls about 1.2–1.5 m high. Loose housing is standard with the exception of housing for sows. Systems where the sows are confined (i.e. not loose) will most likely be abolished or legislatively banned for animal welfare reasons in many countries. Inside a fattening pig house, the room is divided into rows of pens with feeding alleys in between. There is a *batch* of weaners and fattening pigs in each pen, with each batch containing between 100 and 150 pigs.

Pig houses often have forced or mechanical ventilation systems. In temperate regions the air flow is generated by a series of ventilation chimneys along the length of the roof. Mechanical ventilators in the chimneys provide forced air flow where the air enters the house through openings or windows along the long side of the building, or the air may enter through diffuse air inlet or openings in the ceiling (perforated steel plates or wood wool cement boards). The objective is to avoid draughts in the house. In tropical and subtropical climates, tunnel ventilation is used to cool the building. Air is forced out of the building with large gable fans and air is taken in through a controlled tunnel opening at the opposite gable end of the building, with or without a water cooling system. In the tropics, pigs are often cooled by sprinkling a fog of droplets that evaporate, thus no extra water is added to the slurry. Alternatively, the water is sprinkled on the roof of the pig house.

Farrowing sows and sows raising piglets are often housed in combined resting and feeding boxes to reduce the risk of crushing the piglets. Sows with a litter of piglets may be aggressive and may also be affected by aggression from other pigs. However, pigs are generally social animals, so for welfare reasons loose housing systems have been developed where the sows are housed with other pigs. The faeces and urine are typically collected as slurry below a slatted floor at the rear end of sows housed in boxes.

Weaners and fatteners are kept in open pens. The whole floor area in these pens may be slatted (e.g. with openings) and then the floor is *fully slatted*. In contrast, a floor is *partly slatted* when about one-third of the

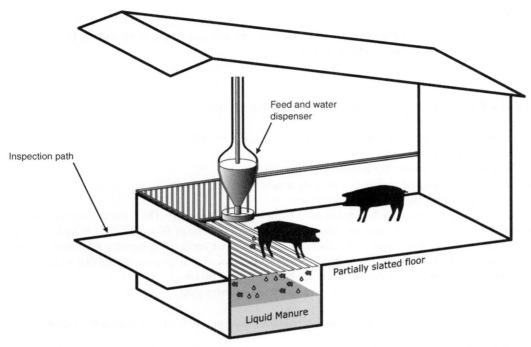

Figure 2.4   *Fattening pigs in loose housing with partly slatted floor where the pig pens have a slatted floor at the back and solid floor at the front. (© University of Southern Denmark.)*

floor has slats and two-thirds is a *solid floor*. The drinking and feeding area may be in the part of the pen with a solid floor, which is also the resting area, and the slatted floor is where the pigs defecate and urinate.

Pig houses may be constructed using deep litter systems, where faeces and urine are deposited in the organic material covering the floor in a similar way to systems seen in beef cattle housing systems. The difference is the behaviour of the pigs, which tend to excrete in a part of the pen and dig and build nests in other parts of the deep litter.

In Asian countries where water is plentiful, shallow water basins are constructed in the rear of the pig pens. The pig uses these for cooling and for excretion. The construction is well adapted to the behaviour of the pigs, which in nature will cool in ponds. From these water basins the liquid is pumped or flows by gravity to lagoons. The liquid manure produced is very dilute and difficult to manage.

2.2.3   Poultry

Houses for intensive broiler production (Text Box – Basic 2.4) are usually simple closed buildings with artificial light and forced ventilated (gable end or chimney ventilation). In warm climates broiler houses may be constructed with open side walls covered with mesh screens and located so that they are exposed to a natural stream of air. Additional ventilation fans may be fitted for use during hot weather. The birds are kept on litter (e.g. chopped straw, wood shavings or shredded paper) spread over the entire floor area. Manure, which is dry poultry litter, is removed at the end of each growing period.

Text Box – Basic 2.4　Poultry categories (Pain and Menzi, 2003)

Poultry: Domesticated birds kept for meat or egg production, the term includes domestic fowl, turkeys, geese and ducks.
Laying hens or *layers*: Chickens kept for table egg production.
Chick: The immature offspring of domesticated birds.
Poult: A chicken less than 8 weeks old. Male chickens are named cockerels.
Pullet: A female chicken in its first egg-laying year between 20 weeks and 18 months old.
Broiler: A chicken reared for meat production, the production period is 5–6 weeks.
Cockerels: A male chicken usually less than 18 weeks old.
Turkey: A large species of poultry kept for meat production.
Duck: Usually denotes a female duck or ducks in general, irrespective of the sex.
Drake: A male duck.
Duckling: A young duck, usually less than 8 weeks old.

Laying hens may be housed in deep litter houses, which are closed insulated buildings with forced ventilation or natural ventilation (Figure 2.5). At least one-third of the floor area must be covered with bedding material and two-thirds arranged as a pit covered with slats to collect droppings. Laying nests, feeders and water supply are placed over the slatted area to keep the litter dry. Below the slatted floor, droppings are collected in a water-tight pit.

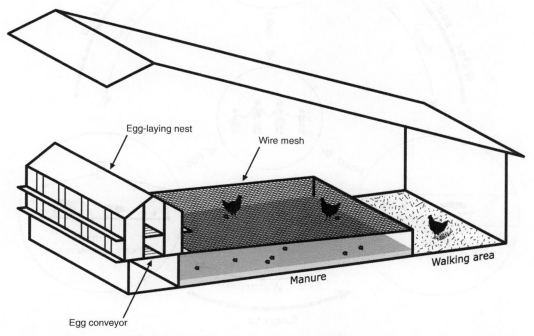

Figure 2.5　*Egg production in nests with a net floor, below which droppings are collected in a pit. (© University of Southern Denmark.)*

In battery cage houses, laying hens are kept in tiered cages, usually made of steel wire, arranged in long rows. Droppings fall through the bottom of the cages and are collected and stored underneath in a deep pit or are removed by a transport belt or scraper system.

2.2.4   Integrated Production Systems

The recycling of animal excreta to fish ponds is an integral part of Asian farming systems, which include crop production, gardening, fish farming and animal rearing (Figure 2.6). While gardening, fish farming and animal rearing provide the main products for family consumption or for sale, the byproducts from one subsystem are used as inputs to the others, reducing the need for external chemicals and minimising pollution. The manure is used directly as feed for the fish or indirectly as feed for phytoplankton, zooplankton and zoobenthos,

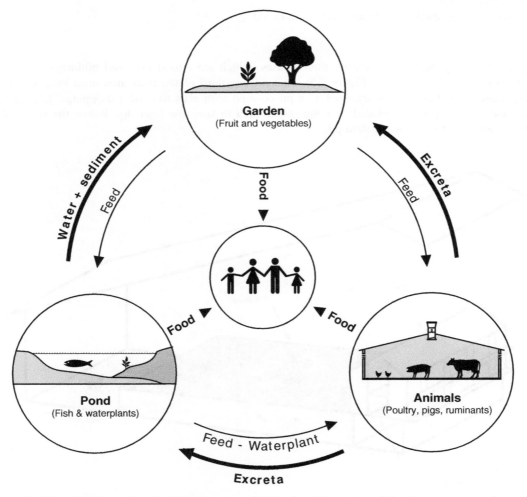

Figure 2.6   *The garden–pond–animal (VAC) system (VAC is the abbreviation of the Vietnamese words* vuon, ao, chuong, *which mean garden, pond, livestock pen). (© University of Southern Denmark.)*

Table 2.1 *Amount of plant nutrients, carbon and water in excreta and straw in manure production in the animal house and in manure transported to the field from dairy cow deep litter management and pig slurry management: one dairy cow produces 15 tons deep litter year^{-1} and one pig place in the pig house (producing three fattening pigs year^{-1}) produces 1.5 tons slurry year^{-1}.*

		Dry matter (kg)	Total-P (kg P)	Total-N (kg N)	TAN (kg N)	K (kg K)	C (kg)	Water (kg)
Dairy cow deep	in house	4500	22.5	126	30	165	1764	10500
litter	to the field	2408	22.5	79	14	165	739	7350
Pig slurry	in house	90	2.7	10.1	7	5.5	36	1410
	to the field	83	2.7	7.8	4.6	5.5	33	1850

It is assumed that in production of deep litter, all N in urine is transformed to TAN ($NH_3 + NH_4^+$).

which are then used to feed herbivorous fish (Vu *et al.*, 2007). An example is systems with pigs housed in pens with a solid floor, where the faeces are scraped off the floor and the urine and water is channelled into fishponds. The solid manure may be added to the ponds without treatment, or composted before being used in fishponds or in cash crop production.

2.3   Management of Manure

Manure may be managed as solid manure or a liquid. Cattle production systems are a major producer of solid manure and excreta from fattening pigs are often collected as slurry. Knowledge about the flow of plant nutrients and dry matter in these systems is used by the farmer when deciding how much plant nutrient is available to fertilise crops and by the authorities to assess the risk of environmental pollution. In slurry management systems where too much dry matter or plant nutrients are produced, the farmers may desludge or separate the slurry. Separators produce a low-nutrient liquid fraction and a solid fraction high in plant nutrients and dry matter.

2.3.1   Deep Litter Management

The input used in the calculation of the nutrient and carbon flow in this system is the deep litter composition at the time when straw, faeces and urine are being mixed (Table 2.1). The manure is produced in the dairy cattle house, removed at 6-month intervals, then stored in a solid manure heap for at least 3 months before being spread on the field (Figure 2.7). It is assumed that straw strewing rates are so high that no liquid drains off from the deep litter in the animal house or during storage outside. Consequently, the only loss pathway

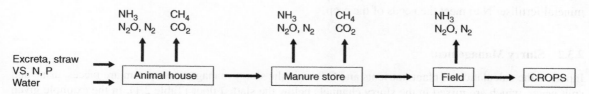

Figure 2.7 *Flow diagram of N and C losses during management of deep litter. (© University of Southern Denmark.)*

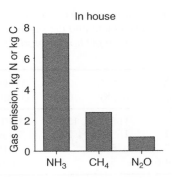 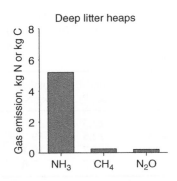 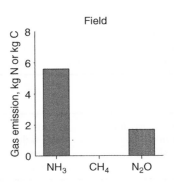

Figure 2.8   *Ammonia, CH_4 and N_2O emissions from deep litter management systems, based on the deep litter collected from one dairy cow production place (Table 2.1). (© University of Southern Denmark.)*

is emission of carbon (C) and nitrogen (N)-containing gases into the atmosphere. Nitrogen is lost due to emission of ammonia (NH_3), as dinitrogen (N_2) or nitrous oxide (N_2O), and carbon is lost as carbon dioxide (CO_2) or methane (CH_4).

Microorganisms transform N between organic and inorganic pools in deep litter, and in the calculations it is assumed the 30% of the organic N is transformed to TAN ($NH_3 + NH_4^+$), which is then oxidised to nitrate (NO_3^-). A significant part of the nitrate is reduced to nitrogen gas (N_2) and some is emitted as N_2O. The farmer must know how much manure has to be transported from the manure heap to a specific field, and also the concentration and amount of plant nutrients in the manure. Therefore, the amount of manure has to be calculated by assessing losses of dry matter and water, and the flow calculations must estimate N, phosphorus (P) and potassium (K) in manure spread on fields. In addition, calculations may provide an estimate of the pollution risks due to the acid rain pollutant NH_3 and the greenhouse gases N_2O and CH_4.

In the management chain for deep litter, most NH_3 is emitted from the animal house (Figure 2.8). Losses from the house are 7.6 kg NH_3-N, out of a total of 18.4 kg NH_3-N lost. In total, 2.8 kg CH_4-C are emitted (0.2% of C), most from in-house storage of the deep litter and some from heaps. CH_4 emissions from the field are insignificant. Nitrous oxide emissions are significant from field-applied manure and total N_2O emissions are 2.1 kg N_2O-N (Webb *et al.*, 2012).

There is a significant loss of carbon and water from the house and the heap. The dry matter in the manure is reduced by 45% due to production and emission of CO_2, and up to 30% of water can be lost due to evaporation from the heap, which is warm due to composting. Much N is lost during storage, mainly due to N_2 production (accounting for 26% of the N), while 18% of N is lost due to NH_3 emissions from deep litter in the animal house, during storage and after application in the field. No potassium (K) or phosphorus (P) is lost (Table 2.1), and the concentration expressed in grams per gram of dry matter is almost doubled due to the significant loss of dry matter. Thus, storage reduces the amount of deep litter the farmer needs to apply to the field to provide sufficient P and K to the crop. Due to loss of TAN, the farmer will probably have to add mineral fertiliser N to meet the needs of the crop.

2.3.2   Slurry Management

Inputs to the calculation of plant nutrients and C flow in the slurry management system are faeces, urine and spilt water, which are mixed in the slurry channels below the slatted floor (Table 2.1). In the example given here the slurry is produced in a fattening pig house, removed at 15-day intervals, then stored in a slurry store (4 m deep) for at least 6 months before being spread on the field (Figure 2.9). It is assumed that the store in

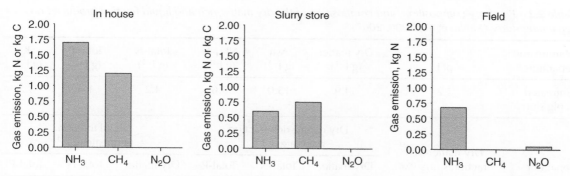

Figure 2.9 *Gaseous emissions from liquid pig manure collected from one pig production place in which three fattening pigs are produced yearly (Table 2.1). (© University of Southern Denmark.)*

the house and outside store are impermeable to liquid, and the slurry is managed so that no liquid is lost in the animal house or during storage outside. For this system the only loss pathway is emissions of N and C gases to the atmosphere in the form of NH_3, N_2, N_2O, CO_2 and CH_4.

In an anaerobic environment the transformation of N between organic and inorganic pools is assumed to be relatively low and is not included in the calculations. No P and K are lost, and all of these plant nutrients collected are spread on fields.

During storage in the house, N may be lost due to NH_3 emissions, which account for 30% of the total N collected. More NH_3 is lost in this system compared with the deep litter system, because much N is in form of TAN. In contrast, N losses due to denitrification are negligible during storage because the slurry is managed anaerobically and no NO_3^- is formed. As a consequence of the storage conditions, N_2O emissions are insignificant from stored slurry, and it is only from field-applied slurry that N_2O is emitted. Methane emissions are significant and account for 2% of the carbon in the manure.

In this system more of the excreted N is applied to the field than in the deep litter system and a greater fraction of the N is in the form of inorganic TAN readily available for plant uptake.

2.3.3 Separation of Slurry

Traditionally, simple procedures have been used to manage manure. However, global livestock production is increasing due to the increasing demand for meat, and production costs are being reduced by intensifying and specialising livestock production, which is increasingly taking place on large farms. Thus, there is a need for new methods for managing manure e.g. solid–liquid separation has been used to reduce the water content of a dry matter-rich fraction of manure with a high content of plant nutrients. This lowers the cost of transportation of nutrients, but also increases the energy density of the manure, which may be used for biogas production or incineration (Chapter 13).

Filtration is the cheapest method for separating solids and plant nutrients from the slurry liquid, but small particles containing P may not be retained on the filter and plant nutrients dissolved in the liquid are passing through the filter. However, the filter technique still retains much dry matter and plant nutrients due to formation of a filter cake, which retains small P-containing particles that ought to pass through. In addition, the filter cake often has a high water content (50–80%-volume), which contains dissolved plant nutrients. The retention of P and small particles can be increased by adding chemicals (coagulants and flocculants) to the manure prior to filtration (See chapter 7 for more information about separation).

Table 2.2   *Pig slurry composition, and composition of the dry matter-rich and liquid fraction produced on separating slurry (Møller et al., 2000, 2007).*

Manure and separators	pH	Dry matter (g l^{-1})	Ash (g l^{-1})	NH$_4$-N (g l^{-1})	Total-N (g l^{-1})	Total-P (g l^{-1})	K (g l^{-1})
Untreated pig slurry	7.2	54.9	15.0	3.6	4.2	1.4	3.2

Separator	Dry matter-rich fraction/slurry (%)	Dry matter-rich fraction, retentate (g l^{-1})			Liquid fraction, permeate (g l^{-1})		
		Dry matter	Total-N	Total-P	Dry matter	Total-N	Total-P
Belt press	17.5	192	6.4	3.5	44	4.4	1.7
Screw press	4.2	344	6.6	2.1	42.5	5.0	1.2

Weight of 1 m^3 of slurry is approximately 1 ton.

In a screw press, the solid fraction is dewatered and thus the dry matter concentration of this fraction is high (Table 2.2). Increasing the applied pressure increases the dry matter concentration in the solid fraction. Although aggregation of particles on the filter may contribute to some degree to the retention of small particles in the screw press, this is not a significant process because the applied pressure forces small particles through the filter pores and a large fraction of small particles are in the liquid fraction after separation. Thus the solid fraction contains little N, P or K. As a consequence, the plant nutrient separation efficiency of the screw press is low, whereas the content of organic materials is high and the water content low. A high dry matter content is important if, for example, the solid fraction has to be incinerated.

2.4   Systems Analysis Method for Assessing Mass Flows

Systems analysis is an efficient and reliable tool for comparing different methods for manure management, and is necessary when assessing new technologies. The core of systems analysis is a clear and standardised method for calculating mass and energy flows of a technology or a combination of technologies.

One of the best tools to get an overview of a complex process is a flowchart using boxes for the process units and arrows for the inputs and outputs. The process unit is the central part of the flowchart. A process is defined as any operation or series of operations that causes a physical or chemical change within the system or its surroundings (separation, biogas reactor, etc.). An example of a flowchart is shown in Figure 2.10.

Materials that enter the process are called the *input* and materials that leave the process are called the *output*. When working with batch processes, it is important to know the quantities of the materials added at

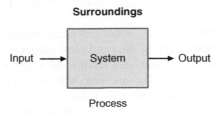

Figure 2.10   *General flowchart of mass input and output (© University of Southern Denmark.)*

the beginning of the process and removed at the end. In continuous processes (Text Box – Basic 2.5), there is a flow of materials to and from the process. Traditionally, masses (m) or mass flow rates (q) have been used for the handling of manure instead of volume or volume flow rates. This is practical because the volume is dependent on temperature and can change during the process, unlike the mass. Furthermore, measurements of mass are often easier and more precise than measurements of volume.

Text Box – Basic 2.5   Definition of engineering terms

Batch process: No input and output to the system during the process.
Continuous process: Both input and output to the system during the process.
Coagulant: Salt added to slurry to aggregate particle and colloids.
Flocculant: Polymer added to flocculate particles and aggregate in slurry.
Process: Operation that causes a change in the system or the surroundings.

Information on the concentration of different components in the input and outputs is also required. However, manure is a complex mixture of organic and inorganic compounds. Therefore, the composition of the input and output flows cannot be characterised in detail, due to the enormous work required to analyse all components; hence key components are identified. Often the dry matter content and plant nutrients (N and P) are measured. When studying the biogas system, interest may centre on the organic compounds, and if manure is to be spread on farmland, then heavy metals may be the key issue. When the key component has been identified, the mass fraction or the molality ($[x]$) must be ensured in all input and output flows. It is often most convenient to use molality, because chemical and biochemical transformation can then be accounted for with equations accounting for reactions.

When the flowchart has been set up, all relevant input and output values have to be known, including the total mass/mass flow and concentrations of the selected compounds. In general, as many as possible of the input and output parameters have to be determined. Values not available because they cannot be measured have to be calculated from mass balances and process-specific equations.

An example of systems analysis is given in Example 2.1.

2.4.1   Mass Balance and Process Specifications

The calculation of mass balance is based on the law of conservation of masses. Several mass balances can be set up such as a total balance as well as component balances for each component.

The integral mass balance is used for components in batch processes, where all the components are fed to an operation at initiation of the processing and all components removed at termination. The mass balance of component x is calculated as:

$$\Delta m(x) = [x]_{\text{in}} \cdot m(\text{tot})_{\text{in}} - [x]_{\text{out}} \cdot m(\text{tot})_{\text{out}} + r_{\text{g}} \cdot \Delta t - r_{\text{c}} \cdot \Delta t \qquad (2.1)$$

where $\Delta m(x)$ is the change in mass of the component x (kg), $[x]$ is the concentration of the component in the streams (kg kg^{-1} total) and $m(\text{tot})$ is the total mass (kg). The "in" and "out" subscripts indicate that the stream is at the start or termination of the process. r_{g} is the mass rate of generation, r_{c} is the mass rate of consumption (kg s^{-1}) and Δt is the duration of the process (s).

The differential mass balance is used for components in continuous processes with a stream of feed to and from the operation. The change d$m(x)$/dt in the amount of component x is assessed as:

$$\frac{dm(x)}{dt} = [x]_{in} \cdot m(tot)_{in} - [x]_{out} \cdot m(tot)_{out} + r_g - r_c \qquad (2.2)$$

At steady state, d$m(x)$/d$t = 0$. At non-steady state, the differential balance has to be solved either analytically or numerically. For more complex systems, it will often only be possible to solve the problem numerically.

Besides the mass balances and input/output parameter, some process specifications are usually required. For biological or chemical processes r_g and r_c have to be known. For unit operations (physical treatment process), both r_g and r_c are zero. Instead, data for the separation process, for example, have to be known. Process specification is often available for the equipment; this may be the fraction of salt removed during reverse osmosis or the size of particles removed during filtration.

Example 2.1   Flowchart for bench-scale solid–liquid experiment

Separation of slurry in a batch test using filter separation in the laboratory after addition of coagulants (FeCl$_2$) and flocculants (polyacrylamide) to the slurry. The pre-treated manure was filtered using a 0.5-mm filter. The material deposited on the filter is called the cake. The amount of slurry, composition of slurry and use of additives is given in Tables 2.3 and 2.4.

Table 2.3   *Data used in calculation of mass balances when separating slurry using additives and batch filtration (Hjorth et al., 2008).*

Feed slurry volume	0.6 l
Slurry characteristic	DM: 56 g kg^{-1}; P: 0.89 g kg^{-1}, density: 1025 kg m^{-3}
Additives	coagulants: 6 ml 1 M FeCl$_3$; flocculant: 60 ml 5% v/v zetag
Cake	weight: 222 g
Cake characteristics	DM: 125 g kg^{-1}
Filtrate characteristics	P: 0.149 g kg^{-1}

Table 2.4   *Additional data used in calculation of mass balances when separating slurry using additives and batch filtration.*

Feed (slurry) weight	m(Manure) = 0.6 l × 1025 kg m^{-3} = 0.62 kg
Feed (flocculant) solution	m(Floc) = 60 ml × 1000 kg m^{-3} = 0.06 kg
Flocculant (5% v/v)	[DM]$_{Floc}$ = 0.050 kg kg^{-1}
Feed (coagulant solution)	m(Coag) = 3 ml × 1000 kg m^{-3} = 3 × 10^{-3} kg
Coagulant (6 ml 1 M FeCl$_3$)[a]	[DM]$_{Coag}$ = 1 mol l^{-1} × $\dfrac{162.2 \text{ g mol}^{-1}}{1000 \text{ kg m}^{-3}}$ = 0.16 kg FeCl$_3$ kg^{-1}

[a] FeCl$_3$ molecular weight 162.2 g mol^{-1}.

A flowchart provides an overview of all the information from the experiment (Figure 2.10), which is divided into two processes: (i) mixing of coagulants and flocculants with slurry and (ii) solid–liquid separation using drainage through a filter. Information is missing, so we have to make assumptions; one important assumption is that the density of additives is 1 kg l^{-1}. Then we can start calculating the mass balance of the system depicted in Figure 2.11.

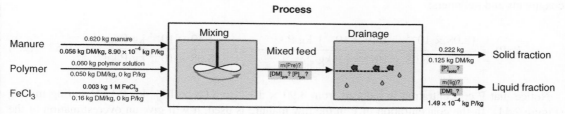

Figure 2.11 *Flowchart used for setting up mass balances for manure dry matter, polymer and FeCl$_3$. On the arrows indicating flows in and out, amount of material is given above the arrow and concentration below. A question mark and shaded background indicate the unknown parameters. (© University of Southern Denmark.)*

Neither the polymer solution nor the FeCl$_3$ solution contains phosphorus. The output from the mixing process can be calculated by setting up a total mass balance as well as a component mass balance for the dry matter and the P. Due to the addition of flocculants and coagulant (FeCl$_3$), the water content of the manure increases by 10%; hence more water has to be removed during the drainage process compared with processes without pre-treatment. After drainage two outputs are observed, the liquid fraction and the solid fraction. Most of the dry matter and the P are found in the solid fraction.

There are six unknown parameters, which can be determined by setting up total and component mass balances for either the mixing process or the drainage process (Table 2.5).

Table 2.5 *Calculation of a mass balance for the separation of animal manure with a simple mechanical separator and using additives.*

Unknown	Mass balance	Calculation
$m(\text{Pre})$	total balance mixing	$m(\text{Pre}) = 0.620\ \text{kg} + 0.060\ \text{kg} + 0.003\ \text{kg} = 0.683\ \text{kg}$
$[\text{DM}]_{\text{pre}}$	dry matter balance, mixing	$[\text{DM}]_{\text{pre}} = \dfrac{0.056\ \text{kg DM kg}^{-1} \times 0.620\ \text{kg} + 0.050\ \text{kg DM kg}^{-1} \times 0.060\ \text{kg} + 0.16\ \text{kg DM kg}^{-1} \times 0.003\ \text{kg}}{0.683\ \text{kg}}$ $= 0.056\ \text{kg DM kg}^{-1}$
$[\text{P}]_{\text{pre}}$	P balance, mixing	$[\text{P}]_{\text{pre}} = \dfrac{8.90 \times 10^{-4}\ \text{kg P kg}^{-1} \times 0.620\ \text{kg}}{0.683\ \text{kg}} = 8.08 \times 10^{-4}\ \text{kg P kg}^{-1}$
$m(\text{Liq})$	total balance, drainage	$m(\text{Liq}) = 0.683\ \text{kg} - 0.222\ \text{kg} = 0.461\ \text{kg}$
$[\text{DM}]_{\text{liq}}$	dry matter balance, drainage	$[\text{DM}]_{\text{liq}} = \dfrac{0.056\ \text{kg DM kg}^{-1} \times 0.683\ \text{kg} - 0.125\ \text{kg DM kg}^{-1} \times 0.222\ \text{kg}}{0.461\ \text{kg}} = 0.0023\ \text{kg DM kg}^{-1}$
$[\text{P}]_{\text{solid}}$	P balance, drainage	$[\text{P}]_{\text{solid}} = \dfrac{8.08 \times 10^{-4}\ \text{kg P kg}^{-1} \times 0.683\ \text{kg} - 1.49 \times 10^{-4}\ \text{kg P kg}^{-1} \times 0.461\ \text{kg}}{0.222\ \text{kg}}$ $= 2.18 \times 10^{-3}\ \text{kg P kg}^{-1}$

In order to compare different separation processes, a removal or separation index can be calculated. The fraction of P transferred from the slurry to the solid fraction can be calculated. This fraction is defined as the removal efficiency. One should use the concentration of P in the pretreated slurry to which is added coagulants and polymers:

$$R\,(\text{P}) = \frac{(8.08 \times 10^{-4} - 1.49 \times 10^{-4})\ \text{kg P kg}^{-1}}{8.08 \times 10^{-4}\ \text{kg P kg}^{-1}} = 1 - \frac{1.49 \times 10^{-4}\ \text{kg P kg}^{-1}}{8.08 \times 10^{-4}\ \text{kg P kg}^{-1}} = 0.82 \quad (2.3)$$

Notice that P concentration is reduced from 8.90×10^{-4} to 8.08×10^{-4} kg P kg^{-1} during mixing (Figure 2.11). If, the concentration of P in the raw manure is used, it will give an overestimation of the removal efficiency ($R = 0.83$) unless corrected for dilution of P.

Furthermore, 88% of the P is retained in the solid fraction (separation efficiency):

$$E_t(\text{P}) = \frac{2.18 \times 10^{-3}\ \text{kg P kg}^{-1} \times 0.222\ \text{kg}}{8.90 \times 10^{-4}\ \text{kg P kg}^{-1} \times 0.620\ \text{kg}} = 0.88 \quad (2.4)$$

When considering the dry matter the calculation is more complicated, because polymer and FeCl$_3$ are added and contribute to the dry matter. It can be calculated that 96% of the dry matter is removed from the liquid:

$$R(\text{DM}) = 1 - \frac{0.0023\ \text{kg DM kg}^{-1}}{0.056\ \text{kg DM kg}^{-1}} = 0.96 \quad (2.5)$$

Furthermore, 73% of the dry matter is retained in the solid fraction (separation efficiency):

$$E_t(\text{DM}) = \frac{0.125\ \text{kg DM kg}^{-1} \times 0.222\ \text{kg}}{0.056\ \text{kg DM kg}^{-1} \times 0.683\ \text{kg}} = 0.73 \quad (2.6)$$

However, the problem here is that it is difficult to compare data with a separation process without pre-treatment. If, for example, much more polymer is added to raw manure, we can easily increase the calculated separation efficiency, but may not remove more of the manure dry matter from the liquid fraction. The best solution for calculating the separation index is to assume that all polymer and FeCl$_3$ ends up in the solid fraction, and then focus on the manure dry matter.

The mass fraction of dry matter can then be split into three groups containing, respectively, the manure dry matter, the polymer and FeCl$_3$ (i.e. $[\text{DM}]_{\text{total}} = [\text{DM}]_{\text{manure}} + [\text{DM}]_{\text{polymer}} + [\text{DM}]_{\text{FeCl}_3}$). By doing this, $[\text{DM}]_{\text{manure}}$ can be calculated to be 0.109 kg DM kg^{-1} and the separation efficiency is:

$$E_t(\text{DM}) = \frac{0.109\ \text{kg DM kg}^{-1} \times 0.222\ \text{kg}}{0.056\ \text{kg DM kg}^{-1} \times 0.620\ \text{kg}} = 0.70 \quad (2.7)$$

The separation efficiency is lower than the calculated value not corrected for addition of polymer and FeCl$_3$.

2.5   Summary

This chapter gives a short presentation of the most commonly used livestock production systems and the related manure management. Terms used to describe manure management are presented and explained. Furthermore, examples are given of plant nutrient and carbon flows through an aerobically managed deep litter system with manure containing much straw and an anaerobically managed liquid manure system with slurry containing little straw. Finally, a systems analysis approach to calculate the flow and concentration of plant nutrients is presented using separation of slurry into a dry matter-rich solid fraction and a liquid fraction as an example.

References

Hjorth, M., Christensen, M.L. and Christensen, P.V. (2008) Flocculation, coagulation and precipitation of manure affecting three separation techniques. *Bioresour. Technol.*, **99**, 8598–8604.

McGinn, S.M., Flesch, T.K., Crenna, B.P., Beauchemin, K.A. and Coates, T. (2007) Quantifying ammonia emissions from a cattle feedlot using a dispersion model. *J. Environ. Qual.*, **36**, 1585–1590.

McGinn, S.M., Flesch, T.K., Chen, D., Crenna, B., Denmead, O.T., Naylor, T. and Rowell, D. (2010) Coarse particulate matter (PM10) emissions from cattle feedlots in Australia. *J. Environ. Qual.*, **39**, 791–798.

Monteny, G.J. and Erisman, J.W. (1998) Ammonia emission from dairy cow buildings: a review of measurement techniques, influencing factors and possibilities for reduction. *Neth. J. Agric. Sci.*, **46**, 225–247.

Møller, H.B., Lund, I. and Sommer, S.G. (2000) Solid liquid separation of livestock slurry – separation efficiency and costs. *Bioresour. Technol.*, **74**, 223–229.

Møller, H.B., Jensen, H.S., Tobiasen, L. and Hansen, M.N. (2007) Heavy metal and phosphorus content of fractions from manure treatment and incineration. *Environ. Technol.*, **28**, 1403–1418.

Pain, B. and Menzi, H. (2003) *Glossary of Terms on Livestock Manure Management*; http://www.ramiran.net/DOC/Glossary2003.pdf.

Seré, C. and Steinfeld, H. (1996) World livestock production systems: current status, issues and trends, *Animal Production and Health Paper 127*, FAO, Rome.

Vu, T.K.V., Tran, M.T. and Dang, T.T.S. (2007) A survey of manure management on pig farms in Northern Vietnam. *Livest. Sci.*, **112**, 288–297.

Webb, J., Sommer, S.G., Kupper, T., Groenestein, K., Hutchings, N.J., Eurich-Menden, B., Rodhe, L., Misselbrook, T.H. and Amon, B. (2012) Emissions of ammonia, nitrous oxide and methane during the management of solid manures. *Sust. Agric. Rev.*, **8**, 67–107.

3

Regulations on Animal Manure Management

Sven G. Sommer[1], Oene Oenema[2], Teruo Matsunaka[3] and Lars S. Jensen[4]

[1]*Institute of Chemical Engineering, Biotechnology and Environmental Technology,*
University of Southern Denmark, Denmark
[2]*Environmental Sciences, Wageningen University, Netherlands*
[3]*Faculty of Dairy Science, Rakuno Gakuen University, Japan*
[4]*Department of Plant and Environmental Sciences, University of Copenhagen, Denmark*

3.1 Introduction

Animal manure is not only a source of valuable plant nutrients, but also a source of air and soil pollution and a threat to aquifers and surface waters unless managed carefully to minimise nutrient losses (Steinfeld *et al.*, 2006; Sutton *et al.*, 2011).

The risk of pollution due to poor management of manure is increasing due to the emerging larger, more regionally intensive and more specialised animal production systems (Menzi *et al.*, 2010) (Figure 3.1). These intensive animal production systems rely to a large extent on the import of animal feed from elsewhere. As a result, these systems often have insufficient areas of land for proper disposal of the animal manure produced, and hence consider the manures to be not a valuable nutrient and organic matter resource, but a waste. This is especially the case in rich countries, where mineral fertilisers are easily available and relatively cheap. In contrast, animal manures are considered a valuable resource in many poor countries, such as those in sub-Saharan Africa.

The total amount of animal manure in the world is huge. Cattle are the greatest producer followed by pigs, poultry, sheep and goats. The total amount of nitrogen (N) in manure produced annually is as large

Figure 3.1 *Spatial distribution of the total amount of P_2O_5 in livestock excreta (kg km^{-2}) in 2005. (Modified from Menzi et al. (2010). © 2010 SCOPE. Reproduced by permission of Island Press, Washington, DC.)*

as the amount of mineral N fertiliser used per year. The amounts of phosphorus (P), potassium (K) and micronutrients in the animal manure are even much larger than the amounts of mineral P, K and micronutrient fertilisers produced per year. However, animal manure is a bulky product due to the high water content, especially in the case of slurry. As a result, the economic costs of collection, storage and transport of animal manures are very high both in absolute terms and per unit of nutrients. However, poor management of animal manure leads to risks of environmental pollution and diseases.

In response, governments in various countries have implemented regulations to minimise the risks of pollution and diseases. This is especially the case for countries in the European Union (EU). The regulations also reflect the fact that an increasing percentage of the human population is no longer familiar with agriculture and is not in close contact with farmers anymore and hence not willing to accept any nuisance from the sector. In addition, the increased economic wealth of people in rich countries enables a larger proportion of the population to pay for products that are produced in a more environmentally sound and animal friendly way. Most of the regulations promote recycling or use of the manure by encouraging environmentally sound manure management and the use of environmental technologies to mitigate environmental problems.

This chapter briefly summarises the pollution risks related to manure management and the government regulations and incentives to reduce this pollution.

3.2   Environmental Issues

Worldwide, obnoxious odour caused by livestock farming and manure management is considered to be a large, but local pollution problem. The odour is a problem for neighbours of livestock farms and also the farmers themselves. The problem is enlarged by the negative impact that odour from livestock production

Table 3.1 *Animal waste management contributes to environmental pollution at local (farm, village), regional (province, country, part of continent) and global scale, depending on the substances involved.*

Substances	Scale of environmental impact		
	Local	Regional	Global
Odour, H_2S	+		
Nitrate	+	+	
Phosphorus	+	+	
Ammonia	+	+	
Greenhouse gases			+
Heavy metals	+	+	
Pathogens	+	+	

units has on the real estate value of nearby dwellings. Consequently, reducing odour emission is an important issue. The odour components are diluted and degraded in the air, and are therefore a local problem (Table 3.1).

Animal production contributed 55% of the global NH_3 emissions during the 1990s (Bouwman *et al.*, 1997). Ammonia emission is a local-to-regional environmental issue, because approximately 50% of the NH_3 gas is deposited close to the source (Sommer *et al.*, 2009), while the formation of ammonium sulfate and ammonium nitrate particles in the atmosphere may transport the NH_3 over long distances. Deposition of NH_3 and particulate NH_4^+ to land or water may cause acidification and eutrophication of natural ecosystems (Text Box – Basic 3.1). Hydrogen sulfide (H_2S) and NH_3 in animal houses is a hazard to the health of farm workers and animals being housed, as it causes lung and respiratory disease. The formation of ammonium sulfate particles in the atmosphere can be a local and regional health hazard (Erisman and Schaap, 2004), and this may contribute significantly to the socioeconomic external costs of air pollution (Brandt *et al.*, 2011).

Text Box – Basic 3.1 Environmental terms

Acidification: Defined as "a net influx of acidifying substances (protons, H^+), that decrease the capacity of a system to neutralise these protons, and that is associated with a decrease of the pH and other unwanted effects". Acidification may affect terrestrial and aquatic environments, especially when the capacity of the soil or water bodies to resist or neutralise acidifying inputs is low. Acidifying compounds may arrive with rain or snow as wet deposition, or in the form of particles or gases as dry deposition. These are either produced from oxides of nitrogen (N) and sulfur (S; derived from combustion of fossil fuels), that subsequently react in the atmosphere to produce acids which are dissolved in precipitation, or acidity is derived from nitrification (oxidation) of deposited NH_3 that largely originates from animal manure.

Greenhouse gas (GHG): Gases that may contribute to "Climate change and global warming". Carbon dioxide (CO_2), methane (CH_4) and nitrous oxide (N_2O) are the major GHGs. These gases do not absorb incoming short-wavelength and high-frequency radiation, which, therefore, passes through the atmosphere reaching soil and plants. Part of the energy is absorbed by the plants and converted to organic material and heat. A fraction is reflected to the atmosphere in the form of long-wavelength and low-frequency radiation. This reflected radiation is partly retained by CO_2, CH_4 and N_2O, and thereby may contribute to the warming of the atmosphere and the Earth. The capacity to absorb the radiation that is reflected from the soil surface is higher for CH_4 and N_2O than for CO_2. The global warming potential of N_2O is 296 times higher than that of CO_2 per kg and that of CH_4 23 times higher. Methane accounts for 30% of the anthropogenic contributions to net global warming and N_2O accounts for 10% (Solomon *et al.*, 2007).

Eutrophication: Defined as "enrichment of a system with nutrients, including the response". Eutrophication occurs when an ecosystem (aquatic or terrestrial) acquires a high concentration of nutrients, especially phosphates and nitrates, either naturally or due to human activity (e.g. excessive fertilisation, runoff and discharge of sewage). In aquatic systems this promotes excessive growth of algae, and as the algae die and decompose it depletes the water of available oxygen (hypoxia), causing the death of other organisms. In natural terrestrial ecosystems eutrophication changes the composition of plant species.

Surface waters and groundwater may be polluted by leaching and runoff of manure nutrients applied to fields (Xiong *et al.*, 2008). In addition, manures may be discharged directly into surface waters. The diffuse sources and direct discharge of manure nutrients to receiving waters give rise to eutrophication (Text Box – Basic 3.1). The increase in plant nutrients necessitates water purification treatment in the provision of safe drinking water supplies. Thus losses of P, N and other plant nutrient elements to the environment are a local and a regional problem.

Animal farming systems are major sources of the GHGs N_2O and CH_4. The global atmospheric CH_4 concentration has increased 2.5-fold and N_2O by approximately 20% from the pre-industrial values (IPCC, 2007). Indirectly, the NH_3 emissions also contribute to N_2O emissions by increasing the N cycling in natural ecosystems (Davidson, 2009).

Animals require some 22 essential nutrient elements, including some trace metals (Suttle, 2010). Most of these elements are present in animal feed in sufficient amounts, but supplementation is done for some elements in some feeds. Therefore, animal diets may contain much higher trace metal concentrations than plants. For example, copper (Cu) is added to the diet of growing pigs in many intensive production systems as a cost-effective method of enhancing performance and is thought to act as an anti-bacterial agent in the gut. Zinc (Zn) is also used in diets for weaned pigs for the control of post-weaning scours. In poultry production, both Zn and Cu are required in trace amounts. Other heavy metals may be present in animal diets as a result of contamination of mineral supplements (e.g. limestone added to the feed may contain relatively high levels of Zn and cadmium (Cd)). A proportion of the metals in livestock manures, as well as those excreted directly onto grazing land, are recycled through the agricultural system in animal feeds grown and fed on-farm (Nicholson *et al.*, 1999).

Health risks from poor manure management include several pathways of pathogen transfer, of which local transfer of zoonotic pathogens to farm staff and neighbours in air, water, crops and vegetables is a well-known problem. Pathogens may spread among and between animal production systems, and the risks of spread of diseases such as foot and mouth disease, salmonella, and so on, should be carefully considered when establishing new manure handling methods (Albihn and Vinnerås, 2007). Pollution of shallow drinking water wells is a considerable risk to human and animal health, as manures may contain numerous pathogens such as bacteria, viruses and parasites. Highly contagious and pathogenic diseases, such as foot and mouth disease, swine fever and Aujeszky's disease, may spread with animal effluent through waterways. When one farm is infected with the disease, then farms downstream will be at considerable risk of infection, if manure is discharged to rivers. Thus, discharge of animal manure to rivers may pose a much greater risk of pathogen transfer than careful recycling on agricultural land with appropriate precautions during spreading. Pathogens may also be transported together with the transfer of animals (and manure) from one farm to another, to markets and to slaughterhouses. Although not definitively proven, poor manure management, the mixing of human and animal excreta and close contact between domestic and animal housing may propagate the spread of recently emerging diseases such as avian flu (Peiris *et al.*, 2007; Gambotto *et al.*, 2008).

3.3   Need for Government Regulations

The environmental problems described in the previous section are most commonly the result of ignorance, negligence or calculated risk; namely, the polluter is: (i) not aware of creating these problems, (ii) does not know how to prevent or mitigate these problems, (iii) does not have the tools and the means to mitigate these problems, and/or (iv) finds it too costly or too complicated to prevent or mitigate these problems. Hence, the environmental problems continue to exist until measures are taken to prevent or mitigate the effects. The question that now arises is "What triggers a polluter to take measures?" This section briefly addresses this fundamental question in societies (*see also* Oenema *et al.*, 2011).

There are four principal drivers in organising and governing societies:

- Culture (human values, traditions, fashion and cultural habits).
- Market power and expertise (the "invisible hand" of the free market).
- Public policy measures (state coercion, i.e. regulation pressure by governments).
- Civic society pressure (pressure from non-government organisations (NGOs), societal pressure and lobby groups).

These drivers collectively determine what people do, although individuals tend "to listen" differently to different drivers. Public or government policy can be seen as a response of governments to a societal problem, where culture, markets and civic society pressure collectively fail to solve that problem. Government policy aims at modifying human individual behaviour so as to achieve societal objectives (i.e. to contribute to the total welfare of society). The fact that the "public policy" addresses societal objectives does not mean that everybody in the society equally accepts this policy and its consequences. There is often a strong divide in societies between those who believe in the cleansing mechanism of the market and in the ability of humans to act responsibly, and who therefore prefer a minimum of government intervention, and those who emphasise the failures of markets and the need to help the less endowed in society and therefore favour more extensive government intervention.

Policy instruments are the tools to implement the policy in practice. There are three types of instruments, the choice of which depends on the nature of the problem, the objectives of the policy and the competences and characteristics of those addressed: (i) regulatory or command-and-control instruments, (ii) economic or market-based instruments and (iii) communicative or persuasive instruments (Table 3.2).

Regulatory instruments (regulation) involve a restriction on the choice of agent, method and action. Regulations are compulsory measures imposing requirements on producers to achieve specific levels and standards of environmental quality, including environmental restrictions, bans, permit requirements, maximum rights or minimum obligations. They are the most common policy instrument used in the EU environmental policy (e.g. the Nitrates Directive).

Table 3.2   *Possible policy instruments, with some examples.*

Regulatory instruments	Economic instruments	Communicative instruments
Public land use planning (zoning/spatial planning)	taxes	extension services
Pollution standards and ceilings	subsidies (including price support)	education and persuasion
Fertiliser limits	import/export tariffs	co-operative approaches
Best available technique requirement	tradable emission rights and quotas	

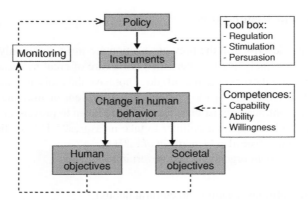

Figure 3.2   *Simple presentation of the intended working of government policy. The tool box is explained further in Table 3.2. (Reprinted with permission from Oenema et al. (2011). © 2011 Cambridge University Press.)*

Economic instruments (stimulation) are intended to stimulate preferred production pathways. These instruments are common in agricultural policy, such as the EU Common Agricultural Policy (CAP), and are also used to stimulate innovative developments. Environmental taxes and tradable rights/quotas have only been implemented in a few countries. Subsidies are increasingly used as a policy instrument to promote environmentally friendly practices and the introduction of new technology.

Communicative instruments (persuasion) include public projects to address environmental issues and measures, to improve information flows, and to promote good practices and environmental objectives. This information can be provided to producers in the form of technical assistance and advice, and to consumers via labelling, for example. Technical assistance and extension services are intended to provide users with information and technical assistance to implement environmentally friendly practices. This category also includes so-called voluntary approaches (e.g. codes of good agricultural practice).

Whether the addressee of the policy changes behaviour and contributes to achieving the objectives depends on the instrument and the decision environment of the addressee. A decision environment can be defined as "the collection of information, alternatives, values and preferences available at the time of decision". An ideal decision environment would include all possible information, all of it accurate, and every possible alternative in time. This is usually not the case. The compliance with a policy will depend on the knowledge and information of the addressee ("capability"), the availability of the appropriate tools and means ("ability") and the persuadability ("willingness") of the addressee to implement the policy (Figure 3.2).

3.4   Global Regulation – Multilateral Environmental Agreements

Intergovernmental organisations (IGOs), such as the United Nations (UN), and multilateral environmental agreements (MEA), such as conventions and protocols, have greatly contributed to tackling many of the known environmental problems and continue to address outstanding issues of international concern (Bull *et al.*, 2011). These IGOs and MEAs have also done much to harmonise the efforts of governments, and have provided important driving forces for international and national action on environmental matters, including manure management.

The UN was founded in 1945 with the aim of maintaining international peace and security, developing friendly relations among nations and promoting social progress, better living standards and human rights. The UN works on a broad range of fundamental issues, including sustainable development and environment protection. The UN has been at the core of the MEA on combating climate change through the UN Framework Convention on Climate Change (UNFCCC) and its Kyoto protocol, which outlines regulations aimed at decreasing greenhouse gas emissions, including those from manure management. Another example is the UN Economic Commission for Europe (UNECE), which hosts five conventions including the Convention on Long-Range Transboundary Air Pollution (CLRTAP). The CLRTAP was established in 1979 to address air pollution problems. It has eight protocols that identify specific measures to be taken by its 51 parties (countries) to cut their emissions of air pollutants including the 1999 Gothenburg Protocol, which addresses emissions of SO_2, O_3, NO_x, NH_3 and volatile organic compounds (VOCs). It sets ceilings on the annual emissions of these pollutants and requires the implementation of measures to decrease emissions. Annex IX of the Gothenburg Protocol contains specific measures to decrease NH_3 emissions from animal manures and fertilisers (UNECE, 2012). Methodology for estimating CH_4 and N_2O emissions from livestock manure based on emissions factors has been devised by the Intergovernmental Panel on Climate Change – better known as the IPCC.

3.5   Regional Regulations – Exemplified with EU Directives and Regulations

As many of the environmental impacts occur at the regional scale (Table 3.1), regulation at the regional scale appears logical, but is often hindered by lack of competent government structures. A good example of joint regional and national environmental regulations can be found in the EU. EU environmental policy is mostly established through means of Directives, which impose environmental objectives to be achieved by the Member States. EU Directives set the framework in which Member States must create legislation directed at industries/civilians in order to attain the environmental quality objectives laid down in the EU Directives. In contrast, EU agricultural policy is mostly established through means of so-called Regulations, which are directly binding for Member States. Implementation of EU Directives by Member States leaves more flexibility to Member States than EU Regulations. Note that EU Directives are commonly based on "regulatory instruments" (Table 3.2) and that EU Regulations are often based on a mixture of "economic instruments" and "regulatory instruments".

Understanding EU policy measures dealing with manure management and its emissions requires insight into the changes in understanding and perception of the cause–effect relationships of these emissions over time by scientists and policy makers. Many current policy measures dealing with emissions reflect a simple "source–receptor/effect" model of understanding (i.e. a source emitting a pollutant and an environment being the receptor affected by the pollutant). In this case, animal manure is seen as sources of pollutants, while atmosphere, surface waters and groundwater are seen as receptors.

3.5.1   EU CAP and its Reforms

The EU Common Alerting Protocol (CAP) was established in 1958, and has contributed greatly to the modernisation and productivity of agriculture and to food security in the EU. Indirectly, it has also contributed to increased use of natural resources and to import of animal feed from outside the EU, as well as to increased environmental effects created by agriculture. Following the recognition and increased awareness of the effects of surpluses of agricultural products and the environmental burden associated with the intensification of agricultural production, the CAP went through a series of reforms, notably in 1984 (implementation of milk quota), 1992 (set-aside regulations), 1997 (Agenda 2000) and 2003 (fundamental change in EU support for agriculture). The reforms of the CAP continue to have a significant influence on agriculture and its environmental effects.

"Cross-compliance" is the main policy vehicle to implement the CAP reform. It is the requirement that farmers in receipt of payments under the CAP must follow other relevant European Community legislation. In June 2003, cross-compliance became an obligatory element of the CAP, thereby coupling existing environmental policies to agricultural income support, as implemented in so-called "Single Farm Payments" to farmers. There are two major aspects of cross-compliance in the Single Farm Payment: (i) compliance with 19 Statutory Management Requirements (SMRs) covering the environment, food safety, animal and plant health and animal welfare, including the provisions of relevant directives, and (ii) compliance with a requirement to maintain land in Good Agricultural and Environmental Condition (GAEC). A few of the SMRs directly or indirectly address animal manure management (see Section 3.5.2). These include, for example, the 1991 Nitrates Directive, but not the Animal Byproducts (ABP) regulation (EC 1774/2002), which sets guidelines for the separation of category 1, 2 and 3 ABP materials at rendering plants.

3.5.2 EU Environmental Directives

The main objective of the 1991 Nitrates Directive is "to reduce water pollution caused or induced by nitrates from agricultural sources and prevent further such pollution". This Directive requires Member States to take the following steps: (i) water monitoring with regard to nitrate concentration and trophic status; (ii) identification of waters that are polluted or at risk of pollution; (iii) designation of vulnerable zones (areas that drain into identified waters); (iv) establishment of codes of good agricultural practices and action programmes (a set of measures to prevent and reduce nitrate pollution); and (v) a review at least every 4 years of the designation of vulnerable zones and action programmes. Waters must be identified as polluted or at risk of pollution, if nitrate concentrations in groundwater and surface waters contain or could contain more than 50 mg NO_3^- l^{-1} if no action is taken, or if surface waters, including freshwater bodies, estuaries, coastal and marine waters, are found to be eutrophic or in the near future may become eutrophic if no action is taken.

The action programmes to reduce pollution must contain mandatory measures relating to: (i) periods when application of animal manure and fertilisers to land is prohibited; (ii) capacity of and facilities for storage of animal manure; and (iii) limits to the amounts of animal manure and fertilisers applied to land, which should ensure balanced fertilisation. The Nitrates Directive has clearly had a huge effect on manure management in practice and is one of the reasons why manure management in the most advanced countries in the EU-27 is more advanced than in, for example, the United States, Japan, and China. All animal manures from confined animals have to be collected and stored in leak-tight and covered manure storage facilities. Manures can be applied to land only in periods when the risk of leaching and surface runoff is small, use low-emissions techniques and the application rate does not exceed the set limit.

The 2008 Directive on Industrial Emissions concerning Integrated Pollution Prevention and Control (IPPC) employs an integrated approach to the management of all types of pollution from industrial installations, including installations for the intensive rearing of poultry or pigs. It requires such installations to have a permit and to minimise all kinds of pollution by using Best Available Techniques (BAT). The IPPC directive also prescribes the manure management of large poultry and pig farms. These farms have to implement BAT for emission control and manure management.

Another EU Directive dealing with atmospheric pollution is the 2001 National Emissions Ceiling (NEC) Directive. This Directive sets upper limits (ceilings) for each Member State for the total emissions in 2010 and 2020 of the four pollutants responsible for acidification, eutrophication and ground-level ozone pollution (SO_2, NO_x, VOCs and NH_3), but leaves it largely to the Member States to decide which measures to take in order to comply. The Directive is basically the EU translation of the Gothenburg Protocol discussed in Section 3.4. It aims at achieving the long-term objectives of not exceeding critical levels and loads by establishing national emission ceilings, taking the years 2010 and 2020 as benchmarks.

The responses of Member States to implement these environmental policies for agriculture in their own laws and regulation have been very variable and in many cases slow. The delays in implementation of EU policies nationally have been ascribed to: (i) The large differences in farming systems and environmental conditions in the EU combined with the complexity of manure management and nutrient cycling;

(ii) varying interpretations by Member States of the targets and measures in environmental directives and regulations; (iii) reluctance to implement measures due to the perceived high costs to farmers and perceived low effectiveness; (iv) reluctance to introduce mechanisms to monitor compliance by farmers, due to the perceived high costs; (v) legislative delays; (vi) failure by farmers to implement measures, due to within-system constraints, perceived and actual costs, and the need for learning time; and (vii) potential antagonisms between measures.

The success of government policies seems to be related to one or a combination of the following factors: (i) Use of economic instruments (subsidies and taxes) to facilitate the implementation of the policy, which results in a high degree of compliance; (ii) availability of relatively straightforward and effective technologies to reduce the emissions; (iii) an efficient advisory system for rapid innovation and knowledge dissemination throughout the management chain – farmers, advisors, technology producers, and so on; (iv) limited number of addressees who must take action to implement the measures; (v) scale of investments required and the degree to which these are shared; (vi) the cost of the compliance measures are relatively small and/or can be transferred to others and (vi) enforcement and control leading to a high degree of compliance with the policy measures.

3.5.3 Reducing Ammonia Emissions from Manure Management in Europe

The effect of the EU Directives is that emissions have been reduced. In some western European (Atlantic) countries, the emissions have been reduced as an effect of new technologies, and the targets set have been achieved in the Netherlands, Belgium, the United Kingdom and Germany (Figure 3.3). In the Baltic Sea countries, Latvia, Lithuania and Estonia, and central European countries, the NH_3 emissions have declined since 1990 primarily due to a reduction in animal production and not so much the implementation of new technologies.

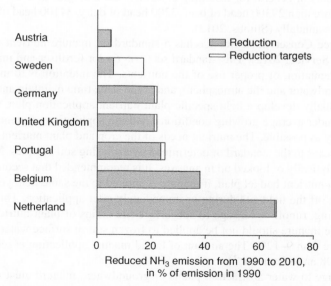

Figure 3.3   *Reduction in NH_3 emissions in eight European countries, presented as the amount reduced in the period 1990–2010 as a percentage of the emissions in 1990 and shown in relation to the reduction target set for the EU countries by the 2010 NEC Directive. (© University of Southern Denmark. Data from EEA (2012).)*

3.6   National Regulations on Agricultural Pollution

The two law systems mentioned in this chapter and much used are civil law and common law (in countries strongly influenced by the British legal system). In the common law tradition, only the frame of regulation is set in the law and within this frame the courts then decide how to judge offences, which means that precedence or earlier decisions made by courts are very important for court decisions. An example of this is given below for US regulation of manure management. In the civil law tradition the government sets the frame for the regulation of citizens' behaviour, often by annexing to the regulation documents that describe in detail the regulation of behaviour. Examples of this system are given below for the Japanese and Danish regulation of manure management.

3.6.1   United States

Livestock production and manure management is regulated by the Clean Water Act (CWA) and the Clean Air Act (CAA). In addition, many States have "State-specific additional regulations".

Concentrated animal feeding operations (CAFOs, i.e. livestock in barns or in feedlots) are defined as "point sources" of water pollution. These units are required to obtain National Pollutant Discharge Elimination System (NPDES) permits under the CWA if they discharge or propose to discharge into waters of the United States (EPA, 2009). The permit to discharge manure contains demands for technologies to treat the manure being discharged. After treatment, manure from CAFOs is, for example, applied to spray fields near the production unit and may also be discharged to recipient waters.

CAFOs may have to follow restrictions on emission of the following gases: Sulfur dioxide (SO_2), nitrogen dioxide (NO_2), particulate matter (PM10 and PM2.5), carbon monoxide (CO), ozone, NH_3 and lead. A ceiling of annual emission of these gases at 100 ton year^{-1} is set at which the farmer needs a pre-construction permit, which normally includes a description of proposed air pollution abatement systems (Grossmann, 2003). Still, there are no strict demands in general for installing abatement technologies at CAFOs (Hebert, 2009). The very large farms (e.g. more than 29300 head of beef, 3200 head of dairy, 34100 head of swine) have to report their emissions of GHGs annually (Stubbs, 2012).

The National Resource Conservation Service has a standard on manure nutrient management (Natural Resource Conservation Service, 2012). This standard sets criteria for fertiliser and manure application, and requirements for documentation of proper use of the nutrients. The intention is to minimise nutrient entry into surface water, groundwater and the atmosphere, and at the same time maintain and improve soil quality.

The farmer must annually develop a field-specific plant nutrient application plan, which is based on the yield goals attainable under average growing conditions, and the application of plant nutrients must match the crop needs as closely as possible. The nutrient needs of the crop and plant nutrient availability in the soil can be looked up in annexes to the standard or determined by analysing soil samples. Manure fertiliser value must be determined analytically or looked up in annexes. It is recommended that second year nutrient credits are included in the plant nutrient budget plan, if manure is applied to the same field year after year.

Manure must not run off the field site during or immediately after application. Therefore measures must be taken to avoid ponding, runoff or leakage to subsurface tile drains of plant nutrients or organic matter. To avoid runoff risks the manure should not be applied to frozen soil in surface water management areas or on slopes inclining more than 9–12%. The amount of liquid manure application is set at between 3000 and 10000 gallons acre^{-1} (28 and 93 ton ha^{-1}).

To avoid direct leakage to water abstraction plants or groundwater, manure must not be applied 50 feet (15 m) from potable water wells and 200 feet (60 m) uphill of conduits to groundwater. Furthermore, special care must be taken when applying manure to fields with high leaching potential or within 1000 feet (305 m) of municipal wells.

An example of regulation of manure management can be found in the state of North Carolina, which has a large pig and poultry production sector concentrated in regions with very high pig and poultry production densities. In the past, liquid pig manure treated in anaerobic lagoons was sprayed on fields at high application rates. In 2007, the State of North Carolina enacted legislation enforcing the five environmental performance standards of environmentally superior waste management technology (EST) as a requirement for the construction of new swine farms or expansion of existing swine farms. These technologies must have been proven to effectively reduce pollution. According to Vanotti *et al.* (2009, p. 5406):

... an EST needs to be technically, operationally and economically feasible and meet the following environmental performance standards: (1) eliminates the discharge of animal waste to surface waters and groundwater through direct discharge, seepage, or runoff; (2) substantially eliminates atmospheric emissions of ammonia; (3) substantially eliminates the emission of odor that is detectable beyond the boundaries of the parcel or tract of land on which the swine farm is located; (4) substantially eliminates the release of disease-transmitting vectors and airborne pathogens; and (5) substantially eliminates nutrient and heavy metal contamination of soil and groundwater. [Citation reproduced with permission from Vanotti *et al.* (2009). © 2009 Elsevier.]

More information about these regulations can be found in the links found in the reference "North Carolina (2012)".

Table 3.3 *Major regulation standards in Japan concerning release of effluent to rivers, lakes and marshes: this table shows only major standards, but there are several additional standards (e.g. for heavy metals); since the concentration of each element in the effluent varies with time of day, the regulation standard is presented as both upper limit and daily average.*

	Upper limit (mg l^{-1})	Daily average (mg l^{-1})
pH		5.8–8.6
Biological oxygen demand	160	120
Chemical oxygen demand	160	120
Suspended solids	200	150
Total-N	120	60
N as nitrate-nitrogenous compound[a]	100	–
Total P	16	8
Coliform bacteria	–	3000[b]

[a]NO_3-N + NO_2-N + 0.4 × NH_4-N, where 0.4 is the empirical conversion factor from NH_4-N to NO_3-N used in Japan. The provisional standard of N as nitrate-nitrogenous compound from the effluent of animal farms is less than 900 mg l^{-1} until 2013.
[b]Number of bacteria cm^{-3}.

3.6.2 Japan

In Japan, animal manure production amounts to about 90 million ton year^{-1}. To prevent serious pollution, farms raising livestock in Japan are controlled by the legislation shown in Figure 3.4 (Haga, 2011). This legislation has been implemented due to increasing complaints over pollution caused by livestock production since the 1970s. The most important laws controlling pollution derived from the animal wastes in Japan are briefly described in this section.

The objective of the law on treatment and utilisation of livestock manure, which was enacted in 1999, is to encourage the appropriate treatment of manure and to promote its use by farmers. The law provides

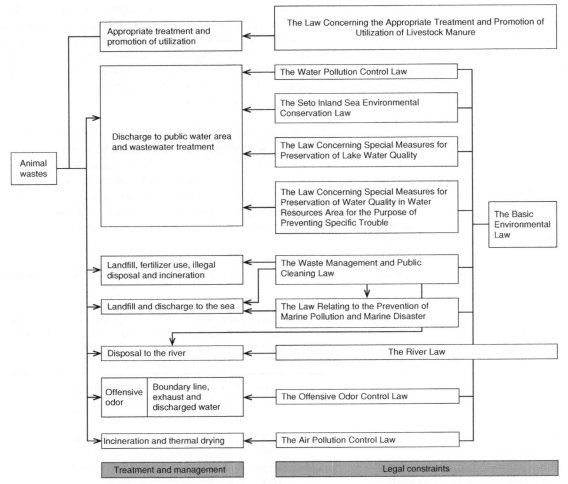

Figure 3.4 *Regulations on animal manure management in Japan. (© University of Southern Denmark; modified after Haga (2011).)*

practical regulation standards for the construction of manure composting facilities (e.g. concrete floor, roof of the facility, side walls, and suitable size and volume for storage of the manures). The law aims at enhancing construction of good manure storage facilities, which do not leak pollutants to the environment. Today, 99.9% of Japanese animal farms have constructed their facilities according to the law (MAFF, 2012). However, the law does not regulate the application methods, application time or rate of manure application to fields.

The water pollution control law enacted in 1970 set limits for the composition of animal wastewater or slurries that are discharged to recipient waters (Table 3.3). The law covers discharge of animal slurries or wastewater from farms with an effluent volume larger than 50 m^3 day^{-1}. The discharge of treated slurries to waterways is also a common removal method in other Asian countries (e.g. Taiwan, Vietnam and Malaysia).

The offensive odour control law enacted in 1971 sets the limit of concentration of 22 offensive odours in exhaust from industries and farms. Nine compounds are considered to contribute to odour caused by manure management (i.e. NH$_3$, methyl mercaptane (methane thiol), hydrogen sulfide, dimethyl sulfide, dimethyl disulfide, propionic acid, *n*-butyric acid, *n*-valeric acid and isovaleric acid). Regulation standards of the odours are set as odour intensity of 2.5 to 3.5. The intensity is qualitatively defined as 0 (no odour) to 5 (severe odour). The intensity level 3 is an odour concentration that is easily detectable by the human nose.

These laws have succeeded in decreasing the pollution caused by composting manures and pollution due to leakage from manure storage and management facilities on livestock farms. However, the regulations do not cover the pollution that may be a consequence of inappropriate manure application to agricultural land. Therefore, at present, livestock farmers can apply manure to fields at any time and rate. Application of excessively high rates of manure to land, corresponding to discharge in many other countries, is not prohibited and farmers do not have to comply with any maximum stocking density (head of livestock per land area) on the farm. Thus, at present, the regulation of farming in Japan does not aim at reducing environmental pollution due to application of manure to agricultural land (i.e. nitrate leaching, NH_3 and N_2O emission).

3.6.3 Denmark

The overall objective of the Danish regulations on livestock production and management of animal manure is to achieve sustainable recycling of manure in order to reduce emissions of plant nutrients and pollutants to the environment. Consequently, the regulations prohibit the discharge of manure to surface waters and also the application of manure to fields in excess of crop nutrient needs. The latter, the so-called "Harmony Regulation", was implemented in response to the EU Nitrate Directive requirements in the early 1990s and specifies a maximum livestock density, corresponding to 140–170 kg total-N in animal manure (in storage) per hectare available land for application. Farms having a livestock herd in excess of this density must have written contracts with neighbouring stockless farms, where the manure can be applied.

Furthermore, the Danish regulation on crop fertilisation, which was implemented at the same time as the Harmony Regulation, requires farmers to carry out detailed fertilisation planning for all fields, and crop N fertilisation standards are limited to only 80–85% of the economically optimal level in order to minimise the risk of overfertilisation (with mineral fertiliser or manure) in individual fields. The regulation caps the maximum overall nutrient use on each farm based on the soil types and the crop rotation, composition and yield levels expected and also implements an accounting procedure to assess the minimum fertiliser efficiency of animal manure applied to land, and the corresponding mandatory reduction in application of mineral fertiliser N.

The Danish regulations on animal manure also focus on reducing leaching and runoff losses of plant nutrients during storage and application of the manure, as well as gaseous emissions of NH_3 and odours from handling of livestock manure. The Danish regulations comply with the EU Regulations or Directives on livestock production, manure or biowaste management and environmental protection. Major guidelines or rules are (Danish Ministry of Environment, 2012):

- Direct discharge of animal manure to surface waters or groundwater is strictly forbidden by law.
- Zoning, which regulates where production units are established (i.e. location of farm and of manure store in relation to surface water, water abstraction plants, natural ecosystems, etc.).
- Construction of livestock production units and of stores for manure (minimum 9 months storage capacity) with the focus on the floor being made of durable and impermeable material, so that liquid cannot leach or flow from the animal houses and manure store to the surroundings.
- The time of year when manures may be applied is restricted (Manure may be applied in the following periods: February–November for solid manures, March–harvest for liquid manures).
- Liquid manure must not be applied by broadspreading, only with trailing hoses or trailing shoes (i.e. placing the slurry on the soil surface in between growing crop plants). On bare soil or grassland, slurries

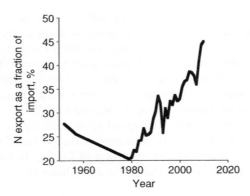

Figure 3.5 *Nitrogen use efficiency presented as N exported in products as a percentage of N imported in fertiliser and feed (Kyllingsbæk, 1995, 2005; Vinther and Olesen, 2011). (© University of Southern Denmark.)*

and other liquid manures must be applied by injection into the soil, be acidified before surface application or applied with a Best available technology (BAT), yielding similar reduction of NH_3 losses as the former.
- Solid manures applied on bare soil must be incorporated within 6 h of application and should not be applied on fields with a potential risk of runoff to streams, lakes or the sea via sloping land, for example.

Control of these regulations is carried out by the farmers keeping mandatory log-books and accounts of the number of animals, cropped land, crops, mineral fertilisers and so on that must be reported in an internet-based accounting system. Thus, the use of manure and fertilisers in the field has to be accounted for in detail, and the documents are inspected both administratively and at farm-level by random site visits.

Since the 1980s, Danish incentives and regulations have reduced losses of N due to nitrate leaching and NH_3 emissions and enhanced efficient use of manure N, which has substituted large amounts of mineral fertilisers. As a consequence, the export of N in plants and animal products as a fraction of N imported in agriculture has increased from 20% in 1980 to about 45% in 2010 (Figure 3.5).

3.7   Summary

Regulations on manure management have been implemented in many rich countries, where animal manure has become a waste rather than a valuable resource of nutrients and organic matter. These regulations define how the manure has to be stored and treated and regulate how the manure has to be applied to land and how much manure can be applied maximally. In contrast, many poor countries do not have government regulations related to manure management, in part because the manure is still treated as a valuable source of nutrients and organic matter.

Government regulations on manure management differ between countries. In the European Union, the Nitrates Directive prescribes how the manure has to be stored and sets limits for the amount of manure applied to land. The NEC Directive set limits for the amounts of NH_3 emitted to air. In the United States, anaerobically treated manure can be spread on small areas of land without accounting for the amount of plant nutrients applied to the land. In Japan and other Asian countries, discharge to surface waters is generally accepted, if the discharged manures have pollutant concentrations below thresholds set by the law. In contrast, discharge of any manure is forbidden in the EU. In Denmark, farmers are only allowed to apply manure to fields with crops, and the amounts applied and the timing are strictly regulated and controlled, the objective being to enhance recycling and the fertiliser value of the plant nutrients in manure.

References

Albihn, A. and Vinnerås, B. (2007) Biosecurity and arable use of manure and biowaste – treatment alternatives. *Livest. Sci.*, **112**, 232–239.

Bouwman, A.F., Lee, D.S., Asman, W.A.H., Dentener, F.J., VanderHoek, K.W. and Olivier, J.G.J. (1997) A global high-resolution emission inventory for ammonia. *Glob. Biogeochem. Cyc.*, **11**, 561–587.

Brandt, J., Silver, J.D., Christensen, J.H., Andersen, M.S., Bønløkke, J.M., Sigsgaard, T., Geels, C., Gross, A., Hansen, A.B., Hansen, K.M., Hedegaard, G.B., Kaas, E. and Frohn, L.M. (2011) Assessment of health-cost externalities of air pollution at the national level using the EVA model system, *CEEH Scientific Report 3*, Centre for Energy, Environment and Health, Copenhagen; www.ceeh.dk/CEEH_Reports/Report_3/CEEH_Scientific_Report3 .pdf.

Bull, K., Hoft, R. and Sutton, M.A. (2011) Coordinating European nitrogen policies between international conventions and intergovernmental organizations, in *The European Nitrogen Assessment* (eds M.A. Sutton, C.M. Howard, J.W. Erisman, G. Billen, A. Bleeker, P. Grennfelt, H. Van Grinsven and B. Grizzetti), Cambridge University Press, Cambridge, pp. 570–584.

Danish Ministry of Environment (2012) *Order on Commercial Livestock, Livestock Manure, Silage etc.*, Danish Ministry of Environment, Copenhagen.

Davidson, E.A. (2009) The contribution of manure and fertilizer nitrogen to atmospheric nitrous oxide since 1860. *Nat. Geosci.*, **2**, 659–662.

Erisman, J.W. and Schaap, M. (2004) The need for ammonia abatement with respect to secondary PM reductions in Europe. *Environ. Pollut.*, **129**, 159–163.

EEA (2012) EEA-32 ammonia (NH_3) emissions; http://www.eea.europa.eu/data-and-maps/indicators/eea-32-ammonia-nh3-emissions; retrieved 6 December 2012.

EPA (2009) EPA targets clean water act violations at livestock feeding operations. *Enforcement Alert*, **10**, 1–4; http://www.epa.gov/compliance/resources/newsletters/civil/enfalert/cafo-alert09.pdf.

Gambotto, A., Barratt-Boyes, S.M., de Jong, M.D., Neumann, G. and Kawaoka, Y. (2008) Human infection with highly pathogenic H5N1 influenza virus. *Lancet*, **371**, 1464–1475.

Grossman, M.R. (2003) *Legal Issues of Livestock Odor and Waste Management*, National Research Council, National Academy of Sciences, Washington, DC.

Haga, K. (2011) Livestock waste composting and wastewater treatment in small-scale farmers, presented at the *International Seminar on Sustainable Resource Management of Livestock and Poultry Wastes for Asian Small-Scale Farmers*, Ho Chi Minh City.

Hebert, T.R. (2009) Agriculture and federal air quality policy outlook, presented at the *National Conference on Mitigating Air Emissions from Animal Feeding Operations: Exploring the Advantages, Limitations, and Economics of Mitigation Technologies*, Iowa State University.

IPCC (2007) Climate change 2007 synthesis report; http://www.ipcc.ch/pdf/assessment-report/ar4/syr/ar4_syr.pdf; retrieved January 2013.

Kyllingsbæk, A. (1995) Kvælstofoverskud i dansk landbrug [Nitrogen surplus in Danish agriculture], *SP Rapport 23*, Landbrugs- og Fiskeriministeriet Statens, Planteavlsforsøg, Copenhagen.

Kyllingsbæk, A. (2005) Næringsstofbalancer og næringsstofoverskud i dansk landbrug 1979–2002 [Plant nutrient balances and surplus in Danish agriculture 1979–2002], *DJF Rapport 116*, Ministry of Food, Agriculture and Fisheries, Danish Institute of Agricultural Sciences, Tjele.

MAFF (2012) *State of Animal Waste Management (in Japanese)*. MAFF, Tokyo; http://www.maff.go.jp/j/chikusan /kankyo/taisaku/t_mondai/02_kanri/index.html; retrieved August 2012.

Menzi, H., Oenema, O., Burton, C., Shipin, O., Gerber, P., Robinson, T. and Franceshini, G. (2010) Impacts of intensive livestock production and manure management on the environment, in *Livestock in a Changing Landscape: Drivers, Consequences and Responses* (eds H. Steinfeld, H. Mooney, F. Schneider and L.E. Neville), Island Press, Washington, DC, pp. 139–164.

Natural Resource Conservation Service (2012) Nutrient management (ACRE) code 590; http://efotg.sc.egov .usda.gov/references/public/WI/590.pdf; retrieved 5 July 2012.

Nicholson, F.A., Chambers, B.J., Williams, J.R. and Unwin, R.J. (1999) Heavy metal contents of livestock feeds and animal manures in England and Wales. *Bioresour. Technol.*, **70**, 23–31.

North Carolina (2012) Standards and regulation of manure management, retrieved November 2012: 15A *NCAC 02T.1307 Swine waste management system performance standards*; http://ncrules.state.nc.us/ncac/title%2015a%20-%20environment%20and%20natural%20resources/chapter%2002%20-%20environmental%20management/subchapter%20t/15a%20ncac%2002t%20.1307.pdf; *15A NCAC 02D.1808 Evaluation of new or modified swine farms*; http://daq.state.nc.us/rules/rules/D1808.pdf; *15A NCAC 02T.1107 Vector attraction reduction requirements*; http://ncrules.state.nc.us/ncac/title%2015a%20-%20environment%20and%20natural%20resources/chapter%2002%20-%20environmental%20management/subchapter%20t/15a%20ncac%2002t%20.1107.pdf; *4   15A   NCAC 02T.1106 Pathogen reduction requirements*; http://ncrules.state.nc.us/ncac/title%2015a%20-%20environment%20and%20natural%20resources/chapter%2002%20-%20environmental%20management/subchapter%20t/15a%20ncac%2002t%20.1106.html; *General Assembly of North Carolina Session Law 2007-523 Senate Bill 1465. An act to (1) codify and make permanent the swine farm animal waste management system performance standards that the general assembly enacted in 1998, (2) provide for the replacement of a lagoon that is an imminent hazard, (3) assist farmers to voluntarily convert to innovative animal waste management systems, and (4) establish the swine farm methane capture pilot program*; http://www.ncleg.net/Sessions/2007/Bills/Senate/HTML/S1465v7.html.

Oenema, O., Bleeker, A., Braathen, N.A., Budnakova, M., Bull, K., Cermak, P., Geupel, M., Hicks, K., Hoft, R., Kozlova, N., Leip, A., Spranger, T., Valli, L., Velthof, G. and Winiwarter, W. (2011) Nitrogen in current European policies, in *The European Nitrogen Assessment* (eds M.A. Sutton, C.M. Howard, J.W. Erisman, G. Billen, A. Bleeker, P. Grennfelt, H. Van Grinsven and B. Grizzetti), Cambridge University Press, Cambridge, pp. 62–81.

Peiris J.S., Malik, de Jong, M.N. and Guan, Y. (2007) Avian influenza virus (H5N1): a threat to human health. *Clin. Microbiol. Rev.*, **20**, 243–261.

Solomon, S., Qin, D., Manning, M., Chen, Z., Marquis, M., Averyt, K.B., Tignor, M. and Miller, H.L. (eds) (2007) *Climate Change 2007: The Physical Science Basis. Contribution of Working Group I to the Fourth Assessment Report of the Intergovernmental Panel on Climate Change*, Cambridge University Press, Cambridge.

Sommer, S.G., Østergård, H.S., Løfstrøm, P., Andersen, H.V. and Jensen, L.S. (2009) Validation of model calculation of ammonia deposition in the neighbourhood of a poultry farm using measured NH_3 concentrations and N deposition. *Atmos. Environ.*, **43**, 915–920.

Steinfeld, H., Gerber, P., Wassenaar, T., Castel, V., Rosales, M. and de Haan, C. (2006) *Livestock's Long Shadow: Environmental Issues and Options*, FAO, Rome.

Stubbs, M. (2012) Environmental regulation and agriculture, *CRS Report for Congress 7-5700, R41622*, Congressional Research Service, Washington, DC.

Suttle, N. (2010) *Mineral Nutrition of Livestock*, 4th edn, CABI, Wallingford.

Sutton, M.A., Oenema, O., Erisman, J.W., Leip, A., van Grinsven, H. and Winiwarter, W. (2011) Too much of a good thing. *Nat. Comment*, **472**, 159–161.

UNECE (2012) Guidance document for preventing and abating ammonia emissions from agricultural sources. United Nations Economic Commission for Europe, Convention on Long-range Transboundary Air Pollution. UNECE, Geneva; www.unece.org/fileadmin/DAM/env/documents/2012/EB/N_6_21_Ammonia_Guidance_Document_Version_20_August_2011.pdf.

Vanotti, M.B., Szogi, A.A., Millner, P.D. and Loughrin, J.H. (2009) Development of a second-generation environmentally superior technology for treatment of swine manure in the USA. *Bioresour. Technol.*, **100**, 5406–5416.

Vinther, F.P. and Olsen, P. (2011) Næringsstofbalancer og næringsstofoverskud i landbruget 1988–2009, *Intern Rapport 102*, Aarhus Universitet, Aarhus.

Xiong, Z.Q., Freney, J.R., Mosier, A.R., Zhu, Z.L., Lee, Y. and Yagi, K. (2008) Impacts of population growth, changing food preferences and agricultural practices on the nitrogen cycle in East Asia. *Nutr. Cyc. Agroecosyst.*, **80**, 189–198.

4

Manure Characterisation and Inorganic Chemistry

Morten L. Christensen[1] and Sven G. Sommer[2]

[1]*Department of Biotechnology, Chemistry and Environmental Engineering, Aalborg University, Denmark*
[2]*Institute of Chemical Engineering, Biotechnology and Environmental Technology,*
University of Southern Denmark, Denmark

4.1 Introduction

Ruminants and monogastric animals excrete plant nutrients and organic matter in the form of urine and faeces. Poultry excreta differ from ruminant and monogastric manure in that only faeces is excreted.

Urine contains the nitrogen (N), phosphorus (P) and potassium (K) from feed that has not been incorporated into animal tissue, milk or eggs, or excreted as faeces. Urine contains components that are surplus to animal needs, and that have been absorbed in the digestive tract of the animal and then transferred to the urine. Faeces contain cell material from the stomach and intestines, microorganisms, and plant nutrients and organic matter in the diet that have not been absorbed by the animal. The plant nutrient content in excreta depends on the nutrient content in the diet eaten by the animals and also on the availability of the components in the diet. A mixture of urine and faeces is called animal manure, but manure usually also contains other materials. The N in excreta from monogastric animals is in the form of urea, a diamide, and proteins. The N compounds excreted by poultry are proteins and uric acid, which is a heterocyclic form of N.

The components of manure in focus depend on how the manure is regarded: as an organic product to ameliorate soil, a fertiliser product, an energy carrier or a risk to the environment. A description of the physico-chemical properties of manure is given here, followed by a discussion of its inorganic components and how the properties and component are determined. Lastly, pH buffering systems, which are very important for the understanding of gas emissions and microbial activity in manure, are presented. The organic content of manure and the structure of the material in manure are discussed in more detail in Chapter 5.

Animal Manure Recycling: Treatment and Management, First Edition. Edited by Sven G. Sommer, Morten L. Christensen, Thomas Schmidt and Lars S. Jensen.

4.2   Livestock Manure Categories

Animal manure from animal houses is a mixture of faeces and urine plus bedding, spilt feed, spilt drinking water and water used for washing. The composition is therefore very variable and affected by manure removal technologies.

The definitions of livestock manure may differ between countries as manure categories are often related to livestock housing design or outdoor production systems where the manure is collected. Manure may either be managed as liquid or solid manure (Text Box – Basic 4.1). Liquid manure is defined as manure with a dry matter content lower than 12%. This can be transported by drainage or pumping. Solid manure is defined as manure that can be stacked and has a dry matter content higher than 12%.

Text Box – Basic 4.1   Manure categories (Pain and Menzi, 2003)

Compost: Solid manure that has been stored in heaps where air replenishment has supported aerobic microbial transformation contributing to a significant temperature increase (composting).

Deep litter: Faeces or droppings and urine mixed with large amounts of bedding (e.g. straw, sawdust paper and wood shavings). Deep litter is produced on the floor of housing where straw, wood chips or paper strewn on the floor retain faeces, urine and spilt water.

Dry matter content (DM): The fraction of manure remaining after drying the manure at 100 °C until no more water evaporates (usually 24 h).

Droppings: Excreta voided by poultry (i.e. poultry do not urinate).

Excreta: Faeces and urine expelled from the body, excreta consist of faeces and urine.

Faeces: Solid waste matter discharged from the body – the terms excrement and excreta are also used to define this manure.

Farmyard manure: Faeces and urine mixed with some bedding material from the floors of cattle or pig housing. The liquid fraction has been drained of the manure.

Liquid manure: A general term that denotes any manure from housed livestock that flows under gravity or can be pumped.

Manure: A general term to denote any organic material that supplies organic matter together with plant nutrients to soils. Organic material from livestock production should be defined as animal manure or livestock manure, but in most texts the term manure defines organic material containing excreta from livestock.

Muck: A colloquial term for livestock manure.

Urine: Wastes removed from the blood stream via the kidneys and voided as a liquid.

Slurry: Faeces and urine produced by housed livestock, usually mixed with some bedding material and some water during management to give liquid manure that can be pumped. Slurry dry matter content is below 12%.

Solid manure: Manure from housed livestock. Solid manure does not flow under gravity and cannot be pumped. The dry matter of solid manure is higher than 12% and solid manure can be stacked in a heap.

Slurries collected from below slatted floors have low dry matter content due to the addition of washing water and low use of bedding materials. In general, the dry matter content of cattle slurry is higher than that of pig slurry, but within each animal category the variation between farms is extreme (Table 4.1). In Asia, water consumption is high and the dry matter of pig slurry is therefore $0.2 \pm 1\%$. In contrast, slurries from Dutch pig houses with restricted water consumption may have dry matter contents close to 10%. New housing systems with slatted floors and slurry collection have been developed with the focus on reducing

Table 4.1 Concentrations of plant nutrients in animal manure: the composition of manures varies widely, so these values should be used normatively when local or measured data are lacking (average ± SE).

Animal category	Manure category	Dry matter (g kg⁻¹ manure)	Total-P (g kg⁻¹ dry matter)	Total-N (g kg⁻¹ dry matter)	TAN (g kg⁻¹ dry matter)	K (g kg⁻¹ dry matter)	Ca (g kg⁻¹ dry matter)
Sows	slurry	23 ± 15	42 ± 18	165 ± 61	114 ± 75	112 ± 60	46 ± 9
Fatteners	slurry	60 ± 26	30 ± 10	112 ± 32	77 ± 26	61 ± 23	38 ± 8
Fatteners – Asia	liquid manure	0.9 ± 0.1					
Dairy cows	slurry	77 ± 21	11	42 ± 24	22 ± 10	59 ± 23	22
Biogas – pig	slurry	30 ± 46	40 ± 24	241 ± 114	186	105	
Biogas – mix[a]	slurry	17 ± 6	6 ± 44	44			
Dairy cows	FYM	211 ± 33	23 ± 7	133 ± 30	25 ± 17	202 ± 31	
	deep litter	418 ± 174	4 ± 3	22 ± 8	6 ± 5	21 ± 12	
Beef cattle	solid manure	231 ± 207		106 ± 123	11 ± 12	21 ± 12	
Poultry	litter	601 ± 114	21	39.0 ± 0.4	7.0 ± 0.5	29.9 ± 0.7	

[a]Pig slurry + 20% biomass from food industry.

water consumption, because water may be a limited resource and water in slurry adds to the costs if it needs transporting over longer distances to its final destination. The dry matter content in these slurries is therefore usually high.

In housing systems with livestock tied in stalls with a solid straw-covered floor, the animals excrete to a gutter behind the resting area. In these systems the solid manure is often scraped off the floor producing a solid manure (farmyard manure (FYM)) mainly containing faeces and straw. The urine and washing water drains off the floor and is collected as liquid manure, which contains washing water, urine and dissolved faecal components. These systems produce a FYM with a high dry matter content and liquid manure with low dry matter content. In Asia, fattening pigs may be raised in pens with solid floors, where the excreta are scraped off the floor manually and the liquid and some excreta are hosed down to a pond.

In loose housing systems straw is spread on the floor producing a deep litter mat, which may reach depths of up to 1–2 m. Faeces and urine are deposited in the deep litter, which has a very high dry matter content and is well structured. In cattle houses the deep litter is compacted, giving it a high density, whereas in pig houses the deep litter tends to be porous because pigs build nests and root around in the litter. This gives good conditions for the composting process, which may start in the animal house. Composting is a microbial aerobic transformation process of the organic matter, which produces heat. In cattle houses litter in the surface layers (10–15 cm) may start to compost, and then with time and with continuous addition of straw and excreta, the litter will be covered and buried beneath new litter and compacted. Thus, about 10–15 cm below the surface there is no composting going on as the oxygen available for the process is consumed in the top layer of 10–15 cm.

Poultry may be housed in loose housing systems where droppings are collected on the floor, which are given a light covering of straw, wood chips or sand. Poultry are also housed in caged systems. Here the droppings are either stored below the cages or collected on a conveyor belt below the cages. The conveyer belt transports the droppings to an external store for dry solid manure (Koerkamp, 1994). Alternatively droppings may be stored as slurry in pits below the cages after addition of water. The slurry and the dry solid chicken manure have a high variation in dry matter content (i.e. from 5% to 27% in slurry and 31% to 67% in dry solid manure). The variation in dry matter content of the solid manure is probably caused by drinking water leaking to the manure (Kroodsma *et al.*, 1988).

Storage and treatment of manure after removal from the animal house affect its composition. The liquid manure and livestock slurries are in most systems transferred to a lagoon or a concrete or steel tank. Little air diffuses into the liquid manure, leading to anaerobic storage conditions. As a consequence, the composition only changes relatively slowly with time. The slurry may be treated with slurry separators to produce a solid and a liquid fraction of the manure.

In stored deep litter and straw-rich FYM, air is transported to the interior of the heap by convection, and oxygen is available for aerobic transformation of the organic matter. The process basically helps to transform organically bound nutrients into inorganic nutrients. FYM from tied cattle houses is a special case, because the heaps are often anaerobic due to a combination of high water content and high density of the organic matter (Forshell, 1993). The density of cattle faeces is high due to the plant material being transformed and mixed in the rumen of the cattle. Chicken droppings have a high level of structure and will start to compost during storage. Thus, the solid manure may naturally or due to active aeration be transformed into compost during storage.

Manure from beef cattle tends to have a higher nutrient content than from dairy cattle. For liquid manure this difference is mainly due to the higher dry matter content of the beef cattle slurry.

Solid and liquid pig manure tends to have a higher plant nutrient content than the equivalent cattle manure. Pigs are monogastric animals and do not absorb plant nutrients, and use the energy in feed as efficiently as ruminant cattle. On the other hand, in warmer climates liquid pig manure tends to get diluted (which lowers the nutrient content), due to hosing or sprinkling the animals with water. Furthermore, in Asia the liquid is low in dry matter and plant nutrients due to the removal of the solids before the liquid is hosed off the floor. The rationale for cleaning the houses is that this reduces the risk of disease and also odour emissions.

The driest manures are from poultry; broiler manure is often very dry owing to the low moisture content in bird droppings, the large amount of litter used and drying effects within the building. The plant nutrient concentration of poultry is consequently highest among manures from different animal categories.

The standard values for the composition within manure categories vary considerably between countries (Table 4.1). Part of this variation is due to differences in livestock production techniques, such as composition of diets, water consumption and manure management.

4.3   Physical Characterisation of Manure

The physical properties of manure (Text Box – Basic 4.2) include slurry density, rheological properties, pH, ionic strength, particle size and surface charge. These characteristics have to be known for the design of pumps, separation equipment and storage tanks. The slurry density and rheological properties of manure are important for estimating the energy requirement for pumping and handling the manure. The particle size influence sedimentation during storage. It is particularly the lower end of the particle size distribution that is important when considering sedimentation of particles in slurry stores, biogas reactors or in the slurry separation process. In addition to particle size, density and liquid viscosity affect the rate of settlement in the technologies that separate particles, P and N from the slurry. Particle size and rheological properties are important for other separation techniques too. Particle surface charge and ionic strength are important for the efficient use of flocculants to remove P and organic matter from liquid waste and when assessing ammonia (NH_3) volatilisation.

Text Box – Basic 4.2   Physical characteristics

Colloids: Particles in the range between approximately 1 nm and 1 μm.
Dissolved suspended solid (DSS): Particles that can pass filters of a pore size of approximately 1 μm and are defined as dissolved as they are assumed not to sediment during storage.
Newtonian fluid: A fluid in which the components of the stress tensor are linear functions of the first spatial derivatives of the velocity components, which means that the liquid at constant temperature and pressure retains constant fluidity at all stresses (e.g. water).
Plastic flow: Steady flow occurring only above a certain finite stress (i.e. the fluid is non-Newtonian).
Suspended solid (SS): Particles retained on filters with a pore size of 1 μm.

4.3.1   Particle Size

In studies of particle size in animal slurry, very different filter sizes have been used, as no standards have been agreed for this analysis, and the characterisations of the biowaste particles are therefore arbitrary. A series of sieves of progressively smaller mesh sizes are usually mounted on top of each other, with the largest mesh size at the top. Manure is added to the top sieves and with a gentle sprinkling of recycled liquid the particles are transported down through the rack until captured on the sieve with a mesh size smaller than the particle size. The bottom sieve should preferably be 1 μm, because particles with diameter less than 1 μm are defined as dissolved suspended solids (or colloids). Colloids are subject to Brownian (diffusive) motion in the liquid, and will not settle or settle only very slowly by gravitation. For that reason, the amount of colloids is important, as colloids are difficult to remove during solid–liquid separation.

Other methods exist for measuring particle size (e.g. different light scattering techniques), but measurement is complicated because of the large particle size distribution and the irregular structure of the particles.

The amount of dry matter in the fraction below 0.025 mm is larger in pig slurry than in cattle slurry, 66–70% and 50–55% of dry matter, respectively. Microbial transformation of the organic pool changes the particle size distribution due to transformation of organic components to carbon dioxide (CO_2), methane (CH_4) and ammonium ($NH_4^+(aq)$), for example. Therefore, anaerobic fermentation during long-term manure storage or in a biogas reactor both reduces particle concentration of animal slurry and changes the partitioning between the fractions of large and small particles. Particles less than 10 μm may account for 64% of dry matter in raw slurry, while the percentage increases to 84% of dry matter in anaerobically digested slurry (Hjorth *et al.*, 2010).

Information about the plant nutrient content of particle size fractions is important when assessing non-homogeneity of the stored animal slurry (e.g. roughly 70% of the undissolved N and P occurs in the 0.45- to 250-μm particle size fraction of cattle slurry). More than 80% of total-N and total-P is in the less than 0.125-mm fraction of cattle slurry (Meyer *et al.*, 2007). At present there are no data available on the amount of plant nutrients in the particle fraction below 1 μm or how this fraction may vary due to manure storage time.

4.3.2   Manure Density and Viscosity

Slurry density is related to dry matter content, and dairy cattle slurry has a lower density than pig slurry at the same dry matter content. The following empirical equations can be used to assess density at a dry matter content below 10% (Thygesen *et al.*, 2012):

- Pig slurry:

$$\rho = \frac{(DM + 279)}{0.28} \tag{4.1}$$

- Cattle slurry:

$$\rho = \frac{(DM + 236)}{0.24} \tag{4.2}$$

where ρ is the slurry density (kg m^{-3}). Landry *et al.* (2004) give this area a thorough review in connection with their own work on concentrated slurry and present algorithms for the calculation of manure density at dry matter contents higher than 12%.

The flow properties for slurry have, in general, been found to be non-Newtonian at dry matter values above 5%, while at lower dry matter values the slurry behaves as a Newtonian fluid (Landry *et al.*, 2004). In general, animal slurry with a higher dry matter content shows pseudoplastic behaviour, but for simple flow considerations an apparent viscosity will often suffice.

Viscosity is a measure of the resistance of a fluid which is being deformed by either shear stress or tensile stress. It is the most important parameter affecting the transport and movement of liquids in porous media such as soil, stored solid manure, stirred liquid manure or slurry transported through pipes. The viscosity can be determined with a stirring viscosimeter, where the force to stir at a specific rate is measured (slurry with a high content of large particles cannot be measured with this technique). Alternatively, the pressure drop in straight tubes can be measured at increasing slurry flow rates and using the values obtained to calculate the slurry viscosity by comparing the pressure drop of water flow through the same system.

Landry *et al.* (2004) correlated their viscosity data, measured for a dry matter between 9 and 14% at 20 °C, as:

- Pig:

$$\mu_{slurry} = 4 \cdot 10^{-6} \cdot DM^{4.6432} \tag{4.3}$$

- Dairy cattle:

$$\mu_{slurry} = 4 \cdot 10^{-5} \cdot DM^{4.4671} \tag{4.4}$$

where μ_{slurry} is the apparent slurry viscosity (Pa s). These data should be used with caution, as also mentioned by Landry *et al.* (2004), but the conclusion that pig slurry is less viscous than cattle slurry at a comparable dry matter content is supported by observations in the field. Pig slurry, therefore, infiltrates the soil more effectively than cattle slurry.

4.3.3 Electrochemical Properties

The concentration of ions, expressed as ionic strength (I), is high in most animal slurry studies. This is revealed as a high electric conductivity (EC), because EC is related to the concentration and species of ions in solution. The EC in animal wastes is in the range from 0.008 to 0.026 S cm^{-1} (Sommer and Husted, 1995a; Christensen *et al.*, 2009). For river water, salt water and slurry, I is linearly related to EC and is given by the following equation, which can be very convenient to use in a solution where not all ionic species concentrations have been measured (Sommer and Husted, 1995a; Griffin and Jurinak, 1978).

$$I = 15.8 \cdot EC_{20} \tag{4.5}$$

where EC_{20} (S cm^{-1}) is the electric conductivity at 20 °C. Also see Text Box – Basic 4.3.

Text Box – Basic 4.3 Activities of ions and ionic strength

The ionic strength is a function of the concentration and valence of ions in the solution. The ionic strength, I, of a solution is a function of the concentration of all ions present in that solution:

$$I = \frac{1}{2} \cdot \sum_{i=0}^{n} \left(c_i z_i^2 \right) \tag{4.6}$$

where c_i is the molar concentration of ion i (M = mol l^{-1}) and z_i is the valence (charge number) of that ion. The sum is taken over all ions in the solution. For a 1 : 1 electrolyte such as NaCl, the ionic strength is equal to the sum of concentration, but for MgSO$_4$ the ionic strength is 4 times higher. The ionic strength plays a central role in the correct calculation of the chemical reactions of ions.

The equilibrium constant is defined as a quotient of activities of the ions involved in the reaction (A) instead of the concentration of the ion [A]. The activity of an ion is given as $(A) = [A] \cdot \gamma_A$ where γ_A is the activity coefficient. Thus, the equilibrium is calculated by expressing the equilibrium constant as:

$$K = \frac{(\gamma_s \cdot [S])^{Z_s} \cdot (\gamma_T \cdot [T])^{Z_T}}{(\gamma_A \cdot [A])^{Z_A} \cdot (\gamma_B \cdot [B])^{Z_B}} \qquad (4.7)$$

where [A] denotes the concentration of A, etc. The activity coefficient can be calculated by use of the extended Debye–Hückel equation, which is valid for an ionic strength up to around 0.01 mol l^{-1}:

$$\log(\gamma_i) = -A \cdot z_i \times \frac{I^{0.5}}{1 + B \cdot d_i \cdot I^{0.5}} \qquad (4.8)$$

where A and B are the temperature-dependent constants, and d is the effective diameter of the hydrated ion (in Å).

The surface charge of a particle, especially colloids, is important for the stability of that particle (i.e. whether the particles aggregate or not). Ions often adsorb to the surface of charged particles, forming the so-called Stern layer (Figure 4.1). Around a charged particle there will be an increased concentration of counter-ions. The concentration of counter-ions decreases from the Stern layer towards the bulk solution (Figure 4.1). This layer is usually called the diffusive layer. The Stern layer and the diffusive layer are together called the electric double layer. The thickness of the diffusive layer depends on the ion strength in the solution and is often characterised using the Derby length $(=1/\kappa)$:

$$\kappa = \sqrt{\frac{F^2 \cdot I}{\varepsilon \cdot RT}} \qquad (4.9)$$

where F is Faraday's constant and R is the gas constant.

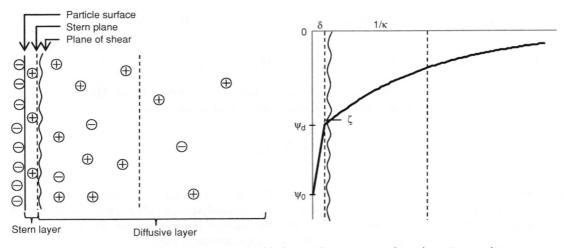

Figure 4.1   *Model of the electric double layer. (© University of Southern Denmark.)*

The Derby length is a key parameter for coagulation and flocculation. At high ionic strength, the Derby length is low and it is easier for charged particles to coagulate than at low ionic strength because the ions shield the charged groups on the particles. A more thorough description can be found in Lyklema (1977).

The electric surface potential (mV) of particles is also important for the stability of the particles. The electrical potential decreases from the surface of the particles to the bulk solution and is usually calculated using the electric double-layer model (Figure 4.1). Near the surface the potential changes linearly due to adsorbed ions (Stern layer); after the Stern layer, the potential decreases exponentially depending on the ionic strength (diffusive layer). It is seldom possible to measure the electric surface potential. Instead, the electrokinetic potential (zeta potential) is used as a measure of the surface potential. The zeta potential is not that at the particle surface, but the potential that exists at the shear plane of the particle. The shear plane of the particle lies in the liquid phase surrounding the particle and is usually located close to the Stern layer. Hence, the zeta potential includes both the charge at the particle surface and the charge of the adsorbed ions. The zeta potential can be calculated from the electrophoretic mobility of the particles, which can be found by measuring the velocity of the particles in a known external electrical field:

$$\mu_E \equiv \frac{v_E}{E} = \frac{\varepsilon \cdot \zeta}{1.5\eta} \tag{4.10}$$

where v_E is the electrophoretic velocity (m s^{-1}), E is the electric field (N C^{-1}), ζ is the zeta potential (V), ε is the permittivity (C^2 N^{-1} m^2) and η is the viscosity (Pa s). The equation is derived for hard particles, so the measured zeta potential must be regarded with caution. At low potentials the charge density is given as $\sigma = \varepsilon \cdot \kappa \cdot \zeta \cdot F^{-1} \cdot S_0^{-1}$ (eq kg^{-1}), where F is the Faraday constant (C mol^{-1}) and S_0 is the specific surface of the particles (m^2 kg^{-1})

An alternative method for determining particle surface charge is polymer titration, where a charged polymer with known charge density is added to the sample until the electrophoretic mobility is zero and the particle neutralised. The polymer needs to have the opposite charge of the particles. Assuming that all polymers adsorb to the particles, the charge density can be calculated as:

$$\sigma = \frac{\sigma_{pol} \cdot C_{pol} \cdot V_{pol}^{\mu_E=0}}{V \cdot C} \tag{4.11}$$

where $V_{pol}^{\mu_E=0}$ is the added volume of cationic polymers (m^3) at $\mu_E = 0$, C_{pol} is the concentration of polymers (kg m^{-3}), V is the sample volume (m^3), C is the concentration of the particle (kg m^{-3}) and $\sigma_{polymer}$ is the charge density of the added polymer (eq g^{-1}) at $\mu_E = 0$ (Christensen *et al.*, 2009).

Studies shows that the organic particles in manure have a negative surface charge. The negative charge has been shown to contribute significantly to the alkalinity of the slurry, being 0.013 M in a slurry containing 1% dry matter. Furthermore, it has been shown that in pig slurry the particle charge density is approximately − 0.18 meq g^{-1} organic solids (Hjorth *et al.*, 2010). Alkalinity and charge density are two interrelated ways to express the negative charge of the organic particles.

4.4   Manure Inorganic Chemistry

In manure a fraction of the inorganic ions is in solution, the remainder being either adsorbed to the charged sites of organic particles, precipitated as crystals or a component of the organic matter. In general, divalent ions (e.g. calcium (Ca^{2+}) and magnesium (Mg^{2+})) have a higher adsorption affinity and a lower solubility than monovalent ions (e.g. sodium (Na^+) and potassium (K^+)) due to the electrostatic interaction forces. Adsorption and crystallisation affect the concentration of inorganic ions in the liquid fraction of the manure, as does microbial transformation of the organic components of manure.

4.4.1   Ions in Solution

In fresh urine or faeces from pigs and cattle there are only traces of liquid total ammoniacal N (TAN = $NH_3(aq) + NH_4^+(aq)$). During storage of manure, urea and organic N or uric acid are transformed to TAN, which is mainly dissolved in the liquid phase. The TAN concentrations are lower in cattle slurry than in pig slurry, because the pigs are fed diets with a high organic N content and the organic N in pig excreta is more readily transformed to TAN than the organic N in cattle excreta. TAN is lower in solid than in liquid manure due to transformation to organic N and larger losses of TAN due to NH_3 and N_2 emissions. Poultry excrete N as uric acid and if the droppings are stored dry, the microbial transformation of uric acid to TAN is very slow and little TAN is dissolved in the liquid phase. The TAN concentration in the liquid phase is also affected by the formation of the crystal struvite and NH_4^+ adsorption to colloids.

Cattle slurry contains two to three times more Na^+, K^+ and Ca^{2+} than pig slurry (Sommer and Husted, 1995a; Massé *et al.*, 2007), due to their intake of roughage, grass and silage with a high content of these cations. The pig diet contains a relatively higher proportion of seeds from cereals, soya and so on with a smaller content of these cations. In cattle slurry, the content of dissolved ions varies by a factor of 2 due to variation in diets (Chapuis-Lardy *et al.*, 2004) and is probably more variable than in pig slurry as pigs are fed a more standardised diet. A relatively small fraction of the Ca^{2+} occurs in solution, because Ca^{2+} precipitates easily with P and carbonate (Bril and Salomons, 1990) and to negative charges of particles composed of organic matter (Masse *et al.*, 2005).

Pigs excrete a relatively high proportion of the P intake, because they are fed additional P to cover their needs. Most of the P in animal manure is in the particulate fraction of slurry and less than 30% is dissolved in the liquid phase (Christensen *et al.*, 2009). Of the dissolved P, more than 80% is in the form of orthophosphate ($PO_4^{3-}(aq)$). The $PO_4^{3-}(aq)$ fraction may vary during storage – increasing immediately after excretion to an optimum and then decreasing slightly.

Pig manure often contains high amounts of copper (Cu^{2+}) and zinc (Zn^{2+}) (Møller *et al.*, 2007a, 2007b; Table 4.2), because these trace metals (heavy metals) are added to the diets with the purpose of reducing

Table 4.2   *Concentration of micronutrients/heavy metals in animal manure: the composition of manures varies widely, so these values should be used normatively when local or measured data are lacking (average ±SE).*

Animal category	Manure category	Dry matter (manure))	Mg (g kg^{-1} (manure))	Cu (g kg^{-1} (manure))	Zn (g kg^{-1} (manure))
Sows	slurry	23 ± 15	11 ± 4	0.5	
Fatteners	slurry	60 ± 26	15 ± 5	0.4 ± 0.2	0.7 ± 0.4
Dairy cows	slurry	77 ± 21	9	0.08 ± 0.02	0.02 ± 0.4
Biogas – pig	slurry	31 ± 3	13	0.7	2
Biogas – mix[a]	slurry	46 ± 17		0.5 ± 0.4	1.3 ± 0.7
Dairy cows	FYM	211 ± 33			
	deep litter	418 ± 174			
Beef cattle	solid manure	231 ± 207		0.02	0.1
Poultry	litter	601 ± 114		0.4 ± 0.3	0.5 ± 0.1

[a]Pig slurry + 20% biomass from food industry.

pathogenic microorganisms in the intestines of piglets. Magnesium concentrations are high in animal manure and originate from the diet, as Mg^{2+} is an essential element in both animal and plant physiological processes as a co-factor for enzymes. Magnesium to a large extent forms crystals with carbonate and phosphate.

4.4.2 pH Buffer System

The most important ion in solution is the proton (H^+) or oxonium (H_3O^+) ion. The proton concentration affects inorganic and microbiological processes in animal manure, so the pH buffering system in manure deserves special attention.

Proton concentrations are usually measured with pH electrodes. In most environments, such measurements are easy to carry out and are considered reliable. This is not the case with animal manure, where pH differs between the bulk and the surface because of evaporation of pH buffers to the atmosphere. Thus, researchers need to consider whether their study or models deals with an average bulk process or a surface process when measuring pH. It should also be borne in mind that the electrodes measure the activity of (H^+); pH is equal to the negative logarithm to the activity of the proton (H^+) concentration. The activity of the proton is defined as (H^+) $= \gamma \cdot [H^+]$ and decreases at increasing ionic strength of the solution (Text Box – Basic 4.3). Most studies and models do not account for this fact and consequently process calculations may be erroneous.

In manure, the calculated $[H^+]$ may be about 20% different from the real proton concentrations $[H^+]$ due to the high ionic strength of the slurry. At very high pH levels small ions may interfere significantly with the measurements, and at $Na^+(aq)$ or $K^+(aq)$ concentrations above 0.1 mol l^{-1} the pH may be overestimated by 0.3 pH units (Sommer and Husted, 2005b). At very low pH the activity of (H_3O^+) and the ionic strength become so high that this also interferes significantly when interpreting the measurements, i.e. assessing the concentrations of the ions.

The main buffer components in animal manure controlling pH are liquid total inorganic carbon (TIC $=$ $CO_2(aq) + HCO_3^-(aq) + H_2CO_3(aq)$), total ammoniacal N (TAN $=$ $NH_3(aq) + NH_4^+(aq)$) and volatile fatty acids (VFAs $=$ C_2–C_5 acids; Vavilin *et al.*, 1998). High-molecular-weight organic matter with carboxyl functional groups may also contribute to the pH buffering capacity (Sommer and Husted, 1995a). The chemical equilibrium reactions for the buffer component are presented in Table 4.3.

The pH buffer components are weak acids or bases, so pH calculations involve several species of the same component. An example is the acid–base pair acetic acid and acetate, which is one of the VFAs. When measuring this acid–base pair in biomass the analytical procedure estimates the concentration of the sum and only by knowing the pH of the solution can the concentration of each of the components be assessed (Figure 4.2).

Weak acid and bases are usually denoted with acronyms: TIC for the CO_2, TAN for the ammonium and VFAs for the organic acid buffer system (Text Box – Basic 4.4).

Table 4.3 Reactions that have to be accounted for when calculating pH in manure liquid (at 25 °C and 1 bar).

Reaction	Equilibrium constant	pK_a	Equation
$NH_4^+(aq) + H_2O(l) \rightleftharpoons NH_3^+(aq) + H_3O^+(aq)$	$K_N = 10^{-9.3}$	9.30	(4.12)
$HCO_3^-(aq) + H_2O(l) \rightleftharpoons CO_3^{2-}(aq) + H_3O^+(aq)$	$K_{HCO_3} = 4.69 \times 10^{-11}$	10.3	(4.13)
$H_2CO_3(aq) + H_2O(l) \rightleftharpoons HCO_3^-(aq) + H_3O^+(aq)$	$K_{CO_2} = 4.45 \times 10^{-7}$	6.32	(4.14)
$CH_3COOH(aq) + H_2O(l) \rightleftharpoons CH_3COO^-(aq) + H_3O^+(aq)$	$K_{CH_3COOH} = 1.9 \times 10^{-5}$	4.72	(4.15)
$2H_2O(l)OH^-(aq) + H_3O^-$	$K_{H_2O} = 10^{-14}$	14	(4.16)

Text Box – Basic 4.4 pH buffer components in biomass

TAC: Total acidity is the amount of hydroxide that has to be added to the biomass to increase pH to 8.3. At pH 8.3, the concentrations of the acids $CH_3COOH(aq)$, $NH_4^+(aq)$, $HCO_3^-(aq)$ and $H_2CO_3(aq)$ are significantly decreased. The units of TAC are meq l^{-1}.

TAL: Total alkalinity is the amount of equivalents of hydrons ($H^+(aq)$) that has to be added to the biomass to reduce pH to 4.5. At pH 4.5, the concentrations of the bases $CH_3COO^-(aq)$, $NH_3(aq)$ and $HCO_3^-(aq)$ are significantly decreased. The units of TAL are meq l^{-1}.

TIC: Total inorganic carbon $= [CO_2] + [HCO_3^-] + [CO_3^{2-}]$.

TAN: Total ammoniacal N $= [NH_3] + [NH_4^+]$.

VFAs: Volatile fatty acids $= [CH_3COOH] + [CH_3CH_2COOH] + [CH_3CH_2CH_2COOH] + [CH_3CH_2CH_2CH_2COOH]$. These acids are often denoted C_2–C_5 acids. Formic acid (methanoic acid $=$ COOH) is not found in the biomass.

The buffer concentration is affected by microbiological transformation of the organic components and emission of volatile acids and bases. The main source of dissolved TAN and TIC in newly excreted manure is hydrolysis of urea, which produces a mixture of $NH_3(aq)$, $NH_4^+(aq)$, $HCO_3^-(aq)$ and $CO_3^-(aq)$. Fresh urine contains high concentrations of urea that are transformed to TAN and TIC. High concentrations of $NH_3(aq)$ and $CO_3^{2-}(aq)$ build up in the urine and as these ions are bases, the pH in fresh urine increases to about 8.5 (Sommer *et al.*, 2006).

The pH is around 7.5 in mixtures of faeces and urine such as fresh slurry, deep litter or FYM, because faeces contain short-chain organic acids (VFAs), which reduce the pH in fresh slurry, deep litter or FYM. VFAs are produced through microbial decomposition of organic components in slurry under anaerobic storage conditions and the predominant VFA in animal slurries is acetic acid (more than 60%; Cooper and Cornforth, 1978). Anaerobic microbial transformation of organic matter transfers approximately 40% of the carbon in the component to $CO_2(aq)$ and aerobic decomposition transfers 100% of the carbon in the component to $CO_2(aq)$, which contributes significantly to the pool of TIC. Microbial transformation may also affect the

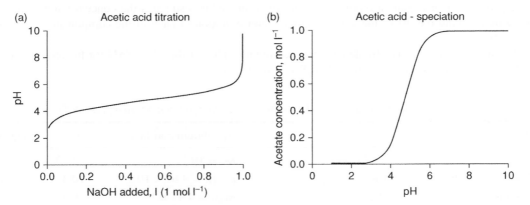

Figure 4.2 *Titration curve of 1 l of 1 M acetic acid (CH_3COOH). (a) Change in pH as affected by addition of 1 M NaOH. (b) CH_4COOH is transformed to CH_4COO^- as pH increases due to NaOH addition. (© University of Southern Denmark.)*

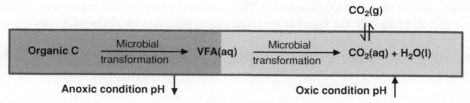

N components. An example is the nitrification of $NH_4^+(aq)$, which produces $H^+(aq)$ in the ratio of 1 : 2 according to:

$$NH_4^+(aq) + 2O_2(aq) + H_2O(l) \rightarrow NO_3^-(aq) + 2H_3O^+(aq) \tag{4.17}$$

and denitrification (Petersen *et al.*, 1996):

$$5(CH_2O(aq)) + 4HNO_3(aq) \rightarrow 5CO_2(aq) + 7H_2O(l) + 2N_2(aq) \tag{4.18}$$

Furthermore, TIC and TAN concentrations may be reduced due to formation of the crystals calcite and struvite. Therefore, the pH in manure varies considerably during storage. This variation is also affected by treatments that influence these reactions and processes. Such treatments include aeration, addition of coagulants to sediment out $PO_4^{3-}(aq)$ or fermentation in biogas plants.

With time, oxic degradation of organic material in manure reduces the content of acids in solution and thereby increases the pH, while anoxic processes contribute to the formation of organic acids and thereby reduce pH (Figure 4.3). Examples are stored slurry where the environment is predominantly anaerobic and organic material is degraded to volatile organic acids (VFAs = C_1–C_5). In contrast, the organic acids are degraded to CO_2 in the aerobic environment of FYM and deep litter, which increases the pH with time. The pH of manure therefore differs between solid manures through which air is moving and anaerobic slurry or compact solid manures, through the bulk of which no air flows. The concentrations of TIC and TAN in manure are also affected by volatilisation of, respectively, $CO_2(aq)$ and $NH_3(aq)$. From the surface of manure, liquid $CO_2(aq)$ is released more readily than $NH_3(aq)$ due to the lower solubility of $CO_2(aq)$ than of $NH_3(aq)$. The greater loss of TIC than of TAN increases the pH (Equations 4.12–4.14).

TIC in manure can be decreased by adding salts, which precipitates carbonate and thus lowers pH and also stops volatilisation of CO_2. Addition of calcium dichloride ($CaCl_2(s)$) or calcium sulfate ($CaSO_4(s)$) has been tested and proven to be most efficient in significantly lowering the pH. At low pH, TAN exists only in the form of $NH_4^+(aq)$, which cannot volatilise and NH_3 emissions are reduced. The reactions contributing to the pH reductions in slurry to which $CaCl_2(s)$ is added are:

$$HCO_3^-(aq) + Ca^{2+}(aq) + 2Cl^-(aq) \leftrightharpoons CaCO_2(s) + H^+(aq) + 2Cl^-(aq) \tag{4.19}$$

Increasing or decreasing the concentration of ionic species in the urine or slurry affects the pH, because the electric charge of the solution has to be neutral (Sommer and Husted, 1995b). At present, soya beans in the diet supply most of the crude protein needed by pigs, and soya contains high concentrations of K^+, which gives a high pH in the manure. Reducing the soya concentration in the diet and supplementing with amino acids decreases the $K^+(aq)$ concentration and increases the $H^+(aq)$ concentration (and reduces the pH). The effect of adding or removing $K^+(aq)$ on slurry pH can be explained by the effect of these ions on the sum of the positive and negative charge of ions in solution. The sum of the charges of the ionic species in the solution has to be zero (i.e. there must be electroneutrality). The electroneutrality for the system is calculated as:

$$Z_{sys} = ([NH_4^+] + [Na^+] + [K^+] + [H^+] + 2[Ca^{2+}] + 2[Mg^{2+}]) - ([HCO_3^-] + 2[CO_3^{2-}] + [CH_3COO^-] + [OH^-])$$

(4.20)

From this equation it can be seen that adding $K^+(aq)$ in feed, which causes increased excretion of $K^+(aq)$, has the consequence of increasing the concentration of $OH^-(aq)$ and the pH (Cahn *et al.*, 1998). Thus, $NH_3(g)$ emissions can be reduced if $K^+(aq)$ in feed can be reduced. This can be achieved by replacing feed components such as oilseed rape with amino acids to ensure that the amino acid requirement of the pigs is fulfilled.

If the concentration of the buffer components and biowaste composition is known, then the change in pH as affected by losses of the pH components can be assessed using the electroneutrality principle. Text Box – Advanced 4.1 describes a model that can be used for this calculation based on (i) the sum of charge of ions in the liquid being zero and (ii) conservation of species of TAN, TIC, VFAs and inorganic ions.

Text Box – Advanced 4.1 Algorithms for assessing the concentration of $NH_3(aq)$ as affected by changes in the pH buffer components TAN, TIC and VFAs in slurry only containing the inorganic ions Na^+ and Cl^-

The concentrations of the three components are: TAN 0.15 mol l^{-1}, TIC 0.14 mol l^{-1} and acetic acid 0.05 mol l^{-1}. The concentrations of Cl^- and Na^+ are, respectively, 0.25 and 0.25 mol l^{-1}. It is assumed that VFAs is only composed of acetic acid (CH_3COOH).

In the liquid the concentrations of the soluble monovalent ions Na^+ and Cl^- are constant and the charge is not affected by changes in the concentration of H_3O^+ or volatilisation. The concentration of the other cations and anions will change due to volatilisation of $CO_2(g)$ and $NH_3(g)$ and to changes in $[H_3O^+]$. The calculations also include the autoprotolysis of water. Listed below are the equations needed to carry out this calculation:

$$[TAN] = [NH_3] + [NH_4^+]$$

(4.21)

$$[TIC] = [H_2CO_2] + [HCO_2^-] + [CO_2^{2-}]$$

(4.22)

$$[NH_4^+] = \frac{[TAN]}{\dfrac{K_N}{[H_3O^+]} + 1}$$

(4.23)

$$[HCO_3^-] = \frac{[TIC]}{\dfrac{[H_3O^+]}{K_{CO_2}} + \dfrac{K_{HCO_3}}{[H_3O^+]} + 1}$$

(4.24)

$$[CO_3^{2-}] = \frac{[TIC]}{\dfrac{[H_3O^+]}{K_{CO_2} \cdot K_{HCO_3}} + \dfrac{[H_3O^+]}{K_{HCO_3}} + 1}$$

(4.25)

$$[CH_3COO^-] = \frac{[VFA] \cdot K_{CH_3COO^-}}{K_{CH_3COO^-} + [H_3O^+]}$$

(4.26)

$$[OH^-] = \frac{K_{H_2O}}{[H_3O^+]} \tag{4.27}$$

$$Z_{sys} = ([NH_4^+(aq)] + [H_3O^+(aq)] + [Na^+(aq)])$$
$$- ([HCO_3^-(aq)] + 2[CO_3^{2-}(aq)] + [Cl^-(aq)] + [OH^-(aq)]) \tag{4.28}$$

The pH iteration to find $Z_{sys} = 0$ can be carried out using various software packages (e.g. Excel (Goal Seek), Matlab, Mathematica, etc.) using the equations given in Table 4.3 and Equations (4.21)–(4.27). Using the above concentrations of TAN, TIC, Na$^+$, Cl$^-$ and acetic acids, the pH of the solution should be 7.7.

If 10% of TAN is lost from this solution due to NH$_3$ volatilisation, the pH will decrease to 7.09. Volatilisation of CO$_2$ causing a 10% reduction in TIC will increase the pH to 8.18.

As mentioned above, the pH of the untreated solution was 7.7. Thus, the NH$_3$(aq) concentration will vary by a factor of 10 (i.e. from 0.001 mol l^{-1} at the low pH of 7.09 to 0.01 mol l^{-1} at the high pH of 8.18, depending on the emission of either NH$_3$(g) or CO$_2$(g)). This reveals that CO$_2$(g) volatilisation increases the potential for NH$_3$(g) volatilisation and NH$_3$(g) volatilisation decreases the potential for NH$_3$(g) volatilisation.

4.4.3 Volatile Components

Animal manure contains components that exchange between the liquid phase and the gas phase in immediate contact with the liquid–air interface. This affects the pH of the slurry because several of the volatile components are acidic or alkaline. The most important of these components are carbon dioxide (CO$_2$(aq)) $\rightleftharpoons$ (CO$_2$(g)) and ammonia (NH$_3$(aq)) $\rightleftharpoons$ (NH$_3$(g)), but other compounds exist (e.g., malodorous hydrogen sulfide (H$_2$S), organic acids and mercaptans).

The equilibrium distribution between the gas and the liquid species of a component can usually be described by Henry's law. Henry's law relates the partial pressure of the gas to the concentration of the liquid phase concentration of the species, which in the equations below is written as the concentration;

$$A(g) \rightleftharpoons A(aq) \tag{4.29}$$

The equilibrium can be calculated using the following equation $H = [A(aq)]/[A(g)]$, where [A(aq)] and [A(g)] are in M and H is dimensionless. This equation is only correct for dilute solution conditions and at low partial pressures of the gas. At higher concentrations one should consider expressing the equilibrium distribution with fugacity for the gas phase and activities for the liquid phase (A(aq)), $K = (A(aq))/f_A$. Fugacity is given units of pressure, and approaches partial pressure at low pressures and high temperature, while K is a constant expressed in mol l^{-1} atm^{-1}. Henry's law equilibria may be presented by both A(g) $\rightleftharpoons$ A(aq) and A(aq) $\rightleftharpoons$ A(g), and authors have used a variety of units when presenting the Henry's law constant H. In the following, the same forms as presented by Stumm and Morgan (1996) are used and the equilibrium reaction A(g) $\rightleftharpoons$ A(aq) (i.e. solubility). Consequently, the equilibrium constant can be expressed as H (no dimensions) and K_H (M atm^{-1}):

$$H = \frac{[A(aq)]}{[A(g)]} \text{ or } K_H = \frac{[A(aq)]}{p_A} \tag{4.30}$$

where p_A is the partial pressure of the gas A (atm). By using the ideal gas law, K_H may be converted to H. An example is given in Text Box – Advanced 4.2.

Text Box – Advanced 4.2 Equilibrium concentration between $NH_3(g)$ in air in close contact with a liquid containing $NH_3(aq)$ and $NH_4^+(aq)$

Animal manure may be contained in covered stores and the system is in principle a closed system. In contrast, manure applied in the field and stored manure without a cover can be considered an open system. Calculations of gas–liquid equilibrium concentration and also concentration of $NH_3(aq)$ and $NH_4^+(aq)$ will differ considerably between the two systems as shown here.

Open system

In this system the NH_3 concentration in the gas phase above the liquid is constant at 0.05 mmol m^{-3}. Consequently, the concentration of $NH_3(aq)$ will be constant assuming a system in chemical equilibrium:

$$[NH_3(aq)] = K_H \cdot p_{NH_3} \tag{4.31}$$

where $K_H = 57$ (M atm^{-1} at 25 °C) and the partial pressure of NH_3 can be calculated using the ideal gas law): 0.05 mmol m^{-3} $R \cdot T = 0.05 \cdot 10^{-6}$ mol l$^{-1} \cdot (0.082057$ l atm K^{-1} mol$^{-1} \cdot 298$ K$) = 1.2 \times 10^{-6}$ atm. Hence, $[NH_3] = 57 \cdot 1.2 \cdot 10^{-6} = 7 \cdot 10^{-5}$ mol l^{-1}.

$NH_4^+(aq)$ can be assessed by the equilibrium equations:

$$NH_4^+(aq) + H_2O(l) \rightleftharpoons NH_3^+(aq) + H_3O^+(aq) \tag{4.32}$$

$$K_a = \frac{[NH_3][H_3O^+]}{[NH_4^+]} \Rightarrow [NH_4^+(aq)] = [NH_3(aq)] \cdot [H_3O^+(aq)] \cdot K_a^{-1} \tag{4.33}$$

In the open system, the $NH_4^+(aq)$ and TAN concentrations decline at decreasing concentrations of $[H_3O^+(aq)]$ (Figure 4.4).

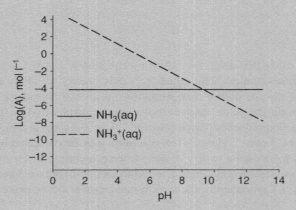

Figure 4.4 *Concentration of ammoniacal species and hydrons in the liquid phase of an open system at chemical equilibrium. $NH_3(g)$ concentrations, $P_{NH_3} = 1.2 \times 10^{-6}$ atm and temperature 25 °C. (© University of Southern Denmark.)*

Closed systems

In the closed system it is assumed that there is no microbiological or other transformation of the compounds, the amount of water (V_1) in the system is $5\,l\,m^{-3}$ and the gas volume V_g is $1000\,l\,m^{-3} - V_1 = 995\,l\,m^{-3}$. The amount of $NH_3(tot)$ is $2 \times 10^{-7}\,mol\,m^{-3}$ of the system and temperature is $25\,^\circ C$:

$$NH_3(tot) = ([NH_4^+(aq)] + [NH_3(aq)]) \cdot V_1 + [NH_3(g)] \cdot V_g \tag{4.34}$$

With the ideal gas law Equation (4.31) can be rearranged to:

$$[NH_3(aq)] = K_H \cdot p_{NH_3} = K_H \cdot [NH_3(g)] \cdot R \cdot T \tag{4.35}$$

Then by combining Equations (4.33) and (4.35), $NH_3(tot)$ can be related to the variable $NH_3(g)$ and H^+:

$$NH_3(tot) = (K_H \cdot [NH_3(g)] \cdot R \cdot T \cdot [H^+] \cdot K_a^{-1} + K_H \cdot [NH_3(g)] \cdot R \cdot T) \cdot V_1 + [NH_3(g)] \cdot V_g$$

$$NH_3(tot) = V_1 \cdot K_H \cdot [NH_3(g)] \cdot R \cdot T \cdot [H^+] \cdot \left(K_a^{-1} + 1\right) + [NH_3(g)] \cdot V_g \tag{4.36}$$

Then $NH_3(g)$ in $mol\,m^{-3}$ can be assessed as:

$$[NH_3(g)] = \frac{NH_3(tot)}{V_1 \cdot K_H \cdot R \cdot T \cdot ([H^+] \cdot K_a^{-1} + 1) + V_g} \tag{4.37}$$

$[NH_3(aq)]$ and $[NH_4^+(aq)]$ can be estimated using Equations (4.35) and (4.33). It can be seen that $[NH_3(aq)]$ increases with pH and $[NH_4^+(aq)]$ decreases with increasing pH (Figure 4.5). Due to increasing concentrations of $NH_3(g)$, $[NH_3(aq)]$ does not increase to the level of the highest concentration of $[NH_4^+(aq)]$.

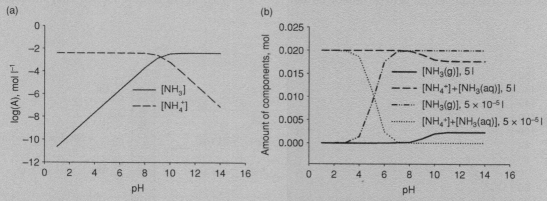

Figure 4.5   *Distribution of NH_3 species between the water and gas phase in a closed system with a volume of 1000 l. The amount of $NH_3(tot)$ ($NH_3(g) + NH_4^+(aq) + NH_3(aq)$) is $2 \times 10^{-2}\,mol\,l^{-1}$ of the system and temperature is $25\,^\circ C$. (a) Concentrations of $NH_4^+(aq)$ and $NH_3(aq)$ in a system with a liquid volume of 5 l. (b) Amount of components in a system with a 5-l liquid phase (V_l) and one with 5×10^{-5} l and a gas volume of, respectively, (V_g) is $995\,l\,m^{-3}$ and $999.99995\,l\,m^{-3}$. (© University of Southern Denmark.)*

In a closed system the concentration of all components changes with pH. In Figure 4.5A the pH increase causes a reduction in the concentration of $[NH_4^+]$ and an increase in $[NH_3]$. In this system where the liquid phase is 5 l out of 1000 l, the amount of $([NH_4^+] + [NH_3(aq)])$ is relatively constant, because $NH_3(g)$ is very soluble in the liquid phase (Figure 4.5 B). In a system where the liquid phase is approximately 0.5% of the gas phase, then $([NH_4^+] + [NH_3(aq)])$ does not decrease much with increasing pH. This changes if the liquid phase is small compared with the gas phase (e.g. a floor in an animal house), in which case most $NH_3(tot)$ is in the gas phase at high pH (Figure 4.5b).

4.4.4   Absorbed Exchangeable Cations

Particles in animal slurry contain organic matter with functional groups of carboxyl, ammonium and thiols. These functional groups contain hydrogen that can be released; the amount released being related to the pH of the liquid surrounding the particle. At high pH a significant amount of H is released and the particles thus become increasingly negatively charged (Figure 4.6). The lowers the zeta potential of the particles (Christensen *et al.*, 2009). Most untreated manures have a pH higher than 7–7.5 and have negatively charged sites where carboxyl groups have released a H atom. The particles therefore function similarly to the resins in ion exchange columns used in chemistry or to the negatively charged functional groups of soil colloids and soil minerals, which contribute to soil cation exchange.

Cations adsorb to these negatively charged sites on particulate organic matter. The adsorbed cations are in dynamic equilibrium with cations dissolved in the manure liquid (Figure 4.6). The term cation exchange capacity (CEC) is used to define the negatively charged sites of the particles.

The relative exchange of the ions in these solutions with dispersed colloids depends on the hydrated size and the electrical potential of the ions. Slurry contains high concentrations of the divalent cations Ca^{2+} and Mg^{2+}, and the monovalent cations K^+, NH_4^+ and Na^+. The divalent cations have a higher affinity for adsorption than the monovalent cations. In slurry, as with soil exchange of ions between dissolution in liquid and adsorption on manure, CEC is therefore affected by the ratio of the concentration of monovalent to divalent cations in the manure. The equilibrium between dissolved and adsorbed ions is in this context defined using the Gapon

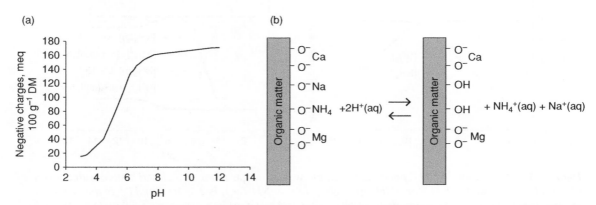

Figure 4.6   *(a) Negative charge of organic manure as affected by manure pH. (b) Example of cation exchange on particles of organic matter in animal manure. (© University of Southern Denmark.)*

equation (Russell, 1977), which is represented here by the exchange of $NH_4^+(aq)$, $Ca^{2+}(aq)$ and $Mg^{2+}(aq)$ with adsorbed NH_4^+, Ca^{2+} and Mg^{2+}:

$$NH_4^+ (aq) + Me_{1/2}X(s) \rightleftharpoons \frac{1}{2}Me^{2+}(aq) + NH_4X(s) \tag{4.38}$$

$$\frac{[NH_4X(s)]}{[Ca_{1/2}X(s)] + [Mg_{1/2}X(s)]} = K_g \frac{[NH_4^+(aq)]}{\sqrt{[Ca^{2+}(aq)] + [Mg^{2+}(aq)]}} \tag{4.39}$$

where Me^{2+} is a divalent cation, here Ca^{2+} or Mg^{2+}. Furthermore, $NH_4X(s)$, $Ca_{1/2}X(s)$ and $Mg_{1/2}X(s)$ are adsorbed ions, $NH_4^+(aq)$, $Ca^{2+}(aq)$ and $Mg^{2+}(aq)$ are dissolved ion compounds, and K_g is the Gapon coefficient. It is assumed that divalent ions have the same affinity for adsorption on the particle surface.

If the concentration of $NH_4^+(aq)$ is changed (e.g. by adding NH_4Cl to the slurry) the exchange process can be depicted through iteration of an extension of the Gapon equation (Fleisher *et al.*, 1987):

$$\frac{[NH_4X(s)]_0 + x}{[Ca_{1/2}X(s)]_0 + [Mg_{1/2}X(s)]_0 - x} = K_g \frac{[NH_4^+(aq)]_0 - x}{\sqrt{([Ca^{2+}(aq)]_0 + [Mg^{2+}(aq)]_0 + 1/2x)}} \tag{4.40}$$

where x represents the cation mass transfer (NH_4^+, Ca^{2+} or Mg^{2+}) between the solution and the particles (M). All concentrations are the concentration after the change of ammonium amount, but before the exchange process. Alternatively, a simple factor may be used, describing the effect of adsorption (Avnimelech and Laher, 1977).

It can be deducted from Equations (4.39) and (4.40) that an increase in CEC increases the NH_4X adsorbed to the manure. Drying the slurry and increasing the concentration of $NH_4^+(aq)$ also increases NH_4X adsorption (Whitehead and Raistrick, 1993). NH_4^+ does not exchange easily with Ca^{2+} and thus high Ca^{2+} concentrations in the manure may reduce the adsorption of NH_4^+.

Both monovalent and divalent cations are dissolved in the manure liquid. As a consequence, the equilibrium between the divalent and monovalent ions in solution and adsorbed is different between manure categories. With increasing water content (i.e. from a compost heap < FYM < slurry) more divalent cations in solution are exchanged with adsorbed NH_4^+ (Chung and Zasoski, 1994). Conversely, if a compost or a FYM is dried and the solution becomes concentrated as water is being removed, then dissolved NH_4^+ exchanges with divalent cations adsorbed to the organic matter and more NH_4^+ is adsorbed.

4.4.5 Crystals and Adsorbed Ions

In manure, the cations $Ca^{2+}(aq)$, $Mg^{2+}(aq)$ and $NH_4^+(aq)$ react with the anions $CO_3^{2-}(aq)$ and $PO_4^{3-}(aq)$, and form crystals of calcite ($CaCO_3(s)$), struvite ($MgNH_4PO_4.6H_2O(s)$) and apatite ($Ca_5OH(PO_4)_3(s)$). Such crystal formation is an important feature of the chemical system, because it will affect the concentration of cations and anions in solution and pH. Thus, due to crystal formation there are only trace amounts of the multivalent ions $Ca^{2+}(aq)$, $Mg^{2+}(aq)$ and $PO_4^{3}(aq)^-$ in the liquid fraction of slurry (Sommer and Husted, 1995a).

The calcite content affects how much acid needs to be added to reduce pH in the manure. Acid is added to slurries, because reducing the pH reduces NH_3 emissions. However, at a slurry pH of 4 or 5 calcite has been known to slowly dissolve, increasing the pH again, so that the effect of adding acid is only temporary.

The formation of calcite ($CaCO_3(s)$) in animal manure is mainly controlled by the concentration of calcium ($Ca^{2+}(aq)$), because carbonate ($CO_3^{2-}(aq)$) is present in large quantities in slurry. Calcite formation not only reduces the concentration of Ca^{2+} in solution, but may also reduce slurry pH because precipitating CO_3^{2-} as calcite releases H_3O^+:

$$Ca^{2+}(aq) + HCO_3^-(aq) + H_2O(aq) \rightleftharpoons CaCO_3(aq) + H_3O^+(aq) \qquad (4.41)$$

$Ca^{2+}(aq)$ and $PO_4^{3-}(aq)$ may also form crystals and precipitate as calcium phosphate (hydroxylapatite). Hydroxylapatite has been seen to adsorb to the surface of calcite crystals and in this way stop the formation of large calcite crystals (Fordham and Schwertmann, 1977a–c).

Minor amounts of hydroxylapatite have been found in manure. Hydroxylapatite is a mineral form of calcium apatite with the formula $Ca_5(PO_4)_3(OH)(s)$ or more correctly $Ca_{10}(PO_4)_6(OH)_2(s)$. In most slurries a high $CO_3^{2-}(aq)$ concentration significantly reduces the $Ca^{2+}(aq)$ available for hydroxylapatite formation, therefore this crystal has not been found in abundance in untreated slurry (Bril and Solomons, 1990; Sommer and Husted, 1995a). Thus, if the intention is to use the hydroxylapatite process to remove phosphorus from a liquid manure, then CO_2 should be released from the slurry by adding acid to significantly reduce the concentration of $CO_3^{2-}(aq)$. As mentioned above, this can only be achieved using large amounts of acid because $CO_3^{2-}(aq)$ bound in calcite crystals will be dissolved due to the reduction in pH (Equation 4.39).

The most significant of the phosphorous crystals in animal manure is struvite ($MgNH_4PO_4{\cdot}6H_2O(s)$). Struvite crystallisation of phosphate is used as a means of settling out phosphorus as particles and has been widely used as a method for separating P from wastewater (Ohlinger *et al.*, 2000; Burton and Turner, 2003). The optimum pH for struvite formation is about 9 (Nelson *et al.*, 2003). The reaction of struvite formation is:

$$Mg^{2+}(aq) + NH_4^+(aq) + PO_4^{3-}(aq) + 6H_2O(l) \rightleftharpoons MgNH_4PO_4{\cdot}6H_2O(s) \qquad (4.42)$$

The equilibrium ion-activity product (K_{str}) is 2.51×10^{-13} and the formation constant (K^0) of struvite is 1.41×10^{13} at 25 °C. The equilibrium ion-activity product is the inverse of the formation constant. In some slurries, the product of $[Mg^{2+}]$, $[NH_4^+]$ and $[PO_4^{3-}]$ is often lower than the conditional formation constant due to low concentrations of dissolved $[Mg^{2+}]$, and therefore little struvite is formed (Nelson *et al.*, 2003). Unfortunately, struvite tends to crystallise on the surfaces of heat exchangers and at outlets of tubes and pipes, due to physical forces and partly due to CO_2 volatilisation increasing pH. This constricts the efficiency of heat exchangers and the flow of slurry through tubes and leads to problems in, for instance, biogas plants where struvite formation reduces heat exchanger performance and thus has to be removed at regular intervals. High temperatures increase the formation constant and thus increase struvite formation. As the heat exchangers cooling the warm slurry from the biogas plant may be covered with struvite, the heat exchangers are constructed so that they can be rinsed with acid, which dissolves the struvite crystals. Also see Text Box – Advanced 4.3.

Text Box – Advanced 4.3   Struvite crystallisation

Struvite formation is one of the most cited methods for removing phosphorus from wastewater and also from slurry. To control the process and assess the need for adding Mg^{2+} and also pH buffer, it is necessary to calculate the conditional stability constants of the formation of struvite crystals and use these to give an idea of the process conditions.

To calculate the solubility of struvite in slurry, the concentration of the species $Mg^{2+}(aq)$, $NH_4^+(aq)$ and $PO_4^{3-}(aq)$ should be known:

$$MgNH_4PO_4 \cdot 6H_2O(s) \rightleftharpoons Mg^{2+}(aq) + NH_4^+(aq) + PO_4^{3-}(aq) + 6H_2O(l) \qquad (4.43)$$

The concentration of the ions in solution is affected by pH, so the side-reaction components of the species have to be calculated (Stumm and Morgan, 1996). These calculations have to include all reactions that affect the concentration of the species contributing to the reaction of struvite formation (i.e. side-reactions).

An example is the equilibrium of $NH_3(aq)$ and $NH_4^+(aq)$, where TAN concentration in the liquid is known. It is then possible to assess the concentration of $[NH_4^+(aq)]$ by knowing the pH, temperature and ionic strength of the liquid. In the calculations the concentration of the ions $Mg^{2+}(aq)$, $NH_4^+(aq)$ and $PO_4^{3-}(aq)$ are assessed as the fraction of the total content of the species (e.g. NH_4^+/TAN), ionic strength is set at 0 and temperature 25 °C). The fraction of the species in relation to total concentration of the component (α) can be calculated (Table 4.4).

Table 4.4   *Fraction of species involved in the formation of struvite.*

$\alpha_{Mg^{2+}}$:	$MgOH^+(aq)$,	$Mg^{2+}(aq) + H_2O(l) \rightleftharpoons MgOH^+(aq) + H^+(aq)$, $\log K_{Mg1} = -11.4$
	$Mg(OH)_2(aq)$	$MgOH^+(aq) + H_2O(l) \rightleftharpoons Mg(OH)_2(aq) + H^+(aq)$, $\log K_{Mg2} = -4.4$

$$\alpha_{Mg^{2+}} = \frac{[Mg^{2+}]}{[Mg^{2+}] + [MgOH^+] + [Mg(OH)_2]}$$

$$\alpha_{Mg^{2+}} = \frac{[Mg^{2+}]}{[Mg^{2+}] + [Mg^{2+}] \cdot K_{Mg1}/[H^+] + [Mg^{2+}] \cdot K_{Mg1} \cdot K_{Mg2}/[H^+]^2}$$

$$\alpha_{Mg^{2+}} = \frac{1}{1 + K_{Mg1}/[H^+] + K_{Mg1} \cdot K_{Mg2}/[H^+]^2}$$

$\alpha_{NH_4^+}$:	$NH_3(aq)$	$NH_4^+(aq) + H_2O(l) \rightleftharpoons NH_3^+(aq) + H_3O^+(aq)$, $K_{N1} = -9.3$, $K_{N1} = -9.3$

$$\alpha_{NH_4^+} = \frac{[NH_4^+]}{[NH_4^+] + [NH_3]}$$

$$\alpha_{NH_4^+} = \frac{1}{1 + K_{N1}/[H^+]}$$

$\alpha_{PO_4^{3-}}$:	$H_3PO_4(aq)$,	$H_3PO_4(aq) + H_2O(l) \rightleftharpoons H_2PO_4^-(aq) + H_3O^+(aq)$, $K_{P1} = -2.1$
	$H_2PO_4^-(aq)$,	$H_2PO_4^-(aq) + H_2O(l) \rightleftharpoons HPO_4^{2-}(aq) + H_3O^+(aq)$, $K_{P2} = -7.2$
	$HPO_4^{2-}(aq)$	$HPO_4^{2-}(aq) + H_2O(l) \rightleftharpoons PO_4^{3-}(aq) + H_3O^+(aq)$, $K_{P3} = -12.3$

$$\alpha_{PO_4^{3-}} = \frac{[Mg^{2+}]}{[H_3PO_4] + [H_2PO_4^-] + [HPO_4^{2-}] + [PO_4^{3-}]}$$

$$\alpha_{PO_4^{3-}} = \frac{1}{[H^+]^3 \cdot \dfrac{1}{K_{P3}} \cdot \dfrac{1}{K_{P2}} \cdot \dfrac{1}{K_{P1}} + [H^+]^2 \cdot \dfrac{1}{K_{P3}} \cdot \dfrac{1}{K_{P2}} + [H^+] \cdot \dfrac{1}{K_{P3}} + 1}$$

The fraction of the different species strongly depends on pH (Figure 4.7a).

The solubility product constant for struvite is 2.51×10^{-13}. Using this, it is possible to calculate the conditional ion activity product constant (K_{str}') for a solution of known contents of TAN, Mg(tot) and PO_4(tot) as affected by pH:

$$K_{str}' = K_{str} \cdot (\alpha_{PO_4^{3-}} \cdot \alpha_{NH_4^+} \cdot \alpha_{Mg^{2+}})^{-1} \qquad (4.44)$$

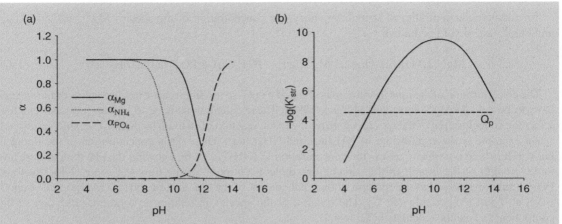

Figure 4.7 *(a) Ratio of the components NH_4^+, Mg^{2+} and PO_4^{3-}. (b) Conditional solubility constants as affected by H^+ (ionic strength is 0 and temperature 25 °C). A line depicts the solubility product (Q_P) of TAN = 0.1 M, Mg(tot) = 0.03 M and PO_4(tot) = 0.01M. (© University of Southern Denmark.)*

Figure 4.7b depicts K'_{str} as a function of pH. Crystals are formed at ion product values higher than K'. In Figure 4.7b it can be seen that between pH 8 and 12 struvite crystals are formed in a solution where the concentration of the components is: TAN = 0.1 M, Mg(tot) = 0.03 M and PO_4(tot) = 0.01 M.

When assessing the effect of PO_4^{3-}(aq) by the side-reactions mentioned above, the contribution of H_3PO_4(aq) need not be considered, because the pH of the slurry is mostly above 7 and at this pH only a trace of trihydrogen phosphate species is in solution. Thus, only dihydrogen phosphate and hydrogen phosphate have to be included when calculating the side-reactions of phosphate formation. Regarding magnesium hydroxide formation at increasing pH, these species should be included when assessing the concentration of Mg^{2+}(aq). Ionic strength should be accounted for through calculations of activities of ions.

In P separation studies, Mg^{2+} has been added to animal slurries to increase the formation product and thus start the formation of struvite. In assessing the potential for struvite formation in slurry as affected by Mg^{2+} addition, the concentration of Mg^{2+}(aq), which is affected by side-reactions forming complexes with Mg^{2+}, must be included.

The phosphate concentration PO_4^{3-}(aq) also has a major impact on struvite crystallisation (Text Box – Advanced 4.3). Phosphate is in equilibrium with HPO_4^{2-}(aq), which is a weak acid. Thus, at low pH HPO_4^{2-}(aq) dominates and consequently the concentration of PO_4^{3-}(aq) is low and struvite crystallisation is reduced (Nelson *et al.*, 2003). In untreated slurry the pH level is between 7.5 and 8, and struvite crystals are not abundant (Sommer and Husted, 1995a). In slurries from biogas treatment the pH is often higher and struvite formation is therefore more likely.

In most slurries, the NH_4^+(aq) concentration is higher than the 1:1:1 ratio ($[Mg^{2+}] : [NH_4^+] : [PO_4^{3-}]$) for the formation of struvite and should not be a limiting factor for the reaction (Nelson *et al.*, 2003). Still, in contrast to how a decreasing H^+(aq) concentration affects the concentration of PO_4^{3-}(aq), a decreasing H^+(aq) reduces the concentration of the weak acid NH_4^+(aq), and consequently struvite formation is limited at a high pH (i.e. a low concentration of H^+(aq)) (Buchanan *et al.*, 1994).

At a pH of about 9–10, the concentrations of PO_4^{3-}(aq) and NH_4^+(aq) are significant in most biomasses and at standard environmental conditions. This pH range therefore provides efficient removal of phosphorus through sedimentation of struvite (Nelson *et al.*, 2003).

4.5 Summary

As demonstrated above, gaseous emissions of NH_3 and also of malodorous compounds are affected by pH, because the charged species of the compounds cannot be released from the liquid phase. Transport of slurry through tubes and when stirred is affected by slurry viscosity, which is related to slurry dry matter, and this is affected by microbial transformation of the organic matter. Again, pH is an important factor, because microbial activity is much affected by pH.

Many other processes can be related to pH or the proton (H^+) concentration of the slurry. Therefore, modelling pH is a most important task, but also very complex. To our knowledge no model has yet been developed that can correctly predict changes in pH for a full-scale system of slurry lagoons or storage or for applied manure. Most models have to be calibrated for each system they describe. It is therefore recommended when considering and assessing novel methods that involve changes of pH that the whole pH buffer system is kept in mind, including the transport of buffer components via volatilisation, and transport in the liquid and solid phase processes. It should be remembered that pH varies during storage due to emission of, for example, NH_3, but also due to the microbial transformation of organic material. Under anoxic condition VFAs are produced and pH decreases. Under oxic conditions VFAs are degraded and pH usually increases.

The particles in manure are mainly negatively charged and the effective charge density changes with pH and ionic strength. The charge density has a high impact on both ion adsorption and the stability of particles. When using solid–liquid separation to remove nutrients from manure, charge density, pH and ionic strength become key parameters that have to be considered in order to choose the right separation strategy. Again, the physico-chemical properties of the manure are important for the design of manure handling equipment.

References

Avnimelech, Y. and Laher, M. (1977) Ammonia volatilization from soils: equilibrium considerations. *Soil Sci. Soc. Am. J.*, **41**, 1080–1084.

Bril, J. and Salomons, W. (1990) Chemical composition of animal manure: a modelling approach. *Neth. J. Agric. Sci.*, **38**, 333–352.

Buchanan, J.R., Mote, C.R. and Robinson, R.B. (1994) Thermodynamics of struvite formation. *Trans. ASAE*, **37**, 617–662.

Burton, C.H. and Turner, C. (2003) *Manure Management: Treatment Strategies for Sustainable Agriculture*, 2nd edn, Silsoe Research Institute, Silsoe.

Cahn, T.T., Aarnink, A.J.A., Verstegen, M.W.A. and Schrama, J.W. (1998) Influence of dietary factors on the pH and ammonia emission of slurry from growing-finishing pigs. *J. Anim. Sci.*, **76**, 1123–1130.

Chapuis-Lardy, L., Fiorini, J., Toth, J. and Dou, Z. (2004) Phosphorus concentration and solubility in dairy feces: variability and affecting factors. *J. Dairy Sci.*, **87**, 4334–4341.

Chung, J.B. and Zasoski, R.J. (1994) Ammonium–potassium and ammonium–calcium exchange equilibria in bulk and rhizosphere soil. *Soil Sci. Soc. Am. J.*, **58**, 1368–1375.

Christensen, M.L., Hjorth, M. and Keiding, K. (2009) Characterisation of swine manure with reference to flocculation and separation. *Water Res.*, **43**, 773–783.

Cooper, P. and Cornforth, I.S. (1978) Volatile fatty acids in stored animal slurry. *J. Sci. Food Agric.*, **29**, 19–27.

Fleisher, Z., Kenig, A., Ravina, I. and Hagin, J. (1987) Model of ammonia volatilization from calcareous soils. *Plant Soil*, **103**, 205–212.

Forshell, L.P. (1993) Composting of cattle and pig manure. *J. Vet. Med. B*, **40**, 634–640.

Fordham, A.W. and Schwertmann, U. (1977a) Composition and reactions of liquid manure (gulle), with particular reference to phosphate: I. Analytical composition and reaction with poorly crystalline iron oxide. *J. Environ. Qual.*, **6**, 133–136.

Fordham, A.W. and Schwertmann, U. (1977b) Composition and reactions of liquid manure (gulle), with particular reference to phosphate: II. Solid-phase components. *J. Environ. Qual.*, **6**, 136–140.

Fordham, A.W. and Schwertmann, U. (1977c) Composition and reactions of liquid manure (gulle), with particular reference to phosphate: III. pH buffering capacity and organic components. *J. Environ. Qual.*, **6**, 140–144.

Griffin, R.A. and Jurinak, J.J. (1973) Estimation of activity coefficients from electrical conductivity of natural aquatic systems and soil extracts. *Soil Sci.*, **116**, 26–30.

Hjorth, M., Christensen, K.V., Christensen, M.L. and Sommer, S.G. (2010) Solid–liquid separation of animal slurry in theory and practice: a review. *Agron. Sustain. Dev.*, **30**, 153–180.

Husted, S., Jensen, L.S. and Storgaard Jørgensen, S. (1991) Reducing ammonia loss from cattle slurry by the use of acidifying additives: the role of the buffer system. *J. Sci. Food Agric.*, **57**, 335–349.

Koerkamp, P.W.G. (1994) Review on emissions of ammonia from housing systems for laying hens in relation to sources, processes, building design and manure handling. *J. Agric. Res.*, **59**, 73–87.

Kroodsma, W., Scholtens, R. and Huis, J. (1988) Ammonia emission from poultry housing systems, in *Volatile Emissions from Livestock Farming and Sewage Operations* (eds V.C. Nielsen, J.H. Voorburg and P. L'Hermite), Elsevier, London, pp. 152–161.

Landry, H., Laguë, C. and Roberge, M. (2004) Physical and rheological properties of manure products. *Appl. Eng. Agric.*, **20**, 277–288.

Lyklema, J. (1977) Water at interfaces: a colloid chemical approach. *J. Colloid Interface Sci.*, **58**, 242.

Masse, L., Masse, D.I., Beaudette, V. and Muir, M. (2005) Size distribution and composition of particles in raw and anaerobically digested swine manure. *Trans. ASAW*, **48**, 1943–1949.

Massé, D.I., Croteau, F. and Massé, L. (2007) The fate of crop nutrients during digestion of swine manure in psychrophilic anaerobic sequencing batch reactors. *Bioresour. Technol.*, **98**, 2819–2823.

Meyer, D., Ristow, P.L. and Lie, M. (2007) Particle size and nutrient distribution in fresh dairy manure. *Appl. Eng. Agric.*, **23**, 113–117.

Møller, H.B., Nielsen, A.M., Nakakubo, R. and Olsen, H.J. (2007a) Process performance of biogas digesters incorporating pre-separated manure. *Livest. Sci.*, **112**, 217–223.

Møller, H.B., Jensen, H.S., Tobiasen, L. and Hansen, M.N. (2007b) Heavy metal and phosphorus content of fractions from manure treatment and incineration. *Environ. Technol.*, **28**, 1403–1418.

Nelson, N.O., Mikkelsen, R.L. and Hesterberg, D.L. (2003) Struvite precipitation in anaerobic swine lagoon liquid: effect of pH and Mg:P ratio and determination of rate constant. *Bioresour. Technol.*, **89**, 229–236.

Ohlinger, K.N., Young, T.M. and Schroeder, E.D. (2000) Postdigestion struvite precipitation using a fluidized bed reactor. *J. Environ. Eng.*, **126**, 361–368.

Pain, B. and Menzi, H. (2003) Glossary of terms on livestock manure management; http://www.ramiran.net/DOC/Glossary2003.pdf.

Petersen, S.O., Nielsen, T.H., Frostegård, Å. and Olesen, T. (1996) O_2 uptake, C metabolism and denitrification associated with manure hot spots. *Soil Biol. Biochem.*, **28**, 341–349.

Russell, E.W. (1977) *Soil Conditions and Plant Growth*, 10th edn, Longman, London, pp. 90–100.

Sawyer, C.N., McCarty, P.L. and Parkin, G.F. (1994) *Chemistry for Environmental Engineering, Civil Engineering Series*, McGraw-Hill International Editions, New York.

Sommer, S.G. and Husted, S. (1995a) The chemical buffer system in raw and digested animal slurry. *J. Agric. Sci.*, **124**, 45–53.

Sommer, S.G. and Husted, S. (1995b) A simple model of pH in slurry. *J. Agric. Sci.*, **124**, 447–453.

Sommer, S.G., Zhang, G.Q., Bannink, A., Chadwick, D., Hutchings, N.J., Misselbrook, T., Menzi, H., Ni, J.Q., Oenema, O., Webb, J. and Monteny, G.-J. (2006) Algorithms determining ammonia emission from livestock houses and manure stores. *Adv. Agron.*, **89**, 261–335.

Stumm, W. and Morgan, J.J. (1996) *Aquatic Chemistry – Chemical Equilibrian and Rates in Natural Waters*, 3rd edn, John Wiley & Sons, Inc., New York.

Thygesen, O., Triolo, J.M. and Sommer, S.G. (2012) Indicators of physical properties and plant nutrient content of animal slurry and separated slurry. *Biol. Eng.*, **5**, 123–135.

Vavilin, V.A., Lokshina, L.Y., Rytov, S.V., Kotsyurbenko, O.R. and Nozhewnikova, A.N. (1998) Modelling low-temperature methane production from cattle manure by an acclimated microbial community. *Bioresour. Technol.*, **63**, 159–171.

Whitehead, D.C. and Raistrick, N. (1993) The volatilization of ammonia from cattle urine applied to soils as influenced by soil properties. *Plant Soil*, **148**, 43–51.

5

Manure Organic Matter – Characteristics and Microbial Transformations

Lars S. Jensen[1] and Sven G. Sommer[2]

[1]Department of Plant and Environmental Sciences, University of Copenhagen, Denmark
[2]Institute of Chemical Engineering, Biotechnology and Environmental Technology,
University of Southern Denmark, Denmark

5.1 Introduction

The content and composition of organic matter in animal manure depend entirely on (i) the animal species, (ii) feeding practices and (iii) the animal production system (including housing and manure management). The first two influence the organic matter excreted from the animals in faeces and urine, as there are major differences between diets fed to ruminant and monogastric animals. Ruminants are usually supplied with roughage-dominated feedstuffs, containing a high proportion of cellulose fibres, while monogastric species are fed more protein-concentrated and easily digestible feeds, which in turn influence the excretion in faeces and urine. Poultry are fed grain with a high concentration of protein and digestible carbohydrate, while the surplus is excreted in the form of faeces as no urine is produced.

The composition of the manure is also affected by the method of collection in the animal house and of the management of the animal manure. During collection in the animal house, organic material is added in the form of spilt feed and any bedding material in the form of straw, wood chips or other absorbent material strewn on the floor. This affects the composition and degradability of the manure entering storage, where it will undergo a number of microbial and biochemical transformation processes, which again change the composition of the organic matter and speciation of the nutrients.

Chapter 4 described the inorganic composition and chemical transformations in manure. This chapter describes the microbial transformations and biochemical characteristics of manure organic matter and nutrients (Text Box – Basic 5.1).

Animal Manure Recycling: Treatment and Management, First Edition. Edited by Sven G. Sommer, Morten L. Christensen, Thomas Schmidt and Lars S. Jensen.

Text Box – Basic 5.1 Characterisation of organic matter in animal manures

Dry matter (DM): The material that remains after drying biowaste at 80–100 °C until there is no weight change, typically for 24 h.

Volatile solids (VS): The material lost when incinerating the dry matter at 550 °C for 1 h. After incineration only the ash remains and volatile solids is determined as the difference between dry matter and ash content. Often volatile solids is given in per cent of dry matter (VS % of DM = 100% – ash % of DM). Volatile solids is synonymous with the term ash-free dry matter often used in plant science.

Biological oxygen demand (BOD): Traditionally determined as the oxygen consumed by microorganisms during 5 days of batch fermentation, the result of this analysis is termed BOD_5 in contrast to the BOD_∞ analysis where the oxygen consumption during 20–30 days of batch fermentation is measured. The classic analytical procedure is to mix the manure or wastewater sample with aerated water in the ratio 1 : 100 or more in a flask. Oxygen content is measured, and the flask encapsulated and stored for 5 days. Then the oxygen content is measured again and BOD_5 is given as the O_2 reduction during fermentation. Ammonia is also oxidised in this analysis, but this can be avoided by inhibiting ammonia oxidation (Henze *et al.*, 1996).

Chemical oxygen demand (COD): An analysis of oxidation of components in wastewater or manure. The oxidation agent is potassium dichromate, and the sample is acidified to ensure complete oxidation. In the process of oxidising the organic substances found in the water sample, potassium dichromate is reduced, forming Cr^{3+}. The amount of Cr^{3+} is determined after oxidisation is complete, and is used as a measure of the organic content of the water sample. The reaction of potassium dichromate with organic compounds is given by:

$$C_nH_aO_bN_c + dCr_2O_7^{2-} + (8d + c)H^+ \rightarrow nCO_2 + \frac{a + 8d - 3c}{2}H_2O + cNH_4^+ + 2dCr^{3+} \quad (5.1)$$

where $d = 2n/3 + a/6 - b/3 - c/2$.

5.2 Manure Organic Matter Composition

Carbon (C), hydrogen (H), oxygen (O), nitrogen (N) and sulfur (S) are all major constituents of the organic matter in the dry matter fraction of the manure. The distribution of especially C and N between the labile and stable organic fractions and between the inorganic and organic phases is strongly affected by microbial transformations of organic matter taking place during storage and management of the organic manure (e.g. if treated by composting or anaerobically digested). In most inventories, C and N losses due to gaseous emissions are assumed to be relative constant. Carbon losses are often expressed as a percentage of dry matter losses (Table 5.1) or loss of C content in dry matter or, more appropriately, volatile solids (Text Box – Basic 5.1) and N losses as a percentage of total ammoniacal N (TAN = $NH_3 + NH_4^+$) or total N.

Monogastric and ruminant manure may contain short-chain volatile organic acids with two to five atoms of carbon (volatile fatty acids (VFAs), C_2–C_5 acids), long-chain fatty acids (LCFAs, six or more carbon atoms), organic lipids, proteins, carbohydrates and lignin (Section 5.2.3). Lignin also contains organic components such as phenols. Transformation of organic matter in livestock and poultry manure follows the same pathways, although the composition of N in the excreta is different. In ruminant and pig manure N is excreted as proteins and urea, whereas poultry excrete N in the form of proteins and uric acid (35–50% of total N; Table 5.2). Uric acid contributes to poultry manure N being less degradable than N in organic manure from ruminants and pigs.

Table 5.1 *Losses of dry matter during storage of solid manure as a percentage of initial content (Chadwick, 2005a, 2005b; Hansen et al., 2006).*

Type of manure	Pre-treatment	Heap weight (ton)	Length of storage (days)	Loss of dry matter (% of initial)
Pig FYM	none	3.4–5.2	180	49
	straw 50%	3.4–5.2	180	46
	covered	3.4–5.2	180	26
Solid fraction from slurry	none	6.9	120	12
separation	covered	6.9	120	5

5.2.1   Carbon

Manure in the form of slurry (for definition of manure categories, see Text Box – Basic 4.1) is an anaerobic matrix in which organic C transformation is relatively slow. The absence of oxygen is a precondition for the production of carbon dioxide (CO_2) and methane (CH_4) via microbial metabolism of organic material in animal manure. During slurry storage inside the animal house, the reduction in organic components is affected by slurry removal frequency (i.e. storage time and temperature). There are few studies of the reduction in dry matter under realistic conditions, but a rough estimate puts dry matter losses from slurry stored in-house at 10%. Outside the animal house, slurry is stored in tanks or lagoons, and in temperate climate zones where temperatures are relatively low and storage periods long the estimated losses may be set at 5% of the dry matter added to the store (Poulsen *et al.*, 2001). Organic matter transformation may be much greater in lagoons with their longer storage periods and often warmer temperatures.

For solid manure produced in deep litter housing systems for cattle, the hooves compact the deep litter, and the temperature may rise to 40–50 °C at 10 cm depth (Henriksen and Olesen, 2000). The deep litter in pig houses is stirred to a greater extent by the pigs and the temperature in the litter may differ from that seen in cattle houses. There is not much information available about the loss of dry matter from deep litter systems, so again expert assumptions have been used to estimate the in-house dry matter losses to 30% in pig and cattle deep litter systems (Poulsen *et al.*, 2001).

Solid manure stored outside livestock houses is typically dominantly aerobic, depending on the aeration rate (i.e. water content and compaction/manure density) and oxygen consumption. If oxygen consumption is higher than the flow of oxygen into the heap, then anaerobic zones will emerge, a process promoted by

Table 5.2 *Composition of faeces and urine from various livestock; range of values as observed in the literature (after Oenema et al., 2008).*

Animal category	Dry matter (g kg^{-1})	Total-N (g kg^{-1} (fw))	Urea (% of total-N)	Uric acid (% of total-N)	Protein-N (% of total-N)	Ammonium (% of total-N)
Dairy cattle						
faeces	100–175	10–17	0	0	90–95	1–4
urine	30–40	4–10	60–95	0–2	0	1
Finishing pigs						
faeces	200–300	7–15	0		90–95	1–7
urine	10–50	2–10	30–90		10–20	5–65
Chicken	200–300	10–20	5–8	35–50	30–50	6–8

compaction and/or a high water content of the manure (Poulsen and Moldrup, 2007). In contrast, a high straw content facilitates air entering the heap (Sommer and Møller, 2000). Consequently, aerobic (composting) conditions are found to prevail in straw-rich heaps of solid manure, whereas a temperature rise due to composting rarely occurs in high density heaps of manure with little bedding material (i.e. cattle farmyard manure (FYM)) (Webb *et al.*, 2012). As a consequence of these patterns, in the Danish manure inventory it is assumed that dry matter losses are 45% from a well-aerated solid manure store of deep litter and swine solid manure, and 10% from high-density cattle FYM (Poulsen *et al.*, 2001). A more recent estimate of dry matter losses during storage of solid manure or the dry matter-rich fraction from separation of livestock slurry is given in Table 5.1.

Carbohydrates constitute the largest fraction of the organic material, followed by proteins, lipids, lignin and VFAs. The organic components include compounds with the functional groups carboxylates, hydroxyls, thiol and phenols (Eriksen 2010), which at the pH interval in the manure will contribute to a negative charge of the organic matter.

In slurry the VFAs constitutes about 15% of the dry matter in pig slurry and about 8% of the dry matter in cattle slurry (Derikx *et al.*, 1994; Sommer and Husted, 1995). In stored solid manure the VFAs is oxidised and the concentration of organic acids in solid manure is therefore low, with the exception of manure heaps where air exchange is impeded by covering or compaction.

The composition of the organic components and the pH influence the charge of the particles and this may affect flocculation and also cation absorption on the dry matter particles in liquid and in solid manure.

5.2.2 Nitrogen

In livestock farming, usually between 5 and 45% of the N in plant protein is transformed into animal protein, depending on animal type and management. The remaining 55–95% of the N is excreted as organically bound N (Table 5.2).

Pigs and ruminants excrete 50–60% of excess N via the urine, where more than 60% is found as urea and the rest is easily degradable organic N components (Table 5.2). After excretion both N fractions are transformed to TAN (i.e. the sum of $NH_3 + NH_4^+$). Roughly half of the N in faeces is undigested and non-absorbed dietary N, whilst the other half is endogenous, resulting from enzymes and mucus excreted into the digestive tract. The undigested dietary N in faeces is poorly degradable, unlike the endogenous N. The excretion of N in faeces is relatively constant; therefore increasing feeding rates and surplus N in diets typically result in a larger fraction of the N excretion ending up in urine.

Following deposition on the floor of animal housing systems or on pasture, all urea and a fraction of the organic N are rapidly hydrolysed into TAN (Oenema *et al.*, 2008). The TAN in pig, cattle or sheep manure originates mainly from hydrolysis of the urea in urine. In traditional livestock buildings or in beef feedlots, transformation of urea is completed before the manure is removed from the house or the feedlot (Voorburg and Kroodsma, 1992).

Unlike mammals, poultry excrete a mixture of organic N and uric acid (Table 5.2), which is excreted in faeces as a dry mass. This involves a complex metabolic pathway that is more energetically costly in comparison to processing of other nitrogenous wastes such as urea (from the urea cycle) or NH_3, but has the advantage of reducing water losses (McCrudden, 2008). Uric acid is a heterocyclic N compound which slowly hydrolyses to urea, which is then hydrolysed to TAN (Groot Koerkamp, 1994). The concentration of TAN and uric acid in chicken manure applied to fields is therefore very variable and related to storage conditions. Furthermore, owing to the uric acid in manure, prediction of NH_3 emissions from poultry manure and its fertiliser efficiency should include transformation of uric acid, which is influenced by water content and temperature (Groot Koerkamp, 1994). If analytical data on N components in poultry manure are not available, then a qualified guess is that the TAN content is about 50% of N in poultry droppings (Sommer and Hutchings, 2001).

Table 5.3 *Organic components constituting volatile solids in slurry (after Møller et al., 2004 and Sommer et al., 2004).*

Volatile solids component	Formula	Amount of component (% of volatile solids)			
		Fattening pigs	Sows	Dairy cows	Organic waste*
Lipids	$C_{57}H_{104}O_6$	13.7	16.3	6.9	50
Proteins	$C_5H_7O_2N$	22.9	20.2	15.0	25
Degradable carbohydrate	$C_6H_{10}O_5$	34.7	38.9	43.4	25
Non-degradable carbohydrate	$C_6H_{10}O_5$	16.6	14.8	19.1	0
Lignin	$C_{10}H_{13}O_3$	4.9	6.8	12.1	
VFAs	$C_2H_4O_2$	7.2	3.0	3.6	
Total		100.0	100.0	100.0	100.0

*: Organic waste from abattoirs.

5.2.3 Characterisation of Manure Organic Matter

The organic matter content of wastewater and manure is commonly characterised by the parameters dry matter, volatile solids, biological oxygen demand (BOD) or chemical oxygen demand COD (Text Box – Basic 5.1). In animal manure, volatile solids is the standard characterisation measure used.

The volatile solids content of biomass is easy to measure and can be used as a key variable when characterising manure as a source of CH_4 emissions (released to the environment), for C addition and retention in soils, for biogas production in anaerobic digestion, and as a factor in the production of nitrous oxide (N_2O) from manure applied to fields. The volatile solids concentration relative to total dry matter is higher in pig slurry than in cattle slurry, mainly due to different feed intake and organic matter degradation in the digestive tract of the animals. As a standard approximation it is assumed that volatile solids is 80% of dry matter in animal manure, but the volatile solids fraction of dry matter in manure can be very variable.

The main components of volatile solids are lipids, proteins and carbohydrates (Table 5.3). The assessment of volatile solids does not include short-chain volatile organic acids (VFAs), because these are not solids suspended in the liquid and they may volatilise during the drying of the manure when determining dry matter. If volatile solids content is deemed to express organic matter in the manure, then these acids should be included in the volatile solids analysis. This can be achieved by adding a base to the manure, so that the acids are in their anionic form, which is not volatile. If volatile acids through the increase in pH are included in the measurement of volatile solids, then they would contribute approximately 10% to the total content of volatile solids in animal slurry (Derikx *et al.*, 1994). In solid manure the organic acid content can be very variable because these are digested almost immediately under aerobic conditions and are only present in anaerobic parts of the solid manure.

In order to characterise the degradability of organic matter in manure, various measures may be applied. BOD (Text Box – Basic 5.1) reflects the microbial consumption of oxygen in the degradation of organic matter in the sample, but is not precise when estimating organic matter in a matrix with a high content of NH_4^+, such as manure. COD is an alternative to BOD, but can be imprecise, because animal slurry has to be greatly diluted for the analysis, which may reduce precision (Angelidaki and Ahring, 1994), and also due to oxidation of inorganic components such as NH_4^+, which may interfere with the analysis. Regarding measurements and use of COD, we recommend the presentation given by Henze *et al.* (1996).

In order to characterise organic matter in manure by dividing the material into organic matter fractions of different digestibility (e.g. rapidly and slowly digestible), other methods must be applied. There are numerous definitions of digestibility of organic material, typically linked to the methods or models for predicting digestibility. The definitions are related to where the organic matter is used and to the models or management tools used to assess how the organic matter will be transformed in soil, in biogas reactors or in compost heaps. The digestibility is typically negatively related to the lignin concentration of the organic matter. Lignin is an important plant cell wall constituent and contributes to plant protection against herbivores and infections, and to the mechanical strength of the plant cell wall. Lignin consists of a mixture of phenyl and aliphatic groups that are covalently linked to hemicellulose, and thereby cross-link different plant polysaccharides.

Thus, the slowly or non-degradable fraction of volatile solids can be defined using the following equation (Møller *et al.*, 2004):

$$VS_{ND;\ carbohydrate} = VS_{crude\ fibre} - VS_{lignin} \tag{5.2}$$

Degradable carbohydrates are defined as:

$$VS_{D;\ carbohydrate} = VS - VS_{protein} - VS_{lipid} - VS_{VFA} - VS_{ND;\ carbohydrates} \tag{5.3}$$

These equations are based on analysis used for assessing the value of plant material as an energy source for ruminants, where the crude fibre content ($VS_{crude\ fibre}$) is defined as the solid material that is not dissolved after treatment of the plant in acids (for further info, see Text Box – Advanced 5.1).

Text Box – Advanced 5.1   Van Soest characterisation of the gross biochemical composition of organic materials of plant origin (including feeds, manures, sludge and other organic residues)

To improve assessment of plants as feedstuffs for livestock, Van Soest (1963) developed an empirical characterisation of plant fibre that could be used in modelling optimum diet formulations. In the digestive tract of animals the microorganisms transform the organic matter into carbohydrates, proteins and short-chain acids that can be transferred from the digestive tract to the bloodstream. The microbial transformation is anaerobic and can be paralleled with the anaerobic transformation of organic matter in animal manure.

The Van Soest characterisation is a three-step dissolution and filtration analysis using enzymes, detergents, acids and heating (Van Soest, 1963).

1. *Neutral detergent fibre (NDF)* is the fibre fraction remaining after 60 min of heating at 100 °C with a "neutral detergent" (soap) for 1 h. In this treatment, sugar, starch, protein and lipids are dissolved. Hemicellulose, cellulose, lignin and silica are retained in the NDF fraction, which can be filtered from the solution. The NDF extraction will not dissolve all starch in the cell wall, so an amylase treatment is often included. After this treatment and treatment with neutral detergents the fibre fraction retained is named amylase neutral detergent fibre (aNDF).
2. *Acid detergent fibre (ADF)* is the fibre fraction remaining after treating the NDF fraction with an acid detergent solution (1 M H_2SO_4 detergent solution). This treatment dissolves hemicelluloses. Retained on the filter will be cellulose, lignin and silica not dissolved. Sometimes the NDF step is skipped, and when the sample is treated directly with acid detergent, pectin is not removed and a fraction of the pectin will precipitate and will contribute to the ADF fraction; this analysis will thus differ from the analysis where the sample has been treated with neutral detergent before treatment with acid detergents.
3. *Acid digested lignin (ACL)* is the fibre fraction remaining after treating the ADF fibre fraction with a strong acid (72% H_2SO_4). In this treatment the cellulose is dissolved and only the lignin will be retained on the filter.
4. *Ash* is the salts and solids remaining after heating the ACL fraction to 550 °C.

The results of the analysis should be given as a fraction of the volatile solids, but often the results are given as a fraction of dry matter. When presenting the analysis it is most important to present this relationship.

The different fractions of dissolved organic components and the cell wall fraction can then be determined as shown in the following (adapted from Hvelplund and Nørgaard, 2003):

Treatment	Measured	Dissolved
Neutral detergent treatment of organic matter (OM)	fibre fraction = NDF + ash	OM – (NDF + ash) = proteins + fat + carbohydrates, etc.
Acid detergent of NDF	fibre fraction = ADF + ash	(NDF + ash) – (ADF + ash) = hemicellulose
Strong acid of ADF Incinerate lignin fraction	fibre fraction = lignin + ash remaining fraction = ash	(ADF + ash) – (lignin + ash) = cellulose

When dissolving the organic matter and fibre fractions in neutral detergents, acid detergents and strong acids, it is assumed that ash is not dissolved. Therefore, the ash content of the organic matter treated has to be subtracted from the results of the sequential analysis of fractions.

5.3 Manure Microbiology

Numerous organisms are present in manure, which are mainly derived from the digestive tract of the animal, although some may also be of environmental origin (e.g. with bedding material), see Text Box – Basic 5.2 for some definitions. A range of different species and subtypes of bacteria, viruses and parasites are found in manure. The type and number vary with the animal species, age of animals, the type of bedding used, the method of storage (liquid or solid) and the storage period (Nodar *et al.*, 1992).

Text Box – Basic 5.2 Taxonomic classification of organisms present in manure and the environment

Several different systems may be used for classification of the microorganisms present in the aquatic, soil or manure environments. Below is a taxonomic classification by phylum, but ecological characteristics, carbon and energy source, electron acceptor or temperature range may also be used for classification (see Text Box – Basic 5.3).

Bacteria: Small single-cell organisms, 0.2–50 μm in size (colloidal) that have spherical through ellipsoidal to rod-like forms, with a negative surface charge under common environmental conditions (deprotonated acidic groups). They proliferate by cell division and when easily degradable organic material is available as substrate, this results in high bacterial growth rates. Bacteria in aerobic environments consist of a broad, versatile and robust range of species capable of degrading many different organic materials under a variety of environmental conditions. A smaller range of bacterial species are capable of degrading organic materials under anaerobic conditions.

Actinomycetes: Special class of unicellular organisms, in the form of fine, branching filaments similar to the fungi. Actinomycetes grow more slowly than most bacteria, but are enzymatically better equipped to degrade more complex substrates. Actinomycetes do not thrive at low oxygen concentrations and at high temperatures. They are active in degrading hemicellulose, cellulose and lignin.

Fungi: Category of microorganisms not commonly found in manure at the stage of excretion from the animal, but since ubiquitously present in the environment, they are often present in stored manure, especially under aerobic, somewhat drier situations. They are found in a wide range of forms and sizes including multicellular species (e.g. mushrooms) readily visible without magnification. Fungi always require pre-synthesised compounds as carbon sources; in using these molecules to grow, they break down the original substance and incorporate the transformed molecules into their own structure (saprophytes). Fungi therefore play a key role in the degradation of organic matter in soil and therefore also in composting, where they are important decomposers of hemicellulose, cellulose and lignin.

Protozoa: Simplest form of animal life; commonly 5–50 μm in size, and present in water, soil and, for some parasitic species, also in the intestines of animals. They are consumers of bacteria and other microorganisms (sometimes referred to as grazers or predators) and hence protozoa regulate the total microbial population (e.g. in a composting process)

Viruses: Small (0.02–0.3 μm), infectious agents that can replicate only inside the living cells of organisms and hence they do not play any direct role in the degradation of organic material. In manure, most viruses are enteric, and many are serious pathogens for animals and humans, and therefore manure management may greatly affect viral disease transmission.

The most numerous group in manure is bacteria, with typical numbers at about 10^{10} bacteria g^{-1} dry manure, dominated by faecal coliforms and streptococci, with *Salmonella* spp. also found occasionally. Some of these are transmitters of disease (pathogens) and many of them also pass between animals and humans (zoonotic), such as *Salmonella*, VTEC (verotoxin-producing *Escherichia coli*) and the protozoan parasite *Cryptosporidium parvum*, causing severe enterohaemorrhagic infections in humans. However, even when present, pathogenic forms rarely exceed 10^5 colony-forming units (CFU) g^{-1} dry manure, so the majority of microorganisms in manure play a role in performing a range of chemical and biological transformation processes of organic matter and nutrients, depending on temperature, oxygen and substrate nutrient level.

After excretion, the microflora of the manure typically changes dramatically, as the microorganisms of the animal intestinal tract are not at all adapted to survival in the environment, due to the radical shift in substrate availability, temperature, aeration and so on (Text Box – Basic 5.3) upon excretion. However, manure management (e.g. the method of storage) affects the environmental conditions and hence the survival of many intestinal microorganisms, including pathogens.

Text Box – Basic 5.3 Classification of microorganisms present in manure and the environment according to ecology, metabolism and temperature preference

Ecological characteristics

Autochthonous microorganisms: A mixed and stable population of microorganisms, indigenous to the environment where they are found, active or dormant, but with great resilience to lack of substrate and energy input. Soil microorganisms are dominantly autochthonous.

Zymogenous microorganisms: A group of microorganisms that can take advantage and proliferate if the environmental situation is changed, such as by an influx of a fresh nutrient supply. The population is then highly active, but fluctuating and rapidly declining when the substrate supply has dissipated.

Carbon and energy source

Autotrophs: Microorganisms capable of growing in a completely inorganic medium using carbonate species (atmospheric CO_2 or aqueous CO_2, HCO_3^- or CO_3^{2-}) as their sole source of carbon to produce carbohydrates, proteins and lipids. Photoautotrophs (e.g. algae) use solar radiation as an energy source for their synthetic processes. *Chemoautotrophs*, including most bacteria, derive energy from chemical oxidation reactions (e.g. by oxidising NH_4^+ to nitrite).

Heterotrophs: Microorganisms that make use of pre-synthesised organic compounds as a carbon source. These organisms are degraders or consumers of other living matter – plant and animal residues in water and soil and many organic waste materials. The heterotroph category includes fungi, actinomycetes, protozoa and most bacteria. When degradation of organic molecules proceeds, often through multiple steps, to the final inorganic products such as CO_2, NH_3 and sulfate, the process is referred to as ultimate degradation or mineralisation.

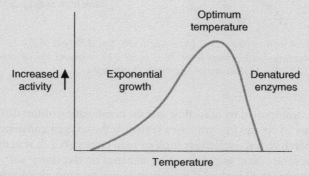

Figure 5.1   *Microbial activity as a function of temperature. (© University of Southern Denmark.)*

Temperature preference: All microorganisms follow an activity versus temperature relationship of the form shown in Figure 5.1. As illustrated, microbial growth rate increases with increasing temperature according to the Arrhenius equation, $\ln k = -E_a/RT + \text{constant}$, where k is the reaction rate constant. However, above a particular optimum temperature, T_{opt}, enzymes within the microorganism become denatured and growth declines precipitously. The position of the maximum along the temperature axis depends on the species of microorganism and species may be defined according to the value of T_{opt} as:

- *Psychrophiles*: $T_{opt} < 20\ °C$ (uncommon, but found in arctic environments)
- *Mesophiles*: $T_{opt} = 20–45\ °C$ (most common in the environment)
- *Thermophiles*: $T_{opt} > 45\ °C$ (found in hot springs, compost piles and biogas reactors)

5.4   Microbial and Biochemical Transformations in Manure

Manure may be collected in various forms, greatly affecting the biological and chemical conditions for microbial activity in the manure. The manure may be stored either as a slurry with a low dry matter content or as a solid manure with a higher dry matter content (for definitions see Text Box – Basic 4.1). The slurry has a low or no oxygen content (i.e. it is anaerobic) because oxygen dissolved in water diffuses very slowly into the matrix and is not replenished at a rate that corresponds to the oxygen consumption of the microorganisms in

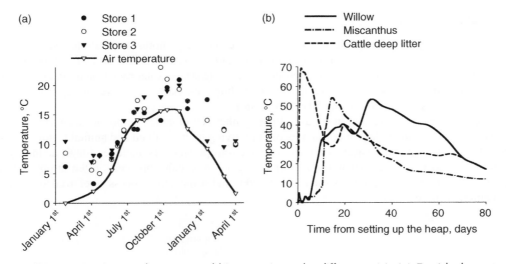

Figure 5.2 *Temperature in stored manure and biowaste (note the different axis). (a) Danish slurry stores. (b) Temperature in heaps containing cattle deep litter (C : N = 21), woodchips of willow (C : N = 123) and Miscanthus (C : N = 105–122). (© University of Southern Denmark.)*

the slurry. In contrast, air is transported by mass flow and the combination of fast diffusion in air-filled spaces into solid manure and influx of air can in most cases replenish the oxygen consumed by microorganisms in the solid manure, which therefore has a relatively high oxygen content (i.e. it is aerobic). Most stored liquid manure is not entirely anaerobic; there may be surface material on the slurry with a high air permeability, supporting an aerobic environment. Stored solid manure may also be a mosaic of aerobic and anaerobic sites, because air flow may be reduced in compacted sites, where the conditions are therefore anaerobic.

Oxygen content (Text Box – Basic 5.4) or partial pressure in the manure will affect the transformation process significantly. In slurry the anaerobic transformation processes are neither exothermic nor endothermic, and do not produce or consume energy from the surroundings. Therefore, slurry stores typically have the same temperature as the ambient temperature of the surroundings. This is not the case in aerobic solid manure, where the organic matter is transformed in oxidation processes which are exothermic. Therefore, the transformation increases the temperature of the solid manure heap (Figure 5.2).

Text Box – Basic 5.4   Definition of oxidation state and electron acceptors for microbial organic matter degradation and transformation processes in manure and the environment

Redox potential

A measure of electrochemical potential or electron availability in soil and aquatic systems. Measurement of redox potential characterises the degree of reduction of an aqueous solution, and enables prediction of stability of various compounds.

Redox potential is often given as $pE = -\log a_E$, where a_E is the electron activity, commonly measured as standard electrode potential E (unit is V). $pE°$ is defined as the redox potential under standard conditions (activities of all components are 1), but in natural aquatic systems, pH is near neutral and hence hydron concentration 10^{-7}, so $pE°(w)$ is defined as the standard redox potential at pH 7.

Organic matter degradation is an oxidation process and the half-reaction for complete oxidation is expressed as:

$$CH_2O + 5H_2O \rightarrow CO_2(g) + 4H_3O^+(aq) + 4e^-, pE°(w) = +8.20 \qquad (5.4)$$

In order for oxidation to occur, this half-reaction must be coupled to an appropriate reduction half-reaction (i.e. an electron accepting process), see Table 5.4. By "appropriate" is meant that a naturally occurring oxidising agent (electron acceptor) must be available in sufficient concentration and the $pE°$ for the overall process must be positive.

The redox potential is diagnostic for whether a system is aerobic or anaerobic. The redox potential is affected not only by dissolved O_2, but also by all other components in the solution that can be reduced or oxidised. The pH is especially important, because the electrical potential is related to H_3O^+ half-cell potential.

Aerobic conditions

Microorganisms use molecular oxygen, either in gaseous form or dissolved in water, as the electron acceptor for their oxidation reactions, in which oxygen is reduced and hydronium ions are consumed. At neutral pH, aerobic conditions are associated with a $pE°(w)$ above 13.

Anaerobic conditions

Anaerobic conditions dominate when air transfer is limited well below the surface in liquid or in compacted manure, so that oxygen becomes depleted and strictly aerobic organisms no longer can function. At neutral pH, anaerobic conditions are associated with a $pE°(w)$ below –3 (Sawyer *et al.*, 1994). Anaerobic microorganisms have the capability of effecting oxidation without molecular oxygen, instead using other electron-poor species, such as sulfate (Table 5.4), to accept electrons from the reduced substrate. The electron acceptor species is converted into a more reduced form, such as sulfide (Table 5.4). Bacteria make up the largest number of anaerobes, and some are obligate anaerobes, for which oxygen is toxic, while others are facultative anaerobes, capable of using either oxygen or other electron acceptors.

Table 5.4 *Half-reactions for relevant oxidation agents (electron acceptors) in natural systems in decreasing order of $pE°$ values; in addition, Gibbs free energy at standard conditions and neutral pH ($\Delta G°(w)$) is given for the full-reaction process when combined with 5.4 (in kJ mol^{-1} CH_2O).*

Oxidation agent	Half-reaction	$pE°(w)$	$\Delta G°(w)$
Oxygen	$O_2 + 4H_3O^+(aq) + 4e^- \rightarrow 6H_2O$	+13.80	–500
Nitrate	$2NO_3^-(aq) + 12H_3O^+(aq) + 10e^- \rightarrow N_2 + 18H_2O$	+12.65	–479
Sulfate	$SO_4^{2-}(aq) + 9H_3O^+(aq) + 8e^- \rightarrow HS^-(aq) + 13H_2O$	–3.75	–102
None[a]	$CO_2 + 8H_3O^+(aq) + 8e^- \rightarrow CH_4(g) + 10H_2O$	–4.13	–93

[a]Self-oxidation of organic material using CO_2 as the electron acceptor (van Loon and Duffy, 2010).

In both slurry and solid manure, the availability of organic matter for microbial transformation is affected by the composition of the excreta and of the bedding material added. The digestibility of the organic matter in cattle excreta is low compared with that of organic matter in pig excreta (Møller *et al.*, 2004). Ruminants can be fed diets of relatively low digestibility, because they ferment the organic matter in the rumen and take up the organic acids and amino acids produced by the rumen microflora. Not only are ruminants fed plant materials of low digestibility, but they also degrade a larger proportion of the degradable fraction in the feed, and hence only the very recalcitrant and slowly degradable organic matter in the feed will be excreted. If ruminants are fed diets containing highly degradable organic matter, then a fraction of this will be pushed through the digestive tract and the excreted organic matter will be more degradable. Monogastric animals (i.e. pigs and poultry) are given a more easily digestible, concentrated feed than cattle diets; in fact, the diets given to pigs could also be considered an adequate diet for humans. In consequence, what is left of organic matter after excretion from pigs and poultry contains relatively more easily degradable organic matter.

5.4.1   Aerobic Decomposition of Organic Matter

Stored solid manure with aerobic sites is found in animal houses and in manure stores. It is only in animal houses with deep litter that the manure is stored for longer periods. The deep litter is often transferred to outside stores and the organic matter is consequently subjected to two storage periods. FYM is removed more frequently, often daily, from the animal house to the outside store. A heap of FYM therefore contains manure of different ages. When deep litter and FYM are added to the soil, the manure components contribute to the soil organic matter, which in most well-drained agricultural soils is transformed in a predominantly aerobic environment.

 If the manure biomass is stacked in a heap with high air-filled porosity, so that the heap is at least initially aerobic, then the following phases of transformation may be defined:

1. The *initial phase*, which includes a lag phase where microorganisms adapt to the environment and the biomass, can be very short if the organic matter is easily degradable or may be of longer duration if the organic matter is lignified and has to be hydrolysed before a significant microbial transformation of the organic material can take place. Still in the initial phase, the temperature then increases to between 60 and 70 °C (Figure 5.2b).
2. This increase in temperature, defined as the *thermophilic phase*, is due to the degradation of organic matter by organoheterotrophic organisms. These microorganisms gain energy from the oxidation of organic compounds to CO_2 and H_2O (Madigan *et al.*, 2003) (Equation 5.5). Aerobic processes generate relatively large amounts of free energy by which the organisms can sustain a high growth rate and are therefore likely to have a competitive advantage in relation to, for example, e.g. NH_4^+-oxidising or sulfide-oxidising processes (chemoautotrophs).

$$CH_2O + O_2 \rightarrow CO_2 + H_2O \; (\Delta G°(w) = -500 \text{ kJ mol}^{-1}) \tag{5.5}$$

 The high-temperature thermophilic phase is often of relatively short duration because the microorganisms die due to the high temperature and because the transport of O_2 to the interior of the heap is too slow to replenish the oxygen consumed by the aerobic microorganisms. Stirring or aeration of the heap facilitates the transport of O_2 into the heap and prolongs the thermophilic period.

 As described in Chapter 4, temperature affects the pK_a value of the NH_4^+ to NH_3 equilibrium, and hence at thermophilic temperatures, a substantial proportion of NH_4^+ in the manure will be in NH_3 form and at risk of being lost with the convective airflow typically created by the strong exothermic reaction and heat flow from the centre of the pile. However, other processes, such as microbial nitrification and immobilisation of the N (see Section 5.5) compete for the NH_4^+, and the risk of N losses as NH_3 during the thermophilic phase depends on a range of different factors.

3. In the following *mesophilic phase* the temperature declines slowly and after 20–30 days it reaches about 20–35 °C. Very often a second increase in temperature of about 5–10 °C is seen (Figure 5.2b, cattle

manure). This increase is related to growth of actinomyces and fungi, which use cellulose and hemicellulose as substrate (Hellman *et al.*, 1997), but are not capable of activity during the high temperatures in the thermophilic phase. They survive the temperature peak as spores or other dormant forms, and then in the mesophilic phase proliferate and start to hydrolyse undecomposed, recalcitrant material, and hence provide a second flush of decomposable substrate for bacteria, causing this second peak of activity. The mesophilic phase may often last for a sustained period of time, provided that the right moisture and oxygen conditions prevail for the fungi and actinomycetes to continue degrading the recalcitrant materials. With this gradual decomposition of carbon, the C : N ratio of the manure decreases, provided that the N is not lost through gaseous emissions.

4. After the mesophilic phase the temperature declines even further to the ambient temperature in the *curing phase*. This decline is partly due to exhaustion of the more easily digestible organic components in the biomass. During this phase the microbiological activity stabilises and most of the mineralised N is nitrified to nitrate.

A mass balance of C and N indicates that between 40 and 50% of the C is lost if the biomass is stored aerobically for a long period and that the reduction in dry matter is more or less at the same level. Nitrogen losses are not as high as the C losses, and losses from deep litter are smaller than losses of N from pig solid manure and cattle FYM (Petersen *et al.*, 1998; Sommer, 2001), probably because the C : N ratio of the pig solid manure and FYM is lower than in deep litter. The C : N ratio in pig solid manure and FYM is between 8 and 10, in contrast to the C : N ratio of about 20 in deep litter, due to the addition and mixing of excreta and urine with straw. The high C : N ratio may enhance immobilisation and thereby reduce nitrification and denitrification (Kirchmann, 1985). The amount of N not accounted for in the mass balance presented in Table 5.5 is low, indicating that little N had been lost through denitrification in the stored deep litter. In contrast, a mass balance study indicated denitrification losses of 13–33% N during storage of pig solid manure and cattle FYM (Petersen *et al.*, 1998).

The organic matter in slurry often floats to the surface and forms a floating dry matter layer, called a surface crust, which is more or less aerobic as it becomes increasingly drier and air-filled during warm, dry periods. Farmers may also actively add straw to stored slurry to create a surface cover, because these surface layers reduce NH_3 emissions. In addition, solid manure heaps may be stratified with an anaerobic core/centre and an aerobic surface layer. Thus, in both categories of manure stores, there is often an anaerobic centre or sites producing reduced organic compounds such as CH_4, H_2S or VFAs. These components may be released from the anaerobic phase where they are produced and then transported into the gaseous phase closer to the surface of the porous organic biomass, from where they may then be released to the environment. However, in the aerobic surface layers they may also be oxidised, because in these layers CH_4 and VFAs are the compounds most likely to serve as electron donors (i.e. components that are oxidised). The oxidation of CH_4 produces

Table 5.5 *Percentage reductions in N, dry matter and carbon content estimated as the change in mass balance after 132 days of deep litter storage (Sommer, 2001).*

Treatment	N leaching (%)	NH_3-N loss (%)	N unaccounted (%)	N reduction (%)	Dry matter reduction (%)	C reduction (%)
Compacted	2.3	14.9	1.0	18	38	39.9
Cut	2.9	7.2	1.4	11.6	41	43.9
Covered	2.6	16.7	0	15.4	34	39.9
Untreated	3.4	ND	ND	27.7	45	48.5

ND, not determined.

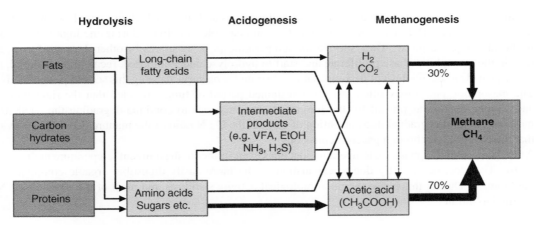

Figure 5.3 *The degradation pathways and end-products in anaerobic decomposition of organic matter or biopolymers in animal manure. This consists of three main phases: hydrolysis, acidogenesis (also called fermentation) and methanogenesis. (© University of Southern Denmark.)*

large amounts of energy and at the same time the CH_4 is a readily available source of C, so chemoautotrophic (Text Box – Basic 5.3) methane-oxidising bacteria are expected to multiply (proliferate) in these sites.

$$CH_4 + 2O_2 \rightarrow CO_2 + 2H_2O \, (\Delta G^{\circ\prime} = -818 \text{ kJ mol}^{-1}) \tag{5.6}$$

The effect of the oxidation is that CH_4 is only emitted from heaps of solid manure at very high CH_4 production rates in the centre of the compost (Sommer and Møller, 2000).

5.4.2   Anaerobic Decomposition of Organic Matter

In animal manure where the organic matter is stored or located in a strictly anaerobic environment, the organic components are transformed through a series of processes starting with the breakdown of the organic material through hydrolysis and microbial metabolism of the organic components through production of long-chain organic acids, acetic acid, CH_4 and CO_2 (Figure 5.3). In addition to CH_4 and CO_2, the end-products of the anaerobic transformation are H_2S and NH_3.

Hydrolysis is the first step in the chain of anaerobic degradation processes (Figure 5.3). In this process organic polymers such as proteins, cellulose, lignin or lipids are transformed to long-chain organic acids, glycerol, dissolved carbohydrates and amino acids. The hydrolysis involves hydrolytic exoenzymes and water. Exoenzymes are extracellular enzymes released by the microorganisms. These enzymes are capable of breaking down biopolymers to units that can be transported over cell membranes. The hydrolysis is a rate-limiting process in the transformation of organic matter and this is especially the case when the manure or the biowaste contains large amounts of straw, which have a high lignin content.

Acidogenesis is the second step in the transformation, where the organic components from the hydrolysis are taken up by heterotrophic, fermenting bacteria, which produce short-chain VFAs, alcohols, amino acids, dihydrogen (H_2) and CO_2. During acidogenesis, H_2S and NH_3 are produced from breakdown of the components containing N and S. The intermediate components produced during hydrolysis and acidogenesis may have a foul smell and, if volatilising from the manure, contribute to emissions. The unpleasant smelling compounds may consist of VFAs, H_2S, dimethyl sulfide (CH_3SCH_3), *p*-cresol, thiols and hundreds of others

(Schiffman *et al.*, 2001; Le *et al.*, 2005; Hudson *et al.*, 2009). The alcohols, amines and VFAs produced are usually further transformed to acetic acid (CH_3COOH), H_2 and CO_2. If amines are transformed, then NH_3 will also be a product of the reaction.

Methanogenesis is the third step in the anaerobic transformation of biopolymers and in this step CH_4 is produced by two entirely different processes: one is the aceticlastic step, which transforms acetic acid to CH_4 and CO_2 and the other combines CO_2 and H_2 to CH_4 through hydrogenotrophic methanogenesis. Some of the reactions are endothermic and others slightly exothermic, as is methane production, for example. In total, the anaerobic breakdown of biopolymers is energetically neutral.

- Aceticlastic methanogenesis:

$$CH_3COO^- + H_2O \rightarrow CH_4 + HCO_3 \ (\Delta G^\circ = -76 \, kJ \, mol^{-1}) \tag{5.7}$$

- Hydrogenotrophic methanogenesis:

$$4H_2 + HCO_3^- + H^+ \rightarrow CH_4 + 3H_2O \ (\Delta G^\circ = -130 \, kJ \, mol^{-1}) \tag{5.8}$$

As Figure 5.3 illustrates, H_2 and CO_2 may combine and produce acetic acid, which is degraded to H_2 and CO_2, the two processes being carried out by two different groups of bacteria. Furthermore, at high H_2 concentrations fatty acids longer than acetic acid increase due to H_2 inhibition of the degradation of long-chain fatty acids.

If the biopolymers are degraded completely, then CH_4, CO_2, H_2S and NH_3 are the end-products produced in the anaerobic transformation of biopolymers. Their quantities can be calculated by Equation (5.9) (Symons and Buswell, 1933):

$$C_nH_aO_bN_cS_dM_{ev} + \left(n - \frac{a}{4} - \frac{b}{2} + \frac{7c}{4} + \frac{d}{2} + \frac{3ev}{4}\right)H_2O \rightarrow \left(\frac{n}{2} - \frac{a}{8} + \frac{b}{4} - \frac{5c}{8} + \frac{d}{4} - \frac{9ev}{8}\right)$$

$$CO_2 + cNH_4HCO_3 + dH_2S + \left(\frac{n}{2} + \frac{a}{8} - \frac{b}{4} - \frac{3c}{8} - \frac{d}{4} + \frac{ev}{8}\right)CH_4 + eM(HCO_3)v \tag{5.9}$$

The kinetics of the combined reactions can be expressed as the production rate of CH_4. The combined reaction kinetics as affected by temperature are expressed as first-order by most authors (Batstone *et al.*, 2002; ADM1 model). Consequently, Arrhenius kinetics are often used to describe the exponential relationship between CH_4 emissions and temperature and this first-order relationship is expressed by the following exponential equation:

$$F(T) = A_o \cdot e^{-E_a/RT} \tag{5.10}$$

where $F(T)$ is the rate of the production per second at temperature T (in K), E_a is the apparent activation energy and R is the gas constant. It can be seen that either increasing the temperature or decreasing the activation energy will result in an increase in the rate of reaction. At temperatures below 10–15 °C, CH_4 production is negligible (Clemens *et al.*, 2006; Sommer *et al.*, 2007). Increasing the temperature above 15 °C will cause an exponential increase in CH_4 production. It is generally assumed that the bacteria consortia contributing to biopolymer degradation can be categorised according to a temperature optimum at 18 °C for psychrophilic, 38 °C for mesophilic and 65 °C for thermophilic microorganisms. However, one could argue that if the microorganisms had enough time to adjust to temperatures between 10 and 65 °C, then peak production rates at 18 and 38 °C might not be observed.

High TAN concentrations may inhibit methanogenesis (Chen *et al.*, 2008), and a high pH interacts with TAN because the toxic NH_3 dominates at high pH (Chapter 4). The inhibition of methanogenesis causes accumulation of VFAs, partly due to inhibition caused by the increased concentration of H_2. The H_2 concentration increases because H_2 is not consumed due to the reduced rate of hydrogenotrophic methanogenesis. With time the increased concentration reduces production of enzymes and the hydrolysis of organic matter (Angelidaki *et al.*, 1993).

A low pH also reduces the degradation of biopolymers due to inhibition, an effect which is partly related to organic acids and sulfide being uncharged and therefore more easily transferred across the bacteria cell membranes. In the bacteria these components inhibit microbial processes.

In fermentation processes, NH_3 and H_2 inhibition of degradation of LCFAs (valerate, butyrate, propionate, acetate) is expressed by the function used in the ADM model (Batstone *et al.*, 2002), in combination with a Monod-type expression. The following Monod equation expresses NH_3 inhibition:

$$I = \frac{1}{1 + \dfrac{[NH_3]}{K_I}} \tag{5.11}$$

where I is an inhibition variable expressing the effect of NH_3, K_I is a constant that can be set at 0.011 for thermophilic processes and 0.0018 for mesophilic processes. As the equation indicates, at increasing concentration of NH_3 the inhibition increases and I declines exponentially with NH_3 concentration.

The rate of the reaction declines linearly with increasing I:

$$dx/dt = k_m \frac{S_I}{K_{S,I} + S_I} \cdot X_1 \cdot I \tag{5.12}$$

where dx/dt is the rate of degradation of the substrate I, X is the concentration of microorganisms degrading substrate I, $K_{S,I}$ is the half-saturation value for the substrate degradation and k_m is the maximum specific uptake rate. It can be seen that I decreases at increasing NH_3 concentration and dx/dt consequently declines.

High loading rates or sudden changes in loading rates of biomass in relation to the amount of slurry stored may also cause an increase in VFAs due to a reduction in CH_4 production. Thus, if a large amount of slurry is added to a store with no or little slurry, the degradation process is insignificant until the microorganisms have recovered (Sommer *et al.*, 2007). The recovery can be partly inhibited by the high concentrations of VFAs.

5.5   Transformations of Nitrogen

In agriculture, manure has been used as a source of N for crops for centuries, as it can be a very valuable fertiliser for supplying crop N requirements. Manure and/or fertiliser supply of N is crucial to maintain high crop production levels in the field. However, only the NH_4^+ and nitrate forms of N can be taken up by the crop from the soil, and hence several different N turnover processes in both manures and soils can affect the availability and loss of manure N during storage, application and in the soil. This section describes the fundamental processes transforming N.

5.5.1 Urea and Uric Acid Transformation

In untreated slurry from pigs, cattle or sheep, about 50–60% of the N is found in the form of TAN. Most of this TAN originates from the hydrolysis of the urea, a diamide, found in urine. This diamide is transformed by the exoenzyme urease to NH_3, NH_4^+ and bicarbonate (HCO_3^-; Krajewska, 2009):

$$CO(NH_2)_2(aq) + 2H_2O \rightleftharpoons NH_3(aq) + NH_4^+(aq) + HCO_3^-(aq) \qquad (5.13)$$

The faeces excreted by livestock contain bacteria-producing urease, so urease is abundant on housing floors and soils in beef feedlots and exercise areas (Elzing and Monteny, 1997; Braam and Swierstra, 1999). In animal houses, the urea deposited on floors is hydrolysed to TAN within 24 h by the exoenzymes that are excreted in faeces and abundant in animal houses. Urease is also found in most soils where urea in urine from grazing animals is transformed to urine, as also seen when mineral fertiliser urea is applied to soil (Sommer *et al.*, 2004).

Hydrolysis of urea is affected by pH (Ouyang *et al.*, 1998) and optimum pH for urease activity has been reported to range from pH 6 to 9. Animal manure pH is buffered to between 7 and 8.4, so hydrolysis of urea is not greatly influenced by pH in manure that has not been treated with acids or bases.

Urease activity is affected by temperature and is often depicted as being exponentially related to temperature (Braam *et al.*, 1997) in the temperature range 10–60 °C. Below 5–10 °C and at temperatures above 60 °C the activity is low (Sommer *et al.*, 2006).

At urea concentrations higher than 3 M, hydrolysis may be inhibited by the substrate urea (Rachhpal-Singh and Nye, 1986), but at concentrations up to this threshold hydrolysis increases with increasing urea concentration on the floor.

Uric acid is a heterocyclic organic compound excreted by poultry (Figure 5.4). In the presence of water, uric acid is transformed to CO_2 and NH_4^+. The reaction is catalysed by the enzyme uricase, present in microorganisms (Lee *et al.*, 1989), which oxidises/hydrolyses uric acid to allatoic acid. Allantoin is hydrolysed to glyoxylic acid ($OCHCO_2H$) and urea, and glyoxylic acid is hydrolysed to urea. Finally, the urea is hydrolysed to NH_4^+.

The enzymes are produced by aerobic and anaerobic microorganisms, but the activity of the anaerobic microorganisms is not high. Furthermore, the reactants, intermediate products and final products are acids. Consequently, the reaction rate is greatly affected by pH and also by water content, which control oxygen availability. Temperature is again the most important reaction variable.

The rate of reaction increases exponentially from 0 to 35 °C, and with pH increasing from 5.5 to 9 the reaction rate also increases (Groot Koerkamp, 1994). Water content affects microbial activity and consequently uric acid degradation, which tends to be fastest at 40–60% water content, where bacteria have optimal growth conditions. At lower water contents degradation is slow, because drying significantly reduces bacterial growth. At higher water contents the conditions become anaerobic and anaerobic transformation of uric acid to urea is slow.

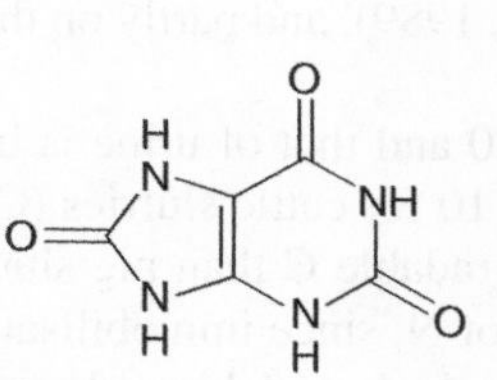

Figure 5.4 *Structural formula of uric acid ($C_5H_4N_4O_3$). (© University of Southern Denmark.)*

5.5.2   Ammonification or Mineralisation – Organic Nitrogen Transformation to Ammonium

Proteins and amines are relatively easily degradable, and TAN is therefore produced during both aerobic and anaerobic degradation processes. Degradation of amino acids takes place via deamination/decarboxylation processes or the Stickland reaction, which is carried out by bacteria belonging to the genus Clostridium (Barker, 1981) that exists in abundance in animal manure (Whitehead and Cotta, 2001; Snell-Castro *et al.*, 2005):

$$\text{Valine} + 2H_2O \rightarrow \text{isobutyric acid} + CO_2 + NH_3 \tag{5.14}$$

$$\text{Alanine} + 2 \text{ glycine} \rightarrow 3 \text{ acetate} + CO_2 + 3NH_3 \tag{5.15}$$

The production is temperature-dependent, as it is related to the transformation of organic matter. For cattle slurry, in one study degradation during the initial approximately 10 days of storage reduced organic N at rates of 12 g N kg^{-1} organic N day^{-1} at 10 °C and 18.4 g N kg^{-1} organic N day^{-1} at 20 °C (Sommer *et al.*, 2007). Between 15 and 150 days, organic N reduction in cattle slurry averaged 5.0 g N kg^{-1} organic N day^{-1} at 20 °C, while at 10 °C the rate between 21 and 114 days incubation was 0.3 g N kg^{-1} organic N day^{-1}. After degradation of biopolymers in biogas digesters for 10–20 days, the organic N constitutes about 20% of the total N in the treated slurry, compared with 50–60% in untreated slurry.

There are currently no appropriate models available that predict TAN concentrations as related to organic N composition of urine and excreta and that can integrate this with the effect of storage duration and temperature. In the present models it is commonly assumed that during in-house storage of the manure, most of the urea is transformed to TAN and about 10% of the organic N is mineralised (Zhang and Day, 1996). During outside storage of slurry, little N is mineralised due to the lower temperature in the storage tank and in Denmark it is assumed that about 5% of the organic N is transformed to inorganic N during a 6- to 9-month storage period (Poulsen *et al.*, 2001). These factors reflect the fact that slurry is constantly added to and removed from the slurry store; if the slurry were to be stored for a longer period in summer and winter temperatures as depicted in Figure 5.2, then more organic N would be transformed to TAN during storage.

5.5.3   Immobilisation

Immobilisation of inorganic N into organically bound N is a microbial process that depends on the C : N ratio of the organic compounds being degraded. When the C : N ratio of the degradable compounds in animal manure is high, the microbes are not capable of obtaining sufficient N for their own synthesis of microbial tissue and hence they obtain this by assimilating inorganic N from the manure into microbial biomass – a process termed immobilisation. Conversely, when the C : N ratio of the degradable compounds in animal manure is low, organically bound N in excess of the microbial demand is transformed (mineralised) into inorganic N. Hence, immobilisation decreases the amount of TAN, whilst mineralisation increases it. The balance between these two processes partly depends on the C : N ratio of degradable organic matter in the animal manure (Kirchmann and Witter, 1989), and partly on the C : N ratio and assimilation efficiency of the microbes.

Typically, the C : N ratio of faeces is 20 and that of urine is in the range 2–5. Slurry mixtures have C : N ratios ranging from 4 for pig slurries to 10 for cattle slurries (Chadwick *et al.*, 2000). Furthermore, cattle slurry has a greater fraction of poorly degradable C than pig slurry (Kirchmann, 1991). This fraction does not contribute as much to immobilisation of N, since immobilisation occurs in connection with degradation processes. The distribution of C between easily degradable and recalcitrant components in manure is therefore important when assessing the potential for immobilisation.

In general, there is no immobilisation of N in slurry mixtures stored in an anaerobic environment, because the C : N ratio of the easily degradable compounds is low (below 15) (Kirchmann and Witter, 1989; Thomsen, 2000). The addition of straw and other bedding material with a high C : N ratio increases the amount of degradable C and induces immobilisation. As a result, FYM (i.e. a mixture of mainly faeces and bedding material with a small amount of urine added) typically has a high C : N ratio and low TAN concentration (Külling *et al.*, 2003). Kirchmann and Witter (1989) estimated an immobilisation potential of 11.2 mg N g^{-1} straw at a C : N ratio between 18 and 24, and 2.2 mg N g^{-1} straw at a ratio between 24 and 36.

As immobilisation of inorganic N in animal manure is considered unlikely, except for FYM containing bedding material, to our knowledge no algorithms have been developed specifically for immobilisation in animal manure. However, for modelling animal manure N immobilisation, use can be made of the algorithms developed for immobilisation in soil.

5.5.4 Nitrification

Nitrification is the two-step oxidation of TAN (NH_4^+ or NH_3) into nitrite (NO_2^-) and then into nitrate (NO_3^-) by predominantly autotrophic microorganisms (Nitrobacteraceae). The first step, the oxidation of TAN into NO_2^-, is conducted by the so-called NH_3 oxidisers or primary nitrifiers, whereas the second step is carried out by NO_2^- oxidisers or secondary nitrifiers. *Nitrosomonas europaea* is the best-studied NH_3 oxidiser, while *Nitrobacter winogradskyi* is one of the most common NO_2^- oxidisers. The Nitrobacteraceae are aerobes and many are obligate autotrophs (i.e. they require oxygen (O_2) and the energy required for growth originates from nitrification). However, NH_4^+, NH_3 and NO_2 are not very effective energy sources, making the Nitrobacteraceae slow growers. They are also highly sensitive to pH; nitrification is negligible at pH values of less than around 4 and increases linearly as pH increases from 4 to 6 (Winter and Eiland, 1996). Currently, there is increased interest in the process of nitrification because of the possible release of the intermediate N_2O during NH_3 oxidation and NO_2^- oxidation. Nitrous oxide is a potent greenhouse gas and nitrification of TAN in animal manure is a possible important source (Chadwick *et al.*, 2011).

Nitrifying activity is absent in slurry and most other manures, because faeces and urine are highly anoxic upon excretion. During storage of animal slurries, nitrifying activity develops only slowly at the interface of atmosphere and slurry, because the diffusion of molecular O_2 into the slurry is slow (Petersen *et al.*, 1996), the biological demand by the host of competing microorganisms is large, and Nitrobacteraceae are slow growers and thus have a competitive disadvantage compared with the native microbial population of the manure. Surface crust drying may accelerate the creation of oxic conditions at the surface and therefore induce nitrifying activity during long-term storage (Nielsen *et al.*, 2010). However, the amount of TAN nitrified in slurries and liquid manures in lagoons and basins is usually very small. Consequently, release of N_2O to the atmosphere from stored slurry is small (Harper *et al.*, 2000; Külling *et al.*, 2003).

In animal manure amended with bedding material, such as that in deep litter stables, feedlots and stacked FYM heaps, significant nitrifying activity can develop during storage. Here, the nitrifying activity results from the much greater aeration of the manure in the surface layer compared with slurry, because the litter-amended manure is rather porous, thus allowing molecular O_2 to be transported more easily into the manure. As a result, measurable quantities of NO_2^- and NO_3^- can be found in the surface layers, and significant emissions of N_2O have also been measured from dung heaps and deep litter animal houses (Groenestein and Van Faassen, 1996; Petersen *et al.*, 1998; Chadwick, 2005b; Webb *et al.*, 2012).

Modelling of nitrification is based either on a mechanistic description of the growth and development of nitrifying populations (Li *et al.*, 1992) or simply on a substrate-dependent process using first-order kinetics (Malhi and McGill, 1982; Gilmour, 1984; Grant, 1994). The microbial growth models consider the dynamics of the nitrifying organisms responsible for the nitrifying activity. The simplified process models are easier to use and do not consider microbial processes and gaseous diffusion. In these simplified models, nitrification rate (d(TAN)/dt) is described as an empirical function of substrate concentration ([TAN]), oxygen partial pressure (pO_2), temperature (T), and pH according to:

$$\int d(TAN)/dt = k_1 \cdot f(TAN) \cdot f(pO_2) \cdot f(T) \cdot f(pH) \tag{5.16}$$

where k_1 is the first-order nitrification coefficient under optimal conditions and $f(TAN) = [TAN]$.

Nitrifying activity is sometimes related to TAN concentration via a Michaelis–Menten-type relationship: $f(TAN) = [TAN]/ (k_2 + [TAN])$. In this case, TAN is limiting for nitrifying activity (*cf.* first-order process) at low TAN concentrations and TAN is not limiting for nitrifying activity (zero-order) at high concentrations. The constant k_2 is the Michaelis–Menten half-saturation constant or the TAN concentration at which $f(TAN) = 0.5$. Like most biological processes, nitrifying activity generally increases exponentially with increasing temperature until a certain temperature, after which the activity decreases with increasing temperature (e.g. composting manure heaps). According to the Arrhenius law, the reduction function for temperature can be described by:

$$f(T) = \exp(K_A(T - T_{ref})/(T_{ref} \cdot T) \tag{5.17}$$

where T is temperature, T_{ref} is the reference temperature where $f(T) = 1$ and K_A is a coefficient characteristic for the environment.

5.5.5   Nitrification–Denitrification Coupling

Denitrifying bacteria are heterotrophic – they require a source of degradable organic carbon (Tiedje *et al.*, 1984). Denitrification is the stepwise conversion of NO_3^- via NO_2^-, NO and N_2O to N_2, each step conducted by a different enzyme. An imbalance between these steps may occur as a result of sub-optimal or rapidly changing conditions (e.g. during rainfall) which can lead to accumulation of N_2O in the pore space or dissolved in the aqueous phase. Nitrous oxide may escape further reduction if the site of production is near an aerobic/anaerobic interface. The enzyme reducing N_2O to N_2, N_2O reductase, is sensitive to oxygen and so even low oxygen conditions may stimulate N_2O emissions by denitrifying bacteria (Firestone *et al.*, 1980).

A coupling can develop around aerobic/anaerobic interfaces between nitrification oxidising NH_4^+ to NO_2^- and NO_3^- in aerobic zones and denitrification reducing the NO_3^- in nearby anaerobic zones. In model systems the opposite fluxes of NH_4^+ and NO_3^- across the aerobic/anaerobic interface have been shown to stimulate both processes (Petersen *et al.*, 1991; Nielsen and Revsbech, 1994).

Nitrification and denitrification are often coupled in manure management systems. The coupling is a consequence of the frequent development of mosaics of aerobic and anaerobic sites in the manure. Thus, ammonium is transformed to nitrate in the aerobic environment and nitrate diffuses into the anaerobic environment where it is denitrified. The coupling of the processes increases the potential for production of the very potent climate gas N_2O.

While the production of N_2O in slurry is generally low, the presence of an organic cover usually results in a significant increase in N_2O emissions over time (Sommer *et al.*, 2000; Berg *et al.*, 2006). Nitrous oxide is produced primarily as a byproduct from the aerobic oxidation of NH_3 carried out by NH_3-oxidising bacteria (AOB) and from the anaerobic reduction of their waste product, NO_2^-, by denitrifying organisms. Emission of N_2O from organic covers is therefore likely to be either a direct or indirect consequence of the activity of AOB and consumption of NH_3.

Nitrite and especially NO_3^- may accumulate in natural and straw covers floating on stored slurry (Petersen and Miller, 2006), which is a strong indication of the activity not only of AOB, but also of nitrite-oxidising bacteria (NOB). These two types of organisms are often found in close association (Schramm, 2003), as NOB benefits from being positioned close to the source of its substrate, NO_2^-, and AOB benefits from the removal of the potentially toxic NO_2^- via its oxidation to NO_3^-:

$$NH_3 + 1.5\,O_2 \rightarrow NO_2^- + H^+ + H_2O \;(\Delta G^{\circ\prime} = -288\;\text{kJ mol}^{-1}) \tag{5.18}$$

$$NO_2^- + 0.5\,O_2 \rightarrow NO_3^- \,(\Delta G^{\circ\prime} = -74\;\text{kJ mol}^{-1}) \tag{5.19}$$

Both AOB and NOB fix CO_2 via the Calvin cycle as their primary carbon source – a process that is estimated to consume up to 80% of the energy budget of autotrophs. Combined with the relatively low energy yield of their metabolic processes (Equations 5.18 and 5.19), these organisms therefore tend to grow relatively slowly compared with many heterotrophs.

This may be of importance in a competitive environment such as organic covers on manure stores, where slow-growing organisms are likely to be inhibited at times due to competition for O_2 or altogether outcompeted. In the case of AOB, however, despite being obligate aerobes, some are known to withstand low O_2 or even anaerobic conditions, and studies indicate that AOB contain cellular mechanisms that help them survive and subsequently regain activity after periods of starvation (Geets, 2006). As such, despite a potentially high competitive pressure, AOB may proliferate within organic covers due to their resilience. Interestingly, recent evidence suggests that AOB can oxidise and consume compounds such as *p*-cresol (Texier and Gomez, 2007; Silva *et al.*, 2009), and the presence of these organisms in organic covers may therefore help reduce odour as well as NH_3 emissions.

While the availability of O_2 is generally considered the limiting factor in highly active, aerobic environments, other compounds may limit microbial activity through inhibition. Free nitrous acid (HNO_2), which exists in a pH-dependent equilibrium with NO_2^-, is toxic to many aerobic organisms, and accumulation of NO_2^- via the activity of AOB could therefore become detrimental to the aerobic community in organic covers and even to AOB themselves.

In addition, high concentrations of NH_3 are known to be inhibitory to many different organisms and especially methane-oxidising bacteria. Hence, even though evidence has been obtained for the possible presence of AOB and NOB in organic covers, very little is known about the overall effect of the many potentially limiting factors inherent in these environments.

5.6   Summary

The composition of the organic components of animal manure and the transformation of these through anaerobic and aerobic microbial activity were presented in this chapter. The main focus was on the effect of oxygen content and the energy outcome of the reactions, which are determining parameters for the rate and direction of these processes. In manure there may be further transformation of C into CO_2 and CH_4, as well as N transformations between the organic pools of N and inorganic pools of NH_4^+ and nitrate (NO_3^-). The rate of transformation is generally related to environmental temperature, oxygen concentration and the composition and degradability of the organic matter in the manure.

References

Angelidaki, I. and Ahring, B.K. (1994) Anaerobic thermophilic digestion of manure at different ammonia loads: effect of temperature. *Water Res.*, **28**, 727–731.

Angelidaki, I., Ellegaard, L. and Ahring, B.K. (1993) A mathematical model for dynamic simulation of anaerobic digestion of complex substrates, focusing on ammonia inhibition. *Biotechnol. Bioeng.*, **42**, 159–166.

Barker, H.A. (1981) Amino-acid degradation by anaerobic bacteria. *Annu. Rev. Biochem.*, **50**, 23–40.

Batstone, D.J., Keller, J., Angelidaki, I., Kalyuzhnyi, S.V., Pavlostathis, S.G., Rozzi, A., Sanders, W.T.M., Siegrist, H. and Vavilin, V.A. (2002) The IWA Anaerobic Digestion Model No 1 (ADM1). *Water Sci. Technol.*, **45**, 65–73.

Berg, W., Brunsch, R. and Pazsiczki, I. (2006) Greenhouse gas emissions from covered slurry compared with uncovered during storage. *Agric. Ecosyst. Environ.*, **112**, 129–134.

Braam, C.R. and Swierstra, D. (1999) Volatilization of ammonia from dairy housing floors with different surface characteristics. *J. Agric. Eng. Res.*, **72**, 59–69.

Braam, C.R., Ketelaars, J. and Smits, M.C.J. (1997) Effects of floor design and floor cleaning on ammonia emission from cubicle houses for dairy cows. *Neth. J. Agric. Sci.*, **45**, 49–64.

Chadwick, D.R. (2005a) Managing techniques to minimise ammonia emissions from solid manure, *Report CSG 15*, Department for Environment, Food and Rural Affairs, London.

Chadwick, D.R. (2005b) Emissions of ammonia, nitrous oxide and methane from cattle manure heaps: effect of compaction and covering. *Atmos. Environ.*, **39**, 787–799.

Chadwick, D.R., John, F., Pain, B.F., Chambers, B.J. and Williams, J. (2000) Plant uptake of nitrogen from the organic nitrogen fraction of animal manures: a laboratory experiment. *J. Agric. Sci.*, **134**, 159–168.

Chadwick, D.R., Sommer, S.G., Thorman, R., Fangueiro, D., Cardenas, L., Amon, B. and Misselbrook, T. (2011) Manure management: implications for greenhouse gas emissions. *Anim. Feed Sci. Technol.*, 166–167, 514–531.

Chen, Y., Cheng, J.J. and Creamer, K.S. (2008) Inhibition of anaerobic digestion process: a review. *Bioresour. Technol.*, **99**, 4044–4064.

Clemens, J., Trimborn, M., Weiland, P. and Amon, B. (2006) Mitigation of greenhouse gas emissions by anaerobic digestion of cattle slurry. *Agric. Ecosyst. Environ.*, **112**, 171–177.

Derikx, P.J.L., Willers, H.C. and Tenhave, P.J.W. (1994) Effect of pH on the behavior of volatile compounds in organic manures during dry-matter determination. *Bioresour. Technol.*, **49**, 41–45.

Elzing, A. and Monteny, G.J. (1997) Modelling and experimental determination of ammonia emissions rates from a scale model dairy-cow house. *Trans. ASAE*, **40**, 721–726.

Firestone, M.K., Firestone, R.B. and Tiedje, J.M. (1980) Nitrous-oxide from soil denitrificaiton – factors controlling its biological production. *Science*, **208**, 749–751.

Geets, J., Boon, N. and Verstraete, W. (2006) Strategies of aerobic ammonia-oxidizing bacteria for coping with nutrient and oxygen fluctuations. *FEMS Microbiol. Ecol.*, **58**, 1–13.

Gilmour, J.T. (1984) The effects of soil properties on nitrification and nitrification inhibition. *Soil Sci. Soc. Am. J.*, **48**, 1262–1266.

Grant, R.F. (1994) Simulation of ecological controls on nitrification. *Soil Biol. Biochem.*, **26**, 305–315.

Groenestein, C.M. and Van Faassen, H.G. (1996) Volatilization of ammonia, nitrous oxide and nitric oxide in deep-litter systems for fattening pigs. *J. Agric. Eng. Res.*, **65**, 269–274.

Groot Koerkamp, P.W.G. (1994) Review of emission of ammonia from housing system for laying hens in relation to sources, processes, building design and manure handling. *J. Agric. Eng. Res.*, **59**, 73–87.

Hansen, M.N., Henriksen, K. and Sommer, S.G. (2006) Observations of production and emission of greenhouse gases and ammonia during storage of solids separated from pig slurry: effects of covering. *Atmos. Environ.*, **40**, 4172–4181.

Harper, L.A., Sharpe, R.R. and Parkin, T.B. (2000) Gaseous nitrogen emissions from anaerobic swine lagoons: ammonia, nitrous oxide and dinitrogen gas. *J. Environ. Qual.*, **29**, 1356–1365.

Hellmann, B., Zelles, L., Palojärvi, A. and Bai, Q. (1997) Emission of climate-relevant trace gases and succession of microbial communities during open-windrow composting. *Appl. Environ. Microbiol.*, **63**, 1011–1018.

Henriksen, K. and Olesen, T. (2000) Kulstof og kvælstof omsætningsprocesser i kvægdybstrøelsesmåtter, in *Husdyrgødning og kompost* (eds S.G. Sommer and J. Eriksen), Forskningscenter for Økologisk Jordbrug, Tjele, pp. 14–21.

Hill, D.T., Taylor, S.E. and Grift, T.E. (2001) Simulation of low temperature anaerobic digestion of dairy and swine manure. *Bioresour. Technol.*, **78**, 127–131.

Henze, M., Harremoës, P., La Cour Jansen, J. and Arvin, E. (1996) *Wastewater Treatment – Biological and Chemical Processes*, 2nd edn, Springer, Berlin.

Hudson, N., Ayoko, G.A., Dunlop, M., Duperouzel, D., Burrell, D., Bell, K., Gallagher, E., Nicholas, P. and Heinrich, N. (2009) Comparison of odour emission rates measured from various sources using two sampling devices. *Bioresour. Technol.*, **100**, 118–124.

Hvelplund, T. and Nørgaard, P. (2003) Kvægets ernæring og fysiologi: næringsstofomsætning og fodervurdering [Textbook about cattle nutrition and physiology], *DJF Rapport Husdyrbrug 53*, Ministry of Food, Agriculture and Fisheries, Danish Institute of Agricultural Sciences, Tjele.

Kirchmann, H. (1985) Losses, plant uptake and utilisation of manure nitrogen during a production cycle. *Acta Agric. Scand.*, Suppl. **24**, 1–77.

Kirchmann, H. (1991) Carbon and nitrogen mineralisation of fresh, aerobic and anaerobic animal manures during incubation with soil. *Swed. J. Agric. Sci.*, **21**, 165–173.

Kirchmann, H. and Witter, E. (1989) Ammonia volatilization during aerobic and anaerobic manure decomposition. *Plant Soil*, **115**, 35–41.

Külling, D.R., Menzi, H., Sutter, F., Lischer, P. and Kreuzer, M. (2003) Ammonia, nitrous oxide and methane emissions from differently stored dairy manure derived from grass- and hay-based rations. *Nutr. Cycl. Agroecosyst.*, **65**, 13–22.

Krajewska, B. (2009) Ureases, I. Functional, catalytic and kinetic properties: a review. *J. Mol. Catal. B*, **59**, 9–21.

Le, P.D., Aarnink, A.J., A., Ogink, N.W.M., Becker, P.M. and Verstegen, M.W.A. (2005) Odour from animal production facilities: its relationship to diet. *Nutr. Res. Rev.*, **18**, 3–30.

Lee, C.C., Caskey, C.T., Wu, X.W. and Muzny, D.M. (1989) Urate oxidase: primary structure and evolutionary implications. *Proc. Natl. Acad. Sci. USA*, **86**, 9412–9416.

Li, C.S., Frolking, S. and Frolking, T.A. (1992) A model of nirous-oxide evolution from soil driven by rainfall events: 1. Model structure and sensitivity. *J. Geo. Res. Atmos.*, **97**, 9759–9776.

Madigan, M., Martinko, J. and Parker, J. (2003) *Brock: Biology of Microorganisms*, 10th edn, Pearson Education International, London.

Malhi, S.S. and McGill, K.S. (1982) Nitrification in three Alberta soils: effect of temperature, moisture and substrate concentration. *Soil Biol. Biochem.*, **14**, 393–399.

McCrudden, F.M. (2008) *Uric Acid*, Bastian Books, Toronto.

Møller, H.B., Sommer, S.G. and Ahring, B.K. (2004) Methane productivity of manure, straw and solid fractions of manure. *Biomembr. Bioenerget.*, **26**, 485–495.

Nielsen, T.H. and Revsbech, N.P. (1994) Diffusion chamber for N-15 determination of coupled nitrification–denitrification around soil manure interfaces. *Soil Sci. Soc. J. Agric.*, **58**, 795–800.

Nielsen, D.A., Nielsen, L.P., Schramm, A. and Revsbech, N.P. (2010) Oxygen distribution and potential ammonia oxidation in floating, liquid manure crusts. *J. Environ. Qual.*, **39**, 1813–1820.

Nodar, R., Acea, M.J. and Carballas, T. (1992) Poultry slurry microbial population: composition and evolution during storage. *Bioresour. Technol.*, **40**, 29–34.

Oenema, O., Bannink, A., Sommer, S.G., van Groenigen, J.W. and Velthof, G. (2008) Gaseous nitrogen emission from livestock farming system, in *Nitrogen in the Environment: Sources, Problems, and Management* (eds J.L. Hatfield and R.F. Follet), Elsevier, Amsterdam, pp. 395–443.

Ouyang, D.S., Mackenzie, A.F. and Fan, M.X. (1998) Ammonia volatilization from urea amended with triple super-phosphate and potassium chloride. *Soil Sci. Soc. Am. J.*, **62**, 1443–1447.

Petersen, S.O. and Miller, D.N. (2006) Perspective: Greenhouse gas mitigation by covers on livestock slurry tanks and lagoons? *J. Sci. Food Agric.*, **86**, 1407–1411.

Petersen, S.O., Henriksen, K. and Blackburn, T.H. (1991) Coupled nitrification denitrification associated with liquid manure in a gel-stabilized model system. *Biol. Fertil. Soils*, **12**, 19–27.

Petersen, S.O., Nielsen, T.H., Frostegård, A. and Olesen, T. (1996) O_2 uptake, C metabolism and denitrification associated with manure hot spots. *Soil Biol. Biochem.*, **28**, 341–349.

Petersen, S.O., Lind, A.-M. and Sommer, S.G. (1998) Nitrogen and organic matter losses during storage of cattle and pig manure. *J. Agric. Sci.*, **130**, 69–79.

Poulsen, T.G. and Moldrup, P. (2007) Air permeability of compost as related to bulk density and volumetric air content. *Waste Manag. Res.*, **25**, 343–351.

Poulsen, H.D., Børsting, C.F., Rom, H.B. and Sommer, S.G. (2001) Kvælstof, fosfor og kalium i husdyrgødning – normtal 2000 [Standard values of nitrogen, phosphorous and potassium in animal manure], *DJF Rapport 36*, Ministry of Food, Agriculture and Fisheries, Danish Institute of Agricultural Sciences, Tjele.

Rachhpal-Singh and Nye, P.H. (1986) A model of ammonia volatilization from applied urea. I. Development of the model. *J. Soil Sci.*, **37**, 9–20.

Sawyer, C.N., McCarty, P.L. and Parkin, G.F. (1994) *Chemistry for Environmental Engineering, Civil Engineering Series*, McGraw-Hill International Editions, New York.

Schiffman, S.S., Bennett, J.L. and Raymer, J.H. (2001) Quantification of odors and odorants from swine operations in North Carolina. *Agric. Forest. Meteorol.*, **108**, 213–240.

Schramm, A. (2003) *In situ* analysis of structure and activity of the nitrifying community in biofilms, aggregates, and sediments. *Geomicrobiol. J.*, **20**, 313–333.

Silva, C.D., Gomez, J., Houbron, E., Cuervo-Lopez, F.M. and Texier, A.C. (2009) *p*-Cresol biotransformation by a nitrifying consortium. *Chemosphere*, **75**, 1387–1391.

Snell-Castro, R., Godon, J.J., Delgenes, J.P. and Dabert, P. (2005) Characterisation of the microbial diversity in a pig manure storage pit using small subunit rDNA sequence analysis. *FEMS Microbiol. Ecol.*, **52**, 229–242.

Sommer, S.G. (2001) Effect of composting on nutrient loss and nitrogen availability of cattle deep litter. *Eur. J. Agron.*, **14**, 123–133.

Sommer, S.G. and Møller, H.B. (2000) Emission of greenhouse gases during composting of dep litter from pig production – effect of straw content. *J. Agric. Sci.*, **134**, 327–335.

Sommer, S.G. and Husted, S. (1995) The chemical buffer system in raw and digested animal slurry. *J. Agric. Sci.*, **124**, 45–53.

Sommer, S.G. and Hutchings, N.J. (2001) Ammonia emission from field applied manure and its reduction – invited paper. *Eur. J. Agron.*, **15**, 1–15.

Sommer, S.G., Petersen, S.O. and Søgaard, H.T. (2000) Emission of greenhouse gases from stored cattle slurry and slurry fermented at a biogas plant. *J. Environ. Qual.*, **29**, 744–751.

Sommer, S.G., Schjoerring, J.K. and Denmead, O.T. (2004) Ammonia emission from mineral fertilizers and fertilized crops. *Adv. Agron.*, **82**, 557–622.

Sommer, S.G., Petersen, S.O. and Møller, H.B. (2004) Algorithms for calculating methane and nitrous oxide emissions from manure management. *Nutr. Cycl. Agroecosyst.*, **69**, 143–154.

Sommer, S.G., Zhang, G.Q., Bannink, A., Chadwick, D., Hutchings, N.J., Misselbrook, T., Menzi, H., Ni, J.Q., Oenema, O., Webb, J. and Monteny, G.-J. (2006) Algorithms determining ammonia emission from livestock houses and manure stores. *Adv. Agron.*, **89**, 261–335.

Sommer, S.G., Petersen, S.O., Sørensen P, Poulsen, H.D. and Møller, H.B. (2007) Greenhouse gas emission and nitrogen turnover in stored liquid manure. *Nutr. Cycl. Agroecosyst.*, **78**, 27–36.

Symons, G.E. and Bushwell, A.M. (1933) The methane fermentation of carbohydrates. *J. Am. Chem. Soc.*, 2028–2036.

Texier, A.C. and Gomez, J. (2007) Simultaneous nitrification and ᴾ-cresol oxidation in a nitrifying sequencing batch reactor. *Water Res.*, **41**, 315–322.

Thomsen, I.K. (2000) C and N transformation in 15-N cross-labelled ruminant manure during anaerobic and aerobic storage. *Bioresour. Technol.*, **72**, 267–274.

Tiedje, J.M., Sextone, A.J., Parkin, T.B., Revsbech, N.P. and Shelton, D.R. (1984) Anaerobic processes in soil. *Plant Soil*, **76**, 197–212.

van Loon, G.W. and Duffy, S.J. (2010) *Environmental Chemistry – A Global Perspective*, 3rd edn, Oxford University Press, Oxford.

Van Soest, P.J. (1963) Use of detergents in the analysis of fibrous feeds. II. A rapid method for the determination of fiber and lignin. *J. Assoc. Off. Agric. Chem.*, **46**, 829–835.

Voorburg, J.H. and Kroodsma, W. (1992) Volatile emissions of housing systems for cattle. *Livest. Prod. Sci.*, **31**, 57–70.

Webb, J., Sommer, S.G., Kupper, T., Groenestein, K., Hutchings, N.J., Eurich-Menden, B., Rodhe, L., Misselbrook, T.H. and Amon, B. (2012) Emissions of ammonia, nitrous oxide and methane during the management of solid manures. *Sust. Agric. Rev.*, **8**, 67–107.

Whitehead, T.R. and Cotta, M.A. (2001) Characterisation and comparison of microbial populations in swine faeces and manure storage pits by 16S rDNA gene sequence analyses. *Anaerobe*, **7**, 181–187.

Winter, F.P. and Eiland, F. (1996) The effect of soil pH on nitrification in coarse sandy soil, in *Progress in Nitrogen Cycling Studies* (ed. O. van Cleemput), Kluwer, Dordrecht, pp. 655–658.

Zhang, R.H. and Day, D.L. (1996) Anaerobic decomposition of swine manure and ammonia generation in a deep pit. *Trans. ASAE*, **39**, 1811–1815.

6
Sanitation and Hygiene in Manure Management

Björn Vinnerås

*Department of Energy & Technology, Swedish University of Agricultural Sciences,
National Veterinary Institute, Sweden*

6.1 Hygiene Risks Associated with Manure Management

Manure management is essential for sustainable use of plant nutrients. The flows and treatments of manure should be as short as possible in order to optimise recycling of the plant nutrients in manure, as each treatment step causes losses of nutrients. However, to minimise the transmission of pathogenic microorganisms the flows should be as long as possible, as each additional step decreases the risk of disease transmission during management of manure and from infected crops.

There are several factors affecting the risk of disease transmission via the flow of plant nutrients in manure from the animal into the food or feed. The first is the health status of the animal, as the risk of disease transmission is lower from healthy animals than from sick animals. During management a risk factor is the age of the manure, as in most cases the pathogen load decreases over time. The disease-causing organisms can either be animal-specific or zoonotic and infect several species. The zoonotic diseases affect manure management by requiring greater awareness, as the risk of transmitting diseases along the food chain is higher than with non-zoonotic diseases. The animal-specific diseases can be controlled in manure management by application to food or feed crops consumed by other species, while the zoonotic diseases require other barriers. A typical example of this is the VAC farming system (VAC is the abbreviation of the Vietnamese words *vuon, ao, chuong*, which mean garden, pond, livestock pen), where the manure is used for production of algae, which in turn are used as fish food (Chapter 2). This practice has been shown to contaminate the skin of fish with *Enterobacteriaceae* (e.g. *Salmonella* spp.) that can affect people along the food management chain as the bacteria are transmitted to the hands of those handling the dead and live fish

Animal Manure Recycling: Treatment and Management, First Edition. Edited by Sven G. Sommer, Morten L. Christensen,
Thomas Schmidt and Lars S. Jensen.

(Yajima and Kurokura, 2008). The fish meat is not affected by this (Son Thi Thanh *et al.*, 2011), but the risk of transmitting pathogenic bacteria to the fish meat during handling is high.

Method of transfer and application of manure also affect the risk of disease transmission (e.g. the risk of transmission to the feed or food crop with direct incorporation into the soil is 99% lower than with surface spreading). In the field, the crop or vegetable to which the manure is applied is a factor affecting the risk of disease transmission. The greatest risks are associated with plants where edible plant parts that are close to the soil are consumed raw (e.g. lettuce and cucumber), while the risks are lower with crops that are processed prior to consumption (e.g. barley for beer production) or even crops that are not consumed at all (e.g. energy crops).

The main barrier to controlling the pathogen level in the manure is treatment prior to application, to ponds or to fields. This lowers the risk of disease transmission later in the manure management chain considerably compared with untreated manure. The focus of current manure management is therefore on treatment before transferring manure to the end use in the field.

See Text Box – Basic 6.1 for definitions used in this chapter.

Text Box – Basic 6.1   Definitions

Animal byproduct (ABP): Term used in EU regulations for organic waste of animal origin that may be hazardous.

Zoonosis: An infectious disease that is transmitted between species (animals to humans).

D_T: Decimal reduction time value. Equal to reduction of 90% of the population, can also be written as 1 $\log_{10}$ reduction.

Pathogen reduction is often written as the number of logarithms it is reduced over a time period/treatment. This is often written as a reduction of 1,000 times is written as 3 $\log_{10}$, and is equal to 99.9% reduction.

Hydraulic retention time (HRT): A measure of the average length of time that a soluble compound remains in a constructed bioreactor:

$$\text{HRT} = \frac{\text{Reactor volume (m}^3\text{)}}{\text{Influent flowrate (m}^3\text{ h}^{-1}\text{)}} \qquad (6.1)$$

Minimum retention time (MRT): A measure of the minimum time between feeding a reactor and detecting any part of the feedstock in the effluent.

6.2   Why Must the Pathogens in Manure be Managed?

When a disease is present in a herd on a farm, the main route of transmission is via animal-to-animal contact. When the disease has decreased in the herd, other transmission routes become important (e.g. fields, infected buildings). It is important to note in manure management that if the disease is zoonotic (i.e. can infect humans as well as animals, such as *Salmonella* spp.), there is a health risk to workers applying the manure to the field.

For the animal-specific diseases there is a barrier to disease transmission as the contaminated manure needs to find its way back to the original host animal. The major risk of transmission here is from animal to animal. However, at the farm level there is a risk of disease transmission via the manure when the disease in the herd has gone, as disease can come back via the manured fields and fodder crops. In addition, the management practice of placing dead animals on the manure heap or in the manure tank increases the risk of maintaining a high contamination level within the manure. However, there is a larger risk with selling the meat of the dead animals, both regarding transmission of zoonotic and non-zoonotic disease, as the animals are transported longer distances and often distributed into multiple homes.

For controlling the spread of disease spread from dead animals, it is important to have proper management. The carcass can be composted at the farm level, as long as there is no risk of further spread of the disease. This can be performed by building a large compost heap that covers the animals, preferably with a large base combined with a plastic lining. This means that a high temperature is generated in the full compost mass and ensures that the fluids produced remain in the compost (Berge *et al.*, 2009). This can also be performed within the animal house if necessary, especially with avian flu, which is so highly contagious that transport of the contaminated dead animals and the manure should be avoided. However, the avian flu virus is readily inactivated even at mesophilic temperatures (Elving *et al.*, 2012).

Additional risks from applying contaminated manure are the risk of disease transmission to a larger animal group, including wild animals, and transmission back to farm animals. The microorganisms applied to fields with manure can be expected to survive for long periods. For example, after application of chicken manure containing *Salmonella* Typhimurium to soil in one study, large numbers of the pathogen were measured for 3 months, after which period the numbers declined by 99.99% (a 10 000-fold or 4 $\log_{10}$ reduction), while in cattle slurry and human urine applied to soil the survival was shorter, 60 and 15 days, respectively (Nyberg *et al.*, 2010). The study also showed that high concentrations of easily available carbon in the manure increased the survival of *S.* Typhimurium in the soil. However, when enrichment methods were used to look for salmonella in 25 g soil, *S.* Typhimurium was still detectable after 180 days in soil with a mean air temperature of 15–20 °C during the first 90 days, followed by decreasing temperatures down to near zero at day 180. This indicates that application of contaminated manure to the soil increases the risk of disease transmission both to crops and to water over a long time.

6.2.1 Manure Treatment

Manure management differs between countries, as does the attitude towards the manure. In the Nordic countries, manure is considered a resource and is used as a fertiliser in agriculture, especially in organic farming where other plant nutrient sources are less available and most often costly. Worldwide, livestock production is becoming more specialised (Steinfeld *et al.*, 2006), and in many parts of the world it is becoming decoupled from plant production and manure is considered to be a waste and is managed accordingly.

Irrespective of how the manure is considered, it has to be managed and either used as a fertiliser or treated as wastewater. Treatment results in production of a solid fraction that is often sold as manure and a water fraction that ends up in the water recipient (Vanotti *et al.*, 2009). In other places the solid manure is just dumped; for example, Komakech *et al.* (2013) found that manure management in and around Kampala city, Uganda, comprises two main pathways: (i) one-third is used as a fertiliser on-farm or sold as a fertiliser (at a price just covering the removal costs), and (ii) between half and two-thirds are discarded locally (mainly for the rain to flush away) or dumped in landfill.

6.2.2 Expression of Pathogen Reduction

The standard commonly used for expressing the reduction in pathogens in animal waste and other organic waste products is the decimal reduction (90%) of the microorganisms over time (D_T). D_T has units of time, which can be seconds, minutes, hours and so on, depending on the units used when measuring the reduction. This value is determined empirically, by relating the reduction in the number of pathogens in a material over time (Figure 6.1), especially in a non-categorised material. However, most materials and organisms have been evaluated in different processes, which may include temperature treatment, addition of chemicals and so on,

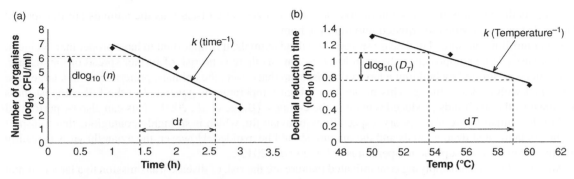

Figure 6.1   *Reduction in number of pathogens as affected by temperature during treatment and treatment time. (a) Log$_{10}$ number of pathogens measured in relation to time. (b) The slope of log$_{10}$ number of pathogens (D$_T$) as a function of temperature, giving Z in degree Celcius.*

and the set-up is often not standardised. Consequently, different times of inactivation have been reported for similar treatments due to the design of the study:

$$k_d = \frac{\mathrm{dlog}_{10}(n)}{\mathrm{d}t}(\text{time}^{-1}) \tag{6.2}$$

$$D_t = \frac{1}{k_d}(\text{time}) \tag{6.3}$$

$$Z = \frac{\mathrm{d}T}{\mathrm{dlog}_{10}(D_T)}(^{\circ}\text{C}) \tag{6.4}$$

The development of a D_T value is based upon the log$_{10}$ reduction (90%) of a single organism over time at a specific temperature. The larger the number of analyses, the more accurate the estimated slope of the decay, which is determined by performing a curve fitting for a log$_{10}$-normal reduction curve (log$_{10}$ number of pathogens versus time; Figure 6.1). The slope should be estimated based on at least three data points and the value of the correlation coefficient for the linearity of the survival curve should not be less than 80%. The k_d value (the slope of the reduction in organisms over time), with units time^{-1}, is calculated according to Equations (6.2) and (6.3). In Equation (6.2), dlog$_{10}(n)$ (the change in number of organisms between time t_0 and t_1) is divided by dt (the difference between time t_0 and t_1). The relationship between the decimal reduction time (D_T) value and the slope (k_d) of the reduction is according to Equation (6.3).

The D_T value is specific for each organism at each temperature and the relationship between the D_T values at different temperatures is called the Z value, which is defined as the number of degrees Celsius required to change the D_T value by a factor of 10 (i.e. a decimal reduction; Figure 6.1b). The correlation between log$_{10}$ D_T and temperature is log-normal according to Equation (6.4).

As with the calculation of the D_T value, the Z value needs three D_T values for calculation and the slope of the curve k_z has the same relationship to the Z value as the D_T value has to k_d (Equation 6.3).

The time required for a decimal reduction can be determined in this way. In the EU regulations on animal byproducts (ABP) (EC1069/2009; EC, 2009) and on treatment of ABP (EC142/2011; EU, 2011), the recommendation for treatment of ABP category 3 and manure to be traded (category 2) is a treatment validated to reduce *Enterococcus faecalis* or *Salmonella* Senftenberg (775W, H$_2$S negative) by five decimals (i.e. by 5 log$_{10}$). If viruses are considered a risk in the system, a thermotolerant animal virus (e.g. parvovirus) should be reduced by 3 log$_{10}$.

For chemical treatment, the relationship between temperature and added chemical is more complex, and in most cases cannot be estimated with the simple linear logarithmic reduction correlation found during high-temperature treatment. The effect of chemical treatment can often have an initial lag phase with little or no reduction before a faster reduction phase that is exponentially related to time, showing a linear logarithmic reduction versus time (Nordin *et al.*, 2009). Most tests of treatments do not provide enough data for development of two-phase models, so instead rather simpler models including both the lag and the log phase are used to present a complete reduction (e.g. the time required for a 3 $\log_{10}$ reduction in *Ascaris* spp. in a treatment). *Ascaris* spp. is often recommended to be used as a model for chemical treatment due to its high tolerance to chemicals; for example, in the ABP regulations (EC1069/2009) and on treatment of ABP (EC142/2011), a 3 $\log_{10}$ reduction in *Ascaris* spp. should be the outcome of a sanitising method using chemical treatment, if the treatment is to be accepted for treatment of the biomass.

6.3 Manure Treatment Alternatives

Treatment of slurry may be anaerobic, or aerobic through aeration of stored slurry, composting of solid manure or through treatment with chemical amendments. For all treatments, time is a most important additional parameter. Furthermore, temperature is a variable included in most studies evaluating the reduction in pathogen content in animal solids and slurries (Table 6.1).

Table 6.1 *Sanitising treatment of manure to be applied in fields according to EU and other regulations and recommendations for safe manure management.*

Management	Treatment	Regulation/recommendation
Transferring and selling manure between farms (e.g. solids from separated manure[a])	1 h at above 70 °C or treatment with similar effect	EC1069/2009 (EC, 2009)
ABP category 3[b]	1 h at above 70 °C or treatment with similar effect	EC1069/2009 (EC, 2009)
Composting	turning compost twice first week and reaching temperature >55 °C, composting >3 months	guidelines for growers to reduce the risks of microbiological contamination of ready-to-eat crops (Hickman *et al.*, 2009)
Composting of human manure	>1 wk at >55°C	WHO (2006)
Composting in reactors	4 h at >55 °C	USEPA (1994)
Storage	batch storage >6 months without addition of fresh manure	guidelines for growers to reduce risks of microbiological contamination of ready-to-eat crops (Hickman *et al.*, 2009)
Liming	addition of quick or slaked lime to reach pH 12 during >2 h	guidelines for growers to reduce risks of microbiological contamination of ready-to-eat crops (Hickman *et al.*, 2009)
Ammonia	addition of urea, 2% treatment for 1 wk	for *Salmonella* removal during outbreaks

[a]Slurry or solid manure that is not mixed with organic waste and not sold may be spread on agricultural land without sanitising treatment.
[b]ABP Animal by-product: Term used in EU regulations for describing waste from animals, including slaughter waste and manure.

6.3.1   Storage

Storage of manure can be considered a pathogen reduction measure, where the main treatment variable is time from initiation of storage to application of manure to the field. This time is affected by the timing of application of manure to ensure optimal nutrient utilisation of the plant nutrients (i.e. the manure should be applied prior to, or early in, the growing season). Therefore, depending on the number of crops that can be grown in 1 year, the storage time varies. For example, in Northern Europe, with one crop per growing season, the average time of storage is 6 months. However, if only one manure tank is used, there will be fresh manure in the tank on the day of spreading to the fields, leading to a minimal retention time of 1 day. In subtropical countries the storage time may be only a few months because three crops are grown per year and storage time may be zero when liquid manure is used to fertilise fish ponds in Asia in the VAC system (Chapter 2).

As storage is the main treatment option performed today, it is important to bear in mind that the main reduction occurs in the manure tank when no fresh manure is added. A constant stream of fresh manure can contaminate the remaining part of the tank with pathogens and provide fresh substrate to support their growth. Even if the count of pathogenic microorganisms in the manure has decreased below the detection limit, it is possible to have regrowth, especially in association with changes in the environment (e.g. during spring when temperatures are increasing) (Gibbs *et al.*, 1997). In addition, when adding fresh manure to the tank the regrowth effect increases with increasing amount of material with a high content of easily available carbon compared with material with little easily available carbon (Elving *et al.*, 2010). The inactivation of pathogenic bacteria is generally faster than inactivation of viruses and parasites, which have a longer survival time. One of the most stable organisms is *Ascaris suum* (intestinal worms in pigs), for which viable individuals have been measured in a pig house that had been empty for 14 years (P. Wallgren, personal communication; Per.Wallgren@SVA.se).

A first step in decreasing the risk of disease transmission is to have more than one slurry tank and thereby increase the minimum storage time prior to field application, as the survival of pathogenic microorganisms is higher in the soil than in the slurry tank. *Salmonella* spp. in stored cattle slurry have been shown to have a decimal reduction in the range of 2–3 weeks (Hutchison *et al.*, 2005), which is considerably faster than the 6- to 7-week decimal reduction time reported for inactivation in manured soil (Nyberg *et al.*, 2011).

6.3.2   Anaerobic Treatment

The degradation of organic matter in an anaerobic treatment does not produce a surplus of heat, as the energy of components ends up in the methane (Chapter 4). Therefore, the anaerobic process requires external heating to reach temperatures above the ambient temperature. It is possible to run the anaerobic treatment process from psychrophilic temperatures around 10 °C up to hyperthermophilic temperatures around 65 °C. The main difference is the speed of the process and some effect on the gas composition and the substances degraded. This treatment is mostly associated with biogas production using digester technology (Chapter 13).

Most anaerobic digesters are run at mesophilic operating temperatures. Pathogenic bacteria such as *Salmonella* spp. can take part in the first degradation step, acidification, and thereby show initial growth (Ottoson *et al.*, 2008b), resulting in higher numbers or a constant outflow of pathogenic *Enterobacteriaceae* (e.g. *Salmonella* spp.) in the effluent of an anaerobic psycrophilic or mesophilic reactor.

The degradation of pathogens in the anaerobic process is the result of three factors. The main effect comes via heat inactivation, as in the case of composting. This can be achieved either via pre-pasteurisation by heating the animal manure fed to the reactor (e.g. 1 h at 70 °C) or from the thermophilic process. Post-pasteurisation of the digestate should be avoided because this treatment is often associated with large risks of regrowth of unwanted microorganisms in the sanitised material and, as a consequence, the level of pathogenic bacteria can be higher than the initial values in the incoming raw material. To ensure inactivation by the thermophilic process a temperature of 50 °C is required, as lower temperatures give a considerably slower reduction (Elving *et al.*, 2012). For estimating the inactivation by temperature, the calculations mentioned in Section 6.2.2 can be used.

The second regulating factor is the ammonia (NH_3) content of the digestate. After a longer adaptation period it is possible to run an anaerobic digester at high NH_3 concentrations (e.g. with high protein load) (Schnürer and Nordberg, 2008). As the microorganisms in the feed biomass are not adapted to the high NH_3 environment, the high NH_3 concentration has a strong lethal effect on incoming microorganisms (Ottoson *et al.*, 2008b).

In a fully mixed continuous anaerobic reactor, which is the most common reactor type found in Europe, the inactivation of pathogens can be considered to be low. The minimum retention time (MRT) in such a reactor is short, because the continuous feeding of the reactor and stirring cause a fraction of the fresh feed to bypass the reactor and be discharged within a short time. The proportion of the material discharged after the minimum retention time is affected by the incoming feed and the hydraulic retention time (HRT) (Equation 6.1). Mesophilic anaerobic treatment generally results in an average inactivation corresponding to a 1–2 $\log_{10}$ reduction in the incoming pathogens. When the hydraulic retention time is increased, the reduction in pathogens in the system increases (Yen-Phi *et al.*, 2009). In a plug flow reactor, the HRT and the MRT (Text Box – Basic 6.1) are closely related, as the mixing of the material is low. This kind of reactor is common in low and middle income countries. One common reactor type is the bag reactor, which is a polyethylene-covered pit or channel containing manure, or the Chinese dome digester, where transport of the material in the reactor is achieved by addition of fresh material, which pushes the old material further into and out of the reactor. In these reactors organic matter and pathogens may sediment and have a long HRT and, as a consequence, the inactivation of pathogens increases. Yen-Phi *et al.* (2009) found that increasing the HRT and MRT from 3 to 30 days resulted in a decrease in the monitored pathogens and indicator organisms by a factor of 2 $\log_{10}$. In such systems with low mixing and laminar flow, larger pathogens such as parasitic eggs sediment and have an even longer retention time.

As with composting, there is a risk of regrowth in the digestate. However, when the material is stabilised the risk of growth of pathogenic bacteria is less than in the raw material (Sidhu *et al.*, 2001; Elving *et al.*, 2010).

6.3.3 Composting

Composting can be performed for a solid or a liquid manure. The main difference is the method of aeration, with the solid manure being composted requiring a dry matter content larger than 35% to have pores for air transport into the compost (Haug, 1993). However, higher water content results in waterlogging of the pores, which impedes aeration. Aeration of liquid manure is treatment of manure with a dry matter content less than 12%, where air is pumped into the slurry to keep the oxygen level high.

Several processes are responsible for the inactivation of pathogenic microorganisms (e.g. competition, nutrient deficiency, etc.). However, the main inactivation effect comes from the heat generated during oxidative degradation of organic matter (Vinnerås *et al.*, 2010). The actual heat released during the build-up phase of composting is less than the potential heat production, since much of the energy released is spent on the growth of the bacterial population.

Some pathogens are easily inactivated at mesophilic temperatures; for example, avian influenza virus undergoes a decimal reduction within less than 30 min at 35 °C, while increasing the temperature by 10 °C more, to 45 °C, has been shown to result in a decimal reduction in this organism in less than 10 min (Elving *et al.*, 2012). However, the recommended temperature for general pathogen inactivation is treatment above 50 °C. The reason for this recommendation is that several organisms show very slow or no reduction at lower temperatures, and in some cases there is even growth of pathogenic bacteria at temperatures close to 50 °C (e.g. *Salmonella* spp. has been reported to grow in the temperature range 6–47 °C) (Mitscherlich and Marth, 1984).

The World Health Organisation recommendation for thermal treatment of faecal compost is to keep the temperature above 55 °C for more than 1 week for safe sanitisation of the material (Table 6.1). The challenge is that the treatment must be performed so that all of the material reaches a high temperature. If this is achieved by aeration of liquid manure in reactors where the temperature is homogeneously distributed, then the treatment time can be shorter. The US Environmental Protection Agency (USEPA, 1994) recommends that slurry treatment in a reactor should be at least four hours at 55 °C to ensure safe sanitisation.

Vinnerås *et al.* (2010) found that applying repeated heating peaks, a process known as Tyndallisation, can have an enhanced effect on inactivation, as the repeated heat peaks add more stress to the organisms. Furthermore, in some cases this treatment has a strong effect on the spore-formers, as the spores can be encouraged to germinate during the periods with lower temperature. Therefore, applying an irregular temperature pattern to compost that is mixed can actually have a positive effect on sanitisation compared with having a constant high temperature.

In a solid composting process, the temperature and the moisture are generally not evenly distributed. The degradation process produces heat that is lost, mainly via evaporation of water (Haug, 1993). The temperature of the compost is determined by the heat production and the amount of heat lost via the surface. To increase the heat of the compost, more easily available organic substances can be added, thereby increasing the heat production, and/or the surface of the compost may be insulated to decrease the heat loss. However, the areas with the highest incoming air flow and areas connected to the surface are still most often colder than the core of the compost (Figure 6.2). The cold zones of a compost pile/windrow can be significant in size and even in compost heaps higher than 1.2 m, up to 35% of the compost can be below the required temperature of 50 °C. As a rule of thumb for composting, a windrow or pile should be mixed five times and have a temperature above 55 °C (USEPA, 1994).

When the material is not degraded but rather preserved (e.g. by high temperature and low pH; Vinnerås *et al.*, 2010), changes in the composition (e.g. buffering the compost) can lead to high regrowth of pathogenic bacteria (Elving *et al.*, 2010). The growth in unmatured compost can actually increase the total number of organisms (e.g. growth of 2 $\log_{10}$ can occur in fresh compost material and this will offset the inactivation in the high-temperature section). However, as the material matures and is mixed several times, the risk of regrowth decreases (Sidhu *et al.*, 2001; Elving *et al.*, 2010).

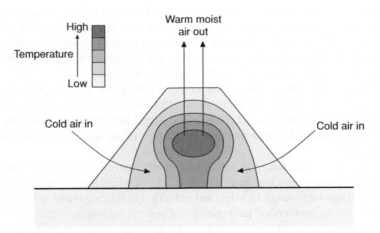

Figure 6.2   *Heat distribution within a windrow compost. (© University of Southern Denmark.)*

Even if there is a risk for growth of pathogenic bacteria in the cold zones, little or no growth of pathogenic bacteria can most often be expected in these zones. However, to achieve homogeneous sanitisation of the composted manure, a large proportion of the material needs to be treated at a high temperature. A general method for management of the temperature distribution in the compost material is mixing during the high-temperature phase. With temperature measurements, it is possible to map the temperature distribution in the compost and obtain a diagram similar to Figure 6.2. This diagram can then be used to determine the proportion of compost that can be considered to have reached sanitising temperature, which can be used to determine the proportion of compost that has not reached sanitisation temperature and inactivation of the pathogens of interest in the remainder of the compost. When several days are left between turnings, it is seldom necessary to calculate the exact inactivation in the warm areas. However, with shorter treatment times (e.g. a reactor with higher temperature prior to a long low-temperature treatment, it can also be interesting to know the inactivation in the high-temperature area) The total survival can be calculated by using the heat distribution between areas with high temperature (f_h) and areas with low temperature (f_l). The inactivation in the high-temperature area can then be calculated by using the inactivation rate (k_d) at a certain temperature over a set time period (d_t; e.g. the period during which this temperature prevails in the zone). The k_d used here is the same k_d as calculated in Equation (6.2). The high-temperature areas can be divided into several areas depending on the number of temperature zones determined in the compost, according to Figure 6.2. The reduction in the initial organisms (n_0) during the treatment time (t) to n_t is set based on the number of turnings of the compost N, plus the initial construction of the compost:

$$n_t = n_0 \left(f_l + f_h \cdot 10^{(-k_d \cdot \Delta t)} \right)^{N+1} \tag{6.5}$$

If the decrease in the number of organisms in the high-temperature zone is large, the reduction in that zone can be removed from Equation (6.5) and the reduction in the compost can be calculated based only on the number of organisms remaining in the cold zone after each complete mixing of the material as:

$$n_t = n_0 \left(f_l \right)^{N+1} \tag{6.6}$$

However, if the organisms are still viable in the cold zone (see above), the total number of microorganisms can actually increase, while if the number of organisms in the cold area decreases, the inactivation of the treatment is underestimated.

6.4 Chemical Treatment

Conventional chemical treatment aims to achieve a high pH that inactivates the microorganisms. However, some organisms are very stable even at high pH (e.g. the recommended lime treatment for sludge in the US is treatment at a pH above 12 during 3 months). If reactive lime (CaO) is used there will also be much production of heat, so in that case the treatment can be performed faster (i.e. if the treatment causes a temperature increase to 55–70 °C) (USEPA, 1994).

6.4.1 Ammonia Treatment

The alternative treatment is with NH_3, which can achieve good inactivation of microorganisms already at pH 8.5 and above (Text Box – Basic 6.2). Ammonia is highly soluble in water (1500 mol NH_3(aq) l^{-1}/mol NH_3(g) l^{-1}) and the dissolved NH_3(aq) is proportional to the partial pressure of NH_3(g) above the solution in a closed container, as given by the Henry's law equilibrium constant (Chapter 8). Ammonia is a weak base, with a pK_a value of 9.25 at 25 °C (where 50% of the molecules are in the charged form (NH_4^+) and the other 50% in uncharged form (NH_3)). The relationship of (NH_4^+) to (NH_3) is:

$$NH_3(g) \rightleftharpoons NH_3(aq) + H_2O(l) \rightleftharpoons NH_4^+(aq) + OH^-(aq) \quad (6.7)$$

Text Box – Basic 6.2 Ammonia (NH_3) chemistry

- Solubility $= 7020$ g l^{-1} (1500 mol $NH_3(aq)$ l^{-1}/mol $NH_3(g)$ l^{-1}).
- pK_a (25 °C) $= 9.25$ when 50% is present as NH_3 and 50% as NH_4^+.
- When used for sanitisation, addition as: $NH_3(aq)$ 25–28%.
- Urea (47% total NH_3-N by weight) enzymatically degraded to NH_3 in material; each urea molecule is degraded into two NH_3 molecules (i.e. 1 mol of urea gives 2 mol of NH_3).

The ammonium ion (NH_4^+) is harmless to microorganisms and the substance that performs the inactivation is NH_3, which has long been known to have this effect (Warren, 1962). The mechanism by which the NH_3 acts on the organisms is not fully described in the literature. However, the NH_3 molecule is small and has a high solubility, in water and in lipids, which allows transport over membranes and other cellular barriers by simple diffusion.

The NH_3 may act as an uncoupler, destroying the membrane potential of bacterial cells, or destroying (denaturing) proteins of the cell, both in the membranes and inside the cell (Bujozek, 2001). As NH_3 can easily be transported across membranes into the cell, it may also cause damage by rapid alkalinisation of the cytoplasm (Diez-Gonzalez *et al.*, 2000). To compensate for this and to maintain optimum internal pH, protons are taken up from outside the cell while at the same time potassium ions (K^+) are released and the loss of this essential element eventually leads to death of the bacterial cell (Bujozek, 2001). Few studies have been performed on the inactivation mechanisms of viruses, but studies on poliovirus have concluded that the inactivation is due to cleavage of the RNA, as the virus cells did not lose their capacity to attach and inject their genome into the host cell, but no reproduction occurred in the host cell (Ward, 1978; Burge *et al.*, 1983). The mechanism for inactivation of larger organisms such as parasites has not yet been identified. The function of NH_3 treatment is still unclear and so far the inactivation is empirically decided for each set of organisms to be used as treatment recommendations for sanitisation.

The inactivation of pathogens and indicator organisms by NH_3 treatment has been evaluated in a number of different substrates, including human urine (Vinnerås *et al.*, 2008), human faeces (Nordin *et al.*, 2009a, 2009b), blackwater (Fidjeland *et al.*, 2013a), sewage sludge (Pecson *et al.*, 2007; Fidjeland *et al.*, 2013b) and cattle manure (Ottoson *et al.*, 2008a) over a temperature range of 4–34 °C. All these studies report decimal reduction data for the organisms at different NH_3 concentrations, temperatures and pH. These data can be useful for a specific setting where the exact conditions of a planned treatment are known. However, when combining these data it is possible to estimate the time required for inactivation according to the ABP treatment regulations (EU142/2011; EU, 2011) at different NH_3 concentrations (Table 6.2). *Ascaris* spp. is included in Table 6.2, as the regulations state that when ABPs are treated chemically, *Ascaris* spp. has to be inactivated by 3 $\log_{10}$ by the treatment. The basic data for the inactivation at the different treatment alternatives include a lag and a log phase for *E. faecalis* and for *Ascaris* spp. (Figure 6.3). When performing the inactivation calculations according to the ABP regulations, it is possible to include both phases in the estimation of the required time. Using the simpler decimal reduction does not include the lag phase of the inactivation, resulting in an inaccurate time of inactivation. To develop the treatment recommendations, the data on inactivation were plotted according to the time for a 5 $\log_{10}$ reduction in the two bacteria species and 3 $\log_{10}$ reduction

Table 6.2 *Time required for fulfilment of EU regulations on ABP (EC142/2011) by chemical treatment of animal manure at different NH_3 concentrations and at temperatures from 4 °C.*

Ammonia concentration [NH₃] (mM)	*Salmonella* spp. (days for 5 log₁₀ reduction)	*E. faecalis* (days for 5 log₁₀ reduction)	*Ascaris* spp.[a] (days for 3 log₁₀ reduction)
50	4	150	200
75	1	80	150
100	0.5	50	100
150	0.5	30	80
200	0.5	20	60
250	0.5	10	40

[a]Data only given for inactivation at temperatures above 20 °C. The reduction in *Ascaris* spp. at lower temperatures is significantly longer, as the lag phase increases in time and there are not sufficient data for generalising the inactivation at this concentration.

in the parasite. From this plot, the time for inactivation was determined at the different NH_3 concentrations, independent of the substrate and the temperature. This provided a conservative estimate for treatments at higher temperatures, as a higher speed of inactivation has been reported for a particular NH_3 concentration with increased treatment temperature (Vinnerås *et al.*, 2008; Nordin *et al.*, 2009a). The lowest temperature given for recommended time of treatment at different NH_3 concentrations is 4 °C for *Salmonella* spp. and *E. faecalis*. However, for *Ascaris* spp. there are only sufficient data for generalisation at temperatures above 20 °C, as most studies of *Ascaris* spp. inactivation at lower temperatures have not lasted long enough to pass the initial lag phase.

The effect of NH_3 treatment depends on the concentration of dissolved NH_3, the uncharged NH_3 molecule. The relationship between $NH_3(aq)$ and TAN ($NH_3 + NH_4^+$) in a solution is quantified by the dissociation constant, K_a (Chapters 4 and 8), which is exponentially related to temperature:

$$[NH_3] = \frac{[TAN] \cdot K}{K + [H^+]}, \ pK_a = \frac{2728.92}{T} + 0.090181 \qquad (6.8)$$

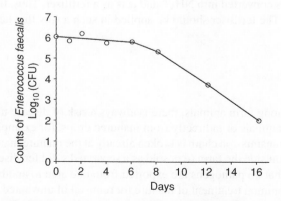

Figure 6.3 *A typical biphasic inactivation curve for NH_3 treatment of* E. faecalis *with an initial lag phase followed by a linear logarithmic decay. This is most clear for treatments with a low concentration of NH_3.*

To determine the concentration of uncharged NH_3 present in the material, the temperature, total ammoniacal concentration and pH of the sample must be determined. Having this information, the $NH_3(aq)$ concentration can be determined and the treatment time of an infected manure to become sanitised can be determined (Table 6.2).

6.4.2   Ammonia Sanitisation at the Farm Level

Ammonia sanitisation can be used during disease outbreaks or as part of the daily regime on farms, especially if the manure or a separated fraction of the manure is intended for sale outside the farm and treatment is required according to the EU ABP regulations (See Text Box – Basic 6.1). Ammonia treatment relies on achieving a sufficiently high concentration of uncharged NH_3 for pathogen inactivation (Table 6.2). According to Equations (6.7) and (6.8), it is possible to increase the NH_3 content by increasing the temperature, increasing the pH or increasing the TAN concentration. The easiest way of increasing the NH_3 concentration is to add NH_3, which increases both the TAN and the pH. The most common method of NH_3 addition is to add a solution of NH_3, 25% $NH_3(aq)$ being the most common commercially available product. The other alternative is to add NH_3 in the form of urea ($CO(NH_2)_2$). The urea itself is not toxic when added to the manure, but the naturally occurring enzymes (urease) transform urea into carbonic acid (H_2CO_3) and NH_3 (Equation 6.9). Urea addition can give a pH increase of up to approximately 9.2 depending on the concentration added and the buffering capacity of the treated material. As the carbonate produced from urea degradation buffers the material, the pH reached after urea addition is lower than that after addition of a solution of $NH_3(aq)$:

$$CO(NH_2)_2 + 2H_2O \rightarrow H_2CO_3 + 2NH_3 \tag{6.9}$$

The treatment needs to be performed under cover to avoid losses of NH_3 (Chapters 8 and 9) As long as the NH_3 concentration is high in the manure there is no risk of recontamination, as the NH_3 keeps the growth of microorganisms low. This probably also results in lower greenhouse gas emissions from the manure, as the general biological activity decreases and thus no CH_4 or N_2O is produced during storage (Chapter 10). Ammonia sanitisation has been proven to function well in liquids (e.g. urine (Vinnerås *et al.*, 2008) and hatchery waste (Emmoth *et al.*, 2011)) and in solids (e.g. sewage sludge (Pecson *et al.*, 2007) and compost (Adamtey *et al.*, 2009)). As the NH_3 is not consumed during the treatment, the sanitisation can continue throughout the treatment and there is no risk of regrowth or recontamination as long as the NH_3 is kept in the material. When applying the treated manure to the soil, the dilution and buffering of the NH_3 leads to decreased pH, and the NH_3 is converted into NH_4^+ and acts as a fertiliser. Thus, the cost of the treatment can be allocated to the fertiliser. The fertiliser should be applied in such a way that large NH_3 losses are avoided (Chapter 9).

6.5   Summary

During disease outbreaks among farm animals, there is always a risk of disease transmission via the manure either directly to humans or animals or indirectly from manured crops, for example. Therefore, it is of major importance that the disease transmission chain is broken already at the manure management level. In addition, when manure is transported outside the farm (e.g. sold as a separated solid fertiliser fraction or soil improver) it must be treated to ensure that no pathogens are exported from one area to another.

Composting is the most common treatment of manure for removal of unwanted microorganisms. The main effect of composting is the heat, which ensures appropriate inactivation of unwanted microorganisms. For efficient removal of pathogenic organisms in the compost, the temperature has to reach above 50 °C, as the efficiency of inactivation decreases considerably at lower temperatures. To ensure that all the material has been exposed to high temperature, the compost heap or windrow compost should be turned 5 times during high temperature treatment. Slurry should be mixed 3 times when treated in aeration reactors with smaller low-temperature zones.

Anaerobic treatment in biogas reactors or during storage of slurry is not a sanitisation treatment per se. Increasing the hydraulic retention time in biogas digesters increases the inactivation of pathogens, especially in plug flow reactors where the minimum retention time and the hydraulic retention time are similar. To ensure full sanitisation effect in anaerobic digestion, the treatment should be performed at temperatures above 50 °C or the incoming material should be pre-pasteurised.

The most efficient chemical treatment for pathogen removal from manure and its solid and liquid fractions is NH_3 treatment. This must be performed at pH above approximately 8.5, the requirement being that sufficient uncharged NH_3 is present in the material for microbial inactivation. With increased temperature and pH, NH_3 treatment becomes more efficient, as a larger proportion of the ammonium and NH_3 is present in the form of NH_3. As NH_3 is a volatile gas, the treatment should be performed in a closed system in order to avoid gaseous losses. The NH_3 is not consumed during the treatment, so there is no risk of recontamination of the treated manure, and upon application to soil the NH_3 acts as a fertiliser.

References

Adamtey, N., Cofie, O., Ofosu-Budu, G.K., Danso, S.K.A. and Forster, D. (2009) Production and storage of N-enriched co-compost. *Waste Manag.*, **29**, 2429–2436.

Berge, A.C.B., Glanville, T.D., Millner, P.D. and Klingborg, D.J. (2009) Methods and microbial risks associated with composting of animal carcasses in the United States. *J. Am. Vet. Medic. Assoc.*, **234**, 47–56.

Bujozek, G. (2001) Influence of ammonia and other abiotic factors on microbial activity and pathogen inactivation during processing of high-solid residues, *Dissertation*, University of Manitoba, Manitoba.

Burge, W.D., Cramer, W.N. and Kawata, K. (1983) Effect of heat on virus inactivation by ammonia. *Appl. Environ. Microbiol.*, **46**, 446–451.

Diez-Gonzalez, F., Jarvis, G.N., Adamovich, D.A. and Russell, J.B. (2000) Use of carbonate and alkali to eliminate *Escherichia coli* from dairy cattle manure. *Environ. Sci. Technol.*, **34**, 1275–1279.

EC (2009) Regulation (EC) No. 1069/2009 of the European parliament and of the council of 21 October 2009. Laying down health rules as regards animal by-products and derived products not intended for human consumption and repealing Regulation (EC) No 1774/2002 (Animal By-products Regulation), *Official Journal of the European Union*, European Community, Brussels.

Elving, J., Ottoson, J., Vinnerås, B. and Albihn, A. (2010) Growth potential of faecal bacteria in simulated psychrophilic/mesophilic zones during composting of organic waste. *J. Appl. Microbiol.*, **108**, 1974–1981.

Elving, J., Emmoth, E., Albihn, A., Vinnerås, B. and Ottoson, J. (2012) Composting for avian influenza virus elimination. *Appl. Environ. Microbiol.*, **78**, 3280–3285.

Emerson, K., Russo, R., Lund, R. and Thurston, R. (1975) Aqueous ammonia equilibrium calculations: effects of pH and temperature. *J. Fish. Res. Board Can.*, **32**, 2379–2383.

Emmoth, E., Ottoson, J., Albihn, A., Belak, S. and Vinnerås, B. (2011) Ammonia disinfection of hatchery waste for elimination of single-stranded RNA viruses. *Appl. Environ. Microbiol.*, **77**, 3960–3966.

EU (2011) EU 142/2011. Regulation (EU) No. 142/2011 of 25 February 2011 implementing Regulation (EC) No 1069/2009 of the European Parliament and of the Council laying down health rules as regards animal by-products and derived products not intended for human consumption and implementing Council Directive 97/78/EC as regards certain samples and items exempt from veterinary checks at the border under that Directive, *Official Journal of the European Union*, European Community, Brussels.

Fidjeland, J., Magri, M.E., Jönsson, H., Albihn, A. and Vinnerås, B. (2013a) The potential for self-sanitization of faecal sludge by intrinsic ammonia. *Water Res.*, in press.

Fidjeland, J., Lalander, C., Jönsson, H. and Vinnerås, B. (2013b) Ammonia sanitisation of sewage sludge using urea. *Water Sci. Technol.*, in press.

Gibbs, R.A., Hu, C.J., Ho, G.E. and Unkovich, I. (1997) Regrowth of faecal coliforms and salmonellae in stored biosolids and soil amended with biosolids. *Water Sci. Technol.*, **35**, 269–275.

Haug, R.T. (1993) *The Practical Handbook of Compost Engineering*, Lewis, Boca Raton, FL.

Hickman, G., Chambers, B. and Moore, T. (2009) *Managing Farm Manures for Food Safety – Guidelines for Growers to Reduce the Risks of Microbiological Contamination of Ready-to-Eat Crops*, Food Standards Agency, London.

Hutchison, M.L., Walters, L.D., Moore, A. and Avery, S.M. (2005) Declines of zoonotic agents in liquid livestock wastes stored in batches on-farm. *J. Appl. Microbiol.*, **99**, 58–65.

Komakech, A., Banadda, N., Gebresenbet, G. and Vinnerås, B. (2013) Feed and manure management for city animals in Kampala Uganda. *Agron. Sustain. Dev.*, in press.

Mitscherlich, E. and Marth, E.H. (1984) *Microbial Survival in the Environment – Bacteria and Rickettsiae Important in Human and Animal Health*, Springer, Berlin.

Nordin, A., Ottoson, J. and Vinnerås, B. (2009a) Sanitation of faeces from source-separating dry toilets using urea. *J. Appl. Microbiol.*, **107**, 1579–1587.

Nordin, A., Nyberg, K. and Vinnerås, B. (2009b) Inactivation of *Ascaris* eggs in source-separated urine and faeces by ammonia at ambient temperatures. *Appl. Environ. Microbiol.*, **75**, 662–667.

Nyberg, K., Vinnerås, B., Ottoson, J., Aronsson, P. and Albihn, A. (2011) Inactivation of *Escherichia coli* O157:H7 and *Salmonella typhimurium* in manure-amended soils studied in outdoor lysimeters. *Appl. Soil Ecol.*, **46**, 398–404.

Ottoson, J., Nordin, A., von Rosen, D. and Vinnerås, B. (2008a) *Salmonella* reduction in manure by the addition of urea and ammonia. *Bioresour. Technol.*, **99**, 1610–1615.

Ottoson, J., Schnürer, A. and Vinnerås, B. (2008b) *In situ* ammonia production as a sanitation agent during anaerobic digestion at mesophilic temperature. *Lett. Appl. Microbiol.*, **46**, 325–330.

Pecson, B.M., Barrios, J.A., Jimenez, B.E. and Nelson, K.L. (2007) The effects of temperature, pH, and ammonia concentration on the inactivation of *Ascaris* eggs in sewage sludge. *Water Res.*, **41**, 2893–2902.

Sidhu, J., Gibbs, R.A., Ho, G.E. and Unkovich, I. (2001) The role of indigenous microorganisms in suppression of salmonella regrowth in composted biosolids. *Water Res.*, **35**, 913–920.

Schnürer, A. and Nordberg, Å. (2008) Ammonia, a selective agent for methane production by syntrophic acetate oxidation at mesophilic temperature. *Water Sci. Technol.*, **57**, 735–740.

Son Thi Thanh, D., Dung Van, T., Madsen, H. and Dalsgaard, A. (2011) Survival of faecal indicator bacteria in treated pig manure stored in clay-covered heaps in Vietnam. *Vet. Microbiol.*, **152**, 374–378.

Steinfeld, H., Gerber, P., Wassenaar, T., Castel, V., Rosales, M. and de Haan, C. (2006) *Livestock's Long Shadow: Environmental Issues and Options*, FAO, Rome.

USEPA (1994) *A Plain English Guide to the EPA Part 503 Biosolids Rule*, US Environmental Protection Agency, Washington, DC.

Vanotti, M.B., Szogi, A.A., Millner, P.D. and Loughrin, J.H. (2009) Development of a second-generation environmentally superior technology for treatment of swine manure in the USA. *Bioresour. Technol.*, **100**, 5406–5416.

Vinnerås, B., Nordin, A., Niwagaba, C. and Nyberg, K. (2008) Inactivation of bacteria and viruses in human urine depending on temperature and dilution rate. *Water Res.*, **42**, 4067–4074.

Vinnerås, B., Agostini, F. and Jönsson, H. (2010) Sanitisation by composting, in *Microbes at Work from Wastes to Resources* (eds H. Insam, I. Franke-Whittle and M. Goberna), Springer, Berlin, pp. 171–191.

Ward, R.L. (1978) Mechanism of poliovirus inactivation by ammonia. *J. Virol.*, **26**, 299–305.

Warren, K.S. (1962) Ammonia toxicity and pH. *Nature*, **195**, 47–49.

WHO (2006) *Guidelines for the Safe Use of Wastewater, Excreta and Greywater. Volume 2: Wastewater Use in Agriculture*, WHO, Geneva.

Yajima, A. and Kurokura, H. (2008) Microbial risk assessment of livestock-integrated aquaculture and fish handling in Vietnam. *Fish. Sci.*, **74**, 1062–1068.

Yen-Phi, V.T., Clemens, J., Rechenburg, A., Vinnerås, B., Lenßen, C. and Kistemann, T. (2009) Hygienic effect of plastic bio-digesters under tropical conditions. *J. Water Health*, **7**, 590–596.

7

Solid–Liquid Separation of Animal Slurry

Morten L. Christensen[1], Knud V. Christensen[2] and Sven G. Sommer[2]

[1]*Department of Biotechnology, Chemistry and Environmental Engineering, Aalborg University, Denmark*
[2]*Institute of Chemical Engineering, Biotechnology and Environmental Technology,
University of Southern Denmark, Denmark*

7.1   Introduction

Animal slurry contains plant nutrients that are essential for crop production. Unfortunately, an expanding and more intensive livestock production has led to a surplus of plant nutrients on these farms. As a consequence of this development, slurry may be applied to fields at nutrient rates higher than those taken up by plants, enhancing the risk of leaching and runoff of the slurry nutrients, which can pollute surface waters and groundwater (Burton and Turner, 2003). The negative effects of heavy applications of slurry may also include salinisation in semi-arid regions and toxic concentrations of heavy metals in soil.

To facilitate environmentally friendly recycling of slurry on these farms, slurry needs to be transported to farms specialising in plant production, for example. The cost of slurry transport may be reduced and its fertiliser value increased by separating the slurry into a liquid fraction intended for on-farm usage and a dry matter and nutrient-rich solid fraction that can be exported to farms with few or no animals (Sørensen *et al.*, 2003). Separation may also contribute to a reduction in odour emissions (Zhang and Westerman, 1997) and in producing an energy-rich biomass that can be used for incineration or biogas production (Møller *et al.*, 2007a).

The nitrogen-to-phosphorus ratio (N : P) of manure often differs from crop requirements. Manure separation may solve this problem, because the manure is separated into an N-rich liquid fraction and a P-rich solids fraction. This would increase the N : P ratio of the liquid fraction, and thereby improve the distribution of nitrogen and phosphorus on the fields near the farm.

As well as the benefits, separation of animal slurry may also create some problems, such as a change in the ratio of plant nutrients to heavy metals in the biomass. Solid–liquid separation with flocculation as pre-treatment may, for example, transfer zinc (Zn), copper (Cu) and cadmium (Cd) to the solid fraction (Møller *et al.*, 2007b). The additives used when optimising separation of slurry (e.g. polymers and aluminium (Al)

Animal Manure Recycling: Treatment and Management, First Edition. Edited by Sven G. Sommer, Morten L. Christensen,
Thomas Schmidt and Lars S. Jensen.

salts) may also constitute an environmental problem (Nahm, 2005). Still, slurry separation and recycling of organic matter and plant nutrients has the potential to mitigate environmental hazards.

To ensure high efficiency of separation, the technology and management of the technology have to be related to the composition of animal manure through an understanding of the physical and chemical processes involved in efficient separation of slurry. This is required if reliable, operational and cheap separation technologies are to be developed, taking into account the actual slurry properties and the end use of the separation product.

7.2   Removal and Separation Efficiency

Several strategies can be used for separating manure, including in-house separation, sedimentation, filtration and centrifugation. When comparing results from different studies it is an advantage to use a single separation parameter to express the efficiency of separation. The removal efficiency (R) expresses the efficiency of removing a specific compound (x) from slurry to the dry-matter-rich solid fraction and is given as:

$$R(x) = 1 - \frac{[x]_{\text{liquid}}}{[x]_{\text{slurry}}} \tag{7.1}$$

where $[x]_{\text{slurry}}$ and $[x]_{\text{liquid}}$ are the concentrations (mol kg^{-1}) of the species in consideration (dry matter, P, N) in, respectively, the slurry being treated in the separator and the liquid fraction that is produced. The larger the removal efficiency, the lower the amount of compound x in the liquid fraction. The equation characterises the efficiency of the separator with respect to the liquid fraction; however, it gives no information on production of the solid fraction.

A separation index (E_t), on the other hand, expresses the distribution of the specific compound between the solid and liquid fraction:

$$E_t(x) = \frac{m(x)_{\text{solid}}}{m(x)_{\text{slurry}}} \tag{7.2}$$

where $m(x)_{\text{slurry}}$ and $m(x)_{\text{solid}}$ are the masses (g) of the compound in consideration in, respectively, the slurry being treated by the separator and the solid fraction being produced. The larger the separation index, the larger the amount of compound x in the solid fraction.

The separation index does not take into account the amount of the solid and liquid fraction. Thus, the simple separation index will in principle become 50%, if a machine separates the slurry in two equally large fractions with similar content of compound x. An improved expression of the separation may be obtained with the reduced separation index (E_t', no units):

$$E_t'(x) = \frac{E_t(x) - \frac{m_{\text{solid}}}{m_{\text{slurry}}}}{1 - \frac{m_{\text{solid}}}{m_{\text{slurry}}}} \tag{7.3}$$

where m_{slurry} and m_{solid} are the total mass (kg) of slurry being treated and the total mass of solids being produced.

It is recommended that the total masses (or mass flow) as well as the concentrations of the relevant compounds in the feed and in the solid fraction and liquid fractions are measured during separation. Only then can the necessary mass balances be calculated, the reduced separation index (E_t') found and the chosen separation process compared with alternative processes.

7.3   In-House Separation

For European and American pig production, in-house separation has been developed for pig houses with partly slatted floors. The faeces are withheld on belts below the slats, and the urine and water from animal drinkers and cleaning is collected in a narrow channel below the belt. The liquid fraction of urine and spilt water is continuously transported to a liquid store. Every 12 or 24 h the solids are removed from the belt by drawing the filter net over a roller in front of the channel under the slats. A brush roller contributes to the removal of faeces from the belt. Collection of fresh solids is very efficient because organic components contributing to the cohesion of the droppings have not been digested by microorganisms. Alternatively, systems have been developed whereby solids deposited on inclining floors in the channels below the slats are scraped to an outside store (Figure 7.1). The liquid runs off to a channel at the bottom, where it is drained off.

The solid fraction from in-house separation of pig manure has a high content of colloids with a low stability, and even though the dry matter content is above 25%, the solid fraction cannot be stacked without adding chopped straw. In a Dutch test of the technology about 35% of the excretion was collected in the solid fraction, which had a high concentration of nutrients and dry matter (Table 7.1). The separation index of the belt filter was more than 90% for P_2O_5, CaO, MgO and Cu. The separation index for the dry matter was 35% and total nitrogen (N) retained was 60%; little potassium (K) and ammonium was retained with the belt filter. The Cu content may restrict the application rate of the solid to crops.

In Asian animal houses with solid floors, the faeces are often scraped off the floor before the pigs are cooled and the floor cleaned with water. Alternatively, the manure is separated in channels under fully slatted floors. The channel sides incline to a narrow gutter in the bottom of the channel for collection of liquid. The

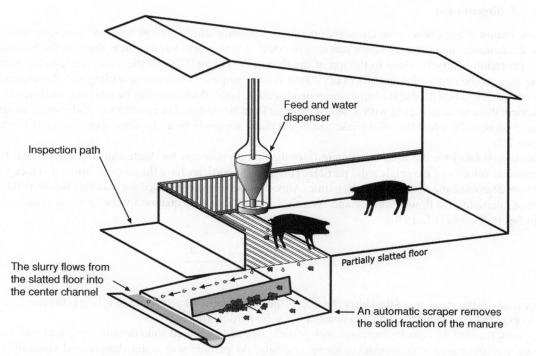

Feed and water dispenser

Inspection path

Partially slatted floor

The slurry flows from the slatted floor into the center channel

An automatic scraper removes the solid fraction of the manure

Figure 7.1   *Example of source-separation of manure. The solids are scraped off the channel floor, and the liquid is channelled and drained to a liquid store. (© University of Southern Denmark.)*

Table 7.1 Average concentration of faeces and urine after in-house separation on filter nets below a partly slatted floor, as a percentage of total amount (Kroodsma, 1986).

	Dry matter (%)	Ash (% of DM)	Total-N (%)	NH$_4$ (%)	P$_2$O$_5$ (%)	K$_2$O (%)	CaO (%)	MgO (%)	Cu (ppm)	pH
Liquid fraction	1.92	63.10	0.34	0.35	0.05	0.62	0.04	0.02	2.5	9.1
Solid fraction	32.5	25.7	1.24	0.34	1.64	0.85	1.45	0.48	189	

height of the channel is so great that farm hands can clean the removed solids from the sides with brooms or scrapers. The collected solids are dried, bagged and sold. The liquid fraction is collected in large lagoons where this fraction is stored and in some systems treated with, for example, aeration.

7.4   Solid–Liquid Separation of Manure Slurry

Different techniques for post-housing separation of slurry into a dry-matter-rich solid fraction and a liquid fraction have been developed and are used on farms. Solid–liquid separation may be carried out in settling tanks where the solids are removed from the bottom of the tank or the settling may be forced using centrifuges. Solids may also be removed mechanically by forced filtration using screw presses or drainage through fabric belts or screens.

7.4.1   Sedimentation

Sedimentation in a thickener is an attractive option for separation due to the low costs and simple technology. Most thickeners consist of a container that is cylindrical at the top and has a conical shape at the bottom. In batch operation, slurry is added to the top of the thickener (Figure 7.2) and the solids settle to the bottom where they can be removed (Loughrin *et al.*, 2006). For the purpose of enhancing settling and also increasing the transfer of solids settled at the top of the conical section, small thickeners can be vibrated, while for larger thickeners this can be achieved with a rake. Thickeners can be operated in continuous mode where slurry is added continuously, while the solid phase and liquid phase are removed at the same speed as slurry is added (Figure 7.2).

The time it takes for the solid to separate from the liquid phase can for dilute slurries be estimated from the terminal velocity of the single solid particle. The smallest particles have the slowest terminal velocity and therefore determine the necessary settling time. Almost no settling is observed for particles below 1–10 μm. For small particles, the flow is laminar and assuming dilute slurry, the equation for the terminal velocity (v_{tg}) simplifies to (Foust *et al.*, 1980):

$$v_{tg}(d_p) = \frac{(\rho_{solid} - \rho_{liquid}) \cdot g \cdot d_p^2}{18 \cdot \mu_{slurry}} \tag{7.4}$$

where μ_{slurry} is the viscosity of the slurry (Pa s), ρ is the density of solid and liquid (g l^{-1}), g is the acceleration due to gravity (m s^{-2}) and d_p is the diameter of the solid particle (m).

As seen, the settling velocity increases with particle size and increased solid density compared with liquid density, and decreases with increasing slurry viscosity. As particle size, solid density and viscosity vary from slurry to slurry, the terminal velocity can seldom be calculated in advance. Furthermore, Equation

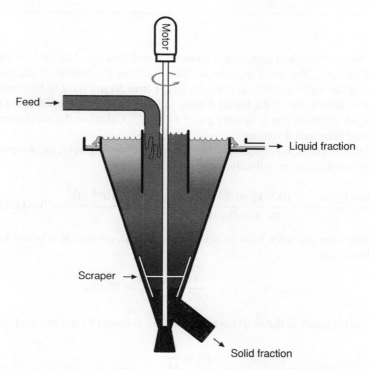

Figure 7.2   *Illustration of a thickener (clarifier) used to sediment particles that are removed from the bottom of the thickener. (© University of Southern Denmark.)*

(7.4) calculates the free settling velocity. The real settling velocity is often lower due to the interaction between the particles (hindered settling, see Text Box – Basic 7.1). Hindered settling is often pronounced for manure. Therefore the terminal velocity as calculated in Equation (7.4) alone cannot be used to determine the dimensions of the thickener. Instead, the calculations or assessments have to be carried out based on laboratory settling experiments. In general, it is possible to dimension the separation equipment when the settling velocity is known. The settling time can be estimated from the settling velocity and the distance from the inlet to the sediment. The retention time of the tank has to be higher than the required settling time (Example 7.1)

Text Box – Basic 7.1   Concepts in mechanical dewatering

Consolidation: Compression of material. During sedimentation processes the settled materials consolidate because of the weight of the overlying layers of materials.

Feed: The slurry being treated in the separator.

Free settling: The free fall of particles through a fluid medium, which is the opposite of hindered settling.

Hindered settling: At high concentrations, the settling of particles is lower than the free settling due to direct and/or indirect (i.e. hydrodynamic) interaction between the individual particles.

Retention time: The average residence time of material within the tank/separation apparatus.

Example 7.1

Manure is pumped into a cylindrical tank with a cross-sectional area equal to 2 m^2, at a distance of 0.5 m from the bottom of the tank. The solid fraction is taken out from the bottom of the tank and the liquid fraction is taken out at the top (Figure 7.2). Particles larger than 50 µm have to be removed. The density of the solid material is 1180 kg m^{-3}, the liquid density is 1000 kg m^{-3}, the viscosity of the fluid is 0.997 $\times$ 10^{-3} Pa s and the acceleration due to gravity $g = 9.81$ m s^{-2}. Calculate the maximum flow of the feed into the tank assuming unhindered settling.

In order to solve the problem, it is necessary to calculate the settling velocity. Assuming free settling, it is possible to calculate the settling velocity for smallest particles:

$$v_{tg} = \frac{\left(1180 \text{ kg m}^{-3} - 1000 \text{ kg m}^{-3}\right) \cdot 9.81 \text{ m s}^{-2} \cdot \left(5 \cdot 10^{-5} \text{ m}\right)^2}{18 \cdot 0.997 \cdot 10^{-3} \text{ Pa s}} = 0.000245 \text{ m s}^{-1}$$

The maximum distance the particles have to settle is 0.5 m (h), so now it is possible to calculate the required settling time (t_{sed}):

$$t_{sed} = \frac{h}{v_{tg}} = \frac{0.5 \text{ m}}{0.000245 \text{ m s}^{-1}} = 2040 \text{ s}$$

The retention time (Rt) is given as the ratio between the tank volume (V) and the feed flow (Q):

$$Rt \equiv \frac{V}{Q}$$

The retention time is the average time a volume element resides in the tank and has to be higher than the settling time. The feed flow can now be calculated as the tank volume is known:

$$Q \leq \frac{V}{t_{sed}} = \frac{0.5 \text{ m} \cdot 2 \text{ m}^2}{2040 \text{ s}} = 0.00049 \text{ m}^3 \text{ s}^{-1}$$

Thus, in order to ensure that particles larger than 50 µm are removed, the feed flow has to be lower than 0.000 49 m^3 s^{-1} or approximately 30 l min^{-1}.

Increasing the settling time increases the separation efficiency, but the settling velocity usually declines at increasing dry matter (Ndegwa *et al.*, 2001). The mechanisms behind the link between increasing dry matter and reduced settling rate are the increased density and viscosity of the slurry. A higher concentration of small (i.e. less than colloidal) particles will increase the viscosity of the slurry liquid and thereby reduce settling. Furthermore, with a high dry matter content, increased hindering of sedimentation will take place. The increased weight of the top sediment particles will also squeeze water out of the thickening zone, causing turbulence that stirs up the particles – a mechanism that may be more common in batch settling systems than with technologies where the sediment is removed continuously (Foust *et al.*, 1980). Fermentation and increased buoyancy of the particles due to trapping of bubbles may reduce settling, if the process takes place over a long time (e.g. in lagoons). Therefore it is recommended that slurry temperature be below 16 °C, if lagoons and long-term treatment are used (Meyer *et al.*, 2007).

Settling of solids in pig slurry is usually completed within 1 h (Converse and Karthikeyan, 2004). The settling may be assumed to follow an exponential curve and thus most of the particles that can settle in pig

Table 7.2 *Separation index at sedimentation: mean of measurements from numerous studies (adapted with permission from Hjorth et al., 2010; © 2009 INRA, EDP Sciences).*

Animal origin	Slurry dry matter (%)	Separation index (%)			
		Dry matter	Total-N	NH$_4$-N	Total-P
Swine	2	57	NA	NA	41
Cattle	2	55	33	28	57

NA: not available.

slurry are deposited at the bottom after 10–20 min (Powers and Flatow, 2002). Plant nutrients are not evenly distributed between particles of different density and size, and consequently the settling of plant nutrients may not be linearly related to the settling of dry matter (Table 7.2). This is indicated by the low P removal after a settling time of 20 min observed in the study by Powers and Flatow (2002).

As for pig slurry, the dry matter settling of cattle slurry increases when the dry matter concentration in the slurry increases (e.g. from 0.1 to 1%). The settling rate of dry matter in cattle slurry decreases exponentially with time and dry matter settling is almost completed after 1.5 h. In contrast, P settling may increase significantly with time (i.e. from approximately 50% settling after 4 h to 75% settling after 48 days) (Hjorth *et al.*, 2010). In contrast, the settling of total-N may not increase with increasing settling time (Converse and Karthikeyan, 2004). Most K and TAN is dissolved in the liquid phase (Massé *et al.*, 2007a), so most K and TAN is recovered in the liquid phase after sedimentation of solids (Masse *et al.*, 2005). The sediment consolidates (or compacts) due to the weight of the overlying layers of sediment; thus, the dry matter content of the solid fraction increases with settling time. The dry matter content has, for example, been found to increase from 1.0% up to 3.2% when the sedimentation time was changed from 4 to 1200 h (Converse and Karthikeyan, 2004).

7.4.2 Centrifugation

Increasing the gravitational force can reduce the settling time required to achieve a given separation efficiency and to remove particles that do not otherwise settle. In practice, this is accomplished in decanter centrifuges (Figure 7.3). In these, a centrifugal force is generated to cause the separation.

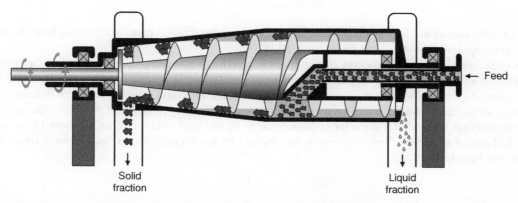

Figure 7.3 *Illustration of a decanter centrifuge, where liquid is continuously drained off to the right and solids are transported to the left. (© University of Southern Denmark.)*

There are vertical and horizontal types of decanter centrifuges. The horizontal decanter centrifuges (Figure 7.3) use a closed cylinder of continuous turning motion. The centrifugal force separates solids and liquids into a layer at the drum wall with high dry matter concentrations, and a layer at a distance from the wall consisting of a liquid containing a suspension of colloids, organic components and salt. The solid and liquid phases are transported to opposite ends of the centrifuge by rotating the entire centrifuge at a high speed and by rotating at the same time the conveyor at a speed that is slightly higher than that of the bowl (outer conical shell). The solid particles are conveyed towards the conical end and extracted there, whereas the supernatant flows towards the larger end in the cylinder formed by the bowl and the flights of the conveyor. During transport of the slurry, the particles are separated from the liquid and a liquid phase is discharged through openings at the wide end of the decanter centrifuge.

For small particles in laminar flow, the terminal velocity (v_{tc}) can be calculated as (Foust *et al.*, 1980):

$$v_{tc}(d_p) = \frac{(\rho_{solid} - \rho_{liquid}) \cdot \omega^2 \cdot r \cdot d_p^2}{18 \cdot \mu_{slurry}} \tag{7.5}$$

where r is the distance of the particle from the centrifuge axis of rotation (m) and ω is the angular velocity (rad s^{-1}), all in SI units. Rotational speed (ω_r) is measured in rotations min^{-1} (rpm) and the angular velocity can be calculated as $\omega = 2 \cdot \pi \cdot \omega_r$.

The only difference between the terminal velocity in a thickener and a centrifuge is the gravitational force. The centrifugal efficiency for simple laboratory centrifuges and, in practice, the performance of the centrifuge for decanting centrifuges can thus be related to the improvement in gravitational force, the G force (no units):

$$G = \frac{\omega^2 \cdot r}{g} \tag{7.6}$$

However, the efficiency of full-scale decanter centrifuges cannot be described so simply, partly because the distance the particle travels in a radial direction is large, so the value of r and the settling velocity vary during the sedimentation, and partly since the geometry of decanter centrifuges is quite complicated.

The performance of a decanting centrifuge is often described by its feed handling capacity (Q, m^3 s^{-1}), which can be calculated as (Foust *et al.*, 1980):

$$Q = \frac{(\rho_{solid} - \rho_{liquid}) \cdot G \cdot d_{pc}^2}{18 \cdot \mu_{slurry}} 2 \cdot \Sigma = v_{t,g} \cdot 2 \cdot \Sigma \tag{7.7}$$

where Q is the volumetric feed rate, d_{pc} is the diameter (m) of the smallest particle separated from the slurry in the centrifuge and Σ is the sigma factor (m^2), all in SI units.

The sigma factor (Σ) is a property of the specific centrifuge geometry and G force and can only be calculated in advance for simple laboratory centrifuges. For full-scale centrifuges Σ has to be found from experiments. As terminal velocity (v_{tg}) is a function of the slurry alone and sigma factor (Σ) a function of the centrifuge alone, the sigma factor can be used to compare the efficiency of different decanter centrifuges.

The dewatering volume of a decanter is considered to be the total volume (V) of the liquid zone in the cylindrical part of the drum. This volume may be changed by level regulators and the retention time (Rt) in seconds can be calculated as:

$$Rt = \frac{V}{Q} \tag{7.8}$$

where V is the dewatering volume of the decanter bowl (m^3).

Table 7.3   *Separation index at centrifugation of animal slurries (adapted with permission from Hjorth et al., 2010; © 2009 INRA, EDP Sciences).*

Animal origin	Slurry dry matter (%)	Separation index (%)			
		Dry matter	Total-N	NH_4-N	Total-P
Swine	5.4	60	25	16	73
Cattle	6.4	63	32	16	69

From Equations (7.7) and (7.8) it is obvious that the volumetric feed rate and therefore the retention time depends on the chosen value of the smallest particle to be separated (d_{pc}). Reducing the feed rate and thereby increasing the retention time automatically leads to better separation, but poorer economic performance of centrifuges.

Increasing the retention time by reducing the volumetric feed rate has been observed to increase the efficiency of the separation of slurry (Møller *et al.*, 2007a). Consequently, separation of, for example, dry matter and P can be high in laboratory studies if the retention time is high, such as 600 s as in the study by Vadas (2006), but will differ significantly from the efficiency seen at full scale at a similar *G* force. Full-scale experiments are required to determine the separation efficiency.

Increasing the dewatering volume within the centrifuge increases the retention time (Equation 7.8), but reduces the thickening zone (conical water-free part); at the same time an increase in the dewatering zone will increase the removal of dry matter from the liquid fraction. However, it will also reduce the draining of water from the solid fraction (Reimann, 1989) and hence the dry matter concentration in the solid fraction will decrease.

Increasing the angular velocity of the decanter centrifuge will increase the dry matter concentration of the solid fraction (Equations 7.6 and 7.7). Thus at a high rotation speed, the decanter centrifuge produces a solid fraction with 40% dry matter. However, increasing the angular velocity has no effect on the separation of P, K and N (Møller *et al.*, 2007a). This indicates that the residual P, K and N are dissolved in the liquid fraction and not adsorbed to particles.

The separation efficiency of dry matter increases at increasing dry matter content of the slurry. One could have expected the contrary, as seen in the sedimentation studies, because the higher viscosity of the slurry may reduce the settling velocity of the small particles. The hypothesis is that the vigorous stirring of the slurry in the decanting centrifuge may enhance attachment of small particles to larger particles and thereby improve settling of the small particles. As a consequence of a higher dry matter content of cattle slurry compared with pig slurry, the dry matter separation index is higher for cattle slurry separation than for pig slurry separation (Table 7.3). For assessment of particle retention of slurry, it is reasonable to assume that decanter centrifuges can retain particles larger than 20–25 μm in the solid fraction (Hjorth *et al.*, 2010).

Organic nitrogen or adsorbed ammonium content is related to the dry matter content of the solid fraction. Therefore, total N separation is related to the dry matter content of the slurry being treated.

7.4.3   Drainage

Separation techniques filtering solids out of slurry use screens and filter belts to retain the solid fractions. In simple screens and belt separators, the liquid is drained by gravity from solids in the separator (Figure 7.4). With a belt separator, the filter cake is continuously removed as the belt rotates and the filters are cleaned continuously. The belt filter can be followed by a press where water is squeezed out of the solid fraction.

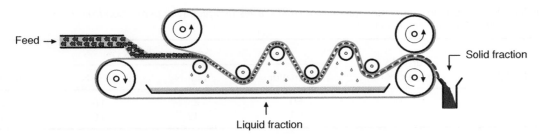

Feed →

Solid fraction

Liquid fraction

Figure 7.4 *In a belt press separator, slurry is moved into the press on a belt from the left and the liquid is squeezed out of the slurry between two belts. (© University of Southern Denmark.)*

The liquid flux through the filter media is determined by the hydraulic resistance of the medium and the hydraulic resistance of the material deposited on the medium (i.e. the filter cake):

$$J = \frac{p}{\mu_{\text{liquid}} \cdot (R_{\text{m}} + \alpha \cdot m)} \tag{7.9}$$

where J is the flux (kg m^{-2} s^{-1}), μ_{liquid} is the viscosity of the filtrate (Pa s), R_{m} is the resistance to flow from the membrane (m^{-1}), p is the effective pressure (Pa) and is related to the effective mass of the slurry (i.e. $p = \rho \cdot g \cdot h \cdot h$), and h is the height of sample above the filter medium (m), which decreases during the process. Furthermore, α is the specific resistance of the filter cake (m kg^{-1}) and m is the mass of cake deposed per area filter medium (kg m^{-2}); the product of these gives the hydraulic resistance within the cake ($R_{\text{cake}} = m \cdot \alpha$). Assuming that all dry material is removed from the filtrate and settling is negligible, the cake height increases proportionally with filtrate volume (i.e. $m = S \cdot V \cdot A^{-1}$, where S is the particle concentration in the feed (kg m^{-3}), A is the media area (m^2) and V is the filtrate volume (m^3)). As a consequence, the filter cake grows and the liquid flux decreases during the drainage process.

Hair in pig slurry may cause an immediate build-up of a filter cake with low specific resistance (α). However, when the filter cake is formed, small particles often clog the filter pores. A mix of particles with a content of particles between 1 and 100 µm will produce a cake with high specific resistance (α), which will reduce the drainage of liquid from the filter cake and the effect will be a solid fraction with a low dry matter concentration. Due to a higher fraction of larger particles, filter technology is more efficient at separating cattle slurry than pig slurry (Table 7.4). Particles can also clog or adhere to the filter media and thereby increase the resistance of the membrane. This may be a greater problem when treating cattle slurry, which tends to "stick" to the filter (Masse *et al.*, 2005).

Table 7.4 *Separation index at filtration without external pressure, screen and filter belt separation (adapted with permission from Hjorth et al., 2010; © 2009 INRA, EDP Sciences).*

Animal origin	Slurry dry matter (%)	Separation index (%)			
		Dry matter	Total-N	NH$_4$-N	Total-P
Swine	3.8	42	24	23	30
Cattle	7.0	47	33	–	40

As small particles are caught within the filter cake or adhere to the medium, screens and filters retain not only particles larger than the size of the mesh or the screen openings, but also smaller particles. Therefore, filtering time and separation index cannot be determined from particle size distribution and plant nutrients in different particle fractions and the size of the screen openings alone.

Increasing the retention time of the filter cake on the screen or the filter fabric will increase the drainage and hence dry matter concentration of the solid fraction produced. Retention time is often longer in laboratory studies, which therefore often give better results than pilot or full-scale tests of a separation technology. The best strategy is to determine specific resistance of the cake (α) in the laboratory and use it for designing pilot or full-scale equipment (Example 7.2).

Example 7.2

Calculate the required belt length of a belt filter for separating pre-treated (flocculated) manure into a liquid and a solid fraction to increase the dry matter content from 60 to 120 kg m^{-3}. The following information for the manure is available: the density of the manure is 1000 kg m^{-3}, gravity $g = 9.81$ m s^{-2}, the filtrate (liquid fraction) viscosity is 1.0×10^3 Pa s and the specific cake resistance (α) is 7×10^8 m kg^{-1}. The effective width of the belt (l_{belt}) is 1.7, the belt speed (v_{belt}) is 2.57 m min^{-1} and the loading rate (Q) is 0.31 m^3 min^{-1}. The filter medium resistance was set at 1×10^6 m^{-1}.

The initial height of the manure on the belt filter is $h_o = Q/(l_{belt} \cdot v_{belt}) = 0.205$ m. The final height on the belt should be $h_f = (60$ kg m$^{-3}/(120$ kg m$^{-3} \times 0.205$ m$) = 0.103$ m. The filtrate flux is given by a $J = -dh/dt$. It is now possible to calculate dh/dt and h as a function of t can be calculated numerically (e.g. by using Euler's method):

$$h \cdot (t + \Delta t) = h(t) + \Delta t \cdot \frac{dh}{dt} \qquad (7.10)$$

where Δt should be small to reduce the numerical error. More advanced methods exist which also lower the numerical error.

From Figure 7.5 we can see that 96 s (or 1.6 min) is enough to ensure the desired dry matter content. Hence, the length should be 1.6 min $\times$ 2.57 m min$^{-1} = 4.1$ m. The calculation can be repeated increasing the belt speed to 4 m min^{-1}.

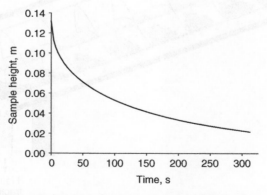

Figure 7.5   *Sample height as function of drainage time.* $\Delta t = 0.000\ 01$ s.

Most P is contained in small particles or is dissolved in the liquid, while little N is found on particles larger than 0.125 μm. As small particles often penetrate the screen or filter fabric, the separation efficiency of P and N is usually low (Meyer *et al.*, 2007). Although K and dissolved P are drained off in liquid, much is retained in the filter cake because this has a high water content (50–80% volume is not uncommon). As a result, the separator solids may still retain a significant proportion of N, P and K (Møller *et al.*, 2000; Table 7.4).

7.4.4   Filtration with Pressure

The dry matter content of the solid fraction filtration can be increased by filtering under pressure. This is usually done by using a screw press or a press auger. In a screw separator the effluent is transported into a cylindrical screen with a screw (Figure 7.6). The liquid passes through the screen and is collected in a container surrounding the screen. The solid fraction is pressed against the plate at the end of the axle and thereby more liquid is pressed out of the solid fraction. The solid fraction drops from the opening between the plate and the opening of the cylindrical mesh.

The liquid flux through the medium can be determined from Equation (7.10) by setting the effective pressure (p) to the applied pressure. It is generally assumed that the specific resistance of the cake (α) is constant during constant pressure filtration. However, for complex organic suspensions, specific resistance often increases during the process. The manure filter cake is compressed during pressure filtration and hence the specific resistance is several orders of magnitude higher for pressure filtration than for gravity drainage (see Example 7.2 for gravity drainage and Example 7.3 for pressure filtration). The advantages of high pressure filtration are therefore seldom a lower filtration time, because α increases with pressure, but instead the higher dry matter content of the solid fraction.

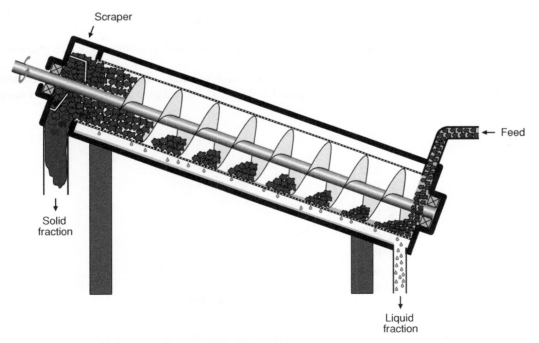

Figure 7.6   *Screw press separator. (© University of Southern Denmark.)*

Example 7.3

Flocculated manure has been filtered using a laboratory-scale piston filtration apparatus with a 45-μm filter paper. The piston pressure was equal to 200 kPa and the diameter of the piston was 5 cm, so the area of the filter medium was 1.96×10^{-3} m². The dry matter concentration of sludge was 56 kg m⁻³ and the viscosity of the filtrate was measured to be 1.0×10^3 Pa s. The experimental filtration data obtained from the laboratory-scale experiments are shown in Table 7.5.

Table 7.5 *Dead-end filtration data.*

t (s)	14	23	32	41	50	59	68	77	86	95
V (ml)	19.2	27.7	33.8	39.0	43.5	47.6	51.3	54.7	57.8	60.5
tV^{-1} (s ml⁻¹)	0.728	0.832	0.947	1.05	1.15	1.24	1.33	1.41	1.49	1.57

Determine the average specific filter cake resistance

The filtrate flux $J = A^{-1}dV/dt$, and combining this equation with Equation (7.9) and integrating gives:

$$\frac{t}{V} = \frac{\mu \cdot S}{2 \cdot A^2 \cdot p} V + \frac{\mu \cdot R_m}{A \cdot p} \qquad (7.11)$$

Thus, plotting $t\, V^{-1}$ versus V gives a straight line and α can be calculated from the slope of the line (Figure 7.7).

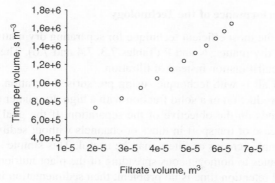

Figure 7.7 *Data plotted as* t V⁻¹ *versus V.*

The slope was found to be 2.15×10^{10} s m⁻⁶ and the specific cake resistance calculated to be 6×10^{11} m kg⁻¹:

$$\alpha = \frac{2 \times 200\ \text{kPa} \cdot \left(1.96 \cdot 10^{-3}\text{m}^2\right)^2 \cdot 2.15 \cdot 10^{10}\text{s m}^{-6}}{1 \cdot 10^{-3}\ \text{Pa s} \cdot 56\ \text{kg m}^{-3}} = 6 \cdot 10^{11}\text{m kg}^{-1} \qquad (7.12)$$

Usually, a slurry where the specific cake resistance is equal to or below 10^9 m kg⁻¹ is easy to filter, at around 10^{11} m kg⁻¹ it is difficult to filter and at around 10^{13} m kg⁻¹ it is extremely difficult to filter. For non-flocculated manure the specific resistance has been measured to be around 10^{15} m kg⁻¹ and it is therefore not possible to filter raw manure with a 45-μm filter paper without flocculation (Section 7.5).

Table 7.6   *Separation index at filtration with external pressure (i.e. screw-press separation) (Hjorth et al., 2008).*

Animal origin	Slurry dry matter (%)	Separation index (%)			
		Dry matter	Total-N	NH_4-N	Total-P
Swine	4.7	35	11	–	20
Cattle	7.0	38	19	–	14

Due to the compression of the cake, the screw press can produce a solid fraction with a high dry matter content – often twice as high as when filtering slurry without applying pressure (Møller *et al.*, 2000). Increasing the pressure applied increases the resulting dry matter concentration in the solid fraction. Although aggregation of particles on the filter may contribute to some degree to the retention of small particles in the screw press, this is not a significant process because the applied pressure forces small particles through the filter pores and a large fraction of small particles ends up in the liquid fraction after separation. Thus, the filter cake contains little N, P or K, which is dissolved in the liquid phase and in the small particles that drain off the filter cake with the filtrate. As a result, the plant nutrient separation efficiency of the screw press is low (Table 7.6).

7.4.5   User Demand on Performance of the Technology

In general, centrifugation is the most efficient technique for separating dry matter and P, and filtration is the least efficient for separating dry matter, N and P (Tables 7.3, 7.4 and 7.6). The separation of NH_4^+ is also more efficient when using centrifugation instead of filtration.

The poorest separation of all is with techniques using pressurised filtration. However, the advantage of pressurised filtration is the production of a solid fraction with a high dry matter concentration.

Choice of separator depends on the objective of the separation. If the goal is to reduce the dry matter content of the slurry for its ease of transport in tubes or channels without sedimentation or blockage of the tubes and channels, then simple screens or filters may be useful. This simple separation will also produce a liquid fraction that contributes to homogeneous spreading of the plant nutrients in the field. If the cost of separation has to be low and retention time is no problem, then sedimentation is a cheap technique that can reduce the plant nutrient composition of the slurry most efficiently (Table 7.2).

A screw press seems to be a good choice if the objective is to produce a biomass with a high dry matter concentration that is usable for biogas production and incineration (Table 7.6). The filtration technologies may retain up to about 25% of the N and P in the slurry, and this may be the only nutrient removal necessary to achieve a better match between the ratio of plant nutrients applied to the field and the crop demand on the livestock farm.

Of the aforementioned techniques, the decanter centrifuge is the most efficient at retaining P and at the same time producing a solid fraction with a relatively high dry matter content. Moreover, this technique can produce a liquid fraction with an N : P : K ratio corresponding to the needs of the crop. However, the investment and operating costs for a decanter centrifuge are usually higher than for the other separation methods.

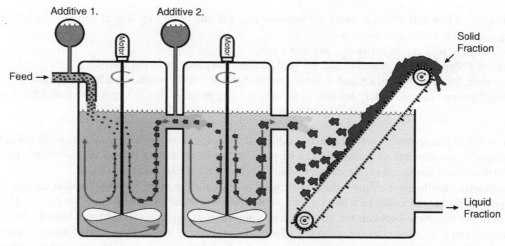

Figure 7.8 *Example of the use of additives and separation of animal manure. The diagram shows (1) slurry pumped into the separator, (2) coagulants added to the slurry, (3) polymers added to slurry, (4) the solid fraction and (5) the liquid fraction. (© University of Southern Denmark.)*

7.5 Pre-Treatment: Chemical Additives

Chemicals can be added prior to the separation to improve the retention of plant nutrients in the solid fraction, and thereby produce a liquid fraction with a composition that better fulfils the needs of both livestock and plant producers.

Additives such as brown coal, bentonite, zeolite, crystals and efficient microorganisms have been examined in numerous studies and are used by some livestock farmers (Hjorth *et al.*, 2010), but are not mentioned further here. In the following, the focus is on methods that have been developed and used as a pre-treatment to sedimentation, centrifugation or filtration (Figure 7.8). These include addition of polymers and multivalent ions, as well as treatment that enhances struvite formation and precipitation.

7.5.1 Precipitation, Coagulation and Flocculation

Flocculation, coagulation and precipitation are chemical pre-treatments that improve mechanical solid–liquid separation of many suspensions (Nowostawska *et al.*, 2005). In most suspensions, colloidal particles do not aggregate spontaneously because the particles are negatively charged and repel each other (Gregory, 1989). However, aggregation is facilitated by adding either multivalent cations that cause coagulation and/or polymers that promote flocculation (Text Box – Basic 7.2). The addition of multivalent cations also triggers the precipitation of P.

Text Box – Basic 7.2 Concepts in coagulation and flocculation

Coagulant: Salt that is added to slurry to lower the electrostatic repulsion between particles and thereby increase the rate of aggregation.

Coagulation: When colloids in a slurry are unstable (i.e. the rate of aggregation is strong), the formation of aggregates is called coagulation.

Colloid: Particle with a size between 1 nm and 1 μm.

Flocculant: Polymer that is added to a slurry to flocculate the particles in the slurry.

Flocculation: Process of contact and adhesion whereby dispersed particles are held together by weak physical interactions ultimately leading to formation of aggregates (flocs) larger than colloidal size.

It is possible to precipitate dissolved ions by adding salt. Precipitation has been used to remove phosphorus from manure: large amounts of phosphate can be precipitated when adding iron or calcium salts (Hjorth *et al.*, 2008) due to the formation of, for example, $FePO_4$, $Fe_5(PO_4)_2(OH)_9$ and $Ca_3(PO_4)_2$.

Salt containing multivalent cations can also be used to coagulate manure. The multivalent cations adsorb to the oppositely charged colloids, whereby they neutralise (or partially neutralise) the surface by lowering the electrostatic repulsion between the colloids, enabling them to aggregate. There is a limit to how much can be added and overdosing occurs when the adsorbed ions reverse the surface charge, which counteracts aggregation (Gregory, 1989). The optimum dose can be found from bench-scale experiments where, for example, the dry matter content has been measured after filtration (Figure 7.9).

Addition of polyelectrolyte polymers (flocculant) to slurry induces flocculation. Different mechanisms are deemed responsible, such as polymer bridging, charge neutralisation and patch flocculation. Polymer bridging is the adsorption of long-chain polymers onto the surface of more than one particle, causing formation of strong aggregates of large flocs (Gregory, 1989). Charge neutralisation occurs when the polymer adsorbs to the particle, whereby the electrostatic repulsion is reduced. Patch flocculation is adsorption to particles of oppositely charged polyelectrolytes, with a charge density much higher than the charge density of the particles. In this way local positively and negatively charged areas on the particle surface are formed (Gregory, 1989). Polymer bridging is believed to be the main reaction mechanism for manure flocculation. Addition of polymers causes flocculation of particles and also of existing flocs that have been produced due to coagulation.

The most common polymers are all based on PAM. They are synthesised by co-polymerising acrylamide with a cationic monomer, because cationic polymers have the highest affinity for the mainly negatively

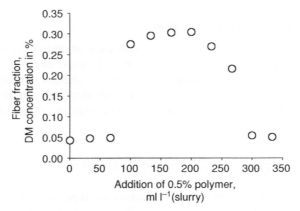

Figure 7.9 *Measured dry matter content of the solid fraction after filtration at 2 bar using a 45-m filter. The raw manure was flocculated adding a high-molecular-weight, cationic polyacrylamide (PAM) polymer. (© University of Southern Denmark.)*

charged particles in animal manure. Polymers with different molecular weight, charge density and molecular structure (linear polymers or branched polymers) exist commercially, and the right choice of polymer depends on the physico-chemical properties of the particles, the ionic strength and the viscosity of the slurry. The ionic strength in manure is high. A high ionic strength lowers the electrostatic repulsion between particles (i.e. particles aggregate more easily if salt is added). However, the cationic polymers also become less efficient due to charge screening. The particle size and surface charge density have a high impact on the stability of the suspension and therefore on the strategy that should be used in relation to the particles (Gregory, 1989). For small particles (colloids), salts or low-molecular-weight polymers are effective, whereas high-molecular-weight polymers have to be used for collection of aggregates or particles larger than 10 μm. The higher the surface charge of the particles, the larger the amount of polymers required (Gregory, 1989).

Multivalent ions, salts and especially polymers need to be carefully added to the slurry in order to obtain satisfactory aggregation of the particles. If both additive types are used, then first the salt is added to the slurry to form small aggregates. Several minutes of stirring are necessary for the charge neutralisation and coagulation to occur. Afterwards, the polymer is slowly added in small doses during vigorous stirring, followed by slow stirring to catch particles and the already formed aggregates. The shear applied by the impeller in terms of time and stirring velocity has a large impact on the formation of the aggregates; too low a shear causes the aggregates be non-uniform and unstable with low particle catchments, while too large a shear causes the aggregates to be destroyed. After the additions, the slurry may be transferred to ordinary solid–liquid separators (Figure 7.8).

As a guideline for manure flocculation, one can assume that (i) a cationic polymer is superior to anionic and neutral polymers, (ii) polymers of medium charge density (20–40 mol%) are efficient, and (iii) linear polymers are better than branched. Optimum separation has been observed by adding 0.5–1.0 g polymer kg^{-1} manure (Hjorth *et al.*, 2008) but experimental test are required to find the optimum (Text Box – Advanced 7.1). The best polymer choice may be different from these guidelines and for fermented slurry tests have shown that anionic polymers have the best flocculation efficiency.

Text Box – Advanced 7.1 Measurement of residual turbidity for optimising pre-treatment of manure

Several types of bench-scale experiments have been used to optimise the added doses of coagulant or flocculant. These methods include simple visual tests of the flocs, residual turbidity, particle dispersion analysis (PDA), capillary suction time (CST) and measurements of specific resistance (*cf.* Examples 7.2 and 7.3 in Section 7.4).

Measurements of the residual turbidity are a simple method to measure the amount of non-coagulated/flocculated particles. Jar test procedures are usually used, where doses of different amounts of coagulant/flocculant are added under standard mixing conditions. The optimum doses can then be identified from a simple settling test (Figure 7.10).

Before measuring residual turbidity, flocculate samples are centrifuged gently (e.g. at 900 *g* for 2 min), after which the turbidity is measured in the supernatant (residual turbidity). Turbidity can be measured in a spectrophotometer and the turbidity calculated as:

$$\tau = \frac{\ln(I/I_0)}{l} \tag{7.13}$$

where I_0 and I are the intensity of a light beam (lux) before and after passage through a layer of the medium of thickness l (m).

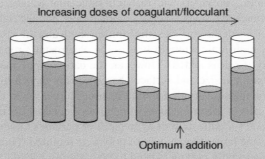

Increasing doses of coagulant/flocculant

Optimum addition

Figure 7.10   *Flocculated pig manure settled for 30 min. Residual turbidity is measured in the liquid phase above the precipitate (© University of Southern Denmark.)*

In dilute slurries, the turbidity increases proportionally to particle concentration. The turbidity decreases with added doses of coagulant/flocculant and is low at optimum doses. As well as small particles, other compounds can be of interest and thus, for example, the concentration of N or P may also be measured in the supernatant.

The solid separation products are often deposited in landfill or applied to cropped fields, and therefore the environmental and health consequences of the applied polymer must be considered. The monomers of PAM used in most slurry separation studies can be toxic, but a study on separated slurry products proved the risk to be minimal (Schechter *et al.*, 1995). Therefore, in the United States the additives are considered "Generally Recognized As Safe" (GRAS) products when added to slurry below a specific application rate related to the end use and "If potentially less toxic polymers are desired, they would first need to be tested". When considering new polymers one must remember to also take into consideration the toxicity of the compounds produced during the degradation of the polymers.

The separation technique used to retain the flocs and remove liquid (permeate) affects the separation efficiency of slurry treated with coagulants and flocculants. As for the combined solid–liquid separation technique, drainage appears to be superior in terms of P and dry matter separation, as pressure, for example, may disintegrate flocs.

The addition of a polymer (flocculant) improves dry matter, N and P separation, because the floc formation will increase the amount of dry matter retained in the solid fraction and also the N and P in organic and inorganic dry matter. Ammonium and K are dissolved in the liquid; hence, the polymer will not catch the TAN or K and improve the separation, and TAN and K cannot be precipitated with coagulants.

A higher slurry dry matter content will in general require an increased amount of coagulant/flocculant to obtain satisfactory floc formation and separation (Zhang and Lei, 1998), because the larger number of charged particles in a larger amount of dry matter have to be neutralised and caught by the additives. Increasing coagulant addition has the effect of lowering the amount of polymer required to achieve satisfactory separation (Krumpelman *et al.*, 2005).

The separation index values for untreated slurry (Tables 7.2–7.4 and 7.6) are much lower than the values obtained by slurry treatment with flocculants and coagulants (Table 7.7). Thus separation is improved by flocculation and further improved when multivalent ions are added to the slurry.

7.5.2   Struvite Crystallisation

Animal slurry contains P partly in the form of crystalline or amorphous struvite ($MgNH_4PO_4 \cdot 6H_2O$) and apatite $Ca_5OH(PO_4)_3$ (hydroxyapatite). Furthermore, struvite formation may be initiated by physical and

Table 7.7 *Separation index for coagulation and flocculation (Hjorth et al., 2008).*

Separation technique	Animal	Manure dry matter (%)	Separation index (%)			
			Dry matter	Total-N	NH$_4$-N	Total-P
Sedimentation, centrifugation or filtration	swine	2.5	63	28	15	79
Sedimentation or filtration (+ applied pressure)	cattle	3.	67	33	20	87

chemical changes. This is a problem in biogas plants, because struvite has a tendency to build up on tube surfaces and reduce the slurry flow, but struvite formation may also be used to remove P and NH$_4^+$ from the slurry (Suzuki *et al.*, 2007).

$$Mg^{2+} + NH_4^+ + PO_4^{3-} + 6H_2O \rightleftharpoons MgNH_4PO_4 \cdot 6H_2O(s) \tag{7.14}$$

In some slurries, the product of [Mg^{2+}], [NH$_4^+$] and [PO$_4^{3-}$] is lower than the conditional formation constant due to low concentrations of dissolved magnesium (Mg^{2+}) (Nelson *et al.*, 2003) and in consequence no struvite is formed. The dissolved Mg is often lower than determined with traditional extraction and measuring techniques, because Mg^{2+} ions form complexes with dissolved organic matter (Bril and Solomon, 1990). The concentration of PO$_4^{3-}$ also has a major impact on struvite precipitation. The concentration of PO$_4^{3-}$ decreases with pH, which reduces struvite crystallisation at low pH (Nelson *et al.*, 2003). In most slurries, the NH$_4^+$ concentration is higher than the 1 : 1 : 1 ratio ([Mg^{2+}] : [NH$_4^+$] : [PO$_4^{3-}$]) required for the formation of struvite and is not the limiting factor in the reaction. However, at pH above 9.5, the NH$_4^+$ concentration will decrease and a low NH$_4^+$ concentration may limit struvite formation. Optimum conditions for sedimentation of struvite therefore occur at about pH 9 (Buchanan *et al.*, 1994; Nelson *et al.*, 2003). Hydroxyapatite is usually not chosen as a means of removing P from slurry and wastewater, because much calcium will have crystallised as calcite (CaCO$_3$) in slurry.

Precipitated struvite is often removed in a thickener or solid–liquid separator. In the pilot plant used by Suzuki *et al.* (2007), the struvite was both formed and removed in a thickener. Alternatively, the struvite may be produced in a psychrophilic anaerobic sequencing batch reactor biogas plant, where the P settles out as struvite and is removed with the sludge (Massé *et al.*, 2007a) or struvite may be settled in anaerobic slurry lagoons and also be removed with sludge (Nelson *et al.*, 2003).

Phosphorus removal can be increased by adding Mg to the slurry. Addition of iron (Fe) and a base to the slurry will enhance dissolution of Mg and thereby increase P removal. On the other hand, P removal may be low even after addition of Mg and NH$_3$ if pH is low. Aeration of slurry or anaerobic digestion of slurry increases the pH and also reduces the content of organic matter in the slurry (Suzuki *et al.*, 2007; Masse *et al.*, 2005). Both processes will greatly enhance crystallisation of struvite due to an increase in the concentration of dissolved Mg^{2+} and PO$_4^{3-}$. Thus, aeration may produce a slurry with a mole ratio optimal for struvite crystallisation and increase struvite crystallisation by a factor of about 10, thereby removing 65–99% of P and 15% of total N in a continuous flow pilot-scale sedimentation plant.

Further struvite formation can be encouraged by improving nucleation, which involves adding nucleation agents such as sand grains to the reactor (Battistoni *et al.*, 2002) or using rakes to scratch on the reactor walls and thereby contribute energy to the nucleation process.

7.6   Post-Treatment: Separation Techniques

Post-treatment mechanical separation techniques need to be included when eco-efficient technical solutions for the processing of animal slurry are developed. Some technologies have been tested at the laboratory scale, while a few have reached farm pilot production scale. Thus, there is still a need for new innovative and sustainable technologies to be developed.

7.6.1   Evaporation of Water and Stripping of Ammonia

Water and volatiles can be removed from the slurry or the liquid fraction by evaporation. The liquid is heated to its boiling point, which for slurry is a little over 100 °C at atmospheric pressure. At this temperature, water, volatile organic compounds such as fatty acids and NH_3 evaporate. Thus, the pH of the slurry has to be reduced by adding acid, so that total ammoniacal N ($TAN = NH_3 + NH_4^+$) is in NH_4^+ form, which cannot volatilise. This vapour phase has to be condensed to retrieve the energy used to evaporate water and volatiles. The energy consumed in the process is high, as the heat of evaporation is roughly 670 kWh ton^{-1} water. The energy (q) consumed in a single step evaporated can thus be estimated as:

$$q = m_{water} \cdot \Delta H^{vap} \tag{7.15}$$

where ΔH^{vap} is the heat of evaporation (J kg^{-1}), m_{water} is the amount of water removed (kg s^{-1}) and q is the energy consumption (W).

To reduce the energy costs, evaporators can be operated either as single evaporators using recompressed steam or in series where the vapour generated in the first evaporator is used as heating steam for the next evaporator. In order to transfer energy from the steam to the liquid in each evaporator, the steam temperature has to be higher than the liquid temperature. Therefore, each consecutive evaporator is operated at a lower temperature than the preceding evaporator. Boiling can thus only be achieved in each evaporator by operating each at a slightly lower pressure than the preceding evaporator. As the boiling point for slurry increases with solids content, the liquid slurry should enter the last evaporator and run counter-current to the vapour. Even though multi-step evaporation increases the investment costs, experiences from industrial applications show that using from three to six evaporators in series is economically sound (Foust *et al.*, 1980). In this way, a 92% volume reduction can be achieved at an energy consumption of 120–130 kWh ton^{-1} slurry treated. Due to the large energy consumption, evaporation will probably not be optimal compared with other options, unless waste heat is available.

Air stripping of NH_3 is a process whereby the liquid fraction after manure separation is brought into contact with air, upon which NH_3 evaporates and is carried away by the gas. Instead of ambient air one can use steam stripping, where steam is used instead of air. Ammonia released from the stripping column is collected using wet scrubbing with an acid solution.

The equilibrium distribution between the gas and the liquid species of NH_3 is described with Henry's constant ($K_H = [NH_3(aq)]/p_{NH_3}$), and the mass transfer of the NH_3 can be calculated by:

$$\frac{1}{V}\frac{dm}{dt} = -K_L \cdot a \cdot ([NH_3(aq)]^* - [NH_3(aq)]) \tag{7.16}$$

where V is the liquid volume (m^3), m is the mass of NH_3 (kg), K_L is the overall liquid mass transfer coefficient (m s^{-1}) and a function of the liquid and gas phase transfer coefficients (it should not be confused with an equilibrium constant), a is the specific interfacial area between gas and liquid (m^2 m^{-3}), $[NH_3(aq)]$ is the average concentration of NH_3 in the liquid phase, (kg m^{-3}), and $[NH_3(aq)]^*$ is the liquid concentration of

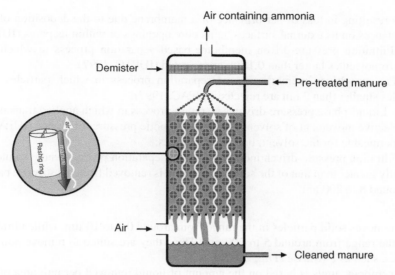

Figure 7.11   *Air stripping for removal of NH₃. (© University of Southern Denmark.)*

NH₃ in equilibrium with the gas phase concentration (kg m⁻³). The product of the mass transfer coefficient and the interfacial area $K_L \cdot a$ is denoted the transfer rate constant.

Since evaporation occurs from the slurry surface, it is advantageous to ensure that the liquid has a large surface area. This can be achieved in a stripping column with structured packing, where the manure spreads over the packing material in a thin film and therefore has a considerably larger surface (Figure 7.11). The mass transport also increases with the concentration of NH₃(aq) in the liquid phase. Hence, if pH is increased, an increasing part of TAN is in NH₃(aq) form and the mass transport of NH₃ also increases.

7.6.2   Membranes

For small particles, membrane filtration (Text Box – Basic 7.3) of the liquid fraction could be an attractive supplement to solid–liquid separation (Section 7.3). Furthermore, membrane separation may be used to separate and concentrate dissolved K, P and N nutrients, producing a nutrient-rich liquid phase and, in principle, pure water for reuse or safe discharge to the environment.

Text Box – Basic 7.3   Concepts in membrane filtration

Concentration polarisation: In pressure-driven membrane processes the concentration of the rejected molecules or particles is higher near the membrane surface than in the feed due to convective transport. This phenomenon is called concentration polarisation.

Cross-flow filtration: Filtration with a tangential flow to the membrane, which reduces membrane clogging.

Cut-off value: The lowest diameter or molecular weight solute (MWCO) in which 90% of the solute is rejected by the membrane.

Dead-end filtration: All fluid passes through the membrane opposite to cross-flow filtration. Particles are deposed on the filter area.

Fouling: Process resulting in loss of performance of a membrane due to the deposition of suspended or dissolved substances on its external surfaces, at its pore openings or within its pores (IUPAC, 1997).

Microfiltration: Filtration pressure-driven membrane-based separation process in which particles and dissolved macromolecules larger than 0.1 μm are rejected (IUPAC, 1997).

Nanofiltration: Pressure-driven membrane-based separation process in which particles and dissolved macromolecules smaller than 2 nm are rejected (IUPAC, 1997).

Reverse osmosis: Liquid-phase pressure-driven separation process in which applied transmembrane pressure causes selective movement of solvent against its osmotic pressure difference (IUPAC, 1997). The membrane is permeable for the solvent, but not the solutes.

Ultrafiltration: Filtration pressure-driven membrane-based separation process where solute of molecular size significantly greater than that of the solvent molecule is removed from the solvent; particles usually range from around 5 to 200 nm.

Microfiltration removes solid particles in the range from around 0.1 to 10 μm, while ultrafiltration retains solid particles in the range from around 5 to 200 nm. Thus, they are suited to remove nutrients associated with small particles, such as P (Massé *et al.*, 2007b).

The design of membrane units is based on the amount of liquid removed per unit area of membrane, the membrane flux. In micro- and ultrafiltration, the flux (J_{liquid}) can be described by the general equation:

$$J_{liquid} = \frac{\Delta P}{\mu_{liquid} \cdot (R_m + R_{rev} + R_{irrev})} \qquad (7.17)$$

where J is the flux (kg m^{-2} s^{-1}), μ_{liquid} is the viscosity of the permeate (liquid fraction), R_m is the resistance to flow from the membrane (m^{-1}), R_{rev} is the reversible fouling resistance often ascribed to filter cake formation and concentration polarisation (m^{-1}), R_{irrev} is the irreversible fouling resistance normally ascribed to blocking of membrane pores and adsorption of materials to the membrane (m^{-1}), and ΔP is the transmembrane pressure over the membrane (Pa).

Micro- and ultrafiltration has to be carried out as cross-flow filtration, where only a fraction of the liquid is removed as permeate through the membrane, while solids and part of the liquid are retained as retentate. The cross-flow removes most of the solids deposited on the surface, but some flow-controlled reversible filter cake formation cannot be avoided. Worse still, irreversible adhesion of minor particles can occur within the membrane pores, partly blocking the path for the liquid. In addition, bacterial growth can occur on the membrane surface, further reducing the flow through the membrane. These kinds of membrane fouling are especially severe for treatment of slurry. Reversible filter cake formation can be removed by flushing with water. In contrast, irreversible fouling can at best only be partly removed by cleaning the membrane in the system ("clean-in-place" (CIP) system) with dilute base (for removal of organic foulant and silica), followed by dilute acid (for removal of inorganic deposits) and, if necessary, enzymatic treatment.

Owing to of fouling, micro- and ultrafiltration membranes can only be used to separate pre-treated slurry, such as effluent streams from biogas reactors or runoff streams from centrifuges.

For a slurry, the microfiltration transmembrane pressure is typically around 100–180 kPa. At this pressure filter cake formation and fouling become rate-determining, and increasing the transmembrane pressure no longer increases the flux significantly. Increasing the feed flow can increase the flux by reducing the filter cake thickness, but at velocities above around 2 m s^{-1} this becomes uneconomical (Owen *et al.*, 1995). Therefore, maximum flux attainable is limited to around 160 l m^{-2} h^{-1}, although long-term fluxes down to 10–40 l m^{-2} h^{-1} should be expected due to fouling. Depending on the size distribution, a retention efficiency of 75% of

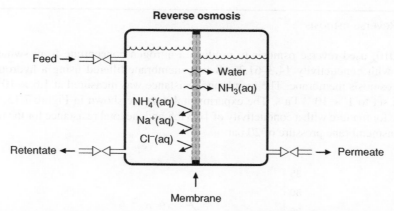

Reverse osmosis

Figure 7.12   *Reverse osmosis filtration for concentration of ions in liquid manure. (© University of Southern Denmark.)*

dry matter should be expected. As most P is contained in particles with diameter between 0.45 and 10 μm retained with membrane filtration, then good P removal can be expected (Massé *et al.*, 2007b), but dissolved N, K and dry matter will not be retained.

For ultrafiltration, transmembrane pressures up to 800 kPa can be used depending on membrane pore size. The larger the pore size, the lower the transmembrane pressure. For filtered pig slurry, fluxes between 10 and $40 \, \mathrm{l \, m^{-2} \, h^{-1}}$ at a transmembrane pressure of 100 kPa have been reported with 100% dry matter removal and a P removal efficiency of up to 87%, depending on operating conditions. For partly digested pig slurry, flushing the membrane every 5 min with permeate leads to a fall in the flux of 25% over a 70-day period compared with a flux decline of 75% if only a single cleaning procedure is performed during the period (maximum flux $16 \, \mathrm{l \, m^{-2} \, h^{-1}}$).

The liquid fraction from an ultrafiltration membrane can contain large amounts of dissolved K^+, NH_4^+ and NH_3. Reverse osmosis and partly nanofiltration membranes can retain dissolved nutrients and deliver a high concentrate retentate and a purified water permeate (Figure 7.12). Depending on the membrane chosen, nanofiltration retains uncharged molecules larger than 200–400 Da and higher valence ions. Nanofiltration can thus be used to remove dry matter and to some extent charged molecules. If a pure water permeate is needed, reverse osmosis can retain NH_4^+ and K^+ and to a lesser extent NH_3 (Massé *et al.*, 2007b).

The water flux through nanofiltration and reverse osmosis membranes can be described by:

$$J_{liquid} = \frac{\Delta P - \Delta \Pi}{\mu_{liquid} \cdot (R_m + R_{rev} + R_{irrev})} \tag{7.18}$$

where $\Delta \Pi$ (Pa) is the osmotic pressure difference between retentate and permeate. The maximum possible retentate concentration is determined by the osmotic pressure difference and the maximum possible transmembrane pressure. Monovalent ions pass through a nanofiltration membrane and hence the osmotic pressure is lower for nanofiltration membranes than reverse osmosis membranes.

As concentration polarisation influences the osmotic pressure on the membrane surface during nanofiltration and reverse osmosis, experimental permeability constants depend on the liquid cross-flow and, because of unavoidable fouling, on time in use (Example 7.4). As for micro- and ultrafiltration membranes, cleaning has to be performed regularly to avoid fouling.

Example 7.4   Reverse osmosis

Massé *et al.* (2010) used reverse osmosis to produce a nitrogen concentrate from swine manure. Pre-filtered manure with conductivity 14.3–61.3 mS was membrane-filtered using a hydrophilic neutrally charged reverse osmosis membrane. The membrane resistance was measured at 1.6×10^{14} m^{-1} and the viscosity can be set to 1×10^{-3} Pa s. The experimental data are shown in Figure 7.13. Determine the osmotic pressure for manure with a conductivity of 14.3 mS and the total resistance for the reverse osmosis setup up to a transmembrane pressure of 20 bar.

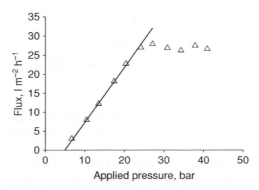

Figure 7.13   *Permeate flux obtained during reverse osmosis membrane filtration of pre-filtered manure. (Reprinted with permission from Massé et al. (2010). © 2010 Elsevier.)*

According to Equation (7.17) the permeate flux is zero when the transmembrane pressure equals the osmotic pressure. Thus, extrapolating the curve to $J = 0$ gives an osmotic pressure of 5 bar. It can be seen that the osmotic pressure increases with conductivity, as both conductivity and osmotic pressure increase with ion concentration.

At 20 bar the flux has been measured to be 24 l m^{-2} h^{-1} = $6.7 \cdot 10^{-6}$ m s^{-1}; hence the slope of the line is equal to $(6.7 \cdot 10^{-6}$ m s$^{-1})/1500$ kPa = $4.4 \cdot 10^{-12}$ m s^{-1} Pa^{-1}:

$$\frac{1}{\mu_{\text{liquid}} \cdot (R_m + R_{\text{rev}} + R_{\text{irrev}})} = 4.4 \cdot 10^{-12} \text{m s}^{-1}\text{Pa}^{-1} \Rightarrow R_m + R_{\text{rev}} + R_{\text{irrev}} = 2.3 \cdot 10^{14}\text{m}^{-1} \quad (7.19)$$

The total resistance is higher than the membrane resistance. Above an applied pressure of 20 bar the flux is almost independent of pressure, which is usually an effect of concentration polarisation. This increases the osmotic pressure, or deposition of materials on the membrane (cake formation), which increases the resistance. Below 20 bar is the pressure-controlled part of the curve. Above 20 bar is the mass transfer coefficient-controlled area, where increased shear (higher cross-flow) can be used to lower the osmotic pressure/cake formation. However, this could be an expensive solution due to the higher energy demand.

For a nanofiltration transmembrane operation, the pressure is around 350 to 3000 kPa. Although not well suited for removal of K or N, nanofiltration membranes have been shown to be able to remove up to 52% NH$_4^+$ and 78% K (Massé *et al.*, 2007b), and, if a proper membrane is chosen, all soluble dry matter with a molecular weight above 200 Da is removed.

For more stringent applications, reverse osmosis has to be used. The transmembrane pressure for reverse osmosis operations is typically around 3.5–6.5 MPa, although up to 150 MPa can be achieved for special design membranes. The retention of K is usually high and independent of pH, but as NH_4^+ is retained better than NH_3, the separation is very pH-dependent and also depends on retentate ionic strength. Therefore, the N separation effect depends on pH, slurry origin and the final volume reduction of the retentate. For pig slurry, the retention of NH_4^+ thus decreased from 90% to 70% and for K from 93% to 87%, when the reduction in retentate volume was increased from 50% to 90% (Massé *et al.*, 2007b).

As the flux for nanofiltration and reverse osmosis is very dependent on fouling and also on the osmotic pressure of the retentate, the flux will change dramatically during a concentration process, with maximum fluxes reaching up to 65 l m^{-2} h^{-1} at initial conditions, but approaching 0 l m^{-2} h^{-1} at the final reduction volume where the osmotic pressure approaches the transmembrane pressure.

7.7 Summary

An overview of separation techniques is given in this chapter. Selection of the most appropriate separator depends on the slurry composition and the end-use of the solid and liquid fractions produced. With mechanical separators significant amounts of dry matter, P and N can be removed from the slurry. The removal of P and some metals can be increased significantly by adding coagulants and flocculants. Membrane technologies can be used to remove K and Na from the liquid fraction, but may be problematic to apply because of fouling of the membranes.

References

Battistoni, P., De Angelis, A., Prisciandaro, M., Boccadoro, R. and Bolzonella, D. (2002) P removal from anaerobic supernatants by struvite crystallization: long term validation and process modelling. *Water Res.*, **36**, 1927–1938.

Bril, J. and Salomons, W. (1990) Chemical composition of animal manure: a modelling approach. *Neth. J. Agric. Sci.*, **38**, 333–351.

Buchanan, J.R., Mote, C.R. and Robinson, R.B. (1994) Struvite control by chemical treatment. *Trans. ASAE*, **37**, 1301–1308.

Burton, C.H. and Turner, C. (2003) *Manure Management: Treatment Strategies for Sustainable Agriculture*, 2nd edn, Silsoe Research Institute, Silsoe.

Christensen, M.L. and Keiding, K. (2007) Filtration model for suspensions that form filter cakes with creep behaviour. *AIChE J.*, **53**, 598–609.

Converse, J.C. and Karthikeyan, K.G. (2004) Nutrient and solids separation of flushed dairy manure by gravity settling. *Appl. Eng. Agric.*, **20**, 503–507.

Foust, A.S., Wenzel, L.A., Clump, C.W., Maus, L. and Andersen, L.B. (1980) *Principle of Unit Operations*, 2nd edn, John Wiley & Sons, New York.

Gregory, J. (1989) Fundamental of flocculation. *Crit. Rev. Environ. Control.*, **19**, 185–229.

Hjorth, M., Christensen, M.L. and Christensen, P.V. (2008) Flocculation, coagulation and precipitation of manure affecting three separation techniques. *Bioresour. Technol.*, **99**, 8598–8604.

Hjorth, M., Christensen, K.V., Christensen, M.L. and Sommer, S.G. (2010) Solid–liquid separation of animal slurry in theory and practice: a review. *Agron. Sustain. Dev.*, **30**, 153–180.

IUPAC (1997) *Compendium of Chemical Terminology*, 2nd edn (the "Gold Book") (compiled by A.D. McNaught and A. Wilkinson), Blackwell, Oxford; http://goldbook.iupac.org.

Kroodsma, W. (1986) Separation and removal of faeces and urine using filter nets under slatted floors in piggeries. *J. Agric. Eng. Res.*, **34**, 75–84.

Krumpelman, B.W., Daniel, T.C., Edwards, E.G., McNew, R.W. and Miller, D.M. (2005) Optimum coagulant and flocculant concentrations for solids and phosphorus removal from pre-screened flushed dairy manure. *Appl. Eng. Agric.*, **21**, 127–135.

Loughrin, J.H., Szogi, A.A. and Vanotti, M.B. (2006) Reduction of malodorous compounds from liquid swine manure by a multi-stage treatment system. *Appl. Eng. Agric.*, **22**, 867–873.

Massé, D.I., Croteau, F. and Massé, L. (2007a) The fate of crop nutrients during digestion of swine manure in psychrophilic anaerobic sequencing batch reactors. *Bioresour. Technol.*, **98**, 2819–2823.

Masse, L., Masse, D.I., Beaudette, V. and Muir, M. (2005) Size distribution and composition of particles in raw and anaerobically digested swine manure. *Trans. ASAE*, **48**, 1943–1949.

Massé, L., Massé, D.I. and Pellerin, Y. (2007b) The use of membranes for the treatment of manure a critical literature review. *Biosyst. Eng.*, **98**, 371–380.

Massé, L., Massé, D.I., Pellerin, Y. and Dubreuil, J. (2010) Osmotic pressure and substrate resistance during the concentration of manure nutrients by reverse osmosis membranes. *J. Membr. Filtrat.*, **348**, 28–33.

Meyer, D., Ristow, P.L. and Lie, M. (2007) Particle size and nutrient distribution in fresh dairy manure. *Appl. Eng. Agric.*, **23**, 113–117.

Møller, H.B., Lund, I. and Sommer, S.G. (2000) Solid liquid separation of livestock slurry – separation efficiency and costs. *Bioresour. Technol.*, **74**, 223–229.

Møller, H.B., Hansen, J.D. and Sørensen, C.A.G. (2007a) Nutrient recovery by solid–liquid separation and methane productivity of solids. *Trans. ASABE*, **50**, 193–200.

Møller, H.B., Jensen, H.S., Tobiasen, L. and Hansen, M.N. (2007b) Heavy metal and phosphorus content of fractions from manure treatment and incineration. *Environ. Technol.*, **28**, 1403–1418.

Nahm, K.H. (2005) Environmental effects of chemical additives used in poultry litter and swine manure. *Crit. Rev. Environ. Sci. Technol.*, **35**, 487–513.

Ndegwa, P.M., Zhu, J. and Luo, A.C. (2001) Effects of solid levels and chemical additives on removal of solids and phosphorus in swine manure. *J. Environ. Eng.*, **127**, 1111–1115.

Nelson, N.O., Mikkelsen, R.L. and Hesterberg, D.L. (2003) Struvite precipitation in anaerobic swine lagoon liquid: effect of pH and Mg:P ratio and determination of rate constant. *Bioresour. Technol.*, **89**, 229–236.

Nowostawska, U., Sander, S.G., McGrath, K.M. and Hunter, K.A. (2005) Effect of trivalent metal sulfates on the coagulation and particle interactions of alumina colloids. *Colloids Surfaces A*, **266**, 200–206.

Owen, G., Bandi, M., Howell, J.A. and Churchouse, S.J. (1995) Economic assessment of membrane processes for water and waste water treatment. *J. Membr. Sci.*, **102**, 77–91.

Powers, W.J. and Flatow, L.A. (2002) Flocculation of swine manure: Influence of flocculant, rate of addition, and diet. *Appl. Eng. Agric.*, **18**, 609–614.

Reimann, W. (1989) Fest-flüssig-trennung von gülle und gülleaufbereitungsprodukten. *Arch. Acker-Pflanzenbau Bodenkd.*, **33**, 617–625.

Schechter, L., Deskin, R., Essenfeld, A., Bernard, B., Friedman, M. and Grube, E. (1995) Evaluation of the toxicological risk associated with the use of polyacrylamides in the recovery of nutrients from food processing waste (II). *Int. J. Toxicol.*, **14**, 34–39.

Sørensen, C.A., Jacobsen, B.H. and Sommer, S.G. (2003) An assessment tool applied to manure management systems using innovative technologies. *Biosyst. Eng.*, **86**, 315–325.

Suzuki, K., Tanaka, Y., Kuroda, K., Hanajima, D., Fukumoto, Y., Yasuda, T. and Waki, M. (2007) Removal and recovery of phosphorous from swine wastewater by demonstration crystallization reactor and struvite accumulation device. *Bioresour. Technol.*, **98**, 1573–1578.

Vadas, P.A. (2006) Distribution of phosphorus in manure slurry and its infiltration after application to soils. *J. Environ. Qual.*, **35**, 542–547.

Zhang, R.H. and Lei, F. (1998) Chemical treatment of animal manure for solid–liquid separation. *Trans. ASAE*, **41**, 1103–1108.

Zhang, R.H. and Westerman, P.W. (1997) Solid–liquid separation of animal manure for odor control and nutrient management. *Appl. Eng. Agric.*, **13**, 657–664.

8

Gaseous Emissions of Ammonia and Malodorous Gases

Sven G. Sommer[1] and Anders Feilberg[2]

[1]*Institute of Chemical Engineering, Biotechnology and Environmental Technology,*
University of Southern Denmark, Denmark
[2]*Department of Engineering, Aarhus University, Denmark*

8.1 Introduction

Animal manure and organic wastes are sources of ammonia (NH_3) and malodorous gases, such as sulfur compounds, organic acids, phenolic compounds and indoles. Of these, (di)hydrogen sulfide (H_2S) is lethal at high concentrations and has been associated with the death of farmers.

Diffusion and convective mass transport are involved in the transport of NH_3 and malodorous gases from animal manure to the free atmosphere. The transport of gas components can be divided into three closely related processes: (i) gas component transport in the source to the surface air–manure boundary layer, (ii) gas transfer over the interface of the manure–air boundary layer and (iii) transport from this interface to the free atmosphere. The transfer over the manure–air interface to the atmosphere is referred to here as "release".

Transport of the gas components in the manure or in the air is strongly affected by the gas concentration gradient in the manure and in the air, because the diffusive nature of part of the transport. The gradient is affected by convection, which is related to natural or forced stirring of the manure and to air movement above the manure surface. In the open air the emitted gas is transported and dispersed during transportation in the atmosphere to the receptor (odour) or sink (NH_3).

Speciation of the gas components in slurry significantly affects the emissions, because it is uncharged gas species that are released from slurry (i.e. electroneutral solutes). The electroneutral species NH_3 and H_2S are a base and an acid, respectively, so release and emission are influenced by the NH_3 to NH_4^+ and H_2S to HS^- reactions, which are affected by the proton (H^+) activity. Proton activity in slurry is buffered by total inorganic carbon ($TIC = CO_2 + HCO_3^- + CO_3^{2-}$), total ammoniacal nitrogen ($TAN = NH_3 + NH_4^+$),

---

Animal Manure Recycling: Treatment and Management, First Edition. Edited by Sven G. Sommer, Morten L. Christensen,
Thomas Schmidt and Lars S. Jensen.

organic acids and organic particles (Chapter 4), an effect that has to be accounted for when assessing pH in the surface layers of an NH_3 or H_2S source, be it stored or applied slurry.

In this chapter, the physics and chemistry of gaseous release, transport and emission to the atmosphere are presented using NH_3 and to some extent H_2S as models for the emission of basic and acidic components. The contribution of manure to the global temperature increase by emission of the greenhouse gases methane (CH_4) and nitrous oxide (N_2O) is presented in Chapter 10.

See Text Box – Basic 8.1 for definitions used in this chapter.

Text Box – Basic 8.1 Definitions used

Convection: Transport of a component due to movement of parcels of air or liquid containing the component.

Diffusion: A molecular form of transport driven by gradients of concentration.

Odour units or D_{50} values: The lowest concentration at which an odour panel can sense the odour of air. Odour units are given in ppb V/V.

NMVOCs: Non-methane volatile organic components.

ppm and ppb: ppm is parts per million concentration of the gas in 10^{-6} l (gas) l^{-1} (air) and ppb is parts per billion in 10^{-9} l (gas) l^{-1}(air).

TAN: Total ammoniacal nitrogen = $NH_3(aq) + NH_4^+(aq)$.

TIC: Total inorganic carbon = $CO_2(aq) + HCO_3^-(aq) + CO_3^{2-}(aq)$

TS: Total sulfide = $H_2S(aq) + HS^-(aq) + S^{2-}(aq)$.

VOCs: Volatile organic compounds including volatile organic acids, phenols and organic sulfur compounds. Different definitions exist. Literally interpreted, methane should be included in VOCs, but this is not always the case. To avoid confusion, NMVOCs can be used to exempt methane.

8.2 Characteristics of Ammonia and Hydrogen Sulfide

The concentration of the gases in the atmosphere ranges from 1 to more than 1000 ppb (Table 8.1). Ammonia is the gas measured in the highest concentrations in animal houses, where concentrations up to 50 000 ppb have been observed. The ambient concentrations is about 1 ppb. Ammonia can be sensed at concentrations above 2500 ppb, which is significantly higher than the detection level of most odorants, which can be sensed at levels of about 1 ppb, apart from acetic acid with a detection level of about 250 ppb (Table 8.1). Thus, in most environments it is not NH_3 in the atmosphere that causes the foul smell, but H_2S, thiols, phenols and so on. Ammonia is a risk to the health of people inhaling the gas or particles formed by the gas, and a component that causes nitrogen (N) enrichment and imbalance of ecosystems (eutrophication). Hydrogen sulfide is an abundant gas in livestock environments; the gas has a low odour detection limit and is a great risk to the farmer at high concentrations.

Ammonia is a colourless gas with a low density (0.73 kg m^{-3} at 1.013 bar and 15 °C). The N atom in the molecule has a lone electron pair, which makes NH_3 a base (i.e. a proton acceptor). The polarity and ability to form hydrogen bonds makes ammonia highly soluble in water. Ammonia dissolved in water is in equilibrium with the acid NH_4^+ with a pK_a value of about 9.3 at 25 °C (Table 8.1). Ammonia is a very small compound that in air and water is transported relatively quickly by diffusion (Table 8.2).

Dihydrogen sulfide is also a colourless but very poisonous and flammable gas, with a density higher than that of NH_3 (1.45 kg m^{-3} at 1.013 bar and 15 °C). It has a foul odour of rotten eggs perceptible at concentrations as low as 1.9 ppb. Dissolved in water, H_2S is an acid in equilibrium with HS^-, with a pK_a

Table 8.1　*Odorants in emissions from manure and livestock facilities: for each compound, the D_{50} value (odour threshold value), typical concentration, Henry's law constant and acid dissociation constant pK_a (if relevant) are presented (the temperature dependency ($dlnK_H/dT$) is included if available for the compound; this can be used to estimate K_H for temperatures other than 298 K).*

Compound	Chemical formula	D_{50}[a] (ppb)	Typical concentration level (ppb)	$pK_{a,298}$	$K_{H,298}$ (M atm^{-1})	$dlnK_H/dT$ (K)
Hydrogen sulfide	H_2S	1.9	10–200	7.1	0.1	2200
Methanethiol	CH_3SH	0.07	1–20	10.4	0.3	3000
Dimethyl sulfide	CH_3SCH_3	4.1	1–20	NA	0.6	3500
Ammonia	NH_3	2500	>1000	9.3[b]	61	4100
Trimethyl amine	$(CH_3)_3N$	2.1	1–20	9.8[b]	10	NA
Acetic acid	CH_3COOH	234	50–500	4.8	6500	6300
Propanoic acid	CH_3CH_2COOH	25	20–200	4.8	4100	NA
Butanoic acid	$CH_3(CH_2)_2COOH$	1.8	20–200	4.8	2800	NA
3-Methyl butanoic acid	$CH_3CH(CH_3)CH_2COOH$	0.6	1–50	4.8	1200	NA
Pentanoic acid	$CH_3(CH_2)_3COOH$	2.1	1–50	4.8	2200	6700
Phenol	$C_6H_5–OH$	54	<1–10	10.0	2100	7100
4-Methyl phenol	$CH_3–C_6H_4–OH$	0.3	1–20	10.3	1300	5900
4-Ethylphenol	$CH_3CH_2–C_6H_4–OH$	1.3	<1–10	NA	NA	NA
Indole		0.4	<1–10	5.7[b]	NA	NA
3-Methyl-1H-indole (skatole)		0.09	<1–10	?[b]	NA	NA

NA, not available.

[a] Odour threshold value (D_{50} for the compound) based on geometric means of detection threshold values from van Gemert (2003).

[b] pK_a values for basic compounds are presented for the corresponding acid (NH_4^+, $(CH_3)_3NH^+$, etc.).

value of 7.1 at 25 °C. However, H_2S is less soluble than NH_3 in water, with a Henry's law coefficient of 0.1 M atm^{-1}. Thus, at a given atmospheric concentration of $H_2S(g)$ and $NH_3(g)$, the concentration of $NH_3(aq)$ is 610 times higher than that of $H_2S(aq)$. Dissolved in water, H_2S reacts with metal ions to form metal sulfides. Hydrogen sulfide is a larger molecule than NH_3 and the diffusion coefficient in water and air is therefore smaller than that of NH_3 (Table 8.2).

8.3　Processes Involved in Emission

The emission of gases from manure is centred around the transfer of gas over the manure–air interface, a process defined as release. This release is affected by the [A(g)]–[A(aq)] gradient over the manure–air interface, where A(g) is the component in gaseous form and A(aq) is the dissolved component (Figure 8.1).

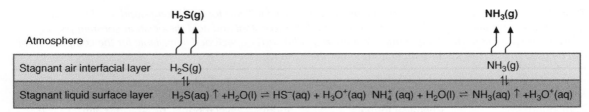

Figure 8.1   *Illustration of emission of the acid species H_2S and the basic species NH_3. At low pH H_2S is emitted and at high pH NH_3 is emitted. (© University of Southern Denmark.)*

The layer of the manure solution containing A(aq) interfaced to air can be at the surface of stored slurry, in urine patches on the floor and on slats in animal houses or in outdoor animal-holding areas (hard standings and feedlots). The source may also be the liquid phase in solid manure containing the component A(aq) (i.e. solid manure stored in heaps, deep litter covering concrete floors or litter on soil surfaces in beef cattle feedlots). For manure applied in the field, the source is the manure–soil mixture. Following release of a compound from the surface, transport from the surface air layers in immediate contact with the liquid to the free atmosphere is affected by turbulent and molecular diffusion. The same physical principles and equations can be used for predicting transport in the air when assessing gaseous emissions from manure in animal houses, manure stores or manure applied in the field.

The component content of a gas in the surface liquid layer is depleted by emission of the gas and components from layers below have to be transported to the surface layer for the emission process to continue. In the surface layers this transport is primarily by diffusion and in the layers below there may be transport affected by both diffusion and convection. Diffusion is a molecular transport driven by gradients of concentration, whereas convection is transport of the component due to movement of parcels of air or liquid containing the component. If the transport is entirely diffusion then the emission rate is low, whereas convection is often a faster transport process than diffusion.

The concentration of gaseous components A(g) in the air just above the liquid surface is assessed using the equilibrium equation for the relationship between gaseous A(g) and liquid A(aq) in the surface liquid layer (Henry's law).

Some of the components emitted are present in equilibrium with charged species depending on the proton concentration (H^+). The charged ions are not released from the liquid. In this case emission is to a great extent affected by the pH of the manure and the equilibrium reactions in the manure surface have to be accounted for (Figure 8.1). In the example given, it is the base NH_3(aq) and the acid H_2S(aq) that are the source of emissions. Consequently, the total concentration of the components and the pH in the surface layers of the manure have to be considered when assessing the potential for gaseous emission of acids or bases. In the two cases mentioned above the literature often uses the nomenclature total ammoniacal nitrogen (TAN = NH_3(aq) + NH_4^+(aq)) and total sulfide (TS = H_2S(aq) + HS^-(aq) + S^{2-}(aq)).

8.3.1   Liquid and Air Diffusion Processes

The rate of transport through the stagnant gaseous layer just above and below the air–liquid interface is considered to be governed by the film thickness and the diffusion coefficient. The rate of diffusive transport of a molecule or a species (A) through a stagnant phase of air or a liquid phase is described by Fick's law of diffusion, which gives the flux (F, mol cm^{-2} s^{-1}) through a layer of thickness Δz (cm):

$$F(A) = -D \cdot \frac{d[A]}{dz} \tag{8.1}$$

$$F(A) = -\frac{D}{\Delta z} \cdot ([A]_2 - [A]_1) \tag{8.2}$$

where [A] is in mol cm^{-3} and D is in cm^2 s^{-1}, and Δz (cm) is the thickness of the phase. The negative sign is because of the convention regarding the orientation of the z-axis (i.e. the component moves from the boundary with a high concentration ([A]$_1$) to that with a low concentration ([A]$_2$) and thus ([A]$_1$) − ([A]$_2$) is negative). By multiplying by −1, the flux of components becomes positive, which is intuitively considered to be correct. Equations (8.1) and (8.2) show that the rate of transport increases with increasing diffusion coefficient and decreases with increasing thickness of the layer through which the component A has to move.

The diffusion coefficient is compound-specific and its magnitude depends mainly on the size of the molecule. In addition, the magnitude of the diffusion coefficient depends on the medium through which the compound diffuses. The coefficient has the units area per unit time (e.g. cm^2 s^{-1}). If extended to three dimensions, the diffusion coefficient is related to the mean free path, λ, and the mean velocity, u, by the following simple relationship:

$$D = \frac{1}{3}\lambda u \tag{8.3}$$

Since diffusion coefficients may not be readily available for all compounds, it is useful to examine methods for estimating diffusion coefficients from tabulated properties of the medium and the diffusing compound (Text Box – Advanced 8.1).

Text Box – Advanced 8.1   Methods for calculation of diffusion coefficient in air and in liquid

Diffusion in air: A convenient empirical relationship can be used to estimate diffusion coefficients in air, depending on the properties of the available compounds (Schwarzenbach *et al.*, 2003):

$$D_i = 10^{-3} \cdot \frac{T^{1.75} \cdot \left(\frac{1}{M_{air}} + \frac{1}{M_i}\right)^{1/2}}{p \cdot \left(\bar{V}_{air}^{1/3} + \bar{V}_i^{1/3}\right)^2} \, (\text{cm}^2 \text{s}^{-1}) \tag{8.4}$$

where M_{air} is the average molecular weight of air (g mol^{-1}), p is pressure (atm), and $\bar{V}_{air}$ and $\bar{V}_i$ are, respectively, the average molar volume of air and the molar volume of the diffusing compound (cm^3 mol^{-1}). The empirical relationship is not dimensionally correct and therefore only gives the correct answer if the specified units are used. The molar volume of a compound can simply be estimated from the density of the pure compound in liquid state and the molecular weight of the compound as: $\bar{V}_i \approx MW_i \cdot \rho_i^{-1}$. This gives a reasonable estimate and is recommended as a first approximation. The approach is not completely accurate since the density is a bulk property and not a molecular property, and it is the effective diffusive volume of the molecule that is of interest. To get a more accurate estimate, an empirical atomic contribution model (see, e.g. Schwarzenbach *et al.*, 2003), which is based on correlations between known diffusion coefficients and molecular structure, can be used. In this way, the volume contributions from the single atoms in the molecule are used to estimate the total molecular (diffusion) volume.

Diffusion in a liquid: The diffusion coefficient of a compound i in water and aqueous media can be assessed by the following empirical equation:

$$D_i = \frac{13.26 \times 10^{-5}}{\eta^{1.14} \cdot \bar{V}_i^{0.589}} \tag{8.5}$$

where η is the solution viscosity (in centipoises (cP); 10^{-2} g cm^{-1} s^{-1}) and $\bar{V}_i$ is the molar volume of the compound (cm^3 mol^{-1}). The temperature dependency is included through the temperature dependency of the viscosity, since molar volume can be considered to be constant with temperature. Thus, the temperature dependency of the diffusion coefficient is included by using the specific viscosity of water at the relevant temperature.

Since diffusivity in both air and water depends on molecular mass and properties, the diffusion coefficient can be estimated from values of related compounds. In both media, a reasonable estimate can be obtained from the molecular weights of the compounds by using the following expression:

$$\frac{D_i}{D_{\mathrm{ref}}} \approx \left(\frac{M_i}{M_{\mathrm{ref}}} \right)^{-1/2} \tag{8.6}$$

where D_{ref} and M_{ref} are the diffusion coefficient in water and the molecular weight of a reference compound, respectively. As can be seen from this relationship, a difference in size/mass between two compounds gives a much smaller difference in diffusivity. For example, H_2S is roughly twice as large as NH_3 but in practice the diffusion coefficient of NH_3 is only a factor of 1.4 higher than that of H_2S.

8.3.2   Air–Water Equilibrium

The release of the dissolved H_2S and NH_3 from the surface of the source to the air phase immediately above the liquid surface (Figure 8.1) is driven by the difference in the atmospheric concentration of the gases, A(g), and the concentration of the components in the surface liquid layer, A(aq). The partitioning of the component between atmosphere and liquid is described by Henry's law. A high Henry's law constant (K_H or H) indicates that the gas is soluble in the liquid and a low constant that the gas is insoluble and has a large fraction in the gaseous phase.

$$A(g) \rightleftharpoons A(aq) \quad K_H = \frac{[A(aq)]}{p_A} \quad \text{or} \quad H = \frac{[A(aq)]}{(A(g))} = \frac{[A(aq)]}{p_A} \cdot R \cdot T = K_H \cdot R \cdot T \tag{8.7}$$

where P_a is the partial pressure of the gas in the air layer, R is the gas constant (8.3144621 J K^{-1} mol^{-1}; 0.08205746 l atm K^{-1} mol^{-1}) and T is the temperature (K). The Henry's law constant (K_H) is often given in M atm^{-1}, which can be transferred to M Pa^{-1} as: (K_H) [M atm^{-1}] $= 101324 \cdot K_H$ [M Pa^{-1}], with H given in mol(aq) l^{-1}/ mol(g) l^{-1}). The Henry's law constant increases exponentially with temperature (Examples 8.1 and 8.2 and Table 8.3). Since the constant is an air–water partitioning equilibrium constant, the temperature dependency can be described by the van't Hoff equation, as with other equilibrium constants (see Text Box – Advanced 8.2 below). For Henry's law constants too, empirical functions have been derived that are not limited to temperature ranges in which the enthalpy of phase exchange can be considered constant, which is the prerequisite for using the van't Hoff function.

Example 8.1   Calculation of fluxes of H_2S and NH_3

Question: Calculate characteristic fluxes of H_2S and NH_3 across the stagnant air boundary layer above a liquid source. The equilibrium concentrations immediately above the liquid surface are: $[H_2S] = 1$ mg m^{-3} and $[NH_3] = 20$ mg m^{-3}. The thickness of the air boundary layer can be assumed to be 3 mm.

Answer: From the diffusion coefficient and Equation (8.2), it is possible to estimate a characteristic flux or transfer rate through a stagnant air layer above a liquid phase for a situation where the air concentration above the stagnant layer is very small compared with the concentration immediately above the surface in equilibrium with the liquid concentration $[A]_{eq}$. In this case, Equation (8.2) can be simplified to:

$$F \approx -\frac{D}{z}\left(-[A]_{eq}\right) \qquad (8.8)$$

The diffusion coefficient of NH_3 is 0.20 cm^2 s^{-1} (Table 8.2) and the diffusion coefficient for H_2S can be estimated from Equation (8.6). The transfer rate ($D/\Delta z$) is thus 0.7 and 0.5 cm s^{-1} for NH_3 and H_2S, respectively. The flux thus becomes 0.14 (NH_3) and 0.005 mg m^{-2} s^{-1} (H_2S).

Table 8.2   *Diffusion coefficients (298 K, 1 atm) of volatile components calculated using Equations (8.4) and (8.5) (Schwarzenbach et al., 2003), including parameters used in the calculations.*

Component	Molecular weight (g mol^{-1})	Density (kg l^{-1})	$V_i{}^a$ (cm^3 mol^{-1})	Diffusion coefficient at 25 °C (cm^2 s^{-1})
NH_3(g)	17.03	0.662	25.725	0.20
CO_2(g)	44.01	0.77	57.27	0.12
Acetic acid (g)	60.05	1.049	57.24	0.11
CO_2(aq)	44.01	0.77	57.27	1.4×10^{-5}
NH_3(aq)	17.03	0.662	25.725	2.2×10^{-5}
H_2S(aq)	34.08	0.972	35.08	1.9×10^{-5}

^{a} Molar volume; estimated from the molecular weight and the density (see text).

Additional parameters used for calculating diffusion coefficients were: $\eta(H_2O) = 0.89$ cP; $V_{air} = 20.1$ cm^3 mol^{-1}; $M_{air} = 28.97$ g mol^{-1}.

Note: This example does not provide the overall flux, but only the flux across the stagnant air layer.

Example 8.2   Solubility of H_2S and NH_3

Based on the empirical equations shown in Table 8.3, the water-to-air distributions of H_2S and NH_3 can be calculated at different temperatures. This is shown in Figure 8.2, where the dimensionless Henry's law constant (H) is used to represent aqueous concentration (mol l$_{aq}{}^{-1}$) relative to gas phase concentration (mol l$_{air}{}^{-1}$). The graph clearly demonstrates the declining aqueous equilibrium concentration as temperature increases.

Table 8.3 *Henry's constant (K_H) for solubility of gases in liquids (Beutier and Renon, 1978).*

Reaction or process	K_H (mol l^{-1} atm^{-1})	Constant at 25 °C
$NH_3(g) \rightleftharpoons NH_3(aq)$	$\ln(K_{H,NH_3}) = -160.559 + 8621.06/T + 25.6767 \cdot \ln(T) - 0.035388 \cdot T$	60.381
$H_2S(g) \rightleftharpoons H_2S(aq)$	$\ln(K_{H,H_2S}) = +403.658 - 7056.07/T - 74.6926 \cdot \ln(T) + 0.14529 \cdot T$	0.105
$CO_2(g) \rightleftharpoons CO_2(aq)$	$\ln(K_{H,CO_2}) = -1082.37 + 34417.2/T + 182.28 \cdot \ln(T) - 0.25159 \cdot T$	0.034

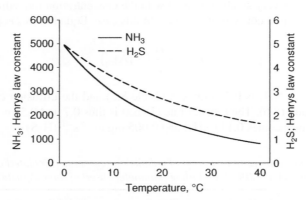

Figure 8.2 *Equilibrium distribution of H_2S and NH_3 between the gaseous and aqueous phases as a function of temperature. The curves are based on the equations presented in Table 8.3. The phase distribution is represented as the dimensionless Henry's law constant (H) with units of mol l_{aq}^{-1}/mol l_{air}^{-1}. (© University of Southern Denmark.)*

Two definitions of Henry's law constants are given in the literature. Here, K_H is defined as liquid concentration divided by pressure or gas phase concentration (Equation 8.7) in units of mol l^{-1} atm^{-1}. However, the opposite definition is frequently used (pressure divided by liquid concentration in units of, for example, bar M^{-1} or Pa M^{-1}). Therefore, one must always take note of the units when using Henry's law constants.

8.3.3 Acid–Base Equilibrium

In manure, the volatile species is the uncharged A(aq) of a component that can volatilise because a species with no charge has a much lower solubility than the ion (i.e. a lower Henry's law constant (K_H)).

In most cases where acids or bases volatilise, the concentration of the uncharged species has to be assessed from knowing the total concentration of the component. Thus, in the case of NH_3 volatilisation, the $NH_3(aq)$ concentration is calculated from knowledge about the concentration of TAN and pH (Equations 8.9 and 8.10). The equilibrium concentration of $NH_3(g)$ in the air phase immediately above the water phase can then be calculated using the Henry's law equation (Equation 8.11).

$$[TAN] = [NH_4^+] + [NH_3] \tag{8.9}$$

$$[NH_3(aq)] = \frac{[TAN(aq)]}{1 + [H^+(aq)]/K_N} \tag{8.10}$$

$$[NH_3(g)] = \frac{1}{H_{NH_3}} \frac{[TAN]}{1 + [H^+]/K_N} \qquad (8.11)$$

where H_{NH_3} is the Henry's law constant (mol(aq) l^{-1}/mol(g) l^{-1}) and K_N is the equilibrium constant for NH_4^+/NH_3 and concentrations are given in mol l^{-1}. For acids and bases with two or more protons, the calculation follows the same principles. So the concentration of uncharged H_2S, which is a diprotic acid, can be assessed using the following equations:

$$[TS] = [H_2S] + [HS^-] + [S^{2-}] \qquad (8.12)$$

$$[H_2S(g)] = \frac{1}{H_{H2S}} \cdot \frac{[TS]}{1 + K_{H2S}^{-1} \cdot K_{HS}^{-1} \cdot [H^+]^{-2} + K_{H2S}^{-1} \cdot [H^+]^{-1}} \qquad (8.13)$$

where H_{H_2S} is the Henry's law constant (mol(aq)l^{-1}/mol(g) l^{-1}), and K_{H_2S} and K_{HS} are the equilibrium constants for the acid–base reactions H_2S/HS^- and HS^-/S^{2-}, respectively. Concentrations are given in mol m^{-3}. The equilibrium constants (K) are exponentially related to temperature (Table 8.4). Taking this into account, the models using equilibrium equations include the effect of temperature in the assessment of emissions.

The potential for emission of acid components is high at low pH, while the potential for emission of basic components is high at high pH, as seen for TS and TAN in Figure 8.3. Emission of the uncharged species (NH_3 or H_2S) that is in equilibrium with the acid or base in the liquid surface layer affects pH. Emission of NH_3 increases H_3O^+ (i.e. decreases pH), whereas emission of the acid H_2S decreases H_3O^+ (i.e. increases pH).

The depletion of acids and bases in the surface layer initiates upward transport of pH buffer components from the layers below. The most important components involved in this process are TAN, TIC and volatile short-chain organic acids (with up to five carbon atoms). Part of the transport is diffusive and part is convective, and the relative importance of the two transport pathways is affected by local conditions and environments, which in each case must be related to a large number of local features that are not always well known. Omitting emission of the three volatile buffer components may introduce significant errors (e.g. in a model calculation of H_2S emissions from a slurry lagoon, the error caused by excluding TIC buffering in the model was a factor of 10; Blanes Vidal *et al.*, 2009).

Table 8.4 *Equilibrium constants of volatile components dissolved in slurry and manure (temperature T is in Kelvin (273.15 + °C): the algorithms are taken from Beutier and Renon (1978).*

Reaction	Constant	pK at 25 °C
$NH_3(aq) + H_2O(l) \rightleftharpoons NH_4^+(aq) + OH^-(aq)$	$\ln(K_{NH_3}) = 191.97 - 8451.61/T - 31.4335 \cdot \ln(T)$ $+ 0.0152123 \cdot T$	4.75
$H_2O(l) \rightleftharpoons H^+(aq) + OH^-(aq)$	$\ln(K_{H_2O}) = 14.01708 - 10294.83/T - 0.039282 \cdot T$	13.99
$NH_4^+(aq) \rightleftharpoons NH_3(aq) + H^+(aq)$	$\ln(K_{NH_4}) = \ln(K_{H_2O}) - \ln(K_{NH_3})$	9.24
$CO_2(aq) + 2H_2O(l) \rightleftharpoons HCO_3^-(aq) + H_3O^+(aq)$	$\ln(K_{CO_2}) = 2767.92 - 80063.5/T - 478.653 \cdot \ln(T)$ $+ 0.714984 \cdot T$	6.35
$HCO_3^-(aq) + H_2O(l) \rightleftharpoons CO_3^{2-}(aq) + H^+(aq)$	$\ln(K_{HCO_3}) = 12.405 - 6286.89/T - 0.050628 \cdot T$	10.33
$H_2S(aq) \rightleftharpoons HS^-(aq) + H^+(aq)$	$\log(K_{H_2S}) = 218.5989 - 12995.40/T - 33.5471 \cdot \ln(T)$	7.01

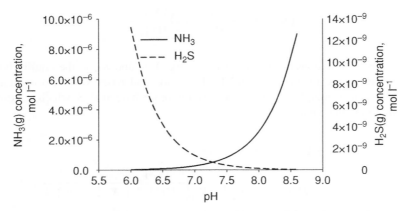

Figure 8.3   *Concentration of $H_2S(g)$ and $NH_3(g)$ as affected by pH of the solution. The concentration of TS is 0.003 M and of TAN is 0.15 M. (© University of Southern Denmark.)*

Microbial transformation of acids and bases also affects the pH (Chapter 4). Nitrification contributes to a reduction in the pH of the slurry/soil mixture through the following relationship:

$$NH_4^+(aq) + 2O_2(aq) \rightarrow NO_3^-(aq) + H_2O(l) + 2H^+(aq) \tag{8.14}$$

Denitrification, on the other hand, may increase pH. Further transformation of organic acids will affect pH. The organic acids can be produced and consumed in anaerobic processes and consumed in aerobic processes (Chapter 5).

Addition of acids and bases affects pH, as does addition of salts where the cation precipitates with CO_3^{2-}. The addition of $CaCl_2$ has been tested as an additive to reduce the pH of manure, because this causes precipitation of $CaCO_3$ (Husted *et al.*, 1991):

$$Ca^{2+}(aq) + HCO_3^-(aq) \rightleftharpoons H^+(aq) + CaCO_3(s) \tag{8.15}$$

In this reaction the weak acid HCO_3^- is removed through crystallisation with Ca_2^{2+}. As a consequence, $H^+(aq)$ increases (pH declines) and NH_3 volatilisation is reduced significantly (see Section 4.3.4).

See Text Box – Advanced 8.2 for temperature dependency of equilibrium constants.

Text Box – Advanced 8.2 Temperature dependency of equilibrium constants: theoretical and empirical relationships

For any equilibrium, the relationship between temperature and the equilibrium constant is described by the van't Hoff Equation. Definite integration of the van' Hoff equation yields the following equations with the assumption that enthalpy is constant (Stumm and Morgan, 1996):

$$\ln\frac{K_2}{K_1} = \frac{-\Delta H}{R} \cdot \left(\frac{1}{T_2} - \frac{1}{T_1}\right) \tag{8.16}$$

$$K_2 = K_1 \cdot \exp\left(\frac{-\Delta H}{R} \cdot \left(\frac{1}{T_2} - \frac{1}{T_1}\right)\right) \tag{8.17}$$

where K_1 is the equilibrium constant at absolute temperature T_1 and K_2 is the equilibrium constant at absolute temperature T_2. ΔH^{θ} is the standard enthalpy change at T_1 (J mol^{-1}) and R is the gas constant (1 bar mol^{-1} K^{-1}). Using the equations, the equilibrium constant (K_2) at a given temperature (T_2) can be determined from a known value (K_1) at another temperature (T_1), if the enthalpy of reaction is known. The temperature must be within a range where the enthalpy can be considered constant (± 20 °C).

Indefinite integration of the van't Hoff equation (Equation 8.16) with the assumption that ΔH is constant yields the following expression:

$$\ln(K) = -\frac{-\Delta H}{R} \cdot \frac{1}{T} + B \tag{8.18}$$

where B is an integration constant that theoretically contains the entropy change of the equilibrium reaction. Experimentally, the temperature dependency of equilibrium constants may be determined from the slope of a plot of $\ln K$ versus $1/T$ with the slope being equal to $\Delta H/R$. This relationship using the parameters given by Hales and Drewes (1979) has been used by many authors to assess NH$_3$ emissions. In order to overcome the limitations of assuming the enthalpy to be constant, more general empirical relationships between $\ln K$ and T can be used (Table 8.4).

8.4   Two-Layer Transport and Release Model

Release of gas from a liquid to the atmosphere is a key process in the chain of processes involved in gas volatilisation and the two-layer diffusion model is an often-used concept to assess this (Liss and Slater, 1974). The model presents the importance of diffusion in a stagnant surface liquid layer and in a laminar boundary air layer at the liquid surface and the transfer of the gas between the two phases. One should bear in mind that the model does not include the chemical equilibria of the species involved in the whole chain of processes.

The model depicts diffusion of the gas from the bulk of slurry through a stagnant liquid layer (liquid film) at the interface boundary, the gas release from the liquid phase to the adjacent air layer and then transport of the gas by diffusion through a stagnant air layer at the interface to the liquid phase (air film or air boundary layer). It is assumed that the liquid phase below the liquid film boundary layer is stirred and homogeneous, and that the air above the air film boundary layer is also mixed and homogeneous (Figure 8.4).

The calculations assume that the barrier (resistances) to emission is diffusion through the stagnant liquid and air film layers and the release from the liquid to the air. This model can provide information on whether the barrier to transport is in the air phase or the liquid phase.

When assessing emission, it is the concentration of the component in the surface layer that is used in the computations. In the following, "w" is water phase, "a" is air phase, "w" is the bulk liquid phase, "w/a" is the liquid in immediate contact with the air and "a/w" is the air in immediate contact with water (Figure 8.4). From the layers below the surface layer the acids and buffer components are transported to the surface layer through a number of layers from below. At a given distance from the surface the concentration of molecules and ions is not significantly affected by the surface processes and a bulk composition of the slurry is assumed.

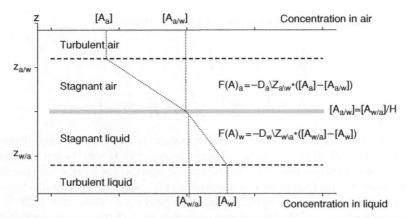

Figure 8.4 *The two-layer theory of gas emission (diagram adapted from Liss and Slater, 1974). Note that the concentration given on the x-axis at the top is the gas concentration [A_a] in the ambient air and the gas concentration [A_{a/w}] in equilibrium with the dissolved component. The bottom x-axis presents the dissolved component in the turbulent layer [A_a] and the dissolved component in the layer where the liquid component is in equilibrium with the air component [A_{a/w}]. If the units are the same on these two axes, then Henrys law constant is 1. (© University of Southern Denmark.)*

The model of Liss and Slater (1974) calculates the following three processes:

- Transfer of the A(aq) through the stagnant surface layer of the liquid as affected by diffusion:

$$F(A)_w = -\frac{D_w}{Z_{w/a}} \cdot ([A_{w/a}] - [A_w]) \tag{8.19}$$

- Release from the liquid to the atmosphere as related to the solubility of the species:

$$[A_{a/w}] \rightleftharpoons [A_{w/a}], \ H = \frac{[A_{w/a}]}{[A_{a/w}]} \tag{8.20}$$

- Diffusive transport through the stagnant air layer immediately above the liquid surface:

$$F(A)_a = -\frac{D_a}{Z_{a/w}} \cdot ([A_a] - [A_{a/w}]) \tag{8.21}$$

Combining the equations given above by setting $F(A)_a = F(A)_w$ and substituting $[A_{a/w}]$ with $[A_{w/a}]/H$ gives the following expression (Text Box – Advanced 8.3):

$$F(A) = \frac{1}{\frac{Z_{w/a}}{D_w} + H \cdot \frac{Z_{a/w}}{D_a}} \cdot ([A_w] - [A_a] \cdot H) \tag{8.22}$$

where the flux F is a rate in mol cm^{-1} s^{-1}, A is the concentration of the component in mol cm^{-3}, z is the distance in cm and D is the diffusion coefficient in cm^2 s^{-1}. Often resistance R is used instead of D_w/z_w and $H \cdot D_a/z_a$:

$$R_w = 1/k_w = \frac{z_{w,a}}{D_w} \quad \text{and} \quad R_a = 1/k_a = \frac{z_{w,a}}{D_a} \tag{8.23}$$

In Equation (8.22), the two parameters k_w and k_a are transfer velocities (cm s^{-1}), which are inversely related to the transfer resistances. Analogously to electrical circuits, resistances are additive, meaning that if it is assumed that overall mass transfer is only determined by transfer through the water and air film, the overall resistance is simply $R_{total} = R_w + R_a$. Introducing Equation (8.22) into Equation (8.21) gives the following equation, which is often used when expressing gas emissions or gas deposition between the atmosphere and water, soil or plants:

$$F(A) = \frac{1}{\frac{1}{k_w} + \frac{H}{k_a}} \cdot ([A_w] - [A_a] \cdot H) \tag{8.24}$$

Equation (8.23) can be used to qualitatively assess which of the barriers to transfer or transport is rate-controlling for emissions.

Text Box – Advanced 8.3   Derivation of the diffusion film model in Equation (8.24)

The diffusion film model is developed from Equations (8.25)–(8.27) and comprises diffusion and solubility:

$$F(A) = -\frac{D_w}{Z_{w,a}} \cdot ([A_{w/a}] - [A_w]) = -\frac{D_a}{Z_{a,w}} \cdot ([A_a] - [A_{a/w}]) \tag{8.25}$$

$$[A_{a/w}] = \frac{[A_{w/a}]}{H} \tag{8.26}$$

$$F(A) = -\frac{D_a}{Z_{a,w}} \cdot ([A_a] - [A_{a/w}]) = -\frac{D_a}{Z_{a,w}} \cdot \left([A_a] - \frac{[A_{w/a}]}{H}\right) \Rightarrow \tag{8.27}$$

$$F(A) \cdot H \cdot \frac{Z_{a,w}}{D_a} = -([A_a] \cdot H - [A_{w/a}]) \Rightarrow \tag{8.28}$$

$$[A_{w/a}] = [A_a] \cdot H + F(A) \cdot H \cdot \frac{Z_{a,w}}{D_a} \tag{8.29}$$

Then the expression for $A_{w/a}(aq)$ is included in Equation (8.26):

$$F(A) = -\frac{D_w}{Z_{w,a}} \cdot \left([A_a] \cdot H + F(A) \cdot H \cdot \frac{Z_{a,w}}{D_a} - [A_w]\right) \Leftrightarrow \tag{8.30}$$

which can be reduced to the following expression:

$$F(A) + F(A) \cdot \frac{H \cdot Z_{a,w} \cdot D_w}{D_a \cdot Z_{w,a}} = -\frac{D_w}{Z_{w,a}} \cdot ([A_a] \cdot H - [A_w]) \Leftrightarrow \tag{8.31}$$

By substituting with the resistance parameters (= inverse transfer velocities) (Equation 8.23), the following is obtained:

$$F(A) + F(A) \cdot \frac{H \cdot k_w}{k_a} = -k_w \cdot ([A_a] \cdot H - [A_w]) \Rightarrow \tag{8.32}$$

$$F(A) \cdot \left(\frac{k_a + H \cdot k_w}{k_a}\right) = k_w \cdot ([A_w] - [A_a] \cdot H) \Rightarrow \tag{8.33}$$

$$F(A) = \frac{k_w}{\left(\frac{k_a + H \cdot k_w}{k_a}\right)} \cdot ([A_w] - [A_a] \cdot H) = \frac{k_w \cdot k_a}{k_a + H \cdot k_w} \cdot ([A_w] - [A_a] \cdot H) \Rightarrow \quad (8.34)$$

$$F(A) = \frac{1}{\frac{1}{k_w} + H \cdot \frac{1}{k_a}} \cdot ([A_w] - [A_a] \cdot H) \quad (8.35)$$

8.4.1   Gas or Liquid Film Controlling Transfer

The two-phase model indicates that the Henry's law constant and the thicknesses of the two stagnant phases affect the emission. Thus, if the solubility of the gas is high (H is large), then the concentration of $[A_a]$ must be very low, as otherwise no gas is emitted. In this situation the air exchange over the liquid surface must be high to enhance gas emission. If the gas is not soluble (H is low), then it is important that much A(aq) is transferred to the surface as otherwise emission will be low (i.e. transfer of A in the liquid must be high).

Stirring the liquid reduces the thickness of the stagnant liquid layer and thereby decreases the liquid-side resistance. Likewise, increasing air velocity above a manure surface decreases the thickness of the stagnant air boundary layer and reduces the air-side resistance. Thus, ventilation strategy in a livestock building or wind speed across an outside manure surface influences the emissions of the compounds with an emission rate mainly affected by air resistance (i.e. those that are very soluble in the liquid and therefore have a high Henry's law constant). For odorants with highly variable H values (Table 8.1), it can be expected that changes in liquid-side or air-side turbulence may affect emissions quite differently.

The resistances and transfer coefficients can be used to identify which of the two stagnant layers mainly affects gas emission. To do this, a critical Henry's law constant can be defined at which resistances are equal. A simple system consisting of stagnant aqueous and air phases is considered. The overall resistance to mass transfer is a combination of liquid resistance and air resistance, which for this system can be expressed with the equation (Schwarzenbach *et al.*, 2003):

$$\frac{1}{K_{total}} = \frac{1}{k_w} + H \cdot \frac{1}{k_a} \quad (8.36)$$

where K_{total} is the overall transfer coefficient (or transfer velocity; $K_{total} = 1/(R_w + R_g)$ in Equation 8.22), k_w and k_a are the liquid and air transfer velocities ($k_w = D_w/z_w$ and $k_a = D_g/z_g$, respectively), and H is the dimensionless Henry's law constant ($H = K_H \cdot R \cdot T$). At the critical H (H^{crit}), $1/k_w$ is by definition equal to H/k_a, which gives:

$$H^{crit} = \frac{k_a}{k_w} = K_H^{crit} \cdot R \cdot T \quad (8.37)$$

To identify a characteristic H^{crit}, the transfer velocity in air (k_a) therefore needs to be estimated relative to the transfer velocity in water (k_w), which depends on film thickness (z) and diffusion coefficient. At a given temperature, the ratio D_a/D_w is relatively independent of the compound. The air film thickness is generally much larger than the water film thickness. For example, an air film thickness of 0.2 cm and a water film thickness of 0.02 cm together with $D_a/D_w = 9000$ (values for NH_3 in Table 8.2) gives $H^{crit} = 900$. In general, air transfer velocities are of the order of 1000 times higher than water transfer velocities (Schwarzenbach *et al.*, 2003), which means that H^{crit} is estimated to be around 1000 and K_H^{crit} is then around 40 (at 298 K and

1 atm). It should be noted that transfer velocities depend on factors such as temperature and air velocity. Thus, K_H^{crit} is not a fixed value, but varies to some extent depending on the conditions of the system. For compounds with K_H close to this value, both liquid-side and air-side resistance are estimated to control emissions. For compounds with significantly lower K_H, liquid-side resistance dominates, whereas for compounds with much higher K_H, air-side resistance dominates.

If we look at typical odorants (Table 8.1), it appears that K_H values are either much higher or much lower than K_H^{crit} for most of the compounds. Sulfur-containing compounds have low K_H and emissions of these from aqueous surfaces are therefore expected to be controlled mainly by liquid-side resistance. Organic acids, NH_3 and phenolic or indolic compounds have high K_H and emissions are therefore expected to be controlled by air-side resistance (Example 8.3).

Example 8.3 Air–water partitioning of dissociating compounds

Two compounds, NH_3 and trimethyl amine, have K_H within an order of magnitude from K_H^{crit}. Both are in equilibrium with corresponding weak acids and at typical manure pH levels the acid form dominates. Since the de-protonation of the weak acids occurs rapidly relative to liquid diffusion, the sum of acid and base must be taken into account. For such compounds, an effective Henry's law constant (K_H^{eff}) should be used (using NH_3 as an example). Thus, the Henry's law constant may be revised by including an estimate of the concentration of $NH_3(aq)$ in equilibrium with NH_4^+ using Equation (8.10) by defining and introducing α:

$$\alpha = \frac{1}{1 + \frac{[H^+]}{K_{NH_3}}} \tag{8.38}$$

$$H^{eff} = \frac{H}{\alpha} = \frac{[TAN]}{[NH_3\,(g)]} \Leftrightarrow \tag{8.39}$$

$$[NH_3\,(g)] = \frac{1}{H^{eff}} \cdot [TAN] = \frac{R \cdot T}{K_H^{eff}} \cdot [TAN] \tag{8.40}$$

For example, K_H for TAN in pure water at 298 K and pH 7 is 11 000 M atm^{-1} compared with 61 M atm^{-1} for NH_3, and a similar relationship is seen for trimethyl amine when the sum of neutral trimethyl amine and charged trimethyl ammonium is considered. As a consequence, both components belong to the group of compounds for which air-side resistance dominates. Analogously to the nitrogen compounds, dissociation of H_2S to HS^- and H^+ needs to be considered and a K_H^{eff} based on $[H_2S]_{aq}$ and $[HS^-]_{aq}$ ($[H_2S]_{aq} + [HS^-]_{aq} = TS$) needs to be used. In this case, the difference is less dramatic because the K_H of H_2S is very low (0.1 M atm^{-1}) and because the pK_a of H_2S is closer to typical manure pH levels. Even at a high pH of 8.5, at which significant dissociation of H_2S occurs, K_H^{eff} is about one order of magnitude lower than K_H^{crit}. Similar considerations are relevant for organic acids, but this does not alter the fact that these belong to the air-side controlled group of compounds.

To summarize, it can be concluded that emissions of sulfur compounds (H_2S, methanethiol and dimethyl sulfide) are expected to be controlled by liquid-side resistance, whereas emissions of other odorants including

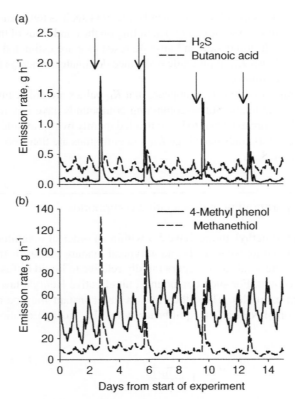

Figure 8.5 *Emissions of odour components from slurry channels emptied every 2 days as indicated with arrows. Stirring of slurry on emptying the channels triggers simultaneous emissions of H_2S and methanethiol. (a) Emissions of H_2S and butanoic acid. (b) Emissions of 4-methyl phenol and methanethiol. (Modified with permission from Liu et al. (2011). © Elsevier.)*

NH_3 are expected to be controlled by air-side resistance. As a consequence, it is predicted that stirring or pumping of manure increases emissions of sulfur compounds, whereas increased air velocity (e.g. by increased ventilation in a livestock building) increases emissions of other odorants.

The effect of liquid turbulence on emissions of sulfur compounds were recently observed experimentally by Liu *et al.* (2011). In a study of emissions from a pig facility it was observed that emissions of H_2S, methanethiol and dimethyl sulfide were strongly increased during events of slurry pit discharging. This is seen in Figure 8.5 in which times of slurry discharging are indicated by arrows. It is seen that emissions of H_2S (upper graph) and methanethiol (lower graph) are strongly elevated at these times. Emissions of, for example, carboxylic acids and phenolic compounds (high solubility) followed a clear diurnal cycle with peak emissions occurring during daytime when ventilation rate is highest, as seen in Figure 8.5 for butanoic acid and 4-methyl phenol. Emissions of these compounds were not affected by discharging.

It has also been experimentally observed that emissions of NH_3 (Lyngbye *et al.*, 2006) and carboxylic acids and phenols (Feilberg *et al.*, 2010) from pig manure increase with increasing ventilation rate, in accordance with the predictions based on K_H^{eff}. This is also reflected in the more pronounced diurnal variation in butanoic acid and 4-methyl phenol emissions compared with H_2S and methanethiol emissions, as seen in Figure 8.5.

8.5 Assessment of Gas Release and Emission

The diffusion two-layer model presented above provides a method for assessing whether the rate of release of the gas is controlled by processes in the stagnant air or stagnant liquid layer at the manure–air interface. This model cannot be used to calculate emissions of the gases.

Simple linear algorithms or complex models can be used to assess gaseous emissions. Often the variables needed to use complex models are not available and a gas release coefficient relating the emission to numbers of animals or to the concentration of the component in the manure may have to be used (Buijsmans *et al.*, 1987; Hutchings *et al.*, 2001). If data on the variables and parameters are available, then emissions can be assessed using increasingly complex chemical and physical models, including the transport of the component in the manure, in air, pH effects and release of the component from the manure (Génermont and Cellier, 1997).

The choice of the best model also depends on the scale of the source and knowledge about environment, chemistry and physics. In inventories of gas emissions from a region or a country, simple emission coefficients are used. More advanced and complex calculations are needed for the development of novel technologies and management practices.

8.5.1 Calculations Using Emission Coefficients

In most national or regional inventories, gaseous emissions are calculated using transfer coefficients and knowledge of number of livestock or amount of TAN or N excreted per year. The most simple calculation assesses the emissions (F_{NH_3}, kg NH_3-N year^{-1}) using the number of animals:

$$F_{NH_3} = K_L \cdot X_L \tag{8.41}$$

where the emissions factor K_L is in kg NH_3-N per livestock entity (L) and X_L is the number of livestock entities of the category L (dairy cows, calves, fattening pigs, etc.). In these inventories, where information about livestock production and manure management is sparse, the emissions calculations are carried out using statistics about numbers of animals (Buijsmans *et al.*, 1987). These calculations may be relatively accurate if emissions factors have been calculated using country-specific data about feeding practices of livestock, livestock production systems and management of manure.

In many inventories the amount of total-N (kg N year^{-1}) or TAN (kg NH_3-N year^{-1}) excreted by the livestock is known, and the emissions from each livestock category are calculated as:

$$F_{NH_3} = K_L \cdot N_L \tag{8.42}$$

where K_L (percent of TAN or of total-N) is related to the category of animal (including their weight and age), composition of the animal feed, type of animal housing and manure storage, method of manure application, meteorological conditions, soil conditions, and agricultural practices (Sommer and Hutchings, 2001). Standard emissions factors for the emissions vary between countries, so annual emissions per animal raised or per ton of manure also differ. As a consequence, different models for calculating the national inventories have been developed (Hutchings *et al.*, 2001; Webb and Misselbrook, 2004). These models can give more precise emissions estimates than estimates from the simple model (Equation 8.41). It must be borne in mind that manure may be transported between the compartments on a farm (i.e. animal houses, store and applied in field). Thus, the emissions are calculated for each compartment of the manure management chain and before calculating emissions from a source downstream, the losses that have taken place already must be deducted.

Process-based models include the variable transfer coefficient, which is affected by environment (temperature and air stability) and the concentration of the neutral component in the surface layers of the source is calculated.

8.5.2   Gas Release and Chemical Equilibrium

The transfer coefficients given in Equations (8.41) and (8.42) encompass the effect of equilibrium reactions, transport in liquid and air phases, and release of the gas from the liquid. This is a simplified description of the gas emission processes that can be improved by taking into account equilibrium reaction of the components and the Henry's law constant. The effect of temperature, wind and equilibrium reactions in the stagnant surface liquid layer is accounted for by the following empirical equation, which can be used when assessing NH_3 emissions from a bare soil supplied with mineral fertiliser, animal slurry or urine (Sherlock *et al.*, 2002):

$$F_A = K \cdot \left([NH_3]_{a/w} - [NH_3]_a\right) \quad \text{and} \quad K = 89 \times u \times 10^{-4} \tag{8.43}$$

where u is the wind speed (m s^{-1}), $[NH_3]_{a/w}$ is the ammonia concentration in air immediately above the liquid surface (mol m^{-3}), $[NH_3]_a$ is the ambient (background) air ammonia concentration (mol m^{-3}) and K is a transfer coefficient (m s^{-1}). Ambient NH_3 concentration is normally negligible compared with $[NH_3]_{a/w}$ and is often set to zero in the calculations. F_{NH_3} is the flux of gases from the surface (mol m^{-2} s^{-1}), $[NH_3]_{a/w}$ is calculated using Equation (8.11) and using source surface pH and TAN concentrations as input to the calculations (Example 8.4).

Example 8.4   Calculating ammonia emissions from slurry

Emissions of NH_3 from slurry applied in the field can be calculated using Equation (8.43). In the surface layers the concentration of TAN is 0.15 M, pH is 7.0, temperature 20 °C, and wind speed 4 m s^{-1}. This is all the information needed for calculating ammonia volatilisation within the first hour after application of slurry using Equation (8.43).

$[NH_3(aq)] = \dfrac{[TAN(aq)]}{1 + [H_3O^+]/K_N}$	$K_N = 10^{-9.40}$ $NH_3(aq) = 0.15/(1 + (10^{-7.0}/10^{-9.40})) = 0.0002$ M	See Table 8.4
$NH_3(g) = NH_s(aq) \cdot 1/H$	$K_H = 93.79$ M atm^{-1} $H = 93.79 \times 1 \times 0.0821 \times 298) = 2294$ $NH_3(g) = 0.0002$ M/0.26 = 0.00077 M	See Table 8.3
$F_{NH_3} = NH_3(g) \cdot K_v(u)$	$K(u) = 89 \times u \times 10^{-4}$ m s^{-1} $K = 89 \times 4 \times 10^{-4} = 0.0356$ m s^{-1} $F_{NH_3} = 0.00077 \times 10^3 \times 0.0356$ mol m^{-2} s^{-1} $= 0.027$ mol m^{-2} s^{-1}	See Equation (8.43)
F_{NH_3}: emissions during 1 h after application	$F = 0.027 \times 14 \times 60 \times 60 = 0.00058$ kg NH_3-N m^{-2} h^{-1} $F_{NH_3} = 5.8$ kg NH_3-N ha^{-1} h^{-1} (1 ha is 10 000 m^2)	

The calculated emissions from the applied manure are 5.8 kg NH_3-N ha^{-1} h^{-1}, which is within the expected range. Assuming that pH in the surface layers of the applied slurry is reduced by acidification to 6.0, then the calculated emissions would be 0.58 kg NH_3-N ha^{-1} h^{-1}. The amount of TAN applied to crops is often within the range 100–250 kg N ha^{-1}.

Equation (8.43) is robust for assessing the emissions within 1 h of the measurements of TAN and pH at the surface, as the emissions will change concentrations and pH over longer time scales. It has been shown to work when assessing NH_3 emissions from slurry applied to the field and from urine patches in situations with fallow land or a low crop (Sommer *et al.*, 2001). The prerequisite for using the algorithm is that: (i) pH and TAN are measured in soil sampled from the top 2- to 3-mm surface layers of the mixture of soil and slurry or urine or fertiliser and soil, (ii) $[NH_3(g)]_{a/w}$ is calculated using Equation (8.11), and (iii) wind speed at 1 m height above the soil surface is known (Example 8.4). According to this simple model, TAN concentration and wind speed affect emissions linearly, while temperature and pH affect emissions exponentially.

8.5.3   Effects of Air Turbulence and Surface Component Concentration on Emissions

The diffusion layer model is in many complex models combined with air physics using micrometeorological models for predicting transport of the gas from the surface of the source. In these models the effect of diffusion and convection in air is included in the assessment of the emissions. The following equation encompasses the processes of release of the component and transfer in the atmosphere:

$$F_A = \frac{1}{R_a + R_b + R_c} \cdot ([A(g)_{a/w}] - [A(g)_a]) \tag{8.44}$$

The transport and chemical reaction processes in the source are included in calculation of $[A_{a/w}]$, which is related to local conditions affecting diffusion and convection in the liquid.

The model in Equation (8.44) is based on the concept that the gas is transported through an interfacial layer, a laminar layer and a turbulent layer from release until it reaches the ambient unaffected atmosphere. Unaffected means that the concentration of the gas is similar to that in a region with no sources of the gas. For each layer a resistance R is calculated, where an interfacial resistance (R_c) represents the resistance to transport within the surface layer of the source of gas, R_b is laminar resistance in the stagnant air layer between the surface layer and the turbulent layer, which is dominated by molecular diffusion, and R_a is the aerodynamic resistance representing the resistance of the turbulent air layer up to the height where the atmosphere is not affected by the emissions.

The transfer coefficient (K) is parameterised by this series of resistances, which are additive and are related to the resistances to transport by diffusion and convection:

$$K_A = \frac{1}{R_a + R_b + R_c} \tag{8.45}$$

These coefficients assess the transport of the gas from the air–liquid interface. The rationale for using resistances is that in serial systems they can be added together. Transfer coefficients or transfer rates ($k = 1/R$) are not additive.

Volatilisation of a gas from stored manure or from land application of manure will take place when the concentration of the gas in air immediately above the surface ($A_{a/w}$) is higher than the concentration in the ambient air (A_a). Emissions of the gas (F_A) are calculated using a transfer coefficient K and the following standard equation, which gives the emissions of the gas in mol m^{-2} s^{-1}:

$$F_A = K_A \cdot ([A(g)_{a/w}] - [A(g)_a]) \tag{8.46}$$

where K_A (m s^{-1}) is a mass transfer coefficient. Different expressions of this coefficient are used to assess the effects of physics when calculating NH_3 transport from different sources.

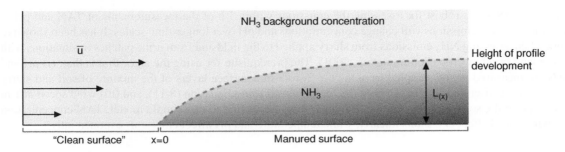

Figure 8.6   *Illustration of the atmospheric gradients of NH$_3$ emitted from a manure surface (e.g. stored manure or a manure–soil mixture in the field). (© University of Southern Denmark.)*

The aerodynamic resistance, R_a, is related to the roughness of the surface of the soil and plant community, average air speed in the turbulent air layer, and stability of the atmosphere. At increasing wind speed or at increasing surface roughness, R_a decreases. In stable atmospheric conditions prevailing at night-time, the boundary layer length is larger than in unstable atmospheric conditions prevailing in daytime, and thus at night R_a is large and emissions may be low.

The resistance to transport across the stagnant boundary layer near the surface, R_b, is dominated by molecular diffusion. Thus, R_b decreases with increasing surface roughness, because of the associated increase in turbulence and R_b also decreases at increasing wind speed.

Interfacial resistance, R_c, represents the resistance to transport within surface layers that may consist of a surface crust on stored slurry or the surface of a soil–slurry mixture where the interface is not well defined. Thus, R_c is related to surface characteristics of the source.

Furthermore, the resistance to transfer from the source of gases to the ambient atmosphere is affected by the size of the source (e.g. R_a increases at increasing distance from the upwind edge of a field). This can be visualised as an increase in the concentration of the gas above the field surface at increasing distance downwind from the edge (Figure 8.6) and an increase in the height of the internal boundary layer (turbulent + stagnant layers).

8.6   Summary

The potential for NH$_3$ and odour release from animal slurry is affected by solubility, as indicated by the Henry's law coefficient, and by the diffusion coefficient of the volatile species in water. Gas transport paths or convection in air and liquid are affected by turbulence. The two-layer model can be used to assess the relative importance of Henry's law coefficient, diffusion coefficient and transport path for the emission process.

Many of the components emitted from animal manure have acid or base functional groups. These components react in water and form cations or anions, which are not released from the liquid. The speciation of the components is affected by pH, so slurry pH is critical for the NH$_3$ or odour emission potential.

In practice, very simple models are used to assess emissions of NH$_3$ and odours. In inventories for assessing NH$_3$ emissions for countries, the emissions are calculated using factors per livestock entity, a fraction of the total-N excreted or a fraction of TAN excreted. Regarding odour emissions, in this case only very simple emissions factors related to livestock number are used. Research or mechanistic models accounting for microbiology, physics and chemistry are being developed.

References

Beutier, D. and Renon, H. (1978) Representation of NH_3–H_2S–H_2O, NH_3–CO_2–H_2O, and NH_3–SO_2–H_2O vapor–liquid equilibria. *Ind. Eng. Chem. Proc. Des. Dev.*, **17**, 220–230.

Blanes-Vidal, V., Sommer, S.G. and Nadimi, E.S. (2009) Modeling surface pH and emissions of hydrogen sulphide, ammonia, acetic acid and carbon dioxide from a swine waste lagoon. *Biosyst. Eng.*, **104**, 510–521.

Buijsman, E., Maas, J.F. and Asman, W.A.H. (1987) Anthropogenic NH_3 emissions in Europe. *Atmos. Environ.*, **21**, 1009–1022.

Feilberg, A., Liu, D.Z., Adamsen, A.P.S., Hansen, M.J. and Jonassen, K.E.N. (2010) Odorant emissions from intensive pig production measured by online proton-transfer-reaction mass spectrometry. *Environ. Sci. Technol.*, **44**, 5894–5900.

Génermont, S. and Cellier, P. (1997) A mechanistic model for estimating ammonia volatilisation from slurry applied to bare soil. *Agric. Forest. Meteorol.*, **88**, 145–167.

Hales, J.M. and Drewes, D.R. (1979) Solubility of ammonia in water at low concentrations. *Atmos. Environ.*, **13**, 1133–1147.

Husted, S., Jensen, L.S. and Jørgensen, S.S. (1991) Reducing ammonia loss from cattle slurry by the use of acidifying additives: the role of the buffer system. *J. Sci. Food Agric.*, **57**, 335–349.

Hutchings, N.J., Sommer, S.G., Andersen, J.M. and Asman, W.A.H. (2001) A detailed ammonia emission inventory for Denmark. *Atmos. Environ.*, **35**, 1959–1968.

Liss, P.S. and Slater, P.G. (1974) Flux of gases over the air sea interface. *Nature*, **247**, 181–184.

Liu, D., Feilberg, A., Adamsen, A.P.S. and Jonassen, K.E.N. (2011) The effect of slurry treatment including ozonation on odorant reduction measured by in-situ PTR-MS. *Atmos. Environ.*, **45**, 3786–3793.

Lyngbye, M., Hansen, M.J., Riis, A.L., Jensen, T.L. and Sørensen, G. (2006) 1000 Olfactometry analyses and 100 TD-GC/MS analyses to evaluate methods for reducing odour from finishing units in Denmark, presented at the *Workshop on Agricultural Air Quality: State of the Science*, North Carolina State University, Raleigh, NC.

Schwarzenbach, R., Gschwend, P.M. and Imboden, D. (2003) *Environmental Organic Chemistry*, 2nd edn, John Wiley & Sons, Hoboken, NJ.

Sherlock, R.R., Sommer, S.G., Rehmat, Z., Khan, R.Z., Wood, C.W., Guertal, E.A., Freney, J.R., Dawson, C.O. and Cameron, K.C. (2002) Emission of ammonia, methane and nitrous oxide from pig slurry applied to a pasture in New Zealand. *J. Environ. Qual.*, **31**, 1491–1501.

Sommer, S.G., Søgaard, H.T., Møller, H.B. and Morsing, S. (2001) Ammonia volatilization from sows on grassland. *Atmos. Environ.*, **35**, 2023–2032.

Stumm, W. and Morgan, J.L. (1996) *Aquatic Chemistry – Chemical Equilibria and Rates in Natural Waters*, 3rd edn, John Wiley & Sons, Inc., New York.

Webb, J. and Misselbrook, T.H. (2004) A mass-flow model of ammonia emissions from UK livestock production. *Atmos. Environ.*, **38**, 2163–2176.

9

Ammonia and Malodorous Gases: Sources and Abatement Technologies

Anders Feilberg[1] and Sven G. Sommer[2]

[1]*Department of Engineering, Aarhus University, Denmark*
[2]*Institute of Chemical Engineering, Biotechnology and Environmental Technology, University of Southern Denmark, Denmark*

9.1 Introduction

Emissions of ammonia (NH_3) and odorous gases constitute an environmental challenge associated with livestock production and the management of animal waste. Since the 1980s there has been an increasing focus on NH_3 emissions from livestock production. At that time, atmospheric deposition of NH_3 emitted from farming was shown to cause damage to natural ecosystems at a level similar to deposition of sulfuric and nitric acid formed from atmospheric oxidation of anthropogenic emissions (Buismans *et al.*, 1987). In addition, NH_3 contributes to particle formation (PM2.5 and PM10), which is a cause of lung disease (Malek *et al.*, 2006). Globally, there is consensus that agriculture and in particular livestock manure is the largest source of NH_3 and ammonium (NH_4^+) in the atmosphere. Furthermore, NH_3 emissions represent a loss of fertiliser that must be replaced by mineral fertiliser nitrogen (N) at some cost to the farmer (Sutton *et al.*, 2011).

Emission of odorous components from animal manure is a large and increasing problem due to intensified livestock production in many parts of the world (East Asia, Europe and North America). Hence, odour emission is becoming an obstacle for increased intensification and productivity of livestock production. Odour is mainly considered to be a nuisance rather than a health hazard, since epidemiological evidence of adverse health effects of odour per se is relatively weak (Schiffman and Williams, 2005). However, lethally high hydrogen sulfide (H_2S) concentrations may accumulate in pits and manure storage facilities.

This chapter presents information on emissions of NH_3 and odorants and on measurement of odour and provides an introduction to abatement technologies.

See Text Box – Basic 9.1 for definitions used in this chapter.

Animal Manure Recycling: Treatment and Management, First Edition. Edited by Sven G. Sommer, Morten L. Christensen, Thomas Schmidt and Lars S. Jensen.
© 2013 John Wiley & Sons, Ltd. Published 2013 by John Wiley & Sons, Ltd.

Text Box – Basic 9.1   Definitions

BOD (biological oxygen demand): the amount of oxygen needed to aerobically metabolise organic compounds, used as a measure of biologically degradable organic material.

C : N ratio: The mass ratio between carbon and nitrogen in a material; important for many microbial processes.

COD (Chemical oxygen demand): the amount of oxygen needed to chemically oxidise organic compounds, used as a measure of chemically degradable organic material.

D_{50}: The level of dilution at which 50% of a population can detect the smell of a sample or the concentration of a specific compound that can just be detected by 50% of a population.

Emission factor: The fraction (or percentage) of a compound in a liquid or solid matrix that is being emitted to the gas phase.

GC/MS: Gas chromatography combined with mass spectrometry – an analytical technique for quantification and identification of volatile organic compounds.

NMVOCs: Non-methane volatile organic compounds; different definitions are used to identify the upper range of volatility.

Olfactometry: A measurement technique for measuring odour. The technique is based on an olfactometer, which is a device for presenting dilutions of smelly gas samples to a panel of humans carrying out a sensory evaluation.

Odour activity value (OAV): The concentration of a specific odorant divided by its D_{50} value.

PM2.5 and PM10: Particles (particulate matter) with an average diameter of less than 2.5 and 10 µm.

Redox potential: A measure of the ability of a chemical compound to attract electrons; measured in units of volts.

Sensory stimulus: The interaction of a gas or mixture of gases with the human senses (here: smell)

Slatted floor: A floor type used in animal buildings, which consists of a solid material (concrete) with openings that enable the collection of waste in pits below the floor. Several differently dimensioned and designed slatted floors are available.

SPME: Solid-phase micro-extraction by which compounds are collected on a sorptive fibre coated with a suitable material and released to an instrument immediately before analysis of odour components.

TAN: Total ammoniacal nitrogen (TAN $= NH_3 + NH_4^+$)

9.2   Measurement Methods

A variety of techniques can be used for measuring gaseous emissions from animal houses, manure stores and applied manure. The concentration measurements are usually combined with an estimate of air exchange in order to determine total gas emissions: $F(A) = [A_g] \cdot Q$, where Q is the air exchange rate. Obtaining Q can be either simple (e.g. as in mechanically ventilated animal buildings) or very complicated (e.g. as in area sources). The methods used encompass use of gas tracers, static and ventilated chambers covering the gas source, and meteorological methods. It is beyond the scope of this chapter to describe these methods, but in-depth presentations can be found in the scientific literature (McGinn and Janzen, 1998; Shah *et al.*, 2006; Ni and Heber, 2008).

Methods for measuring NH_3 concentrations and emissions are well established and agreed, so in the following we focus on measurements of perception of odour and odour components.

9.2.1 Odour Measurement

Odour is a highly subjective sensory perception and quantification of "odour" is by no means a straightforward task, because the sense of odour differs between people and because many components contribute to odour. The most widely used method for odour quantification is dilution-to-threshold olfactometry. Air samples containing odorous compounds are diluted stepwise by clean non-odorous air in order to obtain a wide range of dilutions. Samples are collected in polymer bags made of, for example, poly(vinyl fluoride) (PVF; Tedlar®) or poly(ethylene terephthalate) (PET; Nalophan®). Human panels are exposed to an increasing concentration of the sample, starting with a sample so dilute that the panellists cannot detect a sensory stimulus. The principle is that they should not become adapted to the odour, which would be the case if they were exposed to the highest odour concentration first and thereafter to more diluted samples.

Based on serial dilutions, a threshold level is determined, which is the dilution level that can be detected by 50% of a population (test panel) as a sensory stimulus (D_{50}). The dose, however, is not well defined and has to be determined relative to the undiluted sample. This is accomplished by using the number of dilutions to reach D_{50} as a measure of odour, i.e. odour concentration at D_{50} is defined as the ratio of diluting the sample with clean air (i.e. one dilution is 1 OU (odour unit) m^{-3}). For example, if a sample needs to be diluted 100-fold in order to reach D_{50}, the odour concentration is reported to be 100 OU m^{-3}.

For practical reasons, the number of panellists is limited to four to eight individuals, which is not nearly sufficient to represent a population. In order to overcome this, panellists are selected based on their sensitivity to the well-defined reference compound 1-butanol (CEN, 2003). According to the European standard (CEN 2003), the final result needs to be corrected for the relative sensitivity of the specific panel to 1-butanol. To indicate this, results obtained according to this procedure are reported as OU$_E$ m^{-3}.

Odour measurements by olfactometry are subject to both random variability and systematic bias. Random variability arises from the use of human panellists with variable sensitivity, and both within-panel variability and between-panel variability are important. As a consequence, the uncertainty of a single sample analysed by olfactometry is often a factor of 2. The random variability can be reduced by taking a larger number of samples, which is needed in order to detect significant differences between different sample sources. As an example, the probability of assessing a true difference of 50% between two sets of samples (e.g. from an emission source with and without an abatement technique) has been estimated to be 0.64 based on 10 samples (Clanton *et al.*, 1999).

An even more important problem is that storage in the sampling bags may significantly affect the concentrations of odorants and thus the sensory stimulus (D_{50}) (e.g. Hansen *et al.*, 2011). This is a major problem because the samples may be stored for up to 30 h according to the international standards for measuring odour. The compound recovery of the bags differs significantly between organic components with different functional groups; for example, phenolic and indolic compounds have 24-h recoveries below around 10%, carboxylic acids have 24-h recoveries below around 50% and volatile sulfur compounds (including H$_2$S) have recoveries close to 100% (Hansen *et al.*, 2011). Differences between sampling bag materials have been observed to be relatively small. The largest reductions in compound concentrations occur in the initial stage of storage, thus limiting the possibility of reducing these problems by analysing samples shortly after collection (Hansen *et al.*, 2011). Consequently, the relative impact of volatile sulfur compounds increases strongly with sampling storage and the sample at the time of analysis does not accurately represent the odour composition at the sampling source.

More accurate and reproducible odour measurement techniques based on analytical methods are being developed. More than 100 chemical compounds can be detected in air from livestock facilities and manure (O'Neill and Phillips, 1992; Schiffman *et al.*, 2001). However, even if the many compounds that are present almost everywhere in low concentrations or have a high D_{50} are excluded, it is still a key challenge to identify the specific subset of compounds that are responsible for perceived odour.

In most studies odorants are analysed with gas chromatographs coupled with mass spectrometric detection (GC/MS) (e.g. Trabue *et al.*, 2008). With GC, the compounds are first separated by chromatography according to volatility/vapour pressure. The eluent of the chromatograph is subsequently analysed by MS, which produces a unique fingerprint signal that can be used for both compound quantification and identification. This method is sensitive and gives highly selective results. Samples for GC/MS can be collected either by using sorbent tubes (Trabue *et al.*, 2008), in which compounds are collected in a tube packed with suitable sorbent material, or by using solid-phase micro-extraction (SPME) by which compounds are collected on a sorptive fibre coated with a suitable material (Bulliner *et al.*, 2006). In both cases, compounds are thermally desorbed prior to GC/MS analysis. Sorbent tubes provide quantitative results and SPME qualitative results. Both methods are unsuitable for measuring reactive volatile sulfur compounds containing −SH functional groups (H_2S and thiols, i.e. methanethiol). These compounds can alternatively be analysed by GC combined with one of several available sulfur-specific detectors (Kim *et al.*, 2007), although the sensitivity is lower than for GC/MS. Direct injection mass spectrometry (proton transfer reaction-mass spectrometry (PTR-MS) has recently also been used for odorant analysis (Feilberg *et al.*, 2010).

9.2.2 Relationships Between Odour and Odorants

Most attempts to statistically correlate odour measured by olfactometry with odorant composition have had little success. Good correlations have been obtained using standard odour mixtures (Zahn *et al.*, 2001; Hobbs *et al.*, 2001), but the results have not been satisfactory when using air samples from livestock farms (Hobbs *et al.*, 1999; Gralapp *et al.*, 2001). The possible reasons are: (i) that the analytical methods do not include the effect of all key odorants and (ii) problems with recovery of certain odorants in sampling bags for olfactometry. A recent study (Hansen *et al.*, 2011) has shown that for "pig air", the dominant odorous compounds are predicted to be H_2S, methanethiol, 4-methyl phenol and trimethyl amine (Hansen *et al.*, 2011). The analytical method used (PTR-MS) covers a wider range of chemical compounds than used in earlier studies and with this method concentrations of odorants have been more successfully related to odour (D_{50}).

Alternatively to D_{50}, the relative odour strength of a compound can be assessed by combining both odour concentration and D_{50} of the single compound in order to identify compounds contributing to odour. It is proposed to calculate the compound-specific odour activity values (OAVs) as concentration of a component in a sample divided by the odour threshold value. Studies using this expression show that key odour compounds are H_2S, methanethiol, 4-methylphenol, butanoic acid and 3-methyl butanoic acid, with H_2S and methanethiol seemingly having the highest OAVs (Feilberg *et al.*, 2010). Geometric means of trusted determinations are used to improve accuracy, but lack of precision in the determination of odour threshold values (D_{50}) must be considered when interpreting odour predictions based on OAVs.

Gas chromatography with olfactometric detection has additionally been used to identify key odorous compounds (Table 9.1). The principle is that odorants are separated on a chromatographic column and directed to a nose cone outlet for sensory detection by one or two human panellists. These studies show that in air from pig farms, the components 4-methylphenol, butanoic acid and 3-methyl butanoic acid are important odorants (Bulliner *et al.*, 2006). In these studies, sulfur compounds (H_2S and methanethiol) could not be determined with the analytical techniques used (SPME).

Obviously, more work is needed on identifying the most important odorous compounds and on correlating chemical composition to odour strength. Nevertheless, the results of application of a number of different methods tend to agree that H_2S, methanethiol and 4-methyl phenol can be suitable indicators of odour.

Table 9.1   *Ammonia emission factors for pig buildings (Sommer et al., 2006a).*

Animal category	Pen design	Emission factor (% of total-N)		Emission factor (kg NH_3-N kg^{-1} TAN)	
		Slatted floor and slurry	Littered floor	Slatted floor and slurry	Littered floor
Sows	partially slatted floor and strewed solid floor	12	16	0.16	0.33
	strewed solid floor		16		0.33
	fully slatted floor	20		0.26	
Weaners and fatteners	fully slatted floor	16		0.25	
	partially slatted floor	8–16[a]		0.18	
Cattle	slurry	3		0.06	
	partly slatted floor, 0.4-m deep slurry channel	6		0.12	
	partly slatted floor, 1.2-m deep slurry channel	8		0.17	
	deep litter		6		0.12

[a] Variation due to variation in slatted floor area.

9.3   Ammonia Emissions

The sources of NH_3 emissions from livestock manure management are animal houses or barns, stored manure and manure being applied in the field. The emissions from these sources are affected by the concentration of total ammoniacal N (TAN = NH_3 + NH_4^+), pH, temperature, surface area of source and turbulence in manure and air (see Chapter 8).

9.3.1   Pig and Cattle Houses – Slatted Floor and Slurry Pits

In animal houses with partly or fully slatted floors, slurry is collected in pits beneath the floor (Figure 9.1). The sources of NH_3 are the solid floor, the contaminated animal body, the slatted floor, the walls and the manure pit. The most important sources of NH_3 are the contaminated solid floor, the slatted floor and the slurry beneath the slats (Ni *et al.*, 1999; Kai *et al.*, 2006).

Mean NH_3 losses per amount of excreted N are larger from pigs on slatted floors than from dairy cattle, due to a greater amount of TAN in the slurry and a higher temperature in pig houses. Losses of NH_3 from pig housing systems with slatted floors range from 17% of total-N for piglets to 29% of total-N for rearing pigs (Oenema *et al.*, 2001; Poulsen *et al.*, 2001). Reducing the surface area of the slatted floor reduces NH_3 emissions (Figure 9.2), because the area of the NH_3 source is reduced. As an example, in an animal house with a partly (50%) slatted floor, the NH_3 emissions have been shown to be related to the proportional surface area contaminated with urine and faeces according to the equation (Ni *et al.*, 1999):

$$F(NH_3) = 54.22 \cdot con + 12.02 \text{ and } con = 1.82 \times 10^{-10} \cdot T_i + 1.97^{-19} \cdot W_p \qquad (9.1)$$

where $F(NH_3)$ is the NH_3 emission rate (g h^{-1}), *con* is the floor contamination (dimensionless), T is the inside room temperature (ranging from 2.5 to 22.7 °C) and W_p is the weight of the batch of pigs (0–5095 kg). The model shows that contaminated surface area increases as the pigs grow larger and also with increasing

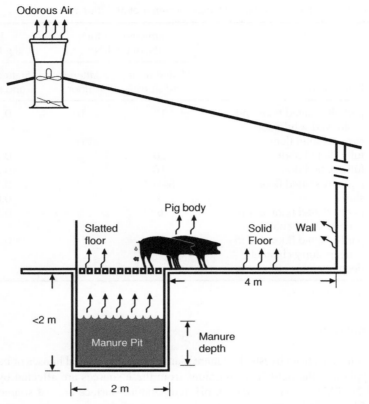

Figure 9.1   *Illustration of sources of NH₃ emissions from a pig house with a partly slatted floor in the pig pen and slurry channels below the slats. (© University of Southern Denmark.)*

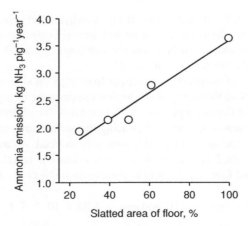

Figure 9.2   *Ammonia emissions related to the area of slatted floor covering slurry stores. (Modified with permission from Sommer et al. (2006a). © Elsevier.)*

temperature. The relationship with temperature reflects the fact that during warm periods, the pigs can change their resting habits and use the slatted floor area for resting and the solid floor for excretion. As a consequence, a pig house with partly slatted floor may periodically emit more NH_3 than a house with a fully slatted floor.

The effect of reducing source area is also the reason for the effect of frequent emptying of slurry channels with inclining (V-shaped) walls. This design reduces NH_3 emissions by up to 50%, because the surface area of the slurry is reduced by lowering the height of the slurry (Groenestein and Van Fassen, 1996).

As for pig houses, the loss of NH_3 from cattle housing systems is dependent on the emission surface area (Table 9.1). For example, little floor area is soiled with excreta in houses with tied dairy cows, and this reduces the NH_3 emissions to 65% of those from dairy cow houses where the cows walk free and excretion is distributed over a larger area (Monteny and Erisman, 1998). In cattle houses, NH_3 emissions may also be reduced by the rapid removal of urine and faeces from the livestock buildings. However, to achieve a significant emissions reduction the liquid manure must be removed efficiently, since even a thin layer is a significant source of NH_3. If the floor is smooth, scraping may reduce emissions by up to 30%, but to the detriment of animal welfare (Sommer *et al.*, 2006a). Very efficient and frequent scraping of a grooved solid floor and cleaning with water may reduce emissions by 50–65%. Thus, it is the combination of cleaning the floor with a scraper and draining the urine to a gutter that reduces the NH_3 release from this floor and reduces NH_3 emissions from the animal building.

In addition, both cooling of manure and slurry acidification to pH $<$ 6 (Kai *et al.*, 2008) have been shown to significantly reduce NH_3 emissions, demonstrating the influence of pH and temperature. However, the effect of such measures is limited to the fraction of NH_3 emissions originating from slurry and they will not reduce emissions from contaminated surfaces.

9.3.2 Pig and Cattle Houses – Solid Floor and Deep Litter

In addition to slatted floors with slurry collection in pits, animals are also kept in houses with a solid floor and bedding material (deep litter), which can be straw or sawdust. Deep litter affects NH_3 emissions in several ways: (i) urine infiltrates into the deep litter and thereby reduces the surface area in contact with the air, (ii) deep litter has a high surface roughness and reduces the air flow over the emitting bedding and (iii) litter material promotes biological conversion of nitrogen either by oxidation of TAN or by immobilisation of inorganic N, producing organic N. Ammonia emissions from fattening pigs on deep litter have been determined to be about 75% of those from fattening pigs on fully slatted floors, while emissions from cattle on straw-bedded systems have been observed to be about 66% of those from cattle in slurry-based systems (Webb *et al.*, 2012). Doubling straw use can reduce the NH_3 emissions factor per pig by 18% (Sommer *et al.*, 2006a). The type of bedding material used influences infiltration rate, air flow over the emitting surface and absorption of liquid effluent (influencing ammonium immobilisation). For example, measured emissions from bulls on bedding of long straw, chopped straw and peat + chopped straw treatments were 58, 46 and 32 g cow^{-1} day^{-1}, respectively (Jeppsson, 1999).

Bedding material is a source of carbon which, depending on the C : N ratio, may enhance loss of N by nitrification and denitrification. This has been observed to account for a loss of 47% of the N excreted in systems where the bedding material was stirred by rakes weekly (Groenestein and Van Faassen, 1996). In addition, a significant fraction of the TAN can be absorbed through cation exchange processes by the straw and immobilised by transformation into organically bound N by microorganisms (Chapter 4).

9.3.3 Poultry Houses

Poultry manure differs from other animal manure because the TAN in poultry manure originates mainly from decomposed uric acid in the droppings. Hydrolysis of uric acid is slow and is affected by storage conditions,

Table 9.2 *Ammonia emissions from poultry houses with cages, conveyer belt, solid floor or access to an outdoor area: the emission coefficient is given in kg NH_3 kg^{-1} N (K_f) (poultry excrete uric acid, which is transformed relatively slowly, and therefore the emissions are given in kg NH_3-N kg^{-1} total-N instead of kg^{-1} TAN) (Sommer et al., 2006b).*

Animal category	Housing system, manure management	Emission coefficient (kg NH_3-N kg^{-1} N)	
		Slurry storage and/conveyor belt	Litter, sand covered floor
Broilers	solid floor	0.25	
Broilers, organic	solid floor and outdoor	0.25	
Turkeys, geese and ducks	solid floor	0.20	
Laying hens	solid floor and manure store	0.40	0.25
	battery cage with conveyer belt and area for scraping behaviour	0.10	0.25
Laying hens – organic	solid floor and outdoor area	0.28	
	solid floor, manure store and outdoor area	0.40	0.25
Poults	battery cages	0.40	
	solid floor		0.25

so the concentration of TAN and uric acid is often more variable than for other manures (Kroodsma *et al.*, 1988). The design of poultry houses and manure management affect transformation of uric acid and thus to a great extent NH_3 emissions (Groot Koerkamp, 1994). Data on ammonia emissions from poultry houses is compiled in Table 9.2.

Emissions are low from houses where excreta are collected dry on a conveyor belt which transports the manure to an external store of dry solid manure (Groot Koerkamp, 1994), because the manure is dried on the conveyer belt and removed to the outside of the poultry house. Drying poultry manure reduces hydrolysis of uric acid and consequently reduces pH, which in dry manure may be 7.3, compared with 9.6 in untreated manure (Groot Koerkamp *et al.*, 1998).

In other poultry houses, the droppings may be stored dry on the floor or, after addition of water, as slurry in pits or channels. A 50% reduction in NH_3 emissions was the reported consequence of reducing water content in the manure from 2400 to 230 g H_2O kg^{-1}, which corresponds to increasing the dry matter content from 30% to 80% (Cabrera and Chiang, 1994). Different systems for drying manure have been developed, one of which involves drying with air flowing through a perforated floor. If the drying is efficient, then this technique reduces NH_3 emissions to 10% of the emissions from a traditional floor (Groenestein, 1993). In housing systems with long-term solid manure storage containers below the cages, transformation may start, pH increases and temperature will often increase due to the microbial activity, and thus the potential for NH_3 emissions is high (Groot Koerkamp, 1994). This is further enhanced if water infiltrates the manure (e.g. due to spillage of drinking water from bell drinkers) and the effect may be a doubling of the NH_3 emissions from these houses (Kroodsma *et al.*, 1988).

The alternative to managing dry manure is to suspend it in water and produce slurry. In the slurry, uric acid is slowly transformed to TAN and temperature does not increase due to anaerobic transformation of organic matter (Groot Koerkamp, 1994). Emissions of NH_3 from slurry systems are therefore low, around 33% of the emissions from houses with manure stored in heaps on the floor.

Table 9.3 *Ammonia emissions from uncovered and covered stored livestock slurry (Sommer et al., 2006a, 2006b).*

Category	Store	Emission (kg NH₃-N m⁻² yr⁻¹)	Emission (% of TAN)		
		No cover	No cover	Covered with surface crust, straw, etc.	Roof
Cattle	concrete store	1.44	9	2	1
Pigs	concrete store	2.18	15	3	
Fermented cattle and pig manure	concrete store	2.33	28	6	1
Pigs	lagoon	0.78			

9.3.4 Ammonia Emissions from Manure Storage

The transfer of NH_3 from stored manure to the ambient atmosphere is not as complex as the transfer of NH_3 from sources in animal houses. The release of NH_3 from stored liquid manure or livestock slurry is primarily of a physical or chemical nature. Since the anaerobic microbial transformation is relatively slow, little TAN is produced during storage and the concentration of organic acids is relatively constant. In contrast, the NH_3 emissions from stored solid manure are related to microbial activity in the manure, which is influenced by air flow through the manure heap.

Ammonia emissions are also larger from stored pig slurry than from cattle slurry, due to a higher TAN concentration. Furthermore, emissions tend to be twice as large from slurry that has been fermented in a biogas plant than from untreated slurry, because fermented slurry has a higher pH and TAN content (Sommer *et al.*, 2006a). Ammonia emissions from slurry in open tanks, silos and lagoons range from 0.78 to 2.33 kg NH_3-N m⁻² year⁻¹ (Table 9.3).

Fitting a cover on the slurry store significantly decreases NH_3 losses (Sommer *et al.*, 2006a). The cover may be a natural surface crust formed by solids floating on the surface, a cover of straw, peat or floating expanded clay particles, or a roof. Covers greatly decrease the air exchange rate between the surface of the slurry and the atmosphere, and an effective cover reduces transport (increases resistance, r_c in Equation (8.44)) and decreases NH_3 losses to less than 10% of those from uncovered slurry (Table 9.3).

In stores of solid manure with a low straw content or high water content (above 50–60%), the diffusion rate of O_2 is low and composting nearly absent (Sommer *et al.*, 2006a; Webb *et al.*, 2012). NH_3 emissions therefore occur exclusively from the outer surface of the stack. The addition of fresh manure to the surface of the stack prevents further emissions from the old outer surface, which is now buried, but creates a new outer surface from which emissions can occur. Each fresh addition of manure creates a new pulse of NH_3 emissions.

In contrast, if self-heating (composting) occurs, then warm air moves through the heap and the potential for NH_3 emissions is large. The decomposition of organic matter results in rapid mineralisation of organic N and an increase in pH due to a reduced concentration of organic acids, which together with high temperatures leads to high concentrations of $NH_3(g)$, and to rapid and substantial emissions. A newly created heap acts as a source of NH_3 for a few weeks, until the moisture content falls sufficiently to halt decomposition or until all the decomposable nitrogen has been emitted as NH_3 or oxidised nitrogen, or has been converted into organic N. Losses of 25–30% of the total-N in stored pig manure and cattle deep litter have been recorded, although

Table 9.4 *Emissions factors for ammonia NH_3 emissions from stored solid manure and the dry matter-rich fraction from slurry separation (Sommer et al., 2006a; Hansen et al., 2008).*

Animal category	Manure category	NH₃-N emissions (% of total-N)		
		Untreated	Compacted	Covered with tarpaulin or PVC
Fattening pigs and sows	solid manure	25	13	13
	deep litter, dry matter-rich fraction from separation of slurry	25	13	13
Dairy cows	solid manure	4	2	2
Cattle	deep litter, dry matter-rich fraction from separation of slurry	8	4	4
Poultry	deep litter	15	7	7

losses as low as 1–10% have also been measured. The low losses may be due to the leaching of TAN with rainwater (Webb *et al.*, 2012).

Addition of straw increases the C:N ratio and promotes immobilisation of TAN, but large amounts of straw are required to reduce NH_3 losses (e.g. a daily addition of 25 kg straw cow^{-1} would be required to reduce NH_3 losses during storage by 50% compared with the standard treatment). Alternatively, losses can be lowered by 50–90% by decreasing the convection of air through the heap with a PVC cover or through compaction of the litter (Table 9.4).

9.3.5   Field-Applied Manure

Ammonia emissions from field-applied manure are affected by the adsorption of NH_4^+ to manure dry matter, the physical processes controlling the movement of manure liquid into and within the soil, and the interaction of slurry liquid with soil cation exchange capacity. It has been estimated that the emissions vary from 0 to more than 50% of TAN, depending on manure type, environmental conditions (temperature, wind speed, rain) and soil properties (calcium content, cation exchange capacity, acidity). Biological N transformation processes have been evaluated to be of relative minor importance due to the short duration of NH_3 emissions from field-applied manure.

Rates of NH_3 emission may be very high immediately after slurry application (12 kg N ha^{-1} h^{-1}; Pain *et al.*, 1989). This high initial loss rate is related to both the initially high TAN concentration in the soil surface and the pH increase taking place immediately after slurry application. The cumulative NH_3 losses increase hyperbolically with time (Figure 9.3). Generally, the rate of NH_3 emission from applied manure is very low after a few days, because the concentration of dissolved NH_4^+ in the soil surface decreases rapidly due to emission and infiltration. Usually 50% of the total NH_3 losses occur within 4–12 h after slurry application (Pain *et al.*, 1989; Moal *et al.*, 1995). Average total emissions of NH_3 may account for up to 43% of the TAN applied in manure (Table 9.5).

Incorporating slurry into the soil is a very effective way of decreasing NH_3 emission. Incorporation of slurry by ploughing or by rotary harrow immediately after surface application of slurry decreases NH_3 losses by 80% (Table 9.5). Shallow direct injection of slurry (3–5 cm) can decrease losses by up to 70%, while deep injection (35 cm) can stop losses almost completely. Furthermore, application of slurry with trailing hoses

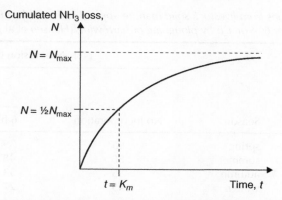

Figure 9.3   *Cumulative NH$_3$ emissions as a function of time following field application of slurry. N$_{max}$ and K$_m$ are the parameters used in the Monod model of the rate of NH$_3$ emissions. (Reprinted with permission from Søgaard et al. (2002). © 2002 Elsevier.)*

on the soil beneath a crop canopy can decrease NH$_3$ volatilisation by more than 50%, the efficiency of this technique increasing with increasing leaf area and height of crop (Thorman *et al.*, 2008).

Losses of NH$_3$ from solid animal manure applied to soil are not as well understood as the emissions from slurry application. The pattern of NH$_3$ volatilisation over time from solid manure is different to that for slurry. The initial rate of loss from solid manure is low, but volatilisation continues for a long period, probably because the NH$_4^+$ from the solid manure infiltrates into the soil more slowly than the NH$_4^+$ from slurry

Table 9.5   *Ammonia emission from surface-spread slurry, slurry injected into the soil and slurry applied to the soil surface and incorporated: the effect of incorporation varies according to the time lag between application and incorporation (adapted from Hansen et al., 2008).*

Season	Soil surface and crop	Application technique	NH$_3$ losses (% of applied TAN) Pig	NH$_3$ losses (% of applied TAN) Cattle
Spring	bare soil	trail hose	17.1	32.6
		trail hose and incorporation[a]	5.0	9.4
		injection 3–5 cm bare soil	1.7	3.3
	cereals	trail hose	14.8	28.1
	grass	trail hose	17.1	32.6
		injection 3–5 cm bare soil	12.8	24.5
Summer	bare soil	trail hose	22.4	42.7
		trail hose and incorporation[a]	6.5	12.4
		injection 3–5 cm bare soil	2.2	4.3
	grass	trail hose	22.3	42.5
		injection 3–5 cm	16.7	31.9
Autumn	grass	trail hose	21.8	41.6
		injection 3–5 cm	16.4	31.2

[a]Incorporation by plough or rotary harrow 6 h after application of manure on soil.

Table 9.6   *Ammonia emissions from livestock solid manure applied in the field and left on the surface or incorporated into the soil after 6, 4 or 1 h by ploughing or harrowing (Hansen et al., 2008).*

Incorporation method	Season	No incorporation	6 h	4 h	1 h
			\multicolumn{3}{}{NH$_3$-N emission (% of NH$_4^+$-N)}		
			\multicolumn{3}{}{Incorporation lag time after application}		
Ploughing	spring	65	39	22	13
	summer	80	48	32	16
	autumn	55	33	12	11
	winter	45	27	7	9
Harrowing	spring	65	41	36	27
	summer	80	54	48	35
	autumn	55	30	27	21
	winter	45	22	20	17

(Chambers *et al.*, 1997). The few studies performed to date on NH$_3$ emissions from solid manure applied to soil indicate that about 50% of the losses occur within 24 h of application and that volatilisation may continue for about 10 days. Ploughing of solid manure into the soil decreases NH$_3$ losses (Table 9.6).

9.4   Odour Emissions

Livestock houses and manure are sources of odorous compounds, and consequently all livestock production facilities cause unpleasant smells in the local surroundings (Nimmermark, 2004; Rappert and Müller, 2005). In general, the odour from livestock facilities is believed to originate almost exclusively from manure excretion (e.g. pen surfaces, slurry pits, manure storage and field-applied manure). Around the world, odour emissions have mainly been measured by olfactometry despite the uncertainties described in Section 9.2.

In order to estimate the impact on the local environment, odour concentration is used together with measurement of air exchange rate to estimate emission rate from a source. Since odour concentration is not a mass unit but is expressed as OU per volume, the flux of odour cannot be expressed in a typical mass per time unit. By multiplication of odour concentration (OU m^{-3}) by air exchange (m^3 s^{-1}), odour emission flux can instead be expressed as OU s^{-1}. These emission rates can be used as input to dispersion models, which predict the odour concentration in the surroundings (e.g. at the setback distance) (Yu *et al.*, 2010). Since odour cannot be represented by a single compound, but is composed of a variable complex mixture of compounds (Rappert and Müller, 2005), it is not possible to determine general mass transfer coefficients of odour as a function of wind speed, temperature and other external factors. Thus, modelling of odour emissions is largely based on empirical data on emission rates.

In the following, examples of data on odour emissions from livestock buildings are presented together with the methods used for estimating these emissions. When comparing these data, it should be kept in mind that odour analysis is associated with bias, as discussed in Section 9.2. Finally, existing knowledge on emissions of volatile organic compounds (VOCs, including odorants) and H$_2$S from agriculture in general is presented. Emissions from manure storage and field application are not included due to the lack of comprehensive data and uncertainties in the methods for measuring emissions from these sources.

Table 9.7 *Danish standard emission rates for odour from livestock houses (Hansen, 2007).*

Animal	Floor type	OE	Range[a] (n)	Unit
Sow (gestation)	partly slatted	16	7–39 (48)	$OU_E\ s^{-1}\ animal^{-1}$
	partly slatted	65	30–170 (48)	$OU_E\ s^{-1}\ 10^{-3}\ kg^{-1}$
Sow (farrowing)	partly slatted	72	40–125 (24)	$OU_E\ s^{-1}\ animal^{-1}$
	partly slatted	240	140–420 (24)	$OU_E\ s^{-1}\ 10^{-3}\ kg^{-1}$
	all other floor types	100	56–280 (24)	$OU_E\ s^{-1}\ animal^{-1}$
	all other floor types	360	200–960 (24)	$OU_E\ s^{-1}\ 10^{-3}\ kg^{-1}$
Finisher pigs (30 kg to end weight)	partly slatted	300	110–810 (24)	$OU_E\ s^{-1}\ 10^{-3}\ kg^{-1}$
	partly slatted	19	8–48 (24)	$OU_E\ s^{-1}\ animal^{-1}$
	other floor types	450	190–1200 (48)	$OU_E\ s^{-1}\ 10^{-3}\ kg^{-1}$
	other floor types	29	13–78 (48)	$OU_E\ s^{-1}\ animal^{-1}$
Piglets (7–30 kg)	partly and fully slatted	380	200–750 (48)	$OU_E\ s^{-1}\ 10^{-3}\ kg^{-1}$
Cattle (all types)	all floor types	170	NA	$OU_E\ s^{-1}\ 10^{-3}\ kg^{-1}$
Laying hens	floor operation with manure pit/channel	900	NA	$OU_E\ s^{-1}\ 10^{-3}\ kg^{-1}$
	cage operation, dry manure	400	NA	$OU_E\ s^{-1}\ 10^{-3}\ kg^{-1}$
Broilers	deep litter	400	NA	$OU_E\ s^{-1}\ 10^{-3}\ kg^{-1}$

[a] The range is represented by the 5 and 95% fractiles with total number of observations in brackets. For pig buildings, the data are based on four production units. NA: not available.

9.4.1 Livestock Buildings

Emissions from livestock buildings are typically determined in OU per unit time per animal or production unit or in OU per unit time per 1000-kg animal. Emissions may also be expressed in OU per unit time per square metre of manure, since a large part of the odorants are expected to be emitted from the manure surface. In mechanically ventilated buildings, the odour emissions can be determined from air flow measured in the ventilation duct either by using a measuring fan or by measuring air velocity in cross-sections of appropriate spatial resolution. For naturally ventilated buildings, it is much more complicated to estimate emission rates: (i) because the air exchange is difficult to determine and (ii) because odour concentrations are relatively low. Therefore, very little data are available on odour emissions from naturally ventilated (mainly dairy and cattle) buildings.

In Denmark, attempts have been made to determine standard emission rates for odour from livestock buildings based on a range of measurements (Table 9.7). For comparison, examples from other countries are given in Table 9.8.

Table 9.8 *Examples of reported ranges of emissions of odour from finisher pigs.*

Location	Emissions range ($OU_E\ s^{-1}\ animal^{-1}$)	Reference
Denmark	8–78[1]	Riis (2006)
Minnesota, USA	4–20	Zhu *et al.* (1999)
The Netherlands	5–85	Ogink and Koerkamp (2001)
Ireland	11–28	Hayes *et al.* (2006)
Minnesota	0.1–1000	Gay *et al.* (2003)

For mechanically ventilated buildings (pig and poultry), emission rates were determined by one of the following expressions:

$$OE \, (OU_E \, s^{-1} \, 10^{-3} \, kg^{-1}) = \frac{OC \, (OU_E \, m^{-3}) \cdot Q \, (m^3 \, h^{-1}) \cdot 1000}{\text{animal weight (kg)} \times 3600} \qquad (9.2)$$

$$OE \, (OU_E \, s^{-1} \, animal^{-1}) = \frac{OC \, (OU_E \, m^{-3}) \cdot Q \, (m^3 \, h^{-1}) \cdot 1000}{\# \, \text{of animals} \cdot 3600} \qquad (9.3)$$

where OE is the emission rate, OC is the odour concentration and Q is the air exchange rate. For the data presented in Table 9.7, air exchange was measured with a calibrated measuring fan in the outlet ventilation duct and number of animals and weight were determined on the measuring days. The standard emission rate for broilers is set to the same rate as for laying hens in cages (400 OU s^{-1} 10^{-3} kg^{-1}) due to lack of relevant data. However, recent measurements have indicated somewhat higher emissions from broilers (summer: 572 OU s^{-1} 10^{-3} kg^{-1}; winter 632 OU s^{-1} 10^{-3} kg^{-1}).

There is a large variation in odour emissions from similar production units, because a number of factors will influence emissions, such as air exchange rate, temperature, manure handling and so on. In mechanically ventilated buildings, odour emissions have been demonstrated to be correlated with air exchange rate (ventilation rate) (Lyngbye *et al.*, 2006). The effect is a combination of a ventilation rate effect and temperature effect, because although ventilation is adjusted to achieve a certain temperature in the animal room (e.g. 18 °C), limitations in the ventilation capacity mean that at high ambient temperatures (at summer time), the ventilation capacity is not high enough to prevent temperatures from increasing. Therefore, fluctuation in ventilation rate is also an indicator of temperature fluctuation. Furthermore, at high ventilation rates the ratio of air exchange in the slurry pit and in the animal room also increases. To complicate matters even further, ventilation rate also increases during a production cycle with increasing animal weight, as well as increasing levels of manure in the pit, which in turn may lead to increasing emissions.

9.4.2   Volatile Organic Compounds and Hydrogen Sulphide Emissions from Livestock Production

Little is known about the contribution of emissions of VOCs and H_2S from livestock production to the overall burden of gaseous emissions of importance for general air quality. These emissions can either influence tropospheric ozone formation or formation of secondary organic or inorganic aerosols.

Emissions of non-methane VOCs (NMVOCs) from livestock farming in the United Kingdom were assessed by scaling measured emissions to emissions of NH_3 for which the total emissions for the United Kingdom has been estimated (Hobbs *et al.*, 2004). The resulting total emissions were 165 ± 56 ktonnes for the year 2002, with important compounds being acetic acid, butanoic acid, 4-methyl phenol and dimethyl sulfide.

In a study of emissions from dairy cows in California (Shaw *et al.*, 2007), it was also found that: (i) total NMVOCs emissions were 6–10 times lower than previously reported values and (ii) the ozone formation potential of the compounds was low compared with that of compounds from other sources (vegetation and combustion). Hence, it was concluded that the impact of dairy farming on atmospheric ozone formation is relatively low, even though California is believed to be home to the largest dairy industry in the world (Shaw *et al.*, 2007). The results of that study are supported by results obtained from measurements of emissions from a German dairy farm (Ngwabie *et al.*, 2008).

Measurements of emissions from pig houses have to some extent demonstrated the presence of similar compounds as in dairy houses. However, the relative composition seems to be different, in that pig house emissions contain a higher proportion of carboxylic acids and a lower proportion of alcohols (Feilberg *et al.*,

2010; Liu *et al.*, 2011), with acetic acid being by far the most abundant NMVOCs. In addition, these studies reported high emissions of H_2S at a level comparable to acetic acid.

The data from California, the UK and Denmark compiled in Table 9.9 are based on carbon emissions (except H_2S). Due to the limited amount of data, the data are reported relative to NH_3, for which detailed emissions inventories are available. Assuming that NH_3 emissions are reasonably well correlated with H_2S and NMVOCs emissions, then total emissions from sources can be estimated if NH_3 emissions are available. In general, there is relatively good agreement between reported data even for different types of animals. It is noteworthy that reported average emissions of H_2S relative to NH_3 are very close. Still due to the limited amount of data, it should be noted that H_2S and NMVOCs emissions are not necessarily highly correlated with NH_3 (Feilberg *et al.*, 2010).

9.5   Technologies and Additives to Reduce NH_3 and Odour Emissions

Techniques and methods for reducing gas emissions from management of animal manure were presented in the previous sections, with the focus on NH_3. This section presents air treatment techniques and additives for reducing NH_3 and odour emissions.

9.5.1   Air Treatment Techniques

Chemical and biological air treatment has been introduced in recent years to reduce emissions of both NH_3 and odour from intensive livestock production. The focus has been mainly on pig production, partly because of the particularly offensive smell from pig production facilities and partly because mechanical ventilation, which is a pre-requisite for installing air treatment systems, is often used in pig facilities.

So far, chemical treatment of ventilation air has mainly been limited to acid scrubbers using dilute sulfuric acid to obtain scrubber solutions in the range of pH 2–4. Such a system efficiently reduces emissions of NH_3 due to the uptake of NH_3 by the acidic scrubber solution. With such systems, removal efficiencies of more than 90% can be achieved (Melse and Ogink, 2005).

Using an acid scrubber is relevant only for basic compounds and many odorants are acidic or neutral. An alternative is biological air treatment, which is based on passing ventilated air through a porous filter matrix in which a biofilm of microorganisms can exist at the surface of the supporting matrix. If the microorganisms are able to metabolise a compound present in the air, the compound can be removed from the air stream provided there is sufficient mass transfer of the compound to the biofilm. In order to maintain conditions suitable for microorganism growth, the filter matrix need to be humidified occasionally or continuously. Three types of biological air treatment systems are used: biofilters, biotrickling filters and bioscrubbers.

In a biofilter (Figure 9.4), the filter matrix is humidified by the moisture present in the ventilated air combined with humidification of the air before it enters the filter. An organic filter matrix such as wood chips, peat or straw is often used in biofilters. Relatively high odour, H_2S and NH_3 reductions have in some cases been obtained by application of biofilters for treating air from livestock facilities (Chen and Hoff, 2009), although in some cases high pressure drops across the filter limit the economic feasibility of the filters.

Biotrickling filters (Figures 9.5 and 9.6) are similar to biofilters except that the moisture content is maintained at a higher level by constant or intermittent irrigation of the filter matrix. The filter matrix (e.g. comprising plastic or inorganic materials) is often less degradable than a biofilter. Relatively high NH_3 reductions have been obtained, whereas variable and relatively low odour reductions (0–50%) have been obtained (Melse and Ogink, 2005). Biotrickling filters typically have lower pressure drops and may therefore be associated with lower running costs than biofilters. The microbial NH_3 oxidation in a biotrickling filter is

Table 9.9 Emissions of H_2S and selected VOCs relative to NH_3 (E_{rel} in units of $g\ C\ g^{-1}\ NH_3$ emitted or $g\ H_2S\ g^{-1}\ NH_3$).

Compound	E_{rel} (Germany) (Ngwabie et al., 2008)	E_{rel} (California) (Shaw et al., 2007)	E_{rel} (UK, cattle) (Hobbs et al., 2004)	E_{rel} (UK, pig) (Hobbs et al., 2004)	E_{rel} (Denmark) (Feilberg et al., 2010)	E_{rel} (S. Korea) (Kim et al., 2008)	E_{rel} (EU/US) (Kim et al., 2008)
H_2S	NA	NA	NA	NA	0.14	0.15	0.17
Acetone	$8\text{–}18 \times 10^{-3}$	NA	$10\text{–}20 \times 10^{-3}$	NA	3.1×10^{-3}	NA	NA
Acetic acid	0.03–0.09	$0.2\text{–}1.2 \times 10^{-3}$	0.1–0.2	0.14	0.08	NA	NA
Propanoic acid	$3\text{–}7 \times 10^{-3}$	$0\text{–}1.5 \times 10^{-3}$	$5\text{–}8 \times 10^{-3}$	1×10^{-2}	3.7×10^{-2}	NA	NA
Butanoic acid	NA	NA	2.4×10^{-3}	7.2×10^{-2}	4.2×10^{-2}	NA	NA
4-Methyl phenol	NA	?	0.16	7.9×10^{-2}	5.9×10^{-3}	NA	NA
C_5-acids	NA	NA	0.02	0.02	0.01	NA	NA

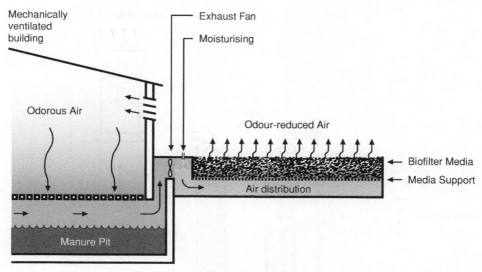

Figure 9.4   *Principle of a biofilter treating air from a pig production facility. (© University of Southern Denmark.)*

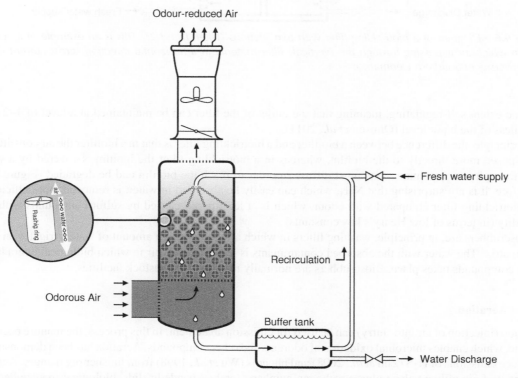

Figure 9.5   *Diagram representing both a biotrickling filter and a bioscrubber with counter-current flow (water flow in opposite direction to air flow). A similar setup can be used for an acid scrubber by substituting the fresh water supply with a supply of sulfuric acid. (© University of Southern Denmark.)*

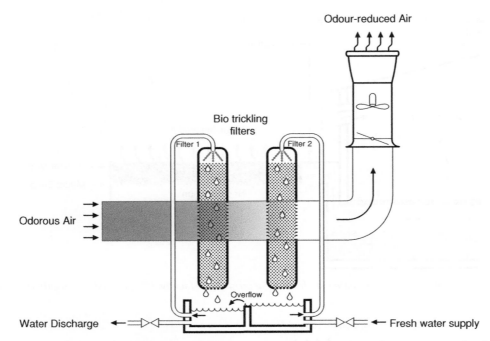

Figure 9.6 *Diagram of a biotrickling filter with two sections (Filter 1 and 2). This is an example of a vertical filter in which air is passing through the (vertical) filter material in a horizontal direction (cross-current flow). (© University of Southern Denmark.)*

to some extent self-regulating, meaning that the outlet of the filter can be maintained at a level of 1–2 ppm regardless of the input level (Ottosen *et al.*, 2011).

In principle, the difference between a biofilter and a biotrickling filter is that in a biofilter the air constituents are exposed more directly to the biofilm, whereas in a biotrickling filter the biofilm is covered by a water layer through which contaminants must diffuse in order to reach the biofilm and be degraded (Figure 9.6). Therefore, it is not surprising that NH_3, which can easily be absorbed in water, is removed more efficiently in a biotrickling filter compared with odour, which is at least partly caused by sulfur compounds with low solubility (in terms of low Henry's law constants).

Bioscrubbers are, in principle, washing filters in which a relatively high amount of water irrigates an inert filter matrix. The water with the absorbed contaminants is led to a reservoir in which biological degradation of the compounds takes place. Bioscrubbers are normally not used for livestock facilities.

9.5.2 Aeration

Aeration (injection of air into slurry) can be used for emission abatement. In this process, the manure becomes aerobic, which enables microbial oxidation of odorants and other compounds. Aeration has been demonstrated to reduce emissions of H_2S (Zhu *et al.*, 2008) and phenols (Wu *et al.*, 1998) from finisher pig manure. Aerobic treatment of pig effluent also reduces pathogen numbers, total suspended solids, biological oxygen demand (BOD), chemical oxygen demand (COD) and content of N (Burton, 1992). When slurry is efficiently aerated, the microbial activity becomes predominantly aerobic and breakdown of organic components, including odorants, and H_2S is accelerated, because aerobic degradation of organic matter is more thermodynamically favourable than anaerobic degradation (Burton, 1992). Although aeration can rapidly reduce odour emissions, persistent effects require more thorough aeration in order to avoid the system quickly returning to the original anaerobic conditions. If volatile solids, BOD and COD are not sufficiently reduced, new odorants will be produced from the residual proteins in the slurry. As an example, it was observed by Ottosen *et al.* (2009) that the combined chemical and biological oxygen consumption rate in untreated slurry was 29 μM min^{-1}. Even with complete saturation by dissolved oxygen resulting in a concentration of around 250 μM, this means that oxygen is depleted within around 10 min. In the same study, a sulfate reduction rate of 0.15 mM day^{-1} was observed. Since total dissolved sulfide concentrations may be in the range of 1–4 mM (Ottosen *et al.*, 2009), H_2S oxidised to sulfate by aeration will be completely regenerated in around 1–3 weeks if the aeration is stopped and significant H_2S emissions can be reached due to sulfate reduction even within a few days after aeration is stopped. If aerobic conditions are not sustained, additional H_2S will be formed from protein degradation if proteins are not aerobically degraded during aeration. As a consequence, one way of using aeration is for short-term treatment prior to handling or application of slurry in order to reduce emissions during these activities. Alternatively, a more complete aerobic treatment can be used to obtain reduced emissions. Measuring BOD and COD can possibly be used as a control of the efficiency of the treatment.

Ozone injection into stored slurry has also been used to reduce emissions of odorants. In aqueous solution, ozone reacts specifically with odorants containing unsaturated carbon–carbon bonds (i.e. phenols and indoles) (Wu *et al.*, 1998), and with sulfur compounds. In the latter case, ozone reacts both with dissolved H_2S and with HS^-, which exist in equilibrium with dissolved H_2S. The latter reaction is particularly fast, which means that the reaction rate of ozone towards dissolved sulfide (H_2S and HS^-) is pH-dependent (Hoigné *et al.*, 1985). In a combined slurry treatment (separation, ozonation, acidification) for slurry from a pit in a pig finisher unit, it has been demonstrated that H_2S emissions can be significantly reduced by around 80% (Liu *et al.*, 2011), whereas emissions of other odorants are less efficiently reduced below around 80%. This reflects the high reactivity of H_2S/HS^-, but may also be related to the source of odorants, which in the case of H_2S is primarily manure collected in the pit, whereas other odorants may also be emitted from surfaces in the animal room.

Aeration and ozonation can be controlled by measuring the redox potential of the slurry (Hjorth *et al.*, 2012). Redox potential is measured as E_h (relative to a standard hydrogen electrode) or as E_{cal} (relative to a standard calomel electrode), both in units of V or mV. The relationship between the two metrics is: $E_h = E_{cal} + 241$ mV. A level of $+250$ mV E_h (corresponding to 2–3% oxygen saturation) has been suggested to reduce COD, BOD and also the perceived odour offensiveness of treated slurry (Evans *et al.*, 1986).

Both for ozonation and aeration, there is a risk of increasing pH as a consequence of the treatment. This is mainly due to release of acidic CO_2 during treatment and may lead to increased emissions of NH_3. If microbial nitrification and denitrification processes are induced, there is also a risk of increasing emissions of N_2O, but this has not been fully investigated.

9.5.3   Additives

Reducing the amount of crude protein fed to pigs by balancing the essential amino acids in the diet with commercial amino acids significantly reduces N excretion, primarily in urine and to a lesser extent in the faecal fraction (Philippe *et al.*, 2011). Faecal N is mainly in the form of organic N compounds that are slowly transformed to TAN in manure and is thus not a significant source of TAN. Urinary N is in the form of urea, which is easily transformed to TAN and total inorganic carbon. The reduced dietary N intake therefore significantly reduces NH_3 emissions.

Reducing the amount of crude protein fed to pigs also reduces the addition of Na^+ and K^+, which in turn lowers the pH. Consequently, this decreases NH_3 emissions, but at the same time the emissions potential of acidic odorants increases, since a larger fraction is present in neutral form.

Salts may be added either in the feed or directly to the slurry to change the pH of the slurry and, as a consequence, gas emissions. Thus, instead of adding lime to pig feed to provide Ca for growth, then $CaCl_2$ or $CaSO_4$ may be added (Chapters 4 and 8). These components reduce the pH of the slurry and thus reduce emissions of NH_3 and affect odour emissions.

Increasing the amount of non-starch carbohydrates in the diet may increase volatile fatty acid production, which decreases pH significantly (Cahn *et al.*, 1998). This is not the only effect, as changing the content of non-starch carbohydrates by altering the composition of the feed may also affect the dietary content of the monovalent ions Na^+, K^+ and Cl^-. The relationship between these components in slurry affects pH, as mentioned above. The content of the ions in feed is defined by pig diet specialists (animal physiologists) as dietary electrolyte balance $dEB = Na^+ + K^+ - Cl^-$ (meq kg^{-1} (DM)). The pH decreases at decreasing dEB due to excretion of H^+ (NH_4^+), which compensates for low Na^+ and K^+ in relation to Cl^- concentration (Philippe *et al.*, 2011).

Addition of acid reduces the pH of the manure (Chapter 4) and NH_3 emissions from livestock houses, manure stores and manure applied in the field (Kai *et al.*, 2008). Treatment of the slurry with acids in the animal house may reduce emissions from all compartments of the manure management. In recently developed technology, slurry in the animal house is acidified to pH below 6. Acidification of the slurry may reduce NH_3 emissions from houses by 70% compared with standard techniques. From stored slurry, the emissions are reduced to less than 10% and the NH_3 emissions from field-applied slurry are reduced by 67%. Alternatively, adding acid during slurry application has been found to be an efficient NH_3 abatement technique.

In addition to the treatments mentioned above, a number of additives have been tested with respect to reduction of NH_3 and odour emissions. Although some positive results have been obtained, this area needs further quality-assured tests in order to identify promising and realistic treatments.

9.6 Summary

Atmospheric emissions of NH_3 and odour are adverse effects of production, handling and application of animal manure. Ammonia is an environmental problem causing eutrophication and acidification, whereas odour is mainly considered to be a nuisance. In terms of measurement, odour is a complex sensory effect caused by a number of odorants and reliable quantification of odour is a challenge.

Ammonia is emitted from all types of animal production facilities, from storage facilities and from field application of manure. Standard emission factors are presented for these sources. Emissions from animal production facilities are affected by building design (e.g. floor type). Emissions from storage are reduced by covering or by formation of a surface crust, whereas emissions from field application can be reduced by incorporation into soil.

Emissions of odour and odorants are less well-described compared with emissions of NH_3 and there is a need for more data on emissions factors for odour and specific odorants.

A number of techniques for reducing gaseous emissions have been developed. These include end-of-pipe air cleaning by chemical or biological methods and treatment of manure (e.g. by acidification or aeration). Reduction efficiency has mainly been obtained for NH_3, whereas abatement of odour is more challenging.

References

Buijsman, E., Maas, J.F. and Asman, W.A.H. (1987) Anthropogenic NH_3 emissions in Europe. *Atmos. Environ.*, **21**, 1009–1022.

Bulliner, E.A., Koziel, J.A., Cai, L.S. and Wright, D. (2006) Characterization of livestock odors using steel plates, solid-phase microextraction, and multidimensional gas chromatography-mass spectrometry-olfactometry. *J. Air Waste Manag. Assoc.*, **56**, 1391–1403.

Burton, C.H. (1992) A review of the strategies in the aerobic treatment of pig slurry: purpose, theory and method. *J. Agric. Eng. Res.*, **53**, 249–272.

Cabrera, M.L. and Chiang, S.C. (1994) Water content effect on denitrification and ammonia emission in poultry litter. *Soil Sci. Soc. Am. J.*, **58**, 811–816.

CEN (2003) *Air Qualitty – Determination of Odour Concentration by Dynamic Olfactometry (EN13725)*. European Committe for Standardization, Brussels, Belgium.

Chambers, B.J., Smith, K. and van der Weerden, T.J. (1997) Ammonia emissions following the land spreading of solid manures, in *Gaseous Nitrogen Emissions from Grasslands* (eds S.C. Jarvis and B.F. Pain), CABI, Wallingford, pp. 275–280.

Chen, L. and Hoff, S.J. (2009) Mitigating odors from agricultural facilities: a review of literature concerning biofilters. *Appl. Eng. Agric.*, **25**, 751–766.

Clanton, C.J., Schmidt, D.R., Nicolai, R.E., Goodrich, P.R., Jacobson, L.D., Janni, K.A., Weisberg, S. and Buckel, J.A. (1999) Dynamic olfactometry variability in determining odor dilutions-to-threshold. *Trans. ASAE*, **42**, 1103–1112.

Evans, M.R., Deans, E.A., Smith, M.P.W., Svoboda, I.F. and Thacker, F.E. (1986) Aeration and control of slurry odors by heterotrophs. *Agric. Wastes*, **15**, 187–204.

Feilberg, A., Liu, D., Adamsen, A.P.S., Hansen, M.J. and Jonassen, K.E.N. (2010) Odorant emissions from intensive pig production measured by online proton-transfer-reaction mass spectrometry. *Environ. Sci. Technol.*, **44**, 5894–5900.

Gay, S.W., Schmidt, D.R., Clanton, C.J., Janni, K.A., Jacobson, L.D. and Weisberg, S. (2003) Odor, total reduced sulfur, and ammonia emissions from animal housing facilities and manure storage units in Minnesota. *Appl. Eng. Agric.*, **19**, 347–360.

Gralapp, A.K., Powers, W.J. and Bundy, D.S. (2001) Comparison of olfactometry, gas chromatography, and electronic nose technology for measurement of indoor air from swine facilities. *Trans. ASAE*, **44**, 1283–1290.

Groenestein, C.M. (1993) Animal-waste management and emission of ammonia from livestock housing systems: field studies, in *Livestock Environment IV* (eds E. Collins and C. Boon), ASAE, St Joseph, MI, pp. 1169–1176.

Groenestein, C.M. and Van Faassen, H.G. (1996) Volatilization of ammonia, nitrous oxide and nitric oxide in deep-litter systems for fattening pigs. *J. Agric. Eng. Res.*, **65**, 269–274.

Groot Koerkamp, P.W.G. (1994) Review on emissions of ammonia from housing systems for laying hens in relation to sources, processes, building design and manure handling. *J. Agric. Res.*, **59**, 73–87.

Groot Koerkamp, P.W.G., Speelman, L. and Metz, J.H.M. (1998) Litter composition and ammonia emission in aviary houses for laying hens. Part I. Performance of a litter drying system. *J. Agric. Eng. Res.*, **70**, 375–382.

Hansen, M.J. (2007) Standardemissioner for lugt, *Artikel 0024*, Dansk Landbrugsrådgivning, Landscentret, Byggeri and Teknik, Aarhus.

Hansen, M.N., Sommer, S.G., Hutchings, N.J. and Sørensen, P. (2008) Emissionsfaktorer til beregning af ammoniakfordampning ved lagring og udbringning af husdyrgødning [Emission factors for calculation of ammonia volatilization by storage and application of animal manure], *DJF Rapport Husdyrbrug 84*, Ministry of Food, Agriculture and Fisheries, Danish Institute of Agricultural Sciences, Tjele.

Hansen, M.J., Adamsen, A.P.S., Feilberg, A. and Jonassen, K. (2011) Storage stability of odorants from pig production in air sampling bags used for olfactometry. *J. Environ. Qual.*, **40**, 1096–1102.

Hayes, E.T., Curran, T.P. and Dodd, V.A. (2006) Odour and ammonia emissions from intensive pig units in Ireland. *Bioresour. Technol.*, **97**, 940–948.

Hjorth, M., Pedersen, C.Ø. and Feilberg, A. (2012) Redox potential as a means to control the treatment of slurry to lower H_2S emissions. *Sensors*, **12**, 5349–5362.

Hobbs, P.J., Misselbrook, T.H. and Cumby, T.R. (1999) Production and emission of odours and gases from ageing pig waste. *J. Agric. Eng. Res.*, **72**, 291–298.

Hobbs, P.J., Misselbrook, T.H., Dhanoa, M.S. and Persaud, K.C. (2001) Development of a relationship between olfactory response and major odorants from organic wastes. *J. Sci. Food Agric.*, **81**, 188–193.

Hobbs, P.J., Webb, J., Mottram, T.T., Grant, B. and Misselbrook, T.M. (2004) Emissions of volatile organic compounds originating from UK livestock agriculture. *J. Sci. Food Agric.*, **84**, 1414–1420.

Hoigné, J., Bader, H., Haag, W.R. and Staehlin, J. (1985) Rate constants of reactions of ozone with organic and inorganic compounds in water – III. Inorganic compounds and radicals. *Water Res.*, **19**, 993–1004.

Jeppsson, K.H. (1999) Volatilization of ammonia in deep-litter systems with diVerent bedding materials for young cattle. *J. Agric. Eng. Res.*, **73**, 49–57.

Kai, P., Kaspers, B. and van Kempen, T. (2006) Modelling sources of gaseous emissions in a pig house with recharge pit. *Trans. ASABE*, **49**, 1479–1485.

Kai, P., Pedersen, P., Jensen, J.E., Hansen, M.N. and Sommer, S.G. (2008) A whole-farm assessment of the efficacy of slurry acidification in reducing ammonia emissions. *Eur. J. Agron.*, **28**, 148–154.

Kim, K.Y., Ko, H.J., Kim, H.T., Kim, Y.S., Roh, Y.M., Lee, C.M., Kim, H.S. and Kim, C.N. (2007) Sulfuric odorous compounds emitted from pig-feeding operations. *Atmos. Environ.*, **41**, 4811–4818.

Kim, K.Y., Ko, H.J., Kim, H.T., Kim, Y.S., Roh, Y.M., Lee, C.M., Kim, H.S. and Kim, C.N. (2008) Quantification of ammonia and hydrogen sulfide emitted from pig buildings in Korea. *J. Environ. Manag.*, **88**, 195–202.

Kroodsma, W., Scholtens, R. and Huis in 't Veld, J.W.H. (1988) Ammonia emission from poultry housing systems, in *Volatile Emissions from Livestock Farming and Sewage Operations* (eds V.C. Nielsen, J.H. Voorburg and P. L'Hermite), Elsevier, London, pp. 152–161.

Liu, D., Feilberg, A., Adamsen, A.P.S. and Jonassen, K.E.N. (2011) The effect of slurry treatment including ozonation on odorant reduction measured by in-situ PTR-MS. *Atmos. Environ.*, **45**, 3786–3793.

Lyngbye, M., Hansen, M.J., Riis, A.L., Jensen, T.L. and Sørensen, G. (2006) 1000 Olfactometry analyses and 100 TD-GC/MS analyses to evaluate methods for reducing odour from finishing units in Denmark, presented at the *Workshop on Agricultural Air Quality: State of the Science*, North Carolina State University, Raleigh, NC.

Malek, E., Davis, T., Martin, R.S. and Silva, P.J. (2006) Meteorological and environmental aspects of one of the worst national air pollution episodes (January, 2004) in Logan, Cache Valley, Utah, USA. *Atmosph. Res.*, **79**, 108–122.

McGinn, S.M. and Janzen, H.H. (1998) Ammonia sources in agriculture and their measurement. *Can. J. Soil Sci.*, **78**, 139–148.

Melse, R.W. and Ogink, N.W.M. (2005) Air scrubbing techniques for ammonia and odor reduction at livestock operations: review of on-farm research in the Netherlands. *Trans. ASAE*, **48**, 2303–2313.

Moal, J.F., Martinez, J., Guiziou, F. and Coste, C.M. (1995) Ammonia volatilization following surface-applied pig and cattle slurry in France. *J. Agric. Sci.*, **125**, 245–252.

Monteny, G.J. and Erisman, J.W. (1998) Ammonia emission from dairy cow buildings: a review of measurement techniques, influencing factors and possibilities for reduction. *Neth. J. Agric. Sci.*, **46**, 225–227.

Ngwabie, N.M., Schade, G.W., Custer, T.G., Linke, S. and Hinz, T. (2008) Abundances and flux estimates of volatile organic compounds from a dairy cowshed in Germany. *J. Environ. Qual.*, **37**, 565–573.

Ni, J.Q. (1999) Mechanistic models of ammonia release from liquid manure: a review. *J. Agric. Eng. Res.*, **72**, 1–17.

Ni, J.Q. and Heber, A.J. (2008) Sampling and measurement of ammonia at animal facilities. *Adv. Agron.*, **98**, 201–269.

Ni, J.Q., Vinckier, C., Coenegrachts, J. and Hendriks, J. (1999) Effect of manure on ammonia emission from a fattening pig house with partly slatted floor. *Livestock Prod. Sci.*, **59**, 25–31.

Nimmermark, S. (2004) Odour influence on well-being and health with specific focus on animal production emissions. *Ann. Agric. Environ. Med.*, **11**, 163–173.

Oenema, O., Bannink, A., Sommer, S.G. and Velthof, G. (2001) Gaseous nitrogen emission from livestock farming system, in *Nitrogen in the Environment: Sources, Problems, and Management* (eds R.F. Follet and B.V. Hatfield), Elsevier, Amsterdam, pp. 255–289.

Ogink, N.W.M. and Groot Koerkamp, P.W.G. (2001) Comparison of odour emissions from animal housing systems with low ammonia emission. *Water Sci. Technol.*, **44**, 245–252.

O'Neill, D.H. and Phillips, V.R. (1992) A review of the control of odor nuisance from livestock buildings. 3. Properties of the odorous substances which have been identified in livestock wastes or in the air around them. *J. Agric. Eng. Res.*, **53**, 23–50.

Ottosen, L.D.M., Poulsen, H.V., Nielsen, D.A., Finster, K., Nielsen, L.P. and Revsbech, N.P. (2009) Observations on microbial activity in acidified pig slurry. *Biosyst. Eng.*, **102**, 291–297.

Ottosen, L.D.M., Juhler, S., Guldberg, L.B., Feilberg, A., Revsbech, N.P. and Nielsen, L.P. (2011) Regulation of ammonia oxidation in biotrickling airfilters with high ammonium load. *Chem. Eng. J.*, **167**, 198–205.

Pain, B.F., Phillips, V.R., Clarkson, C.R. and Klarenbeek, J.V. (1989) Loss of nitrogen through ammonia volatilisation during and following the application of pig or cattle slurry to grassland. *J. Sci. Food Agric.*, **47**, 1–12.

Philippe, F.-X., Cabaraux, J.-F. and Nicks, B. (2011) Review: ammonia emissions from pig houses: influencing factors and mitigation techniques. *Agric. Ecosyst. Environ.*, **141**, 245–260.

Poulsen, H.D., Børsting, C.F., Rom, H.B. and Sommer, S.G. (2001) Kvælstof, fosfor og kalium i husdyrgødning – normtal 2000 [Standard values of nitrogen, phosphorous and potassium in animal manure], *DJF Rapport 36*, Ministry of Food, Agriculture and Fisheries, Danish Institute of Agricultural Sciences, Tjele.

Rappert, S. and Müller, R. (2005) Odor compounds in waste gas emissions from agricultural operations and food industries. *Waste Manag.*, **25**, 887–907.

Riis, A.L. (2006) Standardtal for lugtemission fra Danske svinestalde om sommeren, *Meddelelse 742*, Danish Pig Research Centre, Copenhagen.

Schiffman, S.S. and Williams, C.M. (2005) Science of odor as a potential health issue. *J. Environ. Qual.*, **34**, 129–138.

Schiffman, S.S., Bennett, J.L. and Raymer, J.H. (2001) Quantification of odors and odorants from swine operations in North Carolina. *Agric. Forest Meteorol.*, **108**, 213–240.

Shah, S.B., Westerman, P.W. and Arogo, J. (2006) Measuring ammonia concentrations and emissions from agricultural land and liquid surfaces: a review. *J. Air Waste Manag. Assoc.*, **56**, 945–960.

Shaw, S.L., Mitloehner, F.M., Jackson, W., Depeters, E.J., Fadel, J.G., Robinson, P.H., Holzinger, R. and Goldstein, A.H. (2007) Volatile organic compound emissions from dairy cows and their waste as measured by proton-transfer-reaction mass spectrometry. *Environ. Sci. Technol.*, **41**, 1310–1316.

Søgaard, H.T., Sommer, S.G., Hutchings, N.J., Huijsmans, J.F.M., Bussink, D.W. and Nicholson, F. (2002) Ammonia volatilization from field-applied animal slurry – the ALFAM model. *Atmos. Environ.*, **36**, 3309–3319.

Sommer, S.G., Zhang, G.Q., Bannink, A., Chadwick, D., Hutchings, N.J., Misselbrook, T., Menzi, H., Ni., J.Q., Oenema, O., Webb, J. and Monteny, G.-J. (2006a) Algorithms determining ammonia emission from livestock houses and manure stores. *Adv. Agron.*, **89**, 261–335.

Sommer, S.G., Jensen, B.-E., Hutchings, N., Lundgaard, N.H., Grønkjær, A., Birkmose, T.S., Pedersen, P. and Jensen, H.B. (2006b) Emissionskoefficienter til brug ved beregning af ammoniakfordampning fra stalde, *DJF Rapport Husdyrbrug 70*, Ministry of Food, Agriculture and Fisheries, Danish Institute of Agricultural Sciences, Tjele.

Sutton, M.A., Oenema, O., Erisman, J.W., Leip, A., van Grinsven, H. and Winiwarter, W. (2011) Too much of a good thing. *Nature*, **472**, 159–161.

Thorman, R.E., Hansen, M.N., Misselbrook, T.H. and Sommer, S.G. (2008) Algorithm for estimating the crop height effect on ammonia emission from slurry applied to cereal fields and grassland. *Agron. Sust. Dev.*, **28**, 373–378.

Trabue, S.L., Scoggin, K.D., Li, H., Burns, R. and Xin, H.W. (2008) Field sampling method for quantifying odorants in humid environments. *Environ. Sci. Technol.*, **42**, 3745–3750.

Webb, J., Sommer, S.G., Kupper, T., Groenestein, K., Hutchings, N.J., Eurich-Menden, B., Rodhe, L., Misselbrook, T.H. and Amon, B. (2012) Emissions of ammonia, nitrous oxide and methane during the management of solid manures. *Sust. Agric. Rev.*, **8**, 67–107.

Wu, J.J., Park, S.H., Hengemuehle, S.M., Yokoyama, M.T., Person, H.L. and Masten, S.J. (1998) The effect of storage and ozonation on the physical, chemical, and biological characteristics of swine manure slurries. *Ozone Sci. Eng.*, **20**, 35–50.

Yu, Z., Guo, H. and Laguë, C. (2010) Livestock odor dispersion modeling: a review. *Trans. ASABE*, **53**, 1231–1244.

Zahn, J.A., DiSpirito, A.A., Do, Y.S., Brooks, B.E., Cooper, E.E. and Hatfield, J.L. (2001) Correlation of human olfactory responses to airborne concentrations of malodorous volatile organic compounds emitted from swine effluent. *J. Environ. Qual.*, **30**, 624–634.

Zhu, J., Jacobson, L.D., Schmidt, D.R. and Nicolai, R.E. (1999) Daily variations in odor and gas emissions from animal facilities, *ASAE Paper 994146*, ASAE, St Joseph, MI.

Zhu, J., Zhang, Z. and Miller, C. (2008) Odor and aeration efficiency affected by solids in swine manure during post-aeration storage. *Trans. ASABE*, **51**, 293–300.

10

Greenhouse Gas Emissions from Animal Manures and Technologies for Their Reduction

Sven G. Sommer[1], Tim J. Clough[2], David Chadwick[3] and Søren O. Petersen[4]

[1]*Institute of Chemical Engineering, Biotechnology and Environmental Technology, University of Southern Denmark, Denmark*
[2]*Faculty of Agriculture and Life Sciences, Lincoln University, New Zealand*
[3]*School of Environment, Natural Resources and Geography, Bangor University, Environment Centre for Wales, UK*
[4]*Department of Agroecology, Aarhus University, Denmark*

10.1 Introduction

The greenhouse gases (GHGs) emitted from livestock production include carbon dioxide (CO_2) from fossil fuel consumption, methane (CH_4) from livestock, organic wastes and animal manures, and nitrous oxide (N_2O) from organic wastes (e.g. from abattoirs, food processing industries or organic household waste), mineral fertilisers and animal manure management.

Methane and N_2O are the main targets of mitigation efforts, because these gases have a large capacity to absorb the radiation that is reflected from the Earth. The global warming potential per kilogram emitted of N_2O and CH_4 is around 296 and 23 times higher, respectively, than that of CO_2 over a 100-year time period (Text Box – Basic 3.1 and 10.1). The CH_4 molecule accounts for 30% of the anthropogenic contributions to net global warming and N_2O accounts for 10% (Solomon *et al.*, 2007). The atmospheric CH_4 concentration has increased 2.5-fold and that of N_2O by approximately 20% since the pre-industrial era (IPCC, 2007).

Animal Manure Recycling: Treatment and Management, First Edition. Edited by Sven G. Sommer, Morten L. Christensen, Thomas Schmidt and Lars S. Jensen.

Text Box – Basic 10.1   Terminology and definitions – climate warming and greenhouse gases

Archaea: Archaea and bacteria are quite similar in size and shape. Despite this visual similarity to bacteria, Archaea possess genes and several metabolic pathways that are more closely related to those of eukaryotes, notably the enzymes involved in transcription and translation. Other aspects of Archaean biochemistry are unique, such as their reliance on ether lipids in their cell membranes. Archaea use a much greater variety of sources of energy than eukaryotes: ranging from familiar organic compounds such as sugars, to ammonia, metal ions or even hydrogen gas.

BMP or B_o: The biochemical methane potential (BMP) or the optimal biogas production potential (B_o) measured with batch fermentation until no further biogas or CH_4 is produced.

Concentrate: A feed component comprised mainly of grain and plant material that contains a high density of nutrients, usually low in crude fibre content (less than 18% of dry matter (DM)) and high in total digestible nutrients, used to supply energy and protein sources.

CO_2-equivalent: The global warming effect of GHGs is given in kg CO_2-equivalents and depends on the residence time of the gas in the atmosphere and the energy absorption potential. For CH_4 the global warming potential over a 100-year period is 25 kg CO_2-eq. kg^{-1} CH_4, while that for N_2O is 298 kg CO_2-eq. kg^{-1} N_2O (Forster *et al.*, 2007).

Digestate: Slurry or organic waste (e.g. from abattoirs, dairies, organic household waste) that has been fermented in a biogas reactor. Digestate has a low dry matter (DM) concentration, high total ammoniacal N (TAN) and a low content of digestible volatile solids.

Greenhouse gas (GHG): Gas with the potential to absorb the Earth's radiation and reflect it back into the troposphere.

Intergovernmental Panel on Climate Change (IPCC): The leading international body for the assessment of climate change. The United Nations Environment Programme (UNEP) and the World Meteorological Organization (WMO) established the IPCC to obtain scientific knowledge of climate change and its potential environmental and socio-economic impacts (http://ipcc.ch/index.htm#.UDCq544mYlI).

Livestock unit (LU): The reference livestock unit is considered to be the maintenance of a mature black and white dairy cow yielding an average annual milk yield (Anonymous, 2009).

Methane conversion factor (MCF): The IPCC has proposed that CH_4 emissions from livestock manure be calculated by first estimating the maximum potential CH_4 production (B_o) from a given animal category. Actual emissions are then calculated by multiplying the potential emissions by the MCF factor, which is related to regional conditions and the manure management system. This is the simplest calculation procedure, called Tier One in the accounting system.

Roughage: Feed component mainly consisting of vegetative, less digestible parts of crops (leaves, stalk, roots) which may be dried, ensiled and so on.

Enteric fermentation of organic matter by ruminants is a major source of CH_4 emissions (Figure 10.1), contributing 35–40% of atmospheric CH_4. The CH_4 losses from manure management contribute approximately 20% of the total agricultural CH_4 emissions for most countries (Steinfeld *et al.*, 2006). Between countries, the variation in the percentage contribution of manure to CH_4 emissions reflects differences in the duration of manure storage, the proportion of ruminant livestock relative to other livestock types and livestock production systems (Chadwick *et al.*, 2011). In countries such as Australia with free-roaming cattle, manure contributes only a small proportion of the CH_4 emissions from livestock production. In addition to emitting CH_4, manure from livestock production systems is estimated to contribute as much as 30–50% of the global N_2O emissions from agriculture (Oenema *et al.*, 2005). Most of this N_2O is produced and emitted from manure applied in the field, but some comes from the storage of solid manure.

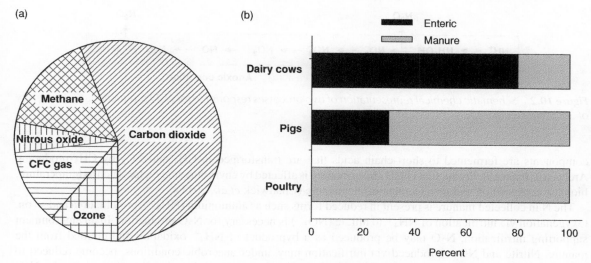

Figure 10.1 *(a) Distribution of greenhouse gases in the atmosphere, CFC = chlorofluorocarbon. (Based on Forster* et al. *(2007). © 2007 Cambridge University Press.) (b) Proportions of CH_4 emitted from animals via enteric fermentation and via animal manure. Animals do not release N_2O, so manure and mineral fertilisers are the only sources of N_2O emissions from livestock production.*

Countries that are signatories to the Kyoto Protocol are moving to regulate emissions of CH_4 and N_2O from livestock production. Consequently, livestock producers and the livestock industry as a whole are developing management systems and technologies for reducing these emissions. In the development of appropriate technologies, it must be borne in mind that manure management represents a continuum that ranges from initial manure excretion by livestock, storage and treatment of manure, and finally spreading of manure on agricultural land (Petersen *et al.*, 2007). Emissions of N_2O and CH_4 may occur at each stage of this continuum. When considering GHG mitigation, it is important to consider GHG emissions on the basis of a whole system approach, because the effects of mitigation methods at one stage may affect emissions in downstream stages (Chadwick *et al.*, 2011). Carbon (C) and nitrogen (N) transformations interact, and this interaction significantly affects N_2O and CH_4 emissions (Petersen and Sommer, 2011). Therefore, potential transformations of both C and N pools must be considered when developing GHG mitigation options for manure management. In addition, indirect losses of GHGs may occur as a result of NH_3 emitted from manure, which is subsequently converted upon deposition into N_2O, further contributing to GHG emissions (IPCC, 2007). In addition, a proportion of any NO_3^- that is leached or lost from land in runoff is subsequently lost as N_2O within the watercourse. This also represents an indirect source of N_2O. However, such indirect N_2O emissions are not considered in this chapter.

10.2 Processes of Methane and Nitrous Oxide Production

Methane is only produced in strictly anaerobic environments, while N_2O is produced in more complex environments with low oxygen concentrations, typically mosaics of aerobic and anaerobic micro-environments or interfaces where aerobic and anaerobic conditions meet (Wrage *et al.*, 2001; Russow *et al.*, 2009).

Methane is only produced in manure when the oxygen consumption rate is higher than the rate of oxygen supply to the site of consumption. The production of CH_4 involves degradation and hydrolysis of organic material to organic compounds, which are then degraded to long-chain acids, proteins or alcohols. These

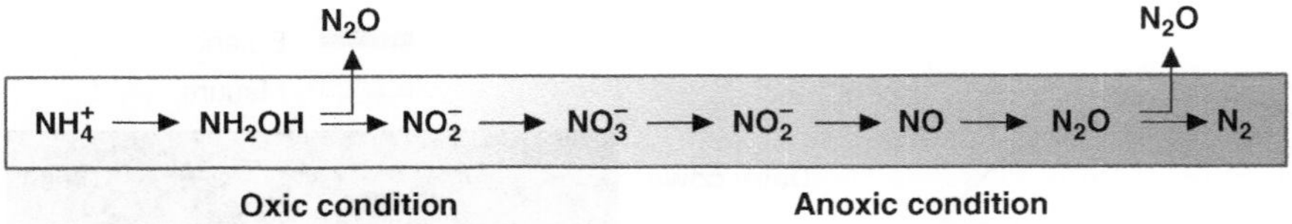

$$N_2O \qquad\qquad\qquad\qquad\qquad\qquad N_2O$$

$$NH_4^+ \longrightarrow NH_2OH \rightleftharpoons NO_2^- \longrightarrow NO_3^- \longrightarrow NO_2^- \longrightarrow NO \longrightarrow N_2O \rightleftharpoons N_2$$

Oxic condition   Anoxic condition

Figure 10.2 *Schematic chemical representation of two processes responsible for N_2O production. (© University of Southern Denmark.)*

components are fermented to short-chain acids that are transformed to CH_4 and CO_2 by CH_4-producing Archaea (Chapter 5). Production of CH_4 from manure is affected by environmental factors such as temperature, biomass composition and management of the manure (Chadwick *et al.*, 2011).

The N in collected manure is present in reduced forms such as ammonium (NH_4^+), protein, urea and so on. Oxygenation via nitrification of NH_4^+ to nitrate (NO_3^-) is necessary for N_2O production. In an environment supporting nitrification, N_2O may be produced as a byproduct of NH_4^+ oxidation and emitted from the manure. Nitrite and NO_3^- produced via nitrification may, under anaerobic conditions, become reduced to gaseous N_2O and N_2. However, in micro-sites with low levels of oxygen there is a greater tendency for N_2O emissions due to incomplete denitrification (Figure 10.2).

The population of microorganisms can be a significant factor affecting GHG production and emissions. For example, the substrates for nitrifiers are NH_4^+ and NO_2^-, which are ineffective energy sources, and therefore the autotrophic nitrifiers grow slowly (Petersen *et al.*, 1991). In environments with relatively few nitrifiers, such as soil, the nitrification of total ammoniacal N ($TAN = NH_4^+ + NH_3$) in manure may become significant only after a lag phase of several days (Petersen *et al.*, 1996).

Methanogenesis in the digestive tract of livestock is mainly due to H_2-utilising methanogens, whereas CH_4 production in the slurry environment probably depends on the activity of slow-growing acetotrophic methanogens (Demirel and Scherer, 2008). If slurry pits or outside stores contain aged slurry with an adapted microflora, CH_4 production is not delayed by the microbial capacity to produce CH_4. In contrast, when slurry channels or a store has been completely emptied, the absence of old manure that may serve as an inoculum can result in low CH_4 production rates for several months (Massé *et al.*, 2003).

10.3 Methane Production from Manure

Organic matter, which is the source of CH_4, contains lignified organic components that are recalcitrant to hydrolysis and therefore only slowly degradable. Consequently, CH_4 production is not related to total organic matter, but rather to the hydrolysable, or digestible, organic matter, which is the fraction that is not lignified (Triolo *et al.*, 2011). Furthermore, the ratio of CH_4 to CO_2 produced varies depending on the composition of the organic matter, with lipid-containing manures producing more CH_4 per C atom than manures rich in carbohydrate, because lipids have few oxygen-containing functional groups (Triolo *et al.*, 2011). Anaerobically stored manure with abundant strawlitter thus produces less CH_4 per C unit in the manure than manure without litter.

In addition, there are upper and lower pH limits for CH_4 production. This is partly due to the effect of hydrogen sulfide (H_2S), organic acids and NH_3, and to the combined effect of environmental conditions on the growth and activity of microorganisms. A pH between 6 and 8 is suitable for CH_4 production, and therefore lowering of the pH reduces CH_4 emissions from manure during storage (Berg *et al.*, 2006; Petersen *et al.*, 2012).

10.3.1   Effect of Temperature

Methane production in anaerobic manure is insignificant at temperatures below 15 °C (Clemens *et al.*, 2006), but above this threshold methanogenesis increases exponentially with temperature. A precondition for significant CH_4 production at a given temperature is the presence of microorganisms adapted to the temperature and manure composition.

In animal houses, manure may be collected as anaerobic slurry in channels or a pit. If no residual slurry containing adapted methanogens has been left in the channel at the beginning of a slurry collection period, little or no CH_4 is initially produced. The length of the period before the onset of significant CH_4 production varies; for example, at 5–15 °C no CH_4 may be produced without inoculum mixed into the manure (Zeeman *et al.*, 1988) or production may not begin for several months (Massé *et al.*, 2003). At 20 °C the production of CH_4 has been found to start after approximately 20 days (Sommer *et al.*, 2007). In contrast, CH_4 production may start immediately when fresh slurry is mixed with more than 7% old slurry (Sommer *et al.*, 2007).

Slurry from biogas reactors contains microorganisms adapted to higher temperatures than those in the slurry tank or lagoon, but CH_4 production may still be high from the stored manure during the cooling phase (Sommer *et al.*, 2000). The temperature decreases only slowly, which may allow the microflora of the digester to adapt to lower ambient temperature. If the temperature of digested slurry is reduced to 20 °C in a heat exchanger at the biogas plant prior to storage, then CH_4 emissions from stored digested slurry (digestate) may be reduced partly due to the microorganisms being poorly adapted to the environment and partly due to the lower content of degradable organic matter in digested slurry.

10.3.2   Manure Storage Methods

The few studies of CH_4 emissions from animal housing conducted to date show that slurry stored in pig houses is a significant, although variable, source of CH_4 (Table 10.1). One source of variation is related to

Table 10.1   *Methane emissions from manure stored in-house and stored outside: emission factors are adapted from reviews by Sommer (2005) and Webb et al. (2012) (note that three different sets of units are used).*

Animal	CH_4 source	Manure	Mean	Maximum	Minimum	SD	Reference
			kg CH_4 animal place^{-1a} year^{-1}				
Dairy	in-house	deep litter	450			360	Webb *et al.* (2012)
		FYM, tied stall	72			7	Webb *et al.* (2012)
Beef	in-house	straw flow	81			7	Webb *et al.* (2012)
Pig	in-house	FYM	3.2			1.8	Webb *et al.* (2012)
			kg CH_4 LU^{-1b} year^{-1}				
Pig	in-house	slurry	40	69	10		Sommer (2005)
Dairy	outside	slurry	107	196	49		Sommer (2005)
			% of C				
Cattle	outside store	FYM	3.5	9.7	0.5	3.3	Webb *et al.* (2012)
		deep litter	0.02	0.03	0.00	0.01	Webb *et al.* (2012)

[a] An animal place is defined as the space needed for breeding one animal. Emissions of CH_4 from in-house stored manure are given by Webb *et al.* (2012) in g CH_4 per animal place per day. We assumed that emissions are constant over the year to give annual emissions per animal place.
[b] LU = livestock units, representing the production of animals defined as 100 kg N in slurry transported from the manure store.

ventilation. For example, in pig houses the indoor temperature may be 20–50% lower with natural ventilation than in houses with forced ventilation. Consequently, CH_4 emissions are significantly lower in the naturally ventilated pig houses. Differences in animal diet also affect CH_4 emissions. For example, dairy cows fed only roughage excrete less digestible volatile solids in the manure, and consequently the CH_4 production potential of the slurry is lower than that of a slurry from dairy cows fed a ration of roughage and concentrates (Møller et al., 2004). Pigs, which are fed rations with readily digestible organic matter, produce slurry that has a higher CH_4 emission potential per organic matter unit than cattle slurry (Massé et al., 2003).

In houses with a solid floor and straw bedding, the hooves of cattle also compact the litter. Aerobic microbial activity in the surface layers of a deep litter of between 0.5 and 2 m thickness may cause the temperature to increase to 40–50 °C at 10 cm depth. In the 0- to 10-cm layer, oxygen entering the mat is depleted and below approximately 10 cm the litter is anaerobic. Measurements in cattle houses with deep litter show that daily CH_4 emissions range from 30 to 70 g C ton^{-1} of litter. The emissions from deep litter mats in pig houses may differ considerably from those measured in cattle houses. In pig houses the deep litter is mixed by the pigs due to nest building and due to the sharper hooves of the pigs. It is assumed that CH_4 emissions are lower from deep pig litter mats than from deep cattle litter mats as a result of litter being mixed by the pigs and possibly also oxidation of CH_4 (Webb et al., 2012). The few studies of emissions from deep litter in pig houses conducted to date indicate that emissions are indeed lower than those from pig houses with slats and slurry tanks.

Outside slurry stores are sources of CH_4 because the environment in these stores, if not actively aerated, favours methanogenesis and because slurry is mostly stored over a long period. As noted above, pig slurry generally has the potential to emit more CH_4 than cattle slurry, because it has a higher content of degradable organic matter than cattle slurry (Møller et al., 2004; Triolo et al., 2011). In outside slurry stores, CH_4 emissions vary over the year due to temperature variations and management practices (Sommer et al., 2009). For example, in countries where slurry stores are emptied in spring, only small amounts of slurry are exposed to high temperatures during summer, whereas in countries where slurry is stored in lagoons for years, emissions may be higher. Emissions may also be higher from lagoons that are not stirred, because dry matter settles out to the bottom and is seldom removed from the lagoons.

The storage of solid manure has also been shown to be a source of CH_4, with losses from cattle manure heaps representing between 0.4 and 9.7% of the total C content of small heaps (less than 6 Mg), and the losses can be higher in larger field-scale heaps (Chadwick, 2005). The large variation is due to the effects of air exchange and aerobic decomposition of the volatile solids, which increases the temperature and produces anaerobic sites in the heap where CH_4 is produced. Deep litter from pig and cattle houses and pig manure with a large proportion of straw decompose aerobically because of the high permeability of the organic material, and little CH_4 is emitted. Cattle manure heaps with high bulk density and low porosity produce little CH_4, because the temperature in the manure remains low (Webb et al., 2012). High CH_4 emissions may be expected from solid manure heaps, with porosities lying between these extremes of high- and low-porosity manures, with the porosity being affected by the amount of straw added to the manure. Manures from open beef feedlots are often so dry that aerobic decomposition will not occur without the addition of water. The gaseous emissions from stored solid manure therefore generally reflect the variation in manure composition (Table 10.1).

10.3.3   Field-Applied Manure

Emissions of CH_4 occur immediately after manure application to land and are usually short-lived, as oxygen diffusion into the manure inhibits CH_4 formation. The disappearance of volatile fatty acids (VFAs) (precursors to CH_4 emissions) within the first few days following manure application indicates an aerobic/oxidising environment (Kirchmann and Lundvall, 1993). Indeed, it is believed that the CH_4 emitted immediately after application of slurry to land is CH_4 trapped within the slurry, having been generated within the slurry store. Accordingly, CH_4 emissions from slurry applied to soil are negligible (Sherlock *et al.*, 2002; Rodhe *et al.*, 2006). As might be expected when slurry is applied via shallow injection, the anaerobic nature of the slot environment results in greater CH_4 emissions than with surface broadcasting (Flessa and Beese, 2000).

10.4 Nitrous Oxide Production from Manure

Nitrogen in manure is a source of N_2O and the emissions increase with N concentration. Therefore, N_2O emissions are calculated using a N_2O emissions factor (EF), which is a proportion of the N in manure (Petersen and Sommer, 2011). Much manure N is in an organic form, which must be mineralised to NH_4^+ before it is a potential source of N_2O, and thus it does not contribute significantly to the short- and medium-term N_2O emissions. It is therefore relevant to calculate N_2O emissions considering the concentration of TAN, together with the fraction of the organic N that can be mineralised within the time that a given source has the potential for N_2O emissions. Nitrifiers increase their activity with temperature, but at temperatures above 40–45 °C they become inactive. Production of N_2O is related to oxygen content and especially to aerobic/anaerobic micro-sites. Therefore, the spatial and temporal distribution of O_2 supply and O_2 demand may be of particular importance for the prediction of N_2O emissions (Petersen and Sommer, 2011).

10.4.1 Stored Manure

In animal houses with slurry collection, the manure remains in a predominantly anaerobic state with little opportunity for the NH_4^+ to be nitrified. Here, N_2O may theoretically be produced at the air/liquid interface of stored slurry or on slats and solid floors where urine and faeces are deposited. In fact, the variation in emissions of N_2O (Table 10.2) is assumed to be related to the area fouled by the animals. Emissions of N_2O may be affected by TAN concentration and pH, since high $NH_3(aq)$ concentrations inhibit nitrification.

The bulk of slurry stored outside is anaerobic, and therefore emissions of N_2O via nitrification and denitrification from slurry without a floating cover are insignificant (Sommer *et al.*, 2000). However, a natural or artificial surface crust on top of the stored slurry can become a mosaic of anaerobic and aerobic sites under drying conditions, thus creating an environment where N_2O can be produced (VanderZaag *et al.*, 2009). For slurry covered with a porous material, N_2O emissions increase during periods when evaporation exceeds rainfall. Dissolved NH_4^+ can be oxidised by nitrifying bacteria in oxic zones of the cover, while in anoxic pockets the products of nitrification can be denitrified. During periods with rain, NH_4^+ in the surface cover/crust is leached downward and there is less air-filled porosity, limiting the potential for nitrification and denitrification.

In houses with deep litter systems, N_2O may be produced in the excreta and litter mixtures. N_2O production in deep cattle litter stored in-house is relatively low, probably because nitrification is inhibited by a combination of low oxygen partial pressure, high temperature and a high NH_3 concentration. In tie-stall systems there may be interfaces between manure and air, which are potential sources of N_2O. There is a greater potential for production of N_2O in deep pig litter than in deep cattle litter owing to air exchange in the deep pig litter, with emissions of approximately 3% of total N having been observed from deep litter left untreated (Chadwick *et al.*, 2011).

Porous solid manure heaps may be a source of N_2O during the initial phase of storage, before the temperature increases. During the composting phase little N_2O is produced, partly because NH_3 volatilisation depletes the pool of NH_4^+, and partly because nitrifying and denitrifying microorganisms are not thermophilic (Chadwick *et al.*, 2011). When the temperature has declined, conditions suitable for nitrification/denitrification may be re-established, which can lead to a secondary increase in N_2O emissions. Nitrous oxide produced at greater

Table 10.2 N_2O emissions from manure stored in-house, in outside stores and applied to land (emission factors are adapted from Chadwick et al. (2011), Sommer (2005) and Webb et al. (2012)).

Animal	Source	Manure	N$_2$O-N Emissions factor, % of total N			kg N$_2$O-N animal place and year*				Reference
			Mean	Minimum	Maximum	Mean	SD	Minimum	Maximum	
Dairy cattle	in-house	deep litter				0.72	0.76			Webb et al. (2012)*
	in-house	FYM				0.29	0.22			Webb et al. (2012)*
Beef cattle	in-house	deep litter				0.03	0.02			Webb et al. (2012)*
Pig	in-house	deep litter				0.90	0.94			Webb et al. (2012)*
	in-house	slurry				0.32		0.57	0.08	Sommer (2005)
Cattle	store outside	deep litter	1.3	4.3	0.1					Chadwick et al. (2011)
		solid manure aerated	0.6	1.1	0.4					Chadwick et al. (2011)
		solid manure, straw addition	0.4	0.5	0.5					Chadwick et al. (2011)
		solid manure, covered/compacted	1.1	2.1	0.6					Chadwick et al. (2011)
		slurry				2.6		5.3	0	Sommer (2005)
Pig	store outside	solid manure	1.7	2.6	0.5					Chadwick et al. (2011)
Poultry	store outside	solid manure	0.5	0.81	0.17					Chadwick et al. (2011)
		solid manure, covered	0.6	0.7	0.6					Chadwick et al. (2011)
Cattle	application	slurry, surface applied	0.5	1.0	0.1					Chadwick et al. (2011)
		slurry, injection	1.1							Chadwick et al. (2011)
		slurry, incorporation	0.9	2.0	0.3					Chadwick et al. (2011)
Pig	application	slurry, surface applied	0.7	1.6	0.1					Chadwick et al. (2011)
		slurry, injection	1.3	3.0	0.1					Chadwick et al. (2011)
Cattle	application	FYM, surface applied	0.2	0.33	0.16					Chadwick et al. (2011)
		FYM, incorporated	0.05	0.12	0.01					Chadwick et al. (2011)
Pig	application	FYM, surface applied	0.02	0.05	0					Chadwick et al. (2011)
		FYM, incorporated	0.20	0.86	0.20					Chadwick et al. (2011)
Poultry	application	FYM, surface applied	0.05							Chadwick et al. (2011)

*Emissions of N_2O from in-house stored manure are given in g N_2O-N per animal place per day. It was assumed that emissions are constant over the year to give annual emissions per animal place.

depth inside the heap may be reduced to N_2 during transport towards the surface and thus not emitted. In general, emissions of N_2O are a function of production and consumption of N_2O and the air exchange rate in the heap. Thus, emissions of N_2O typically range from less than 1% to 4.3% of the total N in stored cattle and pig manure heaps (Table 10.2).

The processes mentioned above for the pig and cattle manure also hold for poultry manure. The N_2O emissions from poultry manure heaps in the one study known to us range between 0.2 and 0.8% of total N content of the heap (Thorman *et al.*, 2006).

10.4.2   Field-Applied Manure

Application of manure to soil allows organic matter in the manure to be mineralised, forming NH_4^+, which may then be subjected to nitrification, with NO_3^- being produced. There is often a delay between manure application and N_2O emissions, generally attributed to the delay in mineralisation/nitrification and the generation of the soil NO_3^- pool. Immediate N_2O emissions following manure application are generally the result of a source of NO_3^- already within the manure (e.g. stored or composted solid manure) or the effect of manure C fuelling denitrification of soil NO_3^-.

Emission factors (i.e. cumulative N_2O-N losses as a proportion of total N applied in the manure) generally range from less than 0.1 to 3% (Tables 10.2), but values up to between 7.3% and 13.9% of total N in manure have been measured during land application of pig slurry (Velthof *et al.*, 2003). The range in the N_2O emission factors following slurry and solid manure application reflects differences in soil type, soil conditions (i.e. temperature, water-filled pore space), manure composition (i.e. NH_4^+-N, C content and availability) and measurement period. N_2O production may be affected by manure application for long periods, and bursts of N_2O emissions have been measured during and after rain events (Sherlock *et al.*, 2002), which have to be accounted for when measuring the emissions from manure applied to soil. Emission factors are lower for land-applied solid manure than from slurry (Table 10.2), which is generally a reflection of the lower available N content of most of these manure types.

In slurry-amended soil the O_2 demand is high, and the O_2 supply is reduced in and around the volume of soil affected by the slurry. This pattern strongly affects the potential for N_2O emissions, because the production is determined by the balance between O_2 demand and O_2 supply, rather than by O_2 supply alone (Figure 10.3). The effect of slurry will depend on soil conditions at the time of application; with a relatively dry soil an

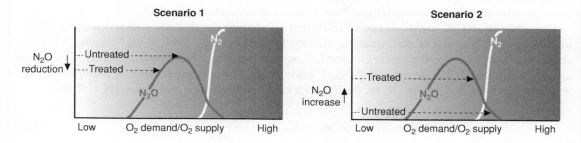

Figure 10.3   *Treatment technologies reducing (degradable) slurry volatile solids, such as separation or anaerobic digestion, influence N_2O emissions differently depending on soil conditions. This is because the balance between emissions of N_2O and N_2 is determined by the ratio between O_2 demand and O_2 supply. Therefore, if untreated slurry is applied to a well-aerated soil, as in scenario 1, this typically stimulates N_2O emissions, but the effect could be a reduction in N_2O emissions if untreated slurry is applied to a wet or compacted soil, as in scenario 2. Application of treated slurry with a lower O_2 demand is predicted to reduce emissions of N_2O relative to untreated slurry in scenario 1, but to increase emissions in scenario 2. (Modified from Petersen and Sommer (2011). © 2011 Elsevier.)*

increase in N_2O can be expected after slurry application, because slurry with a high content of degradable volatile solids increases O_2 demand and much more N_2O is produced. In this case reducing degradable volatile solids in the slurry (e.g. via biogas treatment), or increasing O_2 supply by harrowing to mix slurry and soil, will reduce N_2O emissions after application of slurry. Nitrous oxide emissions do not increase if slurry with a high volatile solids content is applied to a soil with a high potential for N_2O emissions, due to low porosity or high soil organic matter content, because the denitrification process results in complete denitrification to N_2. In this case, treatment of slurry or harrowing the soil could increase the N_2O emissions. Therefore, calculations predicting N_2O emissions from slurry-amended soil must take account of the composition of the slurry, as well as the soil conditions.

10.5   Reduction in Greenhouse Gas Emissions

Greenhouse gases are produced by microorganisms, which are affected by oxygen status, climate and the nature of the organic matter and chemical compounds in manure. GHG emissions can be reduced by management and technologies that induce an environment that is unfavourable for the microorganisms or that removes substrates (N and C) from the manure (Example 10.1).

Example 10.1   GHG emissions as affected by different mitigation techniques

The N_2O emissions factor may be a fraction of total-N or TAN at the time the manure is transferred to the stage of the manure management chain. Chapter 8 presented different methods for using EF for calculating NH_3 emissions. In the example below, the reduction in N as affected by NH_3 emissions is included in calculation of N_2O emissions from applied manure:

$$F_{N_2O:NH_3} = K_L \cdot N_L \qquad (10.1)$$

where $F_{N_2O:NH_3}$ is the emissions of either N_2O or NH_3 in kg N year^{-1} for a given livestock category, N_L is the annual amount of TAN or total-N in the manure excreted by the given livestock category and K_L is a factor in % of TAN or total-N, which is related to the category of animal (including animal weight and age), composition of the animal feed, type of animal housing and manure storage, method of manure application, meteorological conditions, soil conditions, and agricultural practices.

The reduction in CH_4 emissions can be calculated by adapting the methane conversion factor (MCF) to the emissions potential of the manure as affected by the treatment. MCF is used together with data about the potential CH_4 production or biochemical methane production (BMP or B_o, Chapter 13) to assess annual emissions of CH_4 for a given livestock category. In this calculation, MCF is multiplied by the volatile solids excretion rate and B_o for the livestock category in question as follows (IPCC, 2006):

$$F(T) = VS(T) \times 365 \times B_o(T) \times \frac{MCF(S,k)}{100} \times 0.67 \qquad (10.2)$$

where $F(T)$ is the annual CH_4 emissions for livestock category T (kg CH_4 LU^{-1} year^{-1}), $VS(T)$ is the daily volatile solids excreted for livestock category T (kg VS LU^{-1} day^{-1}), 365 is the basis for calculating annual volatile solids production (days year^{-1}), $B_o(T)$ is the ultimate methane production for manure from livestock category T (m^3 CH_4 kg^{-1} VS$_T$) and 0.67 is the conversion factor (kg CH_4 m^{-3} CH_4).

The effect of reducing N intake was not included in these calculations (Mikkelsen *et al.*, 2006). Acidification of slurry (by reducing NH_3 losses) increased N_2O emissions from slurry applied in the field, but significantly reduced CH_4 emissions, so overall GHG emissions were reduced (Figure 10.4).

Anaerobic digestion of slurry reduced both N_2O and CH_4 emissions (fossil fuel substitution not accounted for), while slurry separation increased GHG emissions, because more N_2O was emitted. Covering stored slurry with a crust increased GHG emissions, because N_2O was emitted during storage and covering slurry with a solid cover increased emissions slightly due to more TAN being applied in the field (less NH_3 emissions during storage).

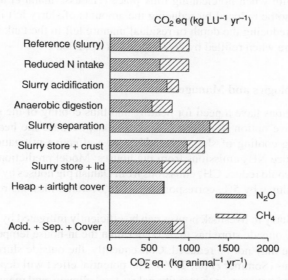

Figure 10.4   *Reduction in annual GHG emissions from a Danish dairy cow calculated using IPCC guidelines and the emissions factors of Table 10.3, and assuming Global Warning Potentials for CH_4 and N_2O of 21 and 310 kg CO_2-eq. (© University of Southern Denmark).*

Table 10.3   *Potential for reducing CH_4 and N_2O emissions from dairy cow manure handling, presented as N_2O emission factors and MCFs.*

| | | \multicolumn{6}{}{Emissions factor} | |
| | | Housing (% of $N_{excr.}$) | | Store (% of $N_{excr.}$) | | Field (% of $N_{appl.}$) | | System MCF (%) |
	Scenario	N_2O	NH_3	N_2O	NH_3	N_2O	NH_3	
Slurry	reference	0	8	0	6	1.25	7.5	10
Slurry	reduced N intake	0	4	0	3	1.25	7.5	10
Slurry	acidification, liquid (no cover)	0	4	0	3	1.25	2.5	4
Slurry	anaerobic digestion (no cover)	0	8	0	21	1.25	7.5	7
Slurry	separation liquid (no cover)	0	8	0	6	1.25	7.5	3
Slurry	separation – solid	NA	8	3	25	1.25	4	3.5
Slurry	storage – floating crust	0	8	0.5	2	1.25	7.5	6
Slurry	storage – solid cover	0	8	0	2	1.25	7.5	10

NA = no solid manure in animal house.

N excretion from a Danish dairy cow is set at 130 kg N year^{-1} at standard feeding and 120 kg N year^{-1} at reduced N intake. Annual potential CH_4 emissions are estimated to be 173 kg CH_4 year^{-1} animal^{-1} (= VS(T) × 365 × B_0(T) × 0.67) and are assumed to be unaffected by a change in diet (i.e. no volatile solids changes) (Mikkelsen *et al.*, 2006; Petersen and Sommer, 2011).

10.5.1   Reduced Inoculum

Removal and flushing of the slurry channel will reduce the pool of methanogenic bacteria adapted to the environment of the pit or slurry channel under slatted floors. In one study in pig houses with removal of slurry after each fattening period, emissions were 40% lower when the slurry channels were cleaned after slurry removal compared with when no cleaning took place (Haeussermann *et al.*, 2006). Emissions from slurry stored outside can also be reduced by reducing the amounts of slurry left in the tank after emptying. For example, in one study, reducing the depth of residual manure left in the tank from 60 to 30 cm reduced CH_4 emissions from the store when refilled by 26% (Massé *et al.*, 2008).

10.5.2   Mitigation Technologies and Management

Pig houses in temperate regions have a need for heating and thus cooling of the slurry combined with heat exchange can be an attractive option to reduce CH_4 emissions, because the heat produced can lower the need for fossil fuel, making cooling of slurry channels a cost-effective mitigation option. A reduction in temperature would also reduce NH_3 emissions from pig houses. Model predictions show that a reduction in slurry temperature of $10\,°C$ could reduce CH_4 emissions from Danish pig houses by 74% while increasing CH_4 from the externally stored slurry by 5%, corresponding to an overall emissions reduction of 31% (Sommer *et al.*, 2004).

Emissions of CH_4 from slurry in livestock houses can be efficiently mitigated by frequent removal of slurry (Massé *et al.*, 2003), but the outside store has to be colder than the in-house temperature, otherwise the total CH_4 emissions from storage may not be reduced. Consequently, the outside slurry storage facility must be emptied during warm seasons (Sommer *et al.*, 2009). The potential effect will depend on the relative storage periods between in-house and outdoors in the traditional system, climate, and management in the new system where slurry is frequently removed from the livestock house. For example, frequent slurry removal from pig houses reduced CH_4 emissions by approximately 40% in a French scenario, but had no effect in Denmark (Sommer *et al.*, 2009). In contrast, reducing in-house storage of slurry in Danish dairy cow houses from 1 month to 1 day reduced CH_4 emissions by up to 49%, including a high reduction in CH_4 emissions from in-house stored cattle slurry and only a modest increase in emissions from the outside store (Sommer *et al.*, 2004). On a Canadian dairy cow farm, slurry was applied to a field before summer and thus the slurry storage unit was empty during the summer period of high temperatures because cattle were outside grazing on grass, a treatment which reduced CH_4 emissions by 9–10% (Massé *et al.*, 2008).

Efforts to reduce CH_4 emissions (e.g. by emptying in-house slurry channels frequently to a colder outside slurry store) will delay decomposition of slurry volatile solids. If slurry is collected during an extended period and applied before crop growth, this may increase N_2O emissions from the slurry applied to soils due to the higher volatile solids content in the applied slurry (Sommer *et al.*, 2009). This is a real risk in well-aerated soils at least. With slurry application on heavy clay soils, higher concentrations of slurry volatile solids may enhance "full denitrification" of nitrogen to N_2 (Figure 10.3). As a rule of thumb, optimising the rate, timing and technique of manure application to crops and grassland with the focus on efficient plant uptake of the N in manure also reduces production and emissions of N_2O (Chadwick *et al.*, 2011).

10.5.3   Reducing Volatile Solids and Nitrogen

Feeding practice affects the amount and digestibility of the manure produced; for example, the substrate for CH_4 production in the form of VFA increases with increasing content of non-starch polysaccharides (NSPs = organic matter – (crude protein + crude lipid + starch + sugar); Canh *et al.*, 1998). Diets low in N, but with an amino acid composition fulfilling the animal demand, will reduce N in manure. The efficient feeding of animals with feed containing readily available volatile solids and an adequate N content can potentially reduce CH_4 and N_2O emissions through the entire manure management chain. This is probably the cheapest and most efficient way to reduce GHG emissions in animal production systems where the diet is not already optimised.

Reducing the volatile solids content of the slurry through fermentation in a biogas digester is an efficient way of reducing CH_4 emissions during outside storage. The emissions from digested slurry during storage have been found to be 30–66% lower than those from untreated slurry (Clemens *et al.*, 2006; Amon *et al.*, 2006). However, Sommer *et al.* (2000) showed that digested slurry must be cooled to ambient temperatures in a post-treatment storage tank with collection of CH_4 from the heated slurry to efficiently reduce CH_4 emissions. Slurry separation results in a liquid fraction and a solid fraction that is similar to solid manure. Cattle slurry separation using the screw-press method reduced CH_4 emissions relative to untreated slurry during winter storage in Portugal by more than 35%, while combining screw-press separation with enhanced separation by addition of additives reduced emissions by up to 50% (Chadwick *et al.*, 2011).

Separation or digestion of manure reduces the volatile solids content of the slurry applied to soil, and this change can either increase or reduce N_2O emissions depending on the soil and environmental conditions at the time of application. The variation is again explained by the conceptual model of the relationship between N_2O/N_2 emissions and O_2 supply and demand in the soil presented in Figure 10.3. Therefore, the effect of separation or digestion of slurry on N_2O emissions from the liquid fraction applied to fields has to be assessed taking local soil characteristics and climate into consideration. When assessing GHG mitigation due to slurry separation, the emissions of N_2O and CH_4 from the solid fraction must be included in the calculations. When this is done, the reduction in total GHG emissions from stored and applied manure is estimated to vary between 0 and 40% (Sommer *et al.*, 2009; Massé *et al.*, 2011).

10.5.4   Additives

Acidification of slurry for the purpose of reducing NH_3 emissions from storage has also been observed to significantly reduce CH_4 emissions from stored slurry or liquid manure (Berg *et al.*, 2006; Petersen *et al.*, 2012). Acidification of slurry or slurry fractions may also delay nitrification and associated N_2O emissions after field application relative to soil amended with non-acidified material. In a study in Portugal, lower amounts of N_2O were released during the growing season from soils amended with acidified liquid fractions from separation, but no effect occurred with acidification of unseparated slurry (Fangueiro *et al.*, 2010), indicating that the effect varies according to manure composition and probably also soil, application technique and climate.

Use of nitrification inhibitors can reduce NO_3^- accumulation for a period and to the extent that a crop or microorganisms take up NH_4^+, this can reduce N_2O emissions from field-applied manure (Chadwick *et al.*, 2011). For example, N_2O emissions from shallow injection slots were reduced by more than 30% when 3,4-dimethylpyrazole phosphate was added to the slurry prior to application. However, no effect or a very low reduction of 0.5–3% was measured when dicyandiamide was added to slurry injected into soil in a Mediterranean climate (Chadwick *et al.*, 2011). However, as their name implies, nitrification inhibitors only slow down the rate of nitrification and thus influence N transformations and losses post-nitrification. If the slurry already contains significant levels of NO_3^-, then the effect of the inhibitor is reduced (McGeough *et al.*, 2012). Nitrification inhibitors have been intensively studied as an N_2O mitigation strategy for grazed pastures with respect to N losses from urine (Clough *et al.*, 2007).

Yamulki (2006) demonstrated that addition of 50% straw (by volume) to small-scale conventional solid cattle manure at the start of the storage period reduced CH_4 emissions by approximately 45%.

The addition of phosphogypsum to cattle feedlot manure has been shown to reduce CH_4 emissions during storage (Hao *et al.*, 2005), perhaps because of the effect of the greater sulfur and NH_4^+ content on the methanogens and the effect of the lower pH on the rate of CH_4 oxidation.

10.5.5   Covers

Covering slurry stores and lagoons is a means to reduce NH_3 emissions. Porous cover materials, including straw, expanded clay pebbles, recycled polyethylene and naturally formed surface crusts, also have the potential to reduce CH_4 emissions via microbial oxidation (Ambus and Petersen, 2005; Petersen *et al.*, 2005). Organic amendments such as straw could theoretically stimulate methanogenesis, but Clemens *et al.* (2006) did not see such an effect under winter or summer storage conditions. Instead, they observed in both seasons a significant reduction when an additional solid cover was added, suggesting that increasing the headspace concentration of CH_4 stimulated CH_4 oxidation. VanderZaag *et al.* (2010) found that a permeable synthetic floating cover delayed the onset of N_2O emissions, possibly because nitrifying and denitrifying bacteria cannot easily establish on synthetic membranes (Petersen and Miller, 2006).

Compacting and covering manure heaps has the potential to markedly reduce N_2O and CH_4 emissions, the key to success being to maintain anaerobic conditions in the manure. Compaction of cattle manure and applying an air-tight cover of plastic has the potential to reduce both CH_4 and N_2O emissions during storage (Hansen *et al.*, 2006; Chadwick *et al.*, 2011). However, if the solid manure is not covered or compacted efficiently the temperature may increase and so will CH_4 production in the centre. At the same time, due to the aerobicity at the surface, N_2O production may be high and the heap may be a significant source of GHG. For example, efficient covering reduced CH_4 emissions, under Danish conditions, from a heap of dry matter-rich separated slurry from 1.6 to 0.2 kg C Mg^{-1}, or from 1.3 to 0.17% of the initial C content (Hansen *et al.*, 2006). In contrast, between 1.8 and 4.4% of the initial C content was emitted as CH_4 during three storage periods of conventionally stored British cattle manure, where covering and compacting the heaps had no consistent effect on CH_4 emissions (Chadwick, 2005).

Alternatively, aerobic treatment of stored liquid and solid manure may be an option. Aeration of liquid manure may reduce CH_4 emissions but brings a great risk of increasing N_2O emissions (Chadwick *et al.*, 2011). Frequent turning of manure heaps can be used to reduce anaerobic zones in the heap. In an Austrian study this technique reduced CH_4 emissions to about 0.5% of the initial C content. At the same time, aeration by turning may either decrease N_2O emissions or turning solid manure heaps can stimulate N_2O production (Chadwick *et al.*, 2011). This indicates that controlling the aerobic process is complicated and there is a risk of increasing GHG emissions.

10.5.6   Whole System Analysis of Technologies to Reduce Greenhouse Gases

Separation of slurry may not always reduce the total GHG emissions from manure under management, because the reduced CH_4 and N_2O emissions from the liquid are negated by the emissions of these two gases from the solid fraction (Sommer *et al.*, 2009). The alternative option may be to separate the slurry and incinerate the dry matter-rich fraction. This procedure would reduce GHG emissions to very low levels. The solid fraction would not be a source of GHGs, and low volatile solids contents in the liquid fraction would mean lower CH_4 emissions during storage and lower N_2O emissions after field application of the liquid fraction. One should consider the energy needed to reduce water content and replace plant nutrients lost by incineration (Chapter 13).

The green energy produced from incineration would replace CO_2 otherwise emitted from a fossil fuel-fired power plant and therefore contribute to reduced emissions. However, incineration would also reduce the amount of C deposited in the soil and therefore sequestration of soil organic C (SOC). This effect would be

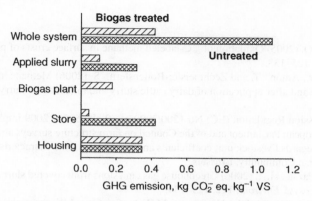

Figure 10.5 *CH_4 and N_2O emissions from slurry as affected by anaerobic digestion in a biogas reactor. (© University of Southern Denmark.)*

particularly high for the cooler climates in Sweden and Denmark, and under some farming conditions the reduction in soil C input from slurry volatile solids might need to be compensated for by other means to maintain SOC stocks (e.g. growing cover crops or adopting reduced tillage practices). Again, the effect of the mitigation option for slurry management has to be set in a regional context.

The fermentation processes in biogas digesters result in CH_4 production at the expense of easily digestible volatile solids, reducing the potential for CH_4 emissions during subsequent storage and for N_2O emissions after field application. This treatment may reduce GHG emissions to approximately one-third of those from untreated slurry (Figure 10.5). Biogas plants are generally only economically sustainable if subsidised (i.e. the power produced is about twice as expensive as the power generated by traditional power plants) and incentives have been given for constructing biogas plants. Some European governments support the plants because they help to contribute to EU and national targets to enhance green energy, and because digestate is considered to emit less odour components, to have a higher fertiliser efficiency, and to reduce risks of spreading diseases and pests.

10.6 Summary

Methane and N_2O emissions are the result of microbial activity and are strongly affected by micro-site environmental conditions, which in turn are affected by manure management system, manure composition, climate, soil and so on. Therefore, care is needed when assessing the effects of introducing mitigation techniques, the effects of which often depend on the environment and management of the source. Depending on where it is implemented, a particular method can reduce emissions efficiently or increase emissions. An efficient mitigation technique for both N_2O emissions and probably also CH_4 emissions may be to change livestock diets in order to achieve greater feed conversion efficiency, which reduces the amount of N and degradable organic C in the excreta. Keeping manure storage units strictly anaerobic, and the temperature as low as possible, efficiently reduces CH_4 and N_2O emissions. Reduction of volatile solids in slurry by biogas treatment reduces the potential for CH_4 emissions and, depending on the environment, may also reduce N_2O emissions from applied manure. Acidification of slurry is a means to reduce CH_4 emissions and in some situations also N_2O emissions, while adding nitrification inhibitors and optimal use of manure as a fertiliser also reduce N_2O emissions.

References

Ambus, P. and Petersen, S.O. (2005) Oxidation of ^{13}C-labeled methane in surface crusts of pig and cattle slurry. *Isotop. Environ. Health Sci.*, **41**, 125–133.

Amon, B., Kryvoruchko, V., Amon, T. and Zechmeister-Boltenstern, S. (2006) Methane, nitrous oxide and ammonia emissions during storage and after application of dairy cattle slurry and influence of slurry treatment. *Agric. Ecosyst. Environ.*, **112**, 153–162.

Anonymous (2009) Commission Regulation (EC) No 1200/2009 of 30 November 2009 implementing Regulation (EC) No 1166/2008 of the European Parliament and of the Council on farm structure surveys and the survey on agricultural production methods, as regards livestock unit coefficients and definitions of the characteristics, *Official Journal of the European Union*, European Community, Brussels.

Berg, W., Brunsch, R. and Pazsiczki, I. (2006) Greenhouse gas emissions from covered slurry compared with uncovered during storage. *Agric. Ecosyst. Environ.*, **112**, 129–134.

Canh, T.T., Sutton, A.L., Aarnink, A.J.A., Verstegen, M.W.A., Schram, J.W. and Bakker, G.C.M. (1998) Dietary carbohydrates alter the fecal composition and pH and the ammonia emission from slurry of growing pigs. *J. Anim. Sci.*, **76**, 1887–1895.

Chadwick, D. (2005) Emissions of ammonia, nitrous oxide and methane from cattle manure heaps: effect of compaction and covering. *Atmos. Environ.*, **39**, 787–799.

Chadwick, D., Sommer, S.G., Thorman, R., Fangueiro, D., Cardenas, L., Amon, B. and Misselbrook, T. (2011) Manure management: implications for greenhouse gas emissions. *Anim. Feed Sci. Technol.*, 166–167, 514–531.

Clemens, J., Trimborn, M., Weiland, P. and Amon, B. (2006) Mitigation of greenhouse gas emissions by anaerobic digestion of cattle slurry. *Agric. Ecosyst. Environ.*, **112**, 171–177.

Clough, T.J., Di, H.J., Cameron, K.C., Sherlock, R.R., Metherell, A.K., Clark, H. and Rys, G. (2007) Accounting for the utilization of a N$_2$O mitigation tool in the IPCC inventory methodology for agricultural soils. *Nutr. Cycl. Agroecosyst.*, **78**, 1–14.

Demirel, B. and Scherer, P. (2008) The roles of acetotrophic and hydrogenotrophic methanogens during anaerobic conversion of biomass to methane: a review. *Rev. Environ. Sci. Biotechnol.*, **7**, 173–190.

Fangueiro, D., Ribeiro, H., Coutinho, J., Cardenas, L., Trindade, H., Queda, C., Vasconcelos, E. and Cabral, F. (2010) Nitrogen mineralization and CO$_2$ and N$_2$O emissions in a sandy soil amended with original and acidified pig slurry or with the relatede fractions. *Biol. Fertil. Soils*, **46**, 383–391.

Flessa, H. and Beese, F. (2000) Laboratory estimates of trace gas emissions following surface application or injection of cattle slurry. *J. Environ. Qual.*, **29**, 262–268.

Forster, P., Ramaswamy, V., Artaxo, P., Berntsen, T., Betts, R., Fahey, D.W., Haywood, J., Lean, J., Lowe, D.C., Myhre, G., Nganga, J., Prinn, R., Raga, G., Schulz, M. and Van Dorland, R. (2007) Changes in atmospheric constituents and in radiative forcing, in *Climate Change 2007: The Physical Science Basis. Contribution of Working Group I to the Fourth Assessment Report of the Intergovernmental Panel on Climate Change* (eds S. Solomon, D. Qin, M. Manning, Z. Chen, M. Marquis, K.B. Averyt, M. Tignor and H.L. Miller), Cambridge University Press, Cambridge, pp. 129–234; http://www.ipcc.ch/pdf/assessment-report/ar4/wg1/ar4-wg1-chapter2.pdf.

Haeussermann, A., Hartung, E., Gallmann, E. and Jungbluth, T. (2006) Influence of season, ventilation strategy, and slurry removal on methane emissions from pig houses. *Agric. Ecosyst. Environ.*, **112**, 115–121.

Hansen, M.N., Henriksen, K. and Sommer, S.G. (2006) Observations of production and emission of greenhouse gases and ammonia during storage of solids separated from pig slurry: effects of covering. *Atmos. Environ.*, **40**, 4172–4181.

Hao, X., Larney, F.J., Chang, C., Travis, G.R., Nichol, C.K. and Bremer, E. (2005) The effect of phosphogypsum on greenhouse gas emissions during cattle manure composting. *J. Environ. Qual.*, **34**, 774–781.

IPCC (2006) *Guidelines for National Greenhouse Gas Inventories. Agriculture, Forestry and Other Land Use*, Vol. **4**. IGES, Kanagawa.

Kirchmann, H. and Lundvall, A. (1993) Relationship between N immobilisation and volatile fatty acids in soil after application of pig and cattle slurry. *Biol. Fertil. Soils*, **15**, 161–164.

Massé, D.I., Croteau, F., Patni, N.K. and Masse, L. (2003) Methane emissions from dairy cow and swine manure slurries stored at 10°C and 15°C. *Can. Biosyst. Eng.*, **45**, 6.1–6.6.

Massé, D.I., Massé, L., Claveau, S., Benchaar, C. and Thomas, O. (2008) Methane emissions from manure storages. *Trans. Am. Soc. Agric. Biol. Eng.*, **51**, 1775–1781.

Massé, D.I., Talbot, G. and Gilbert, Y. (2011) On farm biogas production: a method to reduce GHG emissions and develop more sustainable livestock operations. *Anim. Feed Sci. Technol.*, **166–167**, 436–445.

McGeough, K.L., Laughlin, R.J., Watson, C.J., Muller, C., Ernfors, M., Cahalan, E. and Richards, K.G. (2012) The effect of cattle slurry in combination with nitrate and the nitrification inhibitor dicyandiamide on in situ nitrous oxide and dinitrogen emissions. *Biogeosciences*, **9**, 4909–4919.

Mikkelsen, M.H., Gyldenkærne, S., Poulsen, H.D., Olesen, J.E. and Sommer, S.G. (2006) Emission of ammonia, nitrous oxide and methane from Danish Agriculture 1985–2002. Methodology and estimates, *NERI Research Notes 231*, National Environmental Research Institute, Copenhagen; http://www.dmu.dk/Pub/AR231.pdf.

Monteny, G.-J., Groenestein, C.M. and Hilhorst, M.A. (2001) Interactions and coupling between emissions of methane and nitrous oxide from animal husbandry. *Nutr. Cycl. Agroecosyst.*, **60**, 123–132.

Møller, H.B., Ahring, B.K. and Sommer, S.G. (2004) Methane productivity of manure, straw and solid fractions of manure. *Biomass Bioenergy*, **26**, 485–495.

Oenema, O., Wrage, N., Velthof, G.L., van Groeningen, J.W., Dolfing, J. and Kuikman, P.J. (2005) Trends in global nitrous oxide emissions from animal production systems. *Nutr. Cycl. Agroecosyst.*, **72**, 51–65.

Petersen, S.O. and Miller, D.N. (2006) Greenhouse gas mitigation by livestock waste storage and lagoon covers. *J. Sci. Food Agric.*, **86**, 1407–1411.

Petersen, S.O. and Sommer, S.G. (2011) Ammonia and nitrous oxide interactions – the role of manure organic matter management. *Anim. Feed Sci. Technol.*, 166–167, 503–513.

Petersen, S.O., Henriksen, K. and Blackburn, T.H. (1991) Coupled nitrification–denitrification associated with liquid manure in a gel-stabilized model system. *Biol. Fertil. Soils*, **12**, 19–27.

Petersen, S.O., Nielsen, T.H., Frostegård, Å. and Olesen, T. (1996) Oxygen uptake, carbon metabolism, and denitrification associated with manure hot-spots. *Soil Biol. Biochem.*, **28**, 341–349.

Petersen, S.O., Amon, B. and Gattinger, A. (2005) Methane oxidation in slurry storage surface crusts. *J. Environ. Qual.*, **34**, 455–461.

Petersen, S.O., Sommer, S.G., Béline, F., Burton, C., Dach, J., Dourmad, J.Y., Leip, A., Misselbrook, T., Nicholson, F., Poulsen, H.D., Provolo, G., Sørensen, P., Vinnerås, B., Weiske, A., Bernal, M.-P., Böhm, R., Juhász, C. and Mihelic, R. (2007) Recycling of livestock manure in a whole-farm perspective. *Livest. Sci.*, **112**, 180–191.

Petersen, S.O., Andersen, A.J. and Eriksen, J. (2012) Effects of slurry acidification on ammonia and methane emission during storage. *J. Environ. Qual.*, **41**, 88–94.

Rodhe, L., Pell, M. and Yamulki, S. (2006) Nitrous oxide, methane and ammonia emissions following slurry spreading on grassland. *Soil Use Manag.*, **22**, 229–237.

Russow, R., Stange, C.F. and Neue, H.-U. (2009) Role of nitrite and nitric oxide in the processes of nitrification and denitrification in soil: results from ^{15}N tracer experiments. *Soil Biol. Biochem.*, **41**, 785–795.

Sherlock, R.R., Sommer, S.G., Rehmat, Z., Khan, R.Z., Wesley, C.W., Guertal, E.A., Freney, J.R., Dawson, C.O. and Cameron, K.C. (2002) Emission of ammonia, methane and nitrous oxide from pig slurry applied to a pasture in New Zealand. *J. Environ. Qual.*, **31**, 1491–1501.

Solomon, S., Qin, D., Manning, M., Chen, Z., Marquis, M., Averyt, K.B., Tignor, M. and Miller, H.L. (eds((2007) *Climate Change 2007: The Physical Science Basis. Contribution of Working Group I to the Fourth Assessment Report of the Intergovernmental Panel on Climate Change*, Cambridge University Press, Cambridge.

Sommer, S.G. (2005) Greenhouse gas losses from manure, in *Manure – An Agronomic and Environmental Challenge* (eds M. Stenberg, H. Nilsson, R. Brynjolfsson, P. Kapuinen, J. Morken and T.S. Birkmose), *NJF Seminar 372*, Nordic Association of Agricultural Scientists, Stockholm, pp. 25–34.

Sommer, S.G., Sherlock, R.R. and Khan, R.Z. (1996) Nitrous oxide and methane emissions from pig slurry amended soils. *Soil Biol. Biochem.*, **28**, 1541–1544.

Sommer, S.G., Petersen, S.O. and Søgaard, H.T. (2000) Emission of greenhouse gases from stored cattle slurry and slurry fermented at a biogas plant. *J. Environ. Qual.*, **29**, 744–751.

Sommer, S.G., Petersen, S.O. and Møller, H.B. (2004) Algorithms for calculating methane and nitrous oxide emissions from manure management. *Nutr. Cycl. Agroecosyst.*, **69**, 143–154.

Sommer, S.G., Petersen, S.O., Sørensen, P., Poulsen, H.D. and Møller, H.B. (2007) Greenhouse gas emission and nitrogen turnover in stored liquid manure. *Nutr. Cycl. Agroecosyst.*, **78**, 27–36.

Sommer, S.G., Olesen, J.E., Petersen, S.O., Weisbjerg, M.R., Valli, L., Rohde, L. and Béline, F. (2009) Region-specific assessment of greenhouse gas mitigation with different manure management strategies in four agroecological zones. *Global Change Biol.*, **15**, 2825–2837.

Steinfeld, H., Gerber, P., Wassenaar, T., Castel, V., Rosales, M. and de Haan, C. (2006) *Livestock's Long Shadow: Environmental Issues and Options*, FAO, Rome.

Thorman, R.E., Harrison, R., Cooke, S.D., Chadwick, D.R., Burston, M. and Balsdon, S.L. (2003) Nitrous oxide emissions from slurry- and straw-based systems for cattle and pigs in relation to emissions of ammonia, presented at the *SAC/SEPA Conference on Agriculture, Waste and the Environment*, Edinburgh.

Thorman, E., Chadwick, D.R., Boyles, L.O., Matthews, R., Sagoo, E. and Harrison, R. (2006) Nitrous oxide emissions during storage of broiler litter and following application to arable land. *Int. Cong. Ser.*, **1293**, 355–358.

Triolo, J.M., Sommer, S.G., Møller, H.B., Weisbjerg, M.R. and Xinyua, J. (2011) A new algorithm to characterize biodegradability of biomass during anaerobic digestion: influence of lignin concentration on methane production potential. *Bioresour. Technol.*, **102**, 9395–9402.

Vallejo, A., Garcia-Torres, L., Diez, J.A., Arce, A. and Lopez-Fernandez, S. (2005) Comparison of N losses (NO_3^-, N_2O, NO) from surface applied, injected or amended (DCD) pig slurry of an irrigated soil in a Mediterranean climate. *Plant Soil*, **272**, 313–325.

VanderZaag, A.C., Gordon, R.J., Jamieson, R.C., Burton, D.L. and Stratton, G.W. (2009) Gas emissions from straw covered liquid dairy manure during summer storage and autumn agitation. *Trans. Am. Soc. Agric. Biol. Eng.*, **52**, 599–608.

VanderZaag, A.C., Gordon, R.J., Jamieson, R.C., Burton, D.L. and Stratton, G.W. (2010) Floating covers to reduce gas emissions from liquid manure storages: a review. *Appl. Eng. Agric.*, **26**, 287–297.

Velthof, G.L., Kuikman, P.J. and Oenema, O. (2003) Nitrous oxide emission from animal manures applied to soil under controlled conditions. *Biol. Fertil. Soils* **37**, 221–230.

Webb, J., Sommer, S.G., Kupper, T., Groenestein, K., Hutchings, N.J., Eurich-Menden, B., Rodhe, L., Misselbrook, T.H. and Amon, B. (2012) Emissions of ammonia, nitrous oxide and methane during the management of solid manures. *Sust. Agric. Rev.*, **8**, 67–107.

Wrage, N., Velthof, G.L., van Beusichem, M.L. and Oenema, O. (2001) Role of nitrifier denitrification in the production of nitrous oxide. *Soil Biol. Biochem.*, **33**, 1723–1732.

Yamulki, S. (2006) Effect of straw addition on nitrous oxide and methane emissions from stored farmyard manures. *Agric. Ecosyst. Environ.*, **112**, 140–145.

Zeeman, G., Sutter, K., Vens, T., Koster, M. and Wellinger, A. (1988) Psychrophilic digestion of dairy cattle and pig manure: start-up procedures of batch, fed-batch and CSTR-type digesters. *Biol. Wastes*, **26**, 15–31.

11

Nutrient Leaching and Runoff from Land Application of Animal Manure and Measures for Reduction

Peter Sørensen[1] and Lars S. Jensen[2]

[1]*Department of Agroecology, Aarhus University, Denmark*
[2]*Department of Plant and Environmental Sciences,
University of Copenhagen, Denmark*

11.1 Introduction

Diffuse pollution of groundwater and surface waters with nitrogen (N) and phosphorus (P) is a problem in many regions of the world, especially in areas with high livestock production. Animal manures contain substantial quantities of organic matter, N and P, and if applied to land at excessive rates these may be lost. Nutrient and organic matter losses to aquatic systems mainly occur by leaching through the soil profile and through surface runoff when the infiltration capacity of the soil is exceeded. In surface waters the losses cause problems with eutrophication and algal bloom, and in areas that rely on the use of groundwater high nutrient concentrations can be a problem for the potable water quality. For drinking water the EU limit has been set at a nitrate (NO_3^-) concentration of 50 mg l^{-1} (EU Drinking Water Directive, 98/83/EC). Once leached to surface waters this N may also become a source of emissions of nitrous oxide, which is a potent greenhouse gas. In addition, the loss of valuable nutrient resources is a problem, as P is a limited and essential resource, and both N and P fertilisers are produced using substantial amounts of fossil energy, causing global warming and other environmental emissions. Appropriate management and use of manures is therefore essential for minimising nutrient losses and the environmental impact of agriculture.

The mechanisms involved in leaching and runoff are variable and determine the nutrients that are vulnerable to losses. Transport of water through the soil can occur by matrix or macropore flow (see Text Box – Basic 11.1 for definitions). The nutrients can be in either dissolved or particulate species and in chemically reactive

---

Animal Manure Recycling: Treatment and Management, First Edition. Edited by Sven G. Sommer, Morten L. Christensen, Thomas Schmidt and Lars S. Jensen.

(strongly sorbed to soil particles, relatively immobile) or non-reactive (unbound, and hence mobile) form. Manure nitrogen is mainly present as ammonium N (NH_4-N) and organic N (Text Box – Basic 11.1). Ammonium (NH_4^+) is readily adsorbed in the soil and leaching of NH_4^+ is usually very low in most soils, but it can be leached by macropore flow occurring shortly after manure application (Parkes *et al.*, 1997) and on sandy soils with low cation exchange capacity (CEC, Text Box – Basic 11.1) (Wachendorf *et al.*, 2005). However, NH_4^+ is usually quickly nitrified to NO_3^- in aerobic soils, within a few days or weeks after manure application depending on soil temperature and moisture. Contrary to NH_4^+, NO_3^- is not adsorbed and stays in solution and is therefore more vulnerable to leaching.

Manures contain both soluble and particulate organic N, which may also be leached (Korsaeth *et al.*, 2003), but leaching losses of organic N are usually relatively low. In fresh sheep faeces, 25% of the total N has been found to be water-extractable, including N in microparticles (less than 0.22 µm) (Sørensen *et al.*, 1994). Such organic N is vulnerable to runoff and leaching by macropore flow, whereas it is expected to be adsorbed or physically retained in soil under conditions with matrix flow. Wachendorf *et al.* (2005) observed that 8% of the N in cattle urine was leached as dissolved organic N, while only 0.7% of the N in cattle faeces was leached as organic N on a sandy soil, indicating that it is mainly soluble organic N in urine that is prone to matrix leaching.

Through macropore flow, the transport of particulate and colloidal-bound nutrients (e.g. phosphates) is facilitated, whereas fully soluble and mobile nutrients such as NO_3^- that are present in the soil matrix may be bypassed by the water flow in macropores (Glæsner *et al.*, 2011a). Surface application of manure followed by rainfall on soils with many macropores creates a high risk of fast transport through the soil of both water-soluble and small particle/colloid-bound nutrients present in the manure.

See Text Box – Advanced 11.2 for methods of measuring leaching and runoff.

Text Box – Basic 11.1   Definitions used (see also Figure 11.1)

Ammonium-N (NH_4-N): Agronomists use this term to define how much nitrogen is applied with a compound (e.g. 18 g NH_4^+ contains 14 g NH_4-N; for NO_3^--N, 62 g NO_3^- contains 14 g NO_3-N).

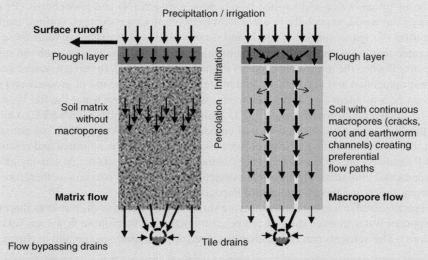

Figure 11.1   *Illustration of major nutrient transport mechanisms in and from soils. (© University of Southern Denmark.)*

Cation exchange capacity (CEC) of soil: Minerals (mostly clays) and organic matter in soil are negatively charged at neutral pH, and hence cations can be adsorbed to these and thus become less prone to losses by leaching or volatilisation. CEC is a measure of the soil's capacity to adsorb cations. Adsorbed cations are not strongly bound and are therefore exchangeable, as they can be replaced by other cations dissolved in the soil water surrounding the soil particles, if the sites are accessible to the soil solution.

Nutrient leaching: The process of nutrient transport with water percolating through the soil profile to below the rooting zone. Nutrients are then normally lost from the agricultural system, but upward movement of water in the soil profile may occur in warmer or drier parts of the year due to the hydraulic lift caused by evapotranspiration.

Transport of solutes: Can be divided into three processes, advection, diffusion and dispersion. *Advection* is the process by which compounds are transported with the flowing water. *Diffusion* processes occur within the flowing water in response to molecular movement of solutes to even out concentration gradients. *Dispersion* spreads solutes in response to water velocity gradients between and within soil pores. During water flow, advection and dispersion are the dominant processes responsible for solute transport, whereas for a matrix with stagnant water, diffusion is the main process by which solutes are assumed to move.

Rooting zone: Vertical depth to which a crop can develop effective roots for water and nutrient uptake. Rooting depth can be highly variable depending on soil type, soil density throughout the soil profile, and crop species and variety, as well as other growing conditions. This influences the determination of leaching, since the lower boundary of measurement used in a specific study may differ from the effective rooting depth of the crop.

Matrix flow: A relatively slow and even movement of water and solutes down the soil profile through all pore spaces obeying convective dispersion theory, which assumes that water follows an average flow path through soil. Matrix flow dominates in coarse-textured (sandy) soils, and mainly transports dissolved matter and soluble nutrients.

Macropore flow (also termed bypass or preferential flow): An uneven and often rapid movement of water, solutes and suspended particulate matter through the larger, more or less continuous soil pores (e.g. cracks, fissures, root channels and earthworm burrows). Occurs when the upper soil layers become water saturated during intensive precipitation events. Macropore flow dominates on more finely textured and structured soils. The faster water transport bypasses most of the soil matrix, having only little contact with the soil matrix and reactive soil surfaces.

Nutrient runoff: The process of nutrient transport when precipitation rate exceeds the infiltration capacity of the soil and the excess is discharged from the area by horizontal flow across the surface, or in the uppermost subsurface layers, without entering the soil profile. The risk of surface runoff increases with slope. Depending on water velocity, runoff may create soil surface erosion, and runoff losses may comprise both dissolved and suspended particulate nutrients.

Text Box – Advanced 11.2 Methods of measuring leaching and runoff (see also Figure 11.2).

Suction cup method: Suction cups are inserted at the bottom of the root zone, usually at 75- to 150-cm depth. The suction cups are made of a porous material (e.g. ceramic) and connected with two tubes to the soil surface (Webster *et al.*, 1993). By adding vacuum to the cups (through tube 1) soil solution is accumulated in the cup, which can then be sampled using tube 2 and analysed for nutrient concentration. Soil solution must be sampled regularly (e.g. every 14 days) and the water flow out of

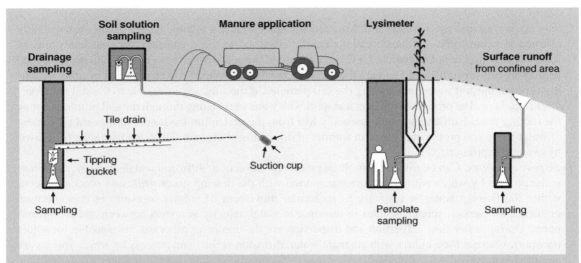

Figure 11.2   *Methods for measuring nutrient leaching and runoff. (© University of Southern Denmark.)*

the rooting zone is estimated using water flow modelling based on measured precipitation and modelled evapotranspiration and soil water storage changes. Nutrient leaching is then estimated from nutrient concentration in soil solution multiplied by the modelled water flow. An advantage of the method is that normal field operations such as tillage, sowing and harvesting are not affected by the installations and can be carried out with normal machines in the field of measurement. However, a critical assumption is that the sampled soil solution is representative of the percolating soil water. It has been questioned whether this is generally valid and the suction cup method is certainly not suitable on structured soils with cracks, where water can bypass the cups.

Lysimeters: These are closed soil systems where all the percolating water can be collected and sampled at the bottom for measurement of nutrient concentration (Webster *et al.*, 1993). Lysimeters can be of variable depth, usually 50–150 cm, and may cover a variable surface area. Normally the walls of lysimeters extend above the soil surface, meaning that normal farm machines cannot be used for tillage and so on. Lysimeters filled with repacked soil or containing intact soil monoliths are both used in studies of nutrient leaching, and the lysimeters may be placed on weighing cells for soil water storage determination.

Drains: Leaching can also be estimated from a tile-drained area, where it is possible to measure or calculate the water flow in the drains and to sample the drainage water regularly for measurement of nutrient concentration. However, to calculate the total nutrient leaching, assumptions and calculations must be made about the proportion of the water flow that bypasses the drains and percolates towards deeper subsoil (Goulding *et al.*, 2000), see also Figure 11.1.

Surface runoff plots: Runoff is usually measured by confining an area of uniformly sloping land within fixed borders (of plastic or metal), extending slightly above the soil surface and down to the depth to which subsurface runoff is to be included, to divert runoff from upslope areas. Along the downslope side, a trough is dug into the ground to collect accumulating runoff (incl. any suspended particulate matter) from the confined area for sampling of runoff. Sampling systems may be equipped with automatic recording of runoff rates during storm events and possibly also time- or flow volume-proportional fraction sampling, for more detailed studies.

11.2   Leaching and Runoff of Manure Nitrogen

Manure application causes an increased risk of N leaching both in the year of application and in many following years. By contrast, runoff of manure N mainly occurs within the first days and weeks after application during heavy rain events. In the following, important factors that influence leaching and runoff of manure N are described.

11.2.1   Leaching of Manure Nitrogen in the First Year after Application and Methods for Reducing the Risk

11.2.1.1   Effects of Application Rate and Timing Related to Temporal Crop Nitrogen Demand

Application of manure or fertiliser at rates with available N in excess of the economically optimal N rate for the crop increases the risk of N leaching significantly (Goulding *et al.*, 2000) and it appears that manure N increases N leaching more than similar levels of fertiliser N (Bergström and Kirchmann, 2006; see also Figure 11.5 below). In contrast to fertiliser, which has a declared N content, a relatively constant availability and is easy to apply accurately, animal manure concentrations of total-N and NH_4^+-N vary significantly, and the N availability and fertiliser value are highly variable, depending on application method, soil, crop and climate conditions (Chapter 15). Finally, most manure and slurry application technology is less capable of applying an accurate and uniform rate than mineral fertiliser spreading technologies. For those reasons, achieving the target available N application rate is much more difficult with manures than with mineral fertilisers and therefore also increases the risk of N leaching with manure N application, simply because of a higher risk of overfertilisation.

Like other fertilisers, manures should only be applied when the crop can utilise the available N, in order to reduce the risk of nutrient losses, in humid climates especially by NO_3^- leaching. For most crops in temperate, humid climates this means in the spring, when both spring- and autumn-sown crops have their major N demand and highest uptake rates. Perennial crops, such as pasture or seed grasses, and some autumn-sown crops, such as winter oilseed rape (canola), are also capable of taking up large amounts of N in the autumn. However, most autumn-sown cereals, such as winter wheat, are relatively inefficient in utilising N applied in the autumn under cool conditions, due to limited autumn growth rate and early dormancy before the winter (Sørensen and Rubæk, 2012).

Application of farmyard manure (FYM) in early autumn (September) to uncropped soil typically results in extra leaching, often corresponding to the NH_4^+ content of the manure under Northern European conditions (Thomsen, 2005). However, on postponing FYM application to December or March, Thomsen (2005) observed no extra leaching in the first year. Similarly, Sørensen and Rubæk (2012) found that when solid manures were applied in autumn to winter wheat extra NO_3^--N leaching occurred, equivalent to the content of NH_4^+-N in the manures (20–30% of the total manure N) under conditions with surplus drainage of 300–400 mm during the winter (Figure 11.3). With spring application, much lower leaching losses occurred in the following winter (less than 4% of solid manure N and 1% of slurry N) and losses were not significantly higher than the leaching from solid manure applied in the previous autumn (Figure 11.3; Sørensen and Rubæk, 2012). Autumn application of animal slurries, normally containing 50–75% of total-N as NH_4^+, should therefore always be avoided in humid climates, since large leaching losses will be inevitable.

On grassland, increases in NO_3^- leaching more or less equivalent to the NH_4^+ content have been observed after FYM application in October (Smith *et al.*, 2002). Thus, even crops with an extended root system may not have effective root N uptake late in the season due to low light intensity and low production, and hence low crop demand for N uptake. Under climate conditions with less surplus precipitation, insignificant losses

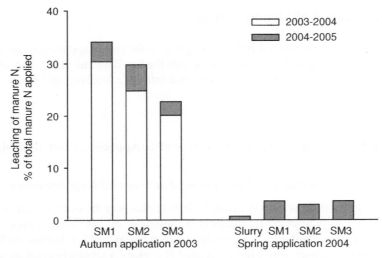

Figure 11.3 *Extra NO_3^- leaching of manure N (above the treatment with only mineral N fertiliser) over 2 years after autumn and spring application of solid manure fractions from animal slurry separation (SM1– SM3, NH_4-N fraction of total-N of 0.34, 0.32 and 0.26, respectively) and pig slurry (spring only) to winter wheat, all supplemented with mineral N fertiliser (to similar amount of available N, 180 kg N ha^{-1} $year^{-1}$). Average of two soil types (loamy sand and sandy loam) in 1.5-m deep lysimeters. (© University of Southern Denmark; based on data from Sørensen and Rubæk (2012).)*

can be observed in the first year after manure application, but under such conditions extra N leaching has been observed in the following years after application (Bergström and Kirchmann, 1999).

On soil types with favourable conditions for deep crop root development to below 2 m depth (e.g. in winter cereals) crops may be able to take up N that has been leached to deep soil layers during the winter period in the following spring/summer period (Jensen *et al.*, 2000).

Nitrification inhibitors, such as dicyandiamide (DCD), may potentially limit the oxidation of NH_4^+ to NO_3^- after manure application and hence reduce the risk of leaching during the wet season because the NH_4^+ cation is adsorbed to the negatively charged soil colloids. However, addition of DCD with manure applications in autumn has shown variable results; for example, it has been found to be ineffective in reducing N leaching after autumn application of both slurry and FYM in UK arable cropping (Beckwith *et al.*, 1998). Similar results have been found for cattle slurry on sandy soils in the Netherlands (Corre and Zwart, 1995). More recently, however, a number of New Zealand studies have found direct field application of DCD to dairy pastures to be effective in reducing NO_3^- leaching significantly (by 44–67%) from deposited cow urine and dung patches (Di *et al.*, 2009).

11.2.1.2 Effect of Nitrogen Mineralisation during Autumn/Winter

The organic N in manures is mineralised continuously after application to soil, but with a declining mineralisation rate over time. Part of the organic N is mineralised during the first autumn/winter period after application, when there are often no plants present to take up mineralised N and the mineralised N can therefore potentially be lost. Furthermore, part of the mineral N initially present in manure is assimilated by soil microbes after application to soil and this has been suggested as a mechanism to retain a significant proportion of manure N over winter (Jensen *et al.*, 2000). Part of the immobilised N present in microbial cells and residues can be remineralised in the autumn following spring application and is thus at risk of being leached. However, there are indications that most of this N is stabilised and only a minor part of it is mineralised in the first autumn (Sørensen and Amato, 2002).

The mineralisation rate of manure organic and immobilised mineral N will depend on soil temperature, which in temperate climates will decline through the autumn season. Several studies have shown that the net N mineralisation from freshly added organic materials is less retarded by low temperatures than older and more recalcitrant soil organic matter, mainly because the gross immobilisation process is restricted more than the gross mineralisation at low temperatures (Andersen and Jensen, 2001). Once mineralised, the NH_4^+ will potentially be nitrified, and although this process is also restricted by low temperatures, several studies have shown that nitrification does proceed at lower but still significant rates over the winter period. Nitrifying bacteria gradually adapt to cold soil temperatures in late autumn and once acclimatised, nitrification may proceed at soil temperatures as low as 0 °C (Jayasundara *et al.*, 2010).

This means that N leaching normally increases in autumn after application of manure. Under humid Northern European conditions, increases in leaching equivalent to 3–10% of the organic N applied have been observed (Thomsen *et al.*, 1997; Thomsen, 2005) and a similar magnitude of leaching has been found by Jayasundara *et al.* (2010) (i.e. 3–5% of applied liquid swine manure N in a silt loam and 8–10% in a sandy loam over 2 years). If the soil is covered by plants in autumn, e.g. an efficient catch crop (see Section 11.2.2.2), this loss can be reduced.

11.2.2 Long-Term Leaching of Manure Nitrogen

11.2.2.1 *Mineralisation Pattern in Relation to Crop Uptake and Precipitation*

Most of the organic N in manures still remains in soil one year after application. Studies with ^{15}N-labelled solid manures have shown that about 70–80% of applied faecal N derived from ruminants remains in soil 1–3 years after application (Jensen *et al.*, 1999). This residual N is mineralised relatively slowly in soil over a period of more than 100 years after application (Petersen *et al.*, 2005). Mineralisation of N occurs over the whole year, and also in periods with surplus precipitation and without plant growth. Therefore, a higher proportion of the mineralised residual manure N can be leached in humid climates and less is taken up by the crop compared with mineral fertiliser N applied at the start of the growing season, as illustrated in Figure 11.4. Thomsen *et al.* (1997) found that in the second year after application, the leaching of manure-derived N was about equal to the uptake of manure-derived N in a spring barley crop (Figure 11.4). That study was conducted under humid Northern European conditions on sandy and sandy loam soils. Similarly, Nielsen and Jensen (1990) found significantly higher NO_3^- leaching in fields with a long history of FYM application than in fields that only received mineral N fertilisers. However, under more continental conditions with yearly drainage below 100 mm, Erhart *et al.* (2007) found no extra N leaching from soil receiving high amounts of organic N in the form of compost over an 11-year period. Thus, under conditions with low yearly drainage the leaching losses from slowly released residual manure N are also expected to be low, as crop roots have better opportunities for taking up mineralised N before it is leached below the root zone.

11.2.2.2 *Effects of Length of Growing Season and Use of Catch Crops*

The long-term leaching losses of residual manure N can be significantly reduced by growing crops with a long growing season (e.g. sugar/fodder beet or silage maize) since these are more capable of taking up mineralised

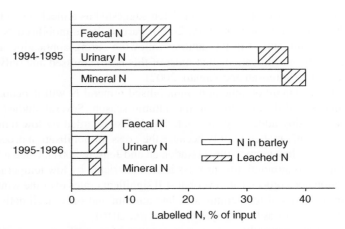

Figure 11.4   *Spring barley crop uptake of labelled N and leaching of labelled NO₃⁻-N during 2 years after application of cross-labelled ruminant slurry containing either ¹⁵N-labelled urine (+ unlabelled faeces) or ¹⁵N-labelled faeces (+ unlabelled urine) and compared with a treatment with labelled mineral N fertiliser. Manures and fertilisers were applied in spring 1994 to a sandy soil in 1.5-m deep lysimeters. (© University of Southern Denmark; based on data from Thomsen* et al. *(1997).)*

residual manure N throughout the autumn season, when cereals or other shorter season crops have already been harvested and when many autumn-sown crops have a limited capacity for N uptake (as described in Section 11.2.1.2).

Alternatively, catch crops can be established in autumn after harvest of the main crop on the area where manures were applied previously. Catch crops are defined as cover crops grown in between main crops, typically over the winter season, to catch available N in the soil and thereby prevent N leaching losses (Thorup Kristensen *et al.*, 2003). In the second year after application of cattle slurry, Sørensen (2004) found that about one-third of the residual manure N that was available for above-ground plant uptake was found in a ryegrass catch crop grown after spring barley, in which the other two-thirds of the available residual N mineralisation were taken up. In the long term, the residual manure N recovered by the catch crop can only be efficiently prevented from losses if a substantial proportion of the recovered N is mineralised and made available to the next main crop, thus potentially substituting fertiliser N inputs. This may be achieved to some extent by intelligent and locally adapted management of catch crop species choice, incorporation time and crop rotation (Thorup Kristensen *et al.*, 2003). However, less optimal management could be expected to result in some leaching of the residual manure N under conditions without plant cover in the autumn in humid areas with significant drainage. This also implies that on areas with a high proportion of perennial grassland with an extensive and efficient root system, the long-term leaching of residual manure N is lower than on areas with mostly arable crops, in which it is difficult to maintain an effective plant cover in all winter seasons.

11.2.2.3   *Effect of Manure Types and Proportion of Ammonium Nitrogen*

Manures with a low content of mineral N, such as composts or manures with a high immobilisation potential, are less vulnerable to NO₃⁻ leaching in the first year of application. However, the risk of long-term leaching is significantly higher, as these manures are mineralised over the years and the mineralised N is at higher risk of leaching due to poorer synchrony between N mineralisation and crop N uptake. Thus, viewed over a period of several decades the accumulated amount of leached manure N can be expected to increase with the proportion of manure N that is present as organic N. This is in contrast to manures with a high proportion of NH_4^+-N, for which appropriate rate, timing and method of application are the dominant factors preventing the risk of N leaching (Section 11.2.1). The main factors for minimising long-term leaching risk from organic N-rich manures are crop rotation and catch crop use.

Treatment of manure by anaerobic digestion for biogas production converts a higher proportion of manure N into NH_4^+, especially for feedstocks with a high degradability (e.g. poultry and pig manures with a diet high in concentrates), which produces a digestate with very high NH_4^+-N proportion of above 80% (Möller and Müller, 2012). If the higher N availability is accounted for in fertiliser planning, then this manure treatment may decrease the risk of N leaching as the manure contains less organic N that is vulnerable to long-term N leaching.

11.2.3   Nitrogen Losses via Runoff and Strategies for Reducing the Risk

Nitrogen may also be lost by surface runoff following the application of animal manures to arable land. Although total-N losses by this route may be insignificant in agronomic terms (e.g. 1–3% of manure mineral N; Ceretta *et al.*, 2010), soluble or suspended N concentrations of NH_4^+ and NO_3^- may be of concern in sensitive catchments, in terms of the potential for contribution to accelerated eutrophication and adverse effects on freshwater biota (e.g. ammonia (NH_3) is toxic to some aquatic organisms). Smith *et al.* (2001a) found cattle slurry and FYM to increase runoff losses of solids and NH_4^+-N, but not NO_3^--N, compared with inorganic fertiliser on a silty clay loam. Rainfall events shortly after manure application were particularly associated with nutrient runoff losses and surface rather than subsurface flow was generally the dominant pathway. Increasing slurry application rate increased solids and NH_4^+ losses via surface runoff, and the threshold above which the risk of losses greatly increased was around 2.5–3.0 tonne slurry solids ha^{-1}, corresponding to approximately 50 m^3 slurry ha^{-1}.

Strategies to reduce the risk of runoff losses of manure N therefore include limiting the manure rate applied per spreading event (e.g. to below 50 m^3 slurry ha^{-1}), but injection or rapid incorporation of the manure will reduce the risk more effectively (Eghball and Gilley, 1999). Furthermore, treatment that reduces the NH_4^+ concentration in the manure, such as composting, which tends to immobilise a fraction of it as organic N (but with an increased risk of NH_3 volatilisation from composting), can also serve to reduce the risk of runoff. Separation of slurry into a liquid fraction with low solids content can promote rapid infiltration in soil and thereby reduce runoff, with a smaller solid fraction to be incorporated.

11.3   Leaching and Runoff of Manure Phosphorus

Phosphorus is normally strongly bound in soil, resulting in generally low P leaching losses from soils, but under specific circumstances significant losses of manure P can occur by both leaching and surface runoff. When manures are directly exposed to moving water shortly after their application in the field, there can be a significant risk of P being lost by runoff, while on soil types with preferential flow manure P can also be leached. When high amounts of manure P are applied repeatedly to the same area over many years, there is also a risk of increased long-term P leaching. In the following, factors that influence leaching and runoff of manure P are described.

11.3.1   Leaching of Manure Phosphorus

11.3.1.1   Manure Phosphorus Forms and Application Method Affect the Phosphorus Leaching Potential

Soil and manure P can be leached in dissolved form, but also in particulate organic or inorganic forms (Schelde *et al.*, 2006; Glæsner *et al.*, 2011b). Livestock manures often contain a high amount of inorganic P that is water-extractable (Kleinman *et al.*, 2005), but they also contain inorganic P that is insoluble and organic P in both soluble and insoluble form (Chapter 4). However, in many cases there is no relationship between manure application rates and P leaching, and Bergström and Kirchmann (2006) even found a negative relationship (Figure 11.5). An explanation for this reduced P leaching with increased slurry applications could possibly be a shift in soil P chemistry, as suggested by Sharpley *et al.* (2004), with long-term additions of manure increasing the pH of the surface soil and shifting P to relative insoluble calcium (Ca) complexes, which could contribute to reduced P leaching.

Sharpley and Moyer (1999) found that water-extractable P gives a good estimate of potential P leaching and P runoff after manure application, and estimated that 16–63% of total P in different manures is water-extractable. By repeated extraction of manure with water, even more manure P can become water-extractable (Dou *et al.*, 2000), indicating that exposure to precipitation and percolating soil water may increase the leaching of dissolved P. Sharpley and Moyer (1999) found after five repeated simulated rainfalls that 58% of total P was leached from dairy manure, 21% from poultry manure, 20% from poultry compost and litter, and 15% from dairy compost and pig slurry. However, manure P brought into solution may easily be adsorbed in soil if it comes into contact with binding sites on the soil particle surfaces and suspended particulate P can be physically retained in soil if transported through small soil pores.

The water flow regime (i.e. whether matrix or macropore flow; Text Box – Basic 11.1) that dominates is therefore determinant for the actual manure P losses, as this controls the contact between manure P and the reactive soil surfaces and sites. In fine-textured soils, structured with continuous macropores, preferential flow may take place, and under such conditions sorption and physical particle retention are lower and significant transport of both soluble and particulate P may take place (e.g. to drains) (Magid *et al.*, 1999; Schelde *et al.*, 2006; Glæsner *et al.*, 2011b). Coarse-textured soils are less prone to macropore flow and most water transport takes place through the soil matrix in small pores. In soils with mainly matrix flow, the risk of P leaching is

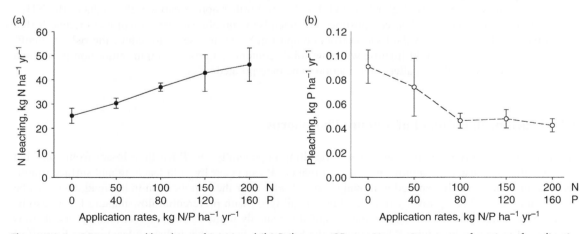

Figure 11.5   *Mean annual leaching of (a) N and (b) P (bars are SE, n = 3) over 3 years as a function of application rates of N and P with animal slurry (only applied first 2 years). (Reprinted with permission from Bergström and Kirchmann (2006); © 2006 Journal of Environmental Quality.)*

low as long as sorption capacity is sufficient (Sharpley *et al.*, 2004; Bergström and Kirchmann, 2006). van Es *et al.* (2004) found that the average P leaching on soil supplied with manure was 39 times higher on a clay loam soil with preferential flow than on a loamy sand soil with mainly matrix flow.

The risk of macropore leaching of manure P is influenced by the manure application method. The leaching of dissolved inorganic P in sandy loam soils can be reduced by manure incorporation instead of surface application (Magid *et al.*, 1999). Similarly, Glæsner *et al.* (2011b) found that P leaching losses were lower after direct injection than after surface-banding of slurry on loam and sandy loam soils, but they found practically no effect on a more sandy soil. When the manure is injected it is placed away from the active flow paths of infiltrating water. This effect can be explained by the fact that many macropores are directly connected with the soil surface in soil that has not been tilled recently (e.g. though earthworm channels). When manures are injected or incorporated such pores are mostly disrupted. Thus, the risk of P leaching through macropores is significantly higher after surface application of manures than after injection or incorporation. When slurry is applied on frozen soils with deep frost penetration, the risk of losses by preferential flow is also increased due to formation of cracks in the soil (Parkes *et al.*, 1997).

11.3.1.2 *Effect of Soil Phosphorus Saturation on Risk of Manure Phosphorus Leaching*

If the supply of N to crops is based entirely on available manure N, this usually results in P application in excess of crop demand, due to the typical range of N : P ratios found in manures being lower than the corresponding ratio of crop demand for N and P (Chapter 15.4.5). Thus, manure application according to this practice for several years will lead to accumulation of P in the manured soils, which may become more or less P saturated, meaning that the potential for further P adsorption becomes significantly reduced. For such more or less P saturated soils, including soils with high repeated manure P application, there is much higher leaching of P (Heckrath *et al.*, 1995). Phosphorus leaching in drainage can be related to a soil P index such as Olsen-P (bicarbonate-extractable P). Leaching of P increases significantly at a certain concentration of Olsen-P in some soils (Heckrath *et al.*, 1995; Figure 11.6) and this concentration is called the "change point". However, studies

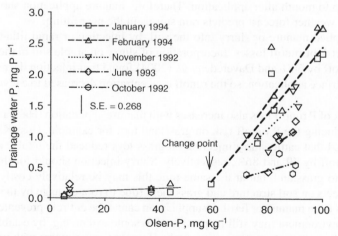

Figure 11.6 *Relationship between total-P (TP) concentration in drainage water and concentration of Olsen P in soil from field plots with differing previous P application (Broadbalk Experiment; Heckrath* et al.*, 1995). The Split-Line Model shown fitted a common initial line and change point (at 57 mg Olsen-P kg⁻¹ soil), but variable slope for the five drainage events at higher P concentrations. (Reprinted with permission from Heckrath* et al. *(1995). © 1995 Journal of Environmental Quality).*

in other soils have shown that there is not always a clear change point and that drainage P concentration may be more linearly related to extractable P in soil, and this relationship is influenced by the soil type (McDowell and Sharpley, 2001). Subsequent studies have quantified various indices for the degree of soil P saturation and have identified that P losses begin to increase when the degree of P saturation is only 25–34% (Butler and Coale, 2005).

11.3.2 Phosphorus Losses via Runoff and Strategies for Reducing the Risk

A significant proportion of the P in FYM and slurry, including the solids fraction from slurry separation, is vulnerable to transport by runoff, but not as vulnerable as similar amounts of soluble mineral P (Heathwaite *et al.*, 1998; Gasser *et al.*, 2012). Although losses of total and soluble P are insignificant in agronomic terms, peak concentrations of P in surface water exert a considerable impact in sensitive freshwater catchments, with accelerated eutrophication especially when combined with high $NO_3{}^-$-N losses.

In general, runoff P losses from soils not recently amended with manure or fertiliser are mainly in the form of soil particle- or colloid-bound P, whereas shortly after application, losses of soluble manure P or suspended manure particulate P may be more dominant. Sealing of the soil surface by slurry solids may also be a major mechanism for initiating surface runoff. Withers and Bailey (2003) studied overland flow of suspended sediment and P from a forage maize field receiving pig slurry and found that surface application of slurry increased runoff losses, with up to 23% of slurry P applied being lost by runoff, accounting for 60% of total plot P export, the majority in soluble bioavailable form. Similarly, Smith *et al.* (2001b) found that cattle FYM and especially slurry increased particulate and soluble P losses in surface water flow from an arable silty clay loam, and estimated a similar threshold for greatly increased risk of P losses as for N losses (Section 11.2.3; Smith *et al.*, 2001a). One strategy to reduce runoff risk can therefore be to split slurry application. For example, Withers and Bailey (2003) found that splitting the application into three doses reduced total P export by 25%.

Increasing time between manure application and the rainstorm events causing runoff also lowers the risk of P runoff. Hahna *et al.* (2012) found that the risk decreased up to 5- to 10-fold when the storm event was more than one week up to month after application. Therefore, manure application should not be planned on very wet soils or if the weather forecast predicts rain storms in the near future.

Incorporation of applied manure or slurry into the soil, by injection or rapid tillage, can be an effective strategy to reduce the risk of runoff losses. Incorporation of slurry on arable soil in the study by Smith *et al.* (2001b) reduced P runoff by 60% and Daverede *et al.* (2004) found that injection decreased total P runoff by 94% compared with surface application, so the runoff concentration decreased to a level similar to that in the non-amended soil.

On grassland the risk of P runoff loss also increases with manure application (Hahna *et al.*, 2012). Injection is quite effective at reducing the P runoff risk on grassland too; for example, Uusi-Kämppä and Heinonen-Tanski (2008) estimated that cattle slurry injection on grass leys reduced the total P and dissolved reactive P losses in surface runoff by 79 and 86%, respectively. Slurry injection should for this reason be preferred when applying slurry to grassland, but at the same time this may be relatively costly (energy demanding), and have negative impacts on soil structure and grass yield (due to root damage by the injection tines).

Even if runoff losses from manure or fertiliser application cannot be entirely prevented, emission of surface runoff to the aquatic environment may still be prevented to some extent, e.g. by establishing buffer strips in between the field and the potential recipient water body. However, experiments with unfertilised grass buffer strips adjacent to runoff areas showed that nearly all the runoff P from inorganic fertiliser was captured in the buffer strip, whereas only about 10% of the runoff P from slurry was retained (Heathwaite *et al.*, 1998). Most of the runoff P from slurry that was not retained in the buffer strip was in particulate or dissolved organic fractions that could not be effectively retained by the grass.

11.4 Leaching and Runoff of Potassium

Potassium (K) is another important plant nutrient in manures that can potentially be lost by runoff and leaching. Potassium losses are not an environmental problem, but K is a valuable resource and K losses should therefore be minimised. Potassium leaching varies with soil texture, clay mineralogy, current and past K inputs, and drainage. In a review, Johnston and Goulding (1992) found a strong correlation between drainage and K leaching in loams and clayey soils with an average loss of 1 kg K (100 mm drainage)$^{-1}$. However, on sandy soils K leaching can be much higher, and Askegaard *et al.* (2004) found K leaching losses of between 0.5 kg K (100 mm drainage)$^{-1}$ in a soil with 24% clay and 7 kg K (100 mm drainage)$^{-1}$ in a soil with only 5% clay.

Most K in manures is present in ionic form as K^+, meaning it is water-soluble, but also easily adsorbed on negatively charged soil particles. Potassium in soil can be divided into four fractions: K in solution, exchangeable K, non-exchangeable K and K bound in soil minerals. These fractions constitute a dynamic system with reversible transfer between the fractions and only a minor part of soil K is in solution (Askegaard *et al.*, 2004). Therefore most manure K is vulnerable to runoff and to leaching by macropore flow immediately after application, whereas leaching losses by matrix transport are usually low, except on sandy soils with a low CEC (Askegaard *et al.*, 2004). Potassium leaching is more influenced by previous K applications and soil type than by a new manure application (Table 11.1).

Leaching losses of K from sandy soils can be reduced by growing a catch crop that is able to retain part of the available K in the upper soil layers and then release it again when the catch crop is incorporated or killed off for other reasons (Thorup-Kristensen *et al.*, 2003; Askegaard and Eriksen, 2008).

Regarding runoff losses of K, Ceretta *et al.* (2010) found K losses from surface-applied pig slurry under no tillage to be higher than corresponding P and N losses. With slurry application rates between 20 and 80 m^3 ha^{-1} they found surface runoff losses of 1–3% of slurry mineral N, 3–6% of available P and 9–17% of available K, indicating the relatively high solubility and mobility of K in manure that has not yet come into contact with the soil, where the K^+ can be adsorbed. The runoff losses of N and K were positively related to the volume of surface runoff, whereas the P losses were more positively related to the slurry rate.

11.5 Summary

Manure application may cause increased nutrient losses by leaching and runoff in both the short and long term, potentially resulting in environmental impacts. Nutrient leaching losses can be measured by use of suction cups, lysimeters or tile drains and runoff can be measured on framed sloping field plots. Nutrient leaching with percolating water takes place by both matrix and macropore flow, and the transport mechanism involved is important for the form of nutrients that are leached. Nutrients that are easily adsorbed on soil

Table 11.1 *Average K leaching from organic crop rotations only supplied with organic manures at different locations on differing soil types (data from Askegaard et al., 2004).*

Soil	Soil clay content (%)	Manure K input (kg ha^{-1} yr^{-1})	Drainage (mm yr^{-1})	K leaching (kg ha^{-1})
Jyndevad	5	30	667	34
Foulum	9	20	366	15
Flakkebjerg	16	18	254	2
Holeby	24	15	270	1

particles and colloids, such as phosphate, NH_4^+ and K, can be quickly leached by macropore flow or lost by runoff if excessive water accumulates at the soil surface after manure surface application. Macropore flow mainly takes place on finely textured soils, whereas on sandy soils leaching mainly takes place by matrix flow and here there is a high potential for NO_3^- leaching. Most animal manures contain a significant proportion of organic N, which is mineralised continuously over many years in soil and therefore vulnerable to NO_3^- leaching due to poor synchrony between N mineralisation and crop N uptake. Such N losses are especially important in humid regions with significant winter excess precipitation and drainage. Important measures to reduce manure nutrient leaching and runoff are appropriate application methods and timing, optimised nutrient application rates and the use of catch crops.

References

Andersen, M.K. and Jensen, L.S. (2001) Low soil temperature effects on short-term gross N mineralisation-immobilisation turnover after incorporation of a green manure. *Soil Biol. Biochem.*, **33**, 511–521.

Askegaard, M. and Eriksen, J. (2008) Residual effect and leaching of N and K in cropping systems with clover and ryegrass catch crops on a coarse sand. *Agric. Ecosyst. Environ.*, **123**, 99–108.

Askegaard, M., Eriksen, J. and Johnston, A.E. (2004) Sustainable management of potassium, in *Managing Soil Quality: Challenges in Modern Agriculture* (eds P. Schjønning, S. Elmholt and B.T. Christensen), CABI, Wallingford, pp. 85–102.

Beckwith, C.P., Cooper, J., Smith, K.A. and Shepherd, M.A. (1998) Nitrate leaching loss following application of organic manures to sandy soils in arable cropping. I. Effects of application time, manure type, overwinter crop cover and nitrification inhibition. *Soil Use Manag.*, **14**, 123–130.

Bergström, L.F. and Kirchmann, H. (1999) Leaching of total nitrogen from nitrogen-15-labeled poultry manure and inorganic nitrogen fertilizer. *J. Environ. Qual.*, **28**, 1283–1290.

Bergström, L. and Kirchmann, H. (2006) Leaching and crop uptake of nitrogen and phosphorus from pig slurry as affected by different application rates. *J. Environ. Qual.*, **35**, 1803–1811.

Butler, J.S. and Coale, F.J. (2005) Phosphorus leaching in manure-amended atlantic coastal plain soils. *J. Environ. Qual.*, **34**, 370–381.

Ceretta, C.A., Girotto, E., Lourenzi, C.R., Trentin, G. and Vieira, R.C.B. and Brunetto, G. (2010) Nutrient transfer by runoff under no tillage in a soil treated with successive applications of pig slurry. *Agric. Ecosyst. Environ.*, **139**, 689–699.

Corre, W.J. and Zwart, K. (1995) Effects of DCD addition to slurry on nitrate leaching in sandy soils. *Neth. J. Agric. Sci.*, **43**, 195–204.

Daverede, I.C., Kravchenko, A.N., Hoeft, R.G., Nafziger, E.D., Bullock, D.G., Warren, J.J. and Gonzini, L.C. (2004) Phosphorus runoff from incorporated and surface-applied liquid swine manure and phosphorus fertilizer. *J. Environ. Qual.*, **33**, 1535–1544.

Di, H.J., Cameron, K.C., Shen, J.P., He, J.Z. and Winefield, C.S. (2009) A lysimeter study of nitrate leaching from grazed grassland as affected by a nitrification inhibitor, dicyandiamide, and relationships with ammonia oxidizing bacteria and archaea. *Soil Use Manag.*, **25**, 454–461.

Dou, Z., Toth, J.D., Ferguson, J.D., Galligan, D.T. and Ramberg Jr, C.F. (2000) Laboratory procedures for characterizing manure phosphorus. *J. Environ. Qual.*, **29**, 508–514.

Eghball, B. and Gilley, J.E. (1999) Phosphorus and nitrogen in runoff following beef cattle manure or compost application. *J. Environ. Qual.*, **28**, 1201–1210.

Erhart, E., Feichtinger, F. and Hartl, W. (2007) Nitrogen leaching losses under crops fertilized with biowaste compost compared with mineral fertilization. *J. Plant Nutr. Soil Sci.*, **170**, 608–614.

Gasser, M.-O., Chantigny, M.H., Angers, D.A., Bittman, S., Buckley, K.E., Rochette, P. and Massé, D. (2012) Plant-available and water-soluble phosphorus in soils amended with separated manure solids. *J. Environ. Qual.*, **41**, 1290–1300.

Glæsner, N., Kjaergaard, C., Rubæk, G.H. and Magid, J. (2011a) Interactions between soil texture and placement of dairy slurry application: I. Flow characteristics and leaching of non-reactive components. *J. Environ. Qual.*, **40**, 337–343.

Glæsner, N., Kjaergaard, C., Rubæk, G.H. and Magid, J. (2011b) Interactions between soil texture and placement of dairy slurry application: II. Leaching of phosphorus forms. *J. Environ. Qual.*, **40**, 344–351.

Goulding, K.W.T., Poulton, P.R., Webster, C.P. and Howe, M.T. (2000) Nitrate leaching from the Broadbalk Wheat Experiment, Rothamsted, UK, as influenced by fertilizer and manure inputs and the weather. *Soil Use Manag.*, **16**, 244–250.

Hahna, C., Prasuhn, V., Stamm, C. and Schulin, R. (2012) Phosphorus losses in runoff from manured grassland of different soil P status at two rainfall intensities. *Agric. Ecosyst. Environ.*, **153**, 65–74.

Heathwaite, A.L., Griffith, P. and Parkinson, R.J. (1998) Nitrogen and phosphorus in runoff from grassland with buffer strips following application of fertilizers and manures. *Soil Use Manag.*, **14**, 142–148.

Heckrath, G., Brookes, P.C., Poulton, P.R. and Goulding, K.W.T. (1995) Phosphorus leaching from soils containing different phosphorus concentrations in the Broadbalk experiment. *J. Environ. Qual.*, **24**, 904–910.

Jayasundara, S., Wagner-Riddle, C., Parkin, G., Lauzon, J. and Fan, M.Z. (2010) Transformations and losses of swine manure ^{15}N as affected by application timing at two contrasting sites. *Can. J. Soil Sci.*, **90**, 55–73.

Jensen, B., Sørensen, P., Thomsen, I.K., Jensen, E.S. and Christensen, B.T. (1999) Availability of nitrogen in ^{15}N-labeled ruminant manure components to successively grown crops. *Soil Sci. Soc. Am. J.*, **63**, 416–423.

Jensen, L.S., Pedersen, I.S., Hansen, T.B. and Nielsen, N.E. (2000) Turnover and fate of ^{15}N-labelled cattle slurry ammonium-N applied in the autumn to winter wheat. *Eur. J. Agron.*, **12**, 23–35.

Johnston, A.E. and Goulding, K.W.T. (1992) Potassium concentrations in surface and ground waters and the loss of potassium in relation to land use, in *Potassium in Ecosystems: Biochemical Fluxes of Cations in Agro-Forest-Systems. Proceedings of the 23rd Colloquium of the International Potash Institute* (ed. International Potash Institute), International Potash Institute, Basel, pp. 35–158.

Kleinman, P.J.A., Wolf, A.M., Sharpley, A.N., Beegle, D.B. and Saporito, L.S. (2005) Survey of water-extractable phosphorus in livestock manures. *Soil Sci. Soc. Am. J.*, **69**, 701–708.

Korsaeth, A., Bakken, L.R. and Riley, H. (2003) Nitrogen dynamics of grass as affected by N input regimes, soil texture and climate: lysimeter measurements and simulations. *Nutr. Cycl. Agroecosyst.*, **66**, 181–199.

Magid, J., Jensen, M.B., Mueller, T. and Hansen, H.C.B. (1999) Phosphate leaching responses from unperturbed, anaerobic, or cattle manured mesotrophic sandy loam soils. *J. Environ. Qual.*, **28**, 1796–1803.

McDowell, R.W. and Sharpley, A.N. (2001) Approximating phosphorus release from soils to surface runoff and subsurface drainage. *J. Environ. Qual.*, **30**, 508–520.

Möller, K. and Müller, T. (2012) Effects of anaerobic digestion on digestate nutrient availability and crop growth: a review. *Eng. Life Sci.*, **12**, 242–257.

Nielsen, N.E. and Jensen, H.E. (1990) Nitrate leaching from loamy soils as affected by crop rotation and nitrogen fertilizer application. *Fertil. Res.*, **26**, 197–207.

Parkes, M.E., Campbell, J. and Vinten, A.J.A. (1997) Practice to avoid contamination of drainflow and runoff from slurry spreading in spring. *Soil Use Manag.*, **13**, 36–42.

Petersen, B.M., Berntsen, J., Hansen, S. and Jensen, L.S. (2005) CN-SIM – a model for the turnover of soil organic matter. I: long-term carbon and radiocarbon development. *Soil Biol. Biochem.*, **37**, 349–374.

Schelde, K., de Jonge, L.W., Kjaergaard, C., Lægdsmand, M. and Rubæk, G.H. (2006) Effects of manure application and plowing on transport of colloids and phosphorus to tile drains. *Vadose Zone J.*, **5**, 445–458.

Sharpley, A.N., Mcdowell, R.W. and Kleinman, P.J.A. (2004) Amounts, forms and solubility of phosphorus in soils receiving manure. *Soil Sci. Soc. Am. J.*, **68**, 2048–2057.

Sharpley, A. and Moyer, B. (1999) Phosphorus forms in manure and compost and their release during simulated rainfall. *J. Environ. Qual.*, **29**, 1462–1469.

Smith, K.A., Beckwith, C.P., Chalmers, A.G. and Jackson, D.R. (2002) Nitrate leaching following autumn and winter application of animal manures to grassland. *Soil Use Manag.*, **18**, 428–434.

Smith, K.A., Jackson, D.R. and Pepper, T.J. (2001a) Nutrient losses by surface run-off following the application of organic manures to arable land. 1. *Nitrogen. Environ. Pollut.*, **112**, 41–51.

Smith, K.A., Jackson, D.R. and Withers, P.J.A. (2001b) Nutrient losses by surface run-off following the application of organic manures to arable land. 2. *Phosphorus. Environ. Pollut.*, **112**, 53–60.

Sørensen, P. (2004) Immobilisation, remineralisation and residual effects in subsequent crops of dairy cattle slurry nitrogen compared to mineral fertiliser nitrogen. *Plant Soil*, **267**, 285–296.

Sørensen, P. and Amato, M. (2002) Remineralisation and residual effects of N after application of pig slurry to soil. *Eur. J. Agron.*, **16**, 81–95.

Sørensen, P. and Rubæk, G.H. (2012) Leaching of nitrate and phosphorus after autumn and spring application of separated solid animal manures to winter wheat. *Soil Use Manag.*, **28**, 1–11.

Sørensen, P., Jensen, E.S. and Nielsen, N.E. (1994) The fate of ^{15}N-labelled organic nitrogen in sheep manure applied to soils of different texture under field conditions. *Plant Soil*, **162**, 39–47.

Thomsen, I.K., Kjellerup, V. and Jensen, B. (1997) Crop uptake and leaching of ^{15}N applied in ruminant slurry with selectively labelled faeces and urine fractions. *Plant Soil*, **197**, 233–239.

Thomsen, I.K. (2005) Crop N utilization and leaching losses as affected by time and method of application of farmyard manure. *Eur. J. Agron.*, **22**, 1–9.

Thorup-Kristensen, K., Magid, J. and Jensen, L.S. (2003) Catch crops and green manures as biological tools in nitrogen management in temperate zones. *Adv. Agron.*, **79**, 69–118.

Uusi-Kämppä, J. and Heinonen-Tanski, H. (2008) Evaluating slurry broadcasting and injection to ley for phosphorus losses and fecal microorganisms in surface runoff. *J. Environ. Qual.*, **37**, 2339–2350.

van Es, H.M., Schindelbeck, R.R. and Jokela, W.E. (2003) Effect of manure application timing, crop, and soil type on phosphorus leaching. *J. Environ. Qual.*, **33**, 1070–1080.

Wachendorf, C., Taube, F. and Wachendorf, M. (2005) Nitrogen leaching from ^{15}N labelled cow urine and dung applied to grassland on a sandy soil. *Nutr. Cycl. Agroecosyst.*, **73**, 89–100.

Webster, C.P., Shepherd, M.A., Goulding, K.W.T. and Lord, E. (1993) Comparisons of methods for measuring the leaching of mineral nitrogen from arable land. *J. Soil Sci.*, **44**, 49–62.

Withers, P.J.A. and Bailey, G.A. (2003) Sediment and phosphorus transfer in overland flow from a maize field receiving manure. *Soil Use Manag.*, **19**, 28–35.

12

Technologies and Logistics for Handling, Transport and Distribution of Animal Manures

Claus A.G. Sørensen[1], Sven G. Sommer[2], Dionysis Bochtis[1] and Alan Rotz[3]

[1]*Department of Engineering, Aarhus University, Denmark*
[2]*Institute of Chemical Engineering, Biotechnology and Environmental Technology,*
University of Southern Denmark, Denmark
[3]*USDA-ARS Pasture Systems and Watershed Management Research Unit, USA*

12.1 Introduction

Animal waste or manure is formed and collected in animal houses, feedlots and exercise areas in solid form or liquid form (Chapters 4 and 5). From the collection site, the manure must be moved to intermediate storage or directly to the site of end use. The latter is often application to fields or fish ponds as organic fertiliser, but the manure may also undergo various treatments for use in energy production, such as biodigestion or incineration.

The systems used for managing manure differ between countries due to climate, farming traditions and the structural development of farm holdings in terms of degree of specialisation and intensity of production (e.g. Burton and Turner, 2003). Short-term storage in animal houses and removing manure reduces the risk of disease spread between animals and batches of animals (Chapter 6). It also reduces ammonia (NH_3), odour and greenhouse gas emissions. Thus, there is normally short-term storage within the animal house and long-term storage outside the house. The latter facilitates distribution of the manure to fields at the right time in relation to crop nutrient requirements and also acts as a buffer to ensure that downstream treatment facilities can handle the pulses of manure coming from animal houses at daily, weekly or monthly intervals.

The amounts of manure produced on medium and large-scale animal farms are considerable (Table 12.1). Removal of manure from the animal house, transport to the field or to other end-use location and field application involve considerable labour inputs and/or high-capacity specialised machines and equipment.

Animal Manure Recycling: Treatment and Management, First Edition. Edited by Sven G. Sommer, Morten L. Christensen,
Thomas Schmidt and Lars S. Jensen.

Table 12.1 *Typical amounts of manure produced by dairy cows, feedlots and fattening pigs.*

Animal category	Manure type	Manure amount	Dry matter ($g\ kg^{-1}$)	Reference
Dairy cow	slurry	21.6 Mg animal^{-1} year^{-1}	100	Poulsen *et al.* (2001)
	FYM	9.9 Mg animal^{-1} year^{-1}	200	Poulsen *et al.* (2001)
	liquid	10.0 Mg animal^{-1} year^{-1}	34	Poulsen *et al.* (2001)
	deep litter, straw	14.8 Mg animal^{-1} year^{-1}	300	Poulsen *et al.* (2001)
Beef (feedlot)	FYM		525	Hao *et al.* (2011)
Fattening pig	slurry	0.48 Mg animal^{-1} (30–100 kg)[a]	61	Poulsen *et al.* (2001)
	FYM	0.09 Mg animal^{-1} (30–100 kg)[a]	23	Poulsen *et al.* (2001)
	liquid manure	0.32 Mg animal^{-1} (30–100 kg)[a]	20	Poulsen *et al.* (2001)
	deep litter, straw	0.17 Mg animal^{-1} (30–100 kg)[a]	330	Poulsen *et al.* (2001)

The data should be used as guidelines, as the amount of manure produced differs widely between countries and management systems. Use of water greatly affects the amount of manure collected, see Table 4.1 for details. Characteristics of the different manure categories are given in Text Box – Basic 4.1.
[a] The time needed for a pig to increase from 30 to 100 kg in weight is about 3 months.

Thus, these operations need to be planned carefully using detailed knowledge of the operational and logistical capabilities of the machinery systems in order to reduce the time requirement and the transport costs. Furthermore, transport and application of manure often cause annoyance to neighbours due to noise and odour, and the heavy machines on small roads pose a hazard to other road users. In addition, field traffic with a heavy tanker and application system can cause soil compaction, impairing crop growth.

The challenge of organising and managing manure transport and application throughout the management chain varies according to the physical form of the manure and end-user demands. Solid and liquid manure must be considered separately, and the need for homogenisation may vary. Concern for the environment differs between countries (Chapter 3), as does the demand for the nutrients and carbon in the manure. Therefore, when making decisions about the technology to be used in a specific manure management system, the decision maker must consider end-user demands, statutory requirements and re-use of resources for economical and environmentally sustainable management.

Management systems for the removal, storage, transport and application of animal manure are presented below by outlining different manure handling chains and associated technological options, planning models, etc.

See Text Box – Basic 12.1 for terminology used in this chapter.

Text Box – Basic 12.1   Terminology

BOD: Biological oxygen demand
C : N ratio: Ratio of the mass of carbon to the mass of nitrogen in a substance.
COD: Chemical oxygen demand
Deep litter: Manure system based on repeated spreading of straw or sawdust in animal houses.
Dry matter (DM) content: The total solids part of the manure.
Farmyard manure (FYM): Solid manure.
NH_3: Ammonia.
NH_4^+: Ammonium.
Solid manure: Manure with dry matter at higher concentrations than 120 g DM kg^{-1}.

Slurry: Manure in liquid form containing a mixture of faeces and urine and produced in intensive livestock systems.

TAN: Total ammoniacal nitrogen ($NH_3(aq) + NH_4^+(aq)$).

12.2   Overview of Manure Systems

The transport of manure begins with removal of the manure from animal houses or from enclosures (e.g. feedlots with a high animal density). Implementation of efficient and detailed management systems for manure handling starts with knowing the amount of manure produced (Table 12.1).

In most systems, the manure is stored outside the animal house or enclosure for a shorter or longer period of time. The storage process is a way to contain the manure until it can be applied to the field during periods where the crop is in need of nutrients. Storage also helps to ensure efficient planning and use of machinery and labour by avoiding too many concurrent operations and enhances manure sanitation (Chapter 6).

In some systems, manure is transported directly from the animal house to the end use (fish ponds or field application). Farmyard manure, deep litter or the solid fraction from separation of slurry may also be transported to other farms, or to companies that produce compost, soil amelioration products, energy and ash.

Transport and spreading of manure must ensure that the amount of nutrients applied with the manure matches crop requirements. If an animal farm produces too much nutrients in the manure in relation to the need of the crop grown on the farm, then some should be transported to other farms in need of plant nutrients (Sørensen, 2003). If the manure produced is in the form of slurry, which contains much water, then the cost of "fertiliser" transport can be reduced by separating the slurry into a liquid fraction intended for on-farm use and a nutrient-rich solid fraction (Chapter 7) that can be exported to farms with fewer or no animals (Sørensen *et al.*, 2003; Sørensen and Møller, 2006).

Farmers may cooperate and deliver manure to a central processing plant for composting, incineration or biogas production (Raven and Gregersen, 2007). As part of the biogas production system, the manure may first be stored on-farm for a few weeks before being transported to the central treatment facility. After biogas treatment, the remaining biosolids may be transported back to the farm and applied on the fields or sold to other end users.

12.3   Animal Manure Characteristics

Physical characteristics such as dry matter content set the requirements on technologies for handling and managing manure (Table 12.2). Manure containing more than 120 g DM kg^{-1} is as a rule of a thumb

Table 12.2   *Manure excretion per animal and year by permanently housed young cattle (Poulsen et al., 2001).*

	Manure (Mg)	Dry matter (g kg^{-1})	Dry matter (kg)
Faeces	3.0	197	600
Urine	1.6	49	80
Straw	1.3	850	1105
Deep litter containing straw	5.9	302	1785

considered solid manure or FYM (Text Box – Basic 12.1) and manure with lower concentrations is referred to as slurry or liquid manure.

Farmyard manure is produced in animal houses where the excreted solids and litter (straw, peat, etc.) strewed on the floor is scraped from the floor and transported out of the house and liquid manure is the urine and excess water that is drained through gutters and various types of channel systems.

Deep litter is manure with a high content of litter, which has been strewed on solid floors to absorb the manure and is typically removed using a front loader. Beef feedlot manure represents FYM or deep litter depending on whether bedding is used. Beef feedlots are mostly situated in dry regions where water evaporates from the surface, producing a high dry matter content in the manure. The deep litter amount is affected by the amount of material (straw, wood chips, peat) strewn on the floor and thus the greater amount of manure to be removed must be weighed against the benefits of adding bedding material to keep the surface dry. Other materials used to cover the floor for the purpose of absorbing feces and urine are peat or wood chips.

In houses with a slatted floor, the slurry is collected in channels below the floor. In houses with a solid floor, the liquid is drained in gutters and the solids are scraped to the outlet of the animal house, where the two fractions may be mixed into a slurry that is pumped to a slurry tank. The amount of liquid produced is given in litres (l) and the density of slurry is generally assumed to be 1 kg l^{-1} (Chapter 4).

Water use in animal houses should be minimised, because water in the slurry adds no value and increases the volume to be stored and transported between different locations on the farm. In North European countries with no need for cooling, the volume of slurry produced by a fattening pig is about 480 kg (Table 12.1) or an average of around 5 kg day^{-1}. In warm climates, where the pigs and pens are cooled and cleaned by hosing, water use is high and the volume of slurry produced by one pig is about $30 \text{ l}^{-1} \text{ day}^{-1}$ (Taiganides, 1992).

Inefficient use of water use for cleaning, water losses from inadequate and leaky drinkers and rainwater ingress lead to a need for a larger storage volume and create greater volumes of effluent that must be treated, transported and applied in the field. By using water-saving techniques (e.g. high pressure cleaners and low-pressure drinkers), the volume of effluent can be reduced. Rain gutters on houses and separate drainage of clean rainwater ensure that the manure is not diluted with rainwater.

12.4   Removal from Animal Houses

Various manure removal technologies exist for different animal types (Table 12.3). Chicken manure is mostly produced as solid manure and transported out of the chicken house to storage by front loader, manually or by installing conveyer belts below the cages.

12.4.1   Solid Manure

Solid FYM produced in cattle houses is scraped from solid (concrete) floors regularly using a tractor with a front or rear scraper, chain scrapers, manually or with a cable-drawn scraper blade. In these systems, the liquid manure is drained from the floor through gutters and transferred by gravity through tubes or channels to the storage tank, which is often below the soil surface. Sand may be strewn on the floor to reduce animal slip.

Solid manure may be removed on a daily basis from houses with a sloping solid floor, where the manure mixed with straw slowly slides down the floor to a collection channel with low walls. From this channel the solid manure may be removed manually or by front loader and placed on a manure heap. Alternatively, the manure may be transported on conveyer belts or scraped off to conveyers, and transported to the manure heap. In other systems, conveyer belts transfer the solid manure to a screw transporter encapsulated in a pipe so that the manure can be pushed into the bottom of the heap (Figure 12.1). This reduces exposure of fresh manure to the air and limits emissions of NH_3 and other gases (Muck *et al.*, 1984).

Table 12.3   *Cattle and pig housing systems and related manure storage (Sommer et al., 2006).*

Animal	Type of housing	Flooring/manure type	Storage time
Cattle (dairy)	cubicle, solid floor	solid floor; slurry or slurry and solid manure	regular removal
	cubicle, partly slatted floor	resting area solid floor; walking alleys with slatted floor; slurry or slurry and solid manure	solid floor regular removal, slatted floor continuous or regular removal, but store always contains some slurry
	tied stalls, liquid/solid manure system	tied concrete standing area; daily removal of solid manure; liquid drained by gutter or stored in gridded channel at rear of animals	channel with continuous or regular removal, but store always contains some slurry
Cattle (dairy heifers)	tied stalls, slurry system	tied concrete standing area with gridded channel at rear of animals	continuous or regular removal but store always contains some slurry
Cattle (beef)	fully slatted	all floor slatted	storage below slats or continuous/regular removal but store always contains some slurry
	deep litter	solid floor with deep litter. solid manure	accumulated for several months, stored before land application or spread directly
	deep litter, sloped floor	deep litter on sloped floor. solid manure	accumulated; regular removal of some solid manure at the bottom of the slope
Pigs (sows, fatteners, piglets)	slurry system	fully or partly slatted floor; flush discharge	1–24 h
	slurry system	fully or partly slatted; pit discharge	4–7 days
	slurry systems	fully or partly slatted; pull plug discharge	7–14 days
	slurry system	fully or partly slatted; deep pit below animals	3–6 months
	deep litter system	solid floor with deep litter; solid manure	3 months
Broilers	deep litter system	solid floor with deep litter; solid manure	35–40 days
Laying hens	solid floor and removal/transport band	solid floor and removal/transport band	7 days

Pigs are not raised in houses with floors cleaned with automatic scrapers, because being curious animals by nature they will examine the equipment and may be injured by movement of the scraper. In Asia, the solids are usually scraped manually from solid floors and FYM is removed from pig houses (Vu *et al.*, 2011).

Deep litter is produced when animals are provided with litter, usually cereal straw but also other absorbent materials (Table 12.4), which make the resulting manure stackable.

Some pig housing systems have partly or fully solid concrete floors strewn with straw or sawdust, to improve pig welfare. The solid manure is typically removed manually or with front loaders at monthly

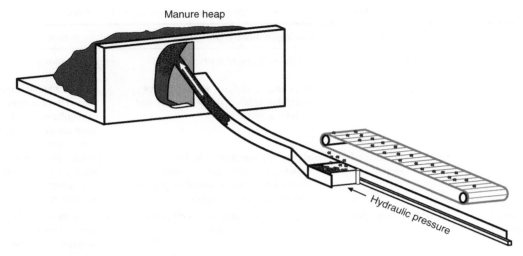

Figure 12.1 *A conveyor transports the deep litter out of the house and a screw in the tube presses the solid manure into the bottom of a heap. (© University of Southern Denmark.)*

intervals. Technologies for turning the deep litter in the pig houses have been developed, the intention being to reduce NH_3 emissions, keep the surface dry and reduce the amount of litter (Groenestein *et al.*, 1993).

In chicken houses, the floors may receive little or no straw, or may be covered with straw or wood chips to produce a deep layer of litter (Groot Koerkamp, 1994). Laying hens in aviary housing systems have access to littered or bare floors or are confined to tiered cages. The cages have wire floors and the droppings are collected on a belt or on the floor and removed from there.

Table 12.4 *Examples of deep litter management systems (Webb et al., 2011).*

	Amount of straw	Area of surface with litter (%)	Type of litter	Removal interval
Dairy	1250–3500 kg straw animal^{-1}	60–85	long straw, chopped straw	3–12 months
Beef (200–640 kg liveweight)	ND	100	long straw	after each group of animals
Fattening pigs (18–55 or 90–146 kg liveweight)	36–395 kg place^{-1} year^{-1}	25–100	straw, sawdust	after each group of animals
Piglets (7.7–12 kg liveweight)	ND	100	straw, sawdust	after each group of animals
Laying hen (2–4 kg end liveweight)	ND	0–100	straw, wood chips	removal; drying and removal, removal depends on system.
Broilers	0.2–10 kg place^{-1} year^{-1}	100	straw, sawdust, rice husks, wood shavings	drying, removal depends on system.

ND: no data.

12.4.2 Liquid Manure and Slurry

Dilute liquid manure containing less than 1% dry matter can be washed out of channels or from solid floors with a low pressure pump or removed with mechanical scrapers. In Asia, it is transported by gravity in open concrete channels. To ensure efficient draining of the dilute liquid manure over floors to "gutters" or through channels, the slope must be more than 1 : 40.

Slurry may be stored for months in deep pits below the floor of animal houses, or frequently removed using scrapers, pull plug, pit recharge or flushing systems depending on the animal category housed (Table 12.3). The frequency of manure removal varies from several times a day in channels up to monthly intervals when stored in deep pits.

In cattle houses, the walking areas may have slatted floors and slurry channels beneath, but in some houses scrapers remove the slurry from the slats, providing an environment that reduces the risk of hoof damage due to infections. The scrapers move slowly so that the cattle can move away or step over them during operation. The floors may be flushed to enhance cleanliness, although this increases the volume of slurry produced. The slurry collected in channels may also be removed with scrapers to ensure that the channels do not fill with sediment. The slurry channels with scrapers fitted are often about 0.5 m deep, so in these systems little slurry is stored in-house in the intervals between slurry removal.

A much used system in cattle houses is a slatted floor over a circular or oval channel, where slurry propellers or pumps in the channel create a flow of slurry. These channels may be relatively deep (2 m) and contain a large volume of slurry. The channels must have no angles where slurry dry matter can settle and eventually block the flow.

In both cattle and pig houses, the slurry collected in the channels may be flushed to the storage tank. The standard method is to regularly pump the liquid from the slurry tank once the solids have precipitated back through the slurry channel at high pressure, to flush residual slurry out of the channels. This method is called back-flushing. Alternatively, fresh water may be used at the cost of increasing water consumption and also increasing the volume of slurry produced.

In pig houses, slurry may be stored below slatted floors in deep pits from where it is pumped out after stirring. Plug flow discharge is used to remove slurry from channels with low height (40–60 cm). In these channels, a gutter with a stopper is placed at the centre of the channel and when the channels are full, the stopper is removed and the slurry flows out of the channel at high speed because of the vacuum. The flow of slurry creates turbulence that stirs up the deposited material, which is therefore transferred with the slurry to the storage tank. There is a limit to the length of channel used, because flow rate and turbulence must be high throughout the entire length of the channel.

It is critical to note that stirring of slurry before removal from channels and pits in closed animal houses releases hydrogen sulfide (H_2S), which can be lethal to animals and humans.

12.5 Manure Storage

Storage of manure results in loss of organic matter and plant nutrients, but also reduce pathogens in the manure. Storage also allows manure application to be matched to crop nutrient requirements. Therefore, storage time varies from 3–6 months in regions where several crops may be planted and harvested during a year or where the crop growing season is long (e.g. Asia) to about 9 months in Northern Europe due to the short crop growing season (Burton and Turner, 2003). Manure may also be stored for several years (e.g. in lagoons).

12.5.1 Solid Manure Stores

Solid manure is usually stored in uncovered heaps on concrete pads or unpaved land near the animal house and the effluent draining from the heap is collected. Solid manure heaps may be turned at regular intervals to enhance microbial turnover of the manure, thereby increasing the temperature of the heap and reducing pathogens and weed seeds (Larney and Hao, 2007). Water evaporates from the heap and organic matter is transformed to gases that are lost to the atmosphere, which can reduce the amount of organic manure by up to 50% of the initial amount (Chapter 5).

An alternative to turning may be compaction or covering with tarpaulin or polyvinyl chloride (PVC) sheets (or clay in Asia) to prevent air entering the manure heap. This stops aerobic microbial activity and thereby reduces emissions of NH_3 from the heap (Webb *et al.*, 2011). Solid manure may also be amended with lime to destroy pathogens or with phosphorus to improve the fertiliser quality of the manure and reduce NH_3 emissions (Tran *et al.*, 2011).

12.5.2 Liquid Manure Stores

Lagoons and lined ponds are the main liquid manure storage system worldwide, but slurry and liquid manure are also stored in tanks made from concrete or enamelled steel sheets located outside the animal houses. In some countries, slurry is stored partly below the slatted floor of the animal building and partly outside in slurry tanks (Burton and Turner, 2003).

Lagoons or ponds for cattle and pig effluents can be constructed at low cost. They may be lined with impermeable material for groundwater protection purposes. The volume of slurry leaching into the soil from unlined lagoons is affected by soil type and lagoon location (e.g. leaching is lower if the lagoon is surrounded by clay soil compared with sandy soil). Sedimentation of organic matter also creates a barrier to leaching. Lagoon depth is between 2 and 6 m depending on soil and groundwater depth at the site, and lagoons should be rectangular, with a 4 : 1 length to width ratio. In regions with much precipitation, extra capacity is needed for collected rainwater; for example, in a region with 800 mm net annual precipitation (rain – evaporation), 80 cm of the lagoon depth is occupied by rainwater. The amount of rainwater collected can be reduced by increasing the depth of the lagoon and reducing the surface area. Furthermore, as solids accumulate in the lagoon the storage capacity is reduced. However, the rate of accumulation can be reduced by solid/liquid separation of the slurry (Chapter 7).

Outside stores made of concrete are often partly buried in the soil at a depth that balances the pressure of the stored slurry on the sides of the tank. About one-third of concrete tank depth is usually below the soil surface and internal tank height can be up to 4–5 m. In countries with heavy rainfall and high groundwater levels, the effect of water in the soil has to be considered, as the empty tank may float on the groundwater and be lifted out of the soil. The liquid manure pumped to slurry tanks is released above or at the bottom of the stored slurry.

Slurry lagoons and tanks are normally not covered, except in countries with this tradition (e.g. Switzerland) or where covers are required by law to reduce emissions of NH_3 and odour (e.g. Denmark, Finland, the Netherlands), or to exclude rainfall.

The liquid manure/slurry is usually homogenised (stirred) in the tank prior to application. If the slurry is not stirred, then sediment builds up and the tank must be desludged periodically to prevent the storage capacity declining considerably with time.

On a farm there may be multiple storage units, often connected in series to reduce BOD and pathogen content. Having multiple stores also reduced the risk of short cuts of fresh slurry through storage and increases retention time. In these systems, solids settle in the first store in the series, which must be desludged at regular intervals. Phosphorus, N, dry matter and COD of the slurry are reduced by sedimentation and microbial transformation (Table 12.5). In warm climates, the first store may be covered and the biogas collected and used as fuel for power production and cooking.

Table 12.5   *Percentage efficiency of a series of lagoons in reducing components in pig effluent: removal as a percentage of original amount of stored pig effluent.*

	P	NH$_4$	Soluble COD	Total solids	Volatile solids
Burton and Turner (2003)	50–90	60	70		
Taiganides (1992)			45	42	33

12.5.3   Stirring – Homogenising Liquid Manure

A crust may develop on the surface of stored slurry and dry matter of higher density accumulates as sludge at the bottom. To maintain the storage capacity the sediment must be removed, and this is costly if the tank is emptied and the sludge removed by mechanical excavation. Therefore, the slurry may be stirred before removal (Cumby, 1990). Mixing slurry ensures that a homogeneous slurry is applied in the field and improves the rheological properties, enabling more constant flow conditions during spreading operations.

During mixing of stored slurry, foul-smelling gases and NH$_3$ are released due to movement of slurry with high amounts of the odorants and TAN to the surface and removal of the surface crust or litter layer (Chapter 5). To avoid odour and NH$_3$ emissions, mixing should be kept to a minimum, with stored slurry preferably only stirred prior to removal.

The slurry may be mixed mechanically with rotating propellers fixed to the storage or lagoon wall and driven by electric motors. Alternatively, a tractor-powered mixer can be lowered into the lagoon using hydraulics. The latter is quite efficient in terms of capacity because the mixer can be moved around the store or lagoon. Slurry can also be mixed by hydraulic jets using pumps set within or outside the slurry store, to which is connected pipes with a diameter greater than 125 mm and ideally adjustable nozzles at the end of the return pipe. Hydraulic mixers can be equipped with vents that add air to the slurry.

12.6   Transport of Manure

12.6.1   Liquid Manure Transport by Gravity

In channels, the flow of the liquid is slow and solids may precipitate, so the liquid transported should be pre-treated with a slurry separator (Chapter 7) or desludged in storage tanks or lagoons. The effluent may be transported in channels on the farm and from the farm to fields close to the farm. Sedimentation of solids in the channels can be reduced by keeping a high flow rate, so channel slope in the direction of transport should be about 1 : 50. The cross-section may be rectangular, with width 0.3 m and height 0.3 m.

12.6.2   Transport of Slurry by Pumping

Pumping is often needed when transporting slurry between different locations on the farm and slurry may also be pumped from the store to the field where it is applied. The pumped distance varies, but effluent with low dry matter content can be pumped for longer distances and at a lower cost per cubic metre than effluent with high dry matter. Thus, per nutrient or dry matter transported, it may be cheaper to pump a dilute, low viscosity slurry with a high volume than a dense slurry with higher pump demand per cubic metre. Instead of diluting slurry with water, slurry can be separated to reduce the dry matter content and thereby the pump requirement. Fermentation of slurry in biogas digesters also reduces the viscosity of slurry and the pumping requirement demand for transport. Thus, when considering the need for pumping capacity, the characteristics of the slurry must be considered (i.e. dry matter content and/or viscosity). Resistance to transport is also affected by the tube and pipe materials, tube bends, valve characteristics and the height difference between the start and the end of the tube. Furthermore, if the slurry is intended to be spread by sprinklers, then the pressure requirement of the sprinklers must be included.

12.6.3 Tanker Transport

The most common way of transporting and spreading slurry is to use a slurry tanker that can perform both functions. This method is sufficient in the case of small transport distances (below 2–3 km), but larger distances require the use of separate transport units, i.e. delivery of the slurry to the spreading unit or an interim storage in the field. The tanker may be tractor-drawn or self-propelled and the load capacity varies between 10 and 35 tons.

12.7 Application of Manure in the Field

12.7.1 Solid Manure Application

Solid manures are spread with manure spreaders in which a chain at the bottom of the spreader box transports the manure to a beater/distributor that can be set horizontally or vertically at the rear or side of the spreader. The beater/distributor is a rotating device that discharges the manure onto the field as homogeneously as possible. Rear-discharge machines generally apply manure more evenly than side-discharge types, but have narrower spreading width (Smith and Baldwin, 1999).

12.7.2 Liquid Manure Application

Effluent transported to the field in channels or pipes may be diluted in river water used to flood the fields or in irrigation water used in furrow irrigation. Furrow irrigation is sometimes used for application of fertiliser in fields, but may not spread the nutrients as homogeneously over the field surface (Benjamin *et al.*, 1997).

Manure effluent can also be spread on the fields through sprinkling. Separation of the manure solids is needed before the liquid portion can be applied. Ammonia is emitted during the spreading event and the emissions increase with the time manure remains on the surface exposed to air (Misselbrook *et al.*, 2004). Furthermore, sprinkling may cause an odour nuisance to neighbours living near the field (Phillips *et al.*, 1991) and aerosols may be formed that can spread diseases. As a consequence, this method of application is not allowed in all countries.

Broad-spreading with a splash plate inclined slightly toward the soil is one of the fastest and cheapest methods of applying slurry and other liquid manures and remains the most commonly used method in Europe. The slurry is dropped onto the plate where it splashes in a half circle to the field. However, this type of application is also being phased out in many countries in an effort to reduce NH_3 and odour emissions.

To reduce gaseous emissions, slurry may be applied using a trailing hose or trailing shoe band technique (Thorman *et al.*, 2008). In trailing hose application, slurry is delivered to the soil surface via a series of hoses mounted typically at 30-cm spacing on a horizontal 12- to 24-m wide boom. This technique is most suited to application in row-grown arable crops. For application of slurry to dense grassland, the trailing shoe application is more efficient, because the slurry is delivered via hoses just behind forward facing "shoes", which are designed to part the crop canopy to ensure that slurry is delivered to the soil surface. Trailing hose and shoe spreaders are equipped with pumps and devices that ensure homogeneous spreading of slurry across the working width of the machine.

Emissions of NH_3 can be reduced by incorporation of slurry into the soil after application. In the case of bare soil, the slurry may be incorporated by ploughing, harrowing and so on (Chapter 9). The more effective the incorporation, the greater the reduction in emissions. Thus, ploughing is a very efficient technique for reducing emissions, but is time-consuming compared with harrowing and slurry may therefore be exposed for longer on the soil surface, causing greater emissions compared with harrowing (Huijsmann and de Mol, 1999). Alternative direct injection methods using tubes in slits in the soil created with discs or tines (Figure 12.2)

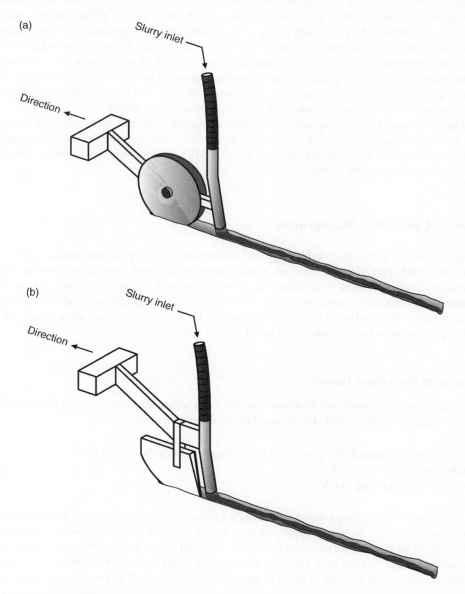

Figure 12.2 *Disc (a) and tines (b) used when injecting slurry into the soil. (© University of Southern Denmark.)*

are efficient in reducing NH_3 and odour emissions, but are energy-demanding. Much work has been devoted to developing injection devices that effectively reduce emissions with low energy consumption (i.e. with low resistance in soil and wide booms for higher capacity). A combination of a set of double discs can inject slurry to 3–4 cm depth (i.e. shallow injection) and a simple tine can go to 9–10 cm depth (i.e. deep injection at a relative low traction demand) (Figure 12.2b). The double-disc injectors can work in cropped fields without damaging the crop significantly, but the tine can only be used in fallow soil. With this technology, injectors can be mounted on a boom up to 18 m wide and pulled using a standard tractor (Nyord *et al.*, 2009).

Slurry application in the field is often the one operation that requires the highest traffic intensity and therefore gives the highest risk of soil compaction (Håkansson and Danfors, 1988). The best use of slurry nutrients is normally achieved by applying slurry to a growing crop, but this requires much labour during periods with many other field operations. Further, the use of traditional heavy slurry tankers on a relative humid soil in the beginning of a growth period creates increased risk of soil compaction. Alternatively, the slurry can be transported through pipes to a self-propelled spreader (with no tanker) applying slurry with trailing hoses. This equipment can spread slurry at distances of up to 2–3 km from the slurry store, reduces soil compaction and the labour requirement, and increases application capacity.

Attempts have been made to develop decision support systems for optimal route planning, so that manure application equipment carries smaller loads on sensitive soils, while driving with heavier loads on less compaction-sensitive areas, for example (Bochtis *et al.*, 2012). Preliminary results show that the risk of damaging soil compaction can be reduced by 23–61%.

12.8   Manure Operations Management

Planning and organising manpower, technology and operations for transport and application of manure is an important and costly part of overall manure management on the farm. Operations must be planned and executed carefully to ensure there is sufficient manpower in a busy period of the year where concurrent tasks on a farm also compete for manpower, and the planning must also involve strategic decisions on the technology to be used. Information is needed on the time and cost requirements of all the stages of handling and transferring the manure from house to end use (Figure 12.3).

12.8.1   Emptying the Animal House

The first operation in the manure supply chain involves transferring the manure from the animal house to intermediate storage, normally outside the house. This involves three steps:

Step 1 Removal from the animal area to a slurry pit or channel under the slatted floor
Step 2 Transfer to a pre-storage tank
Step 3 Transfer to a main storage tank

The labour requirement is estimated based on the daily labour requirement in Step 1 and the periodic work in Step 2. As a general rule, the labour requirement in Step 2 for most slurry systems is minimal, but there may be manual work involved in supplemental cleaning of slatted floors. There is frequent emptying of slurry from the house to the pre-storage tank (Step 1). Removal of slurry by use of mechanical scraper systems may be carried out 4–8 times per day in order to reduce NH_3 emissions. The pre-storage tank can typically

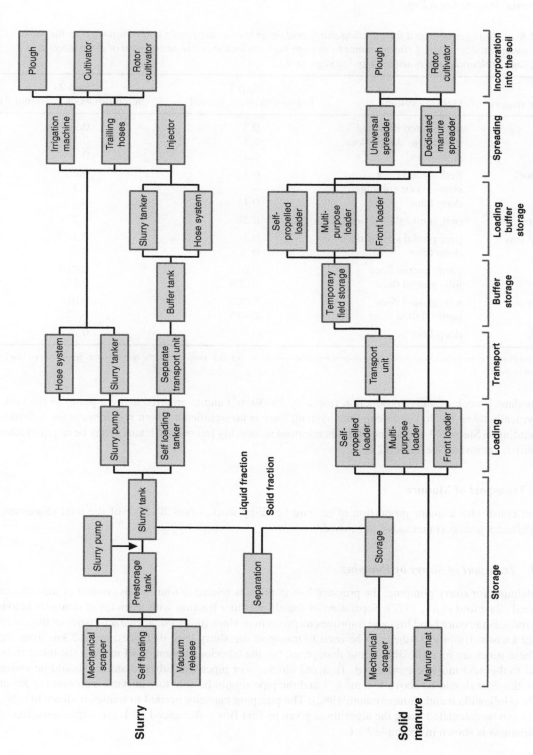

Figure 12.3 Schematic diagram of handling chains for liquid and solid manure. (© University of Southern Denmark.)

Table 12.6 *Labour requirement for handling slurry and deep litter in the house and removing it to the main outside storage facility: Steps 1 (from animal to storage tank under the house or removal of deep litter) and 2 (from storage tank under the house to a pre-storage tank).*

Animal category	Housing system	Step 1 (minimum day^{-1} animal^{-1})	Step 2 (minimum week^{-1} animal^{-1})
Cows	tied, slatted floor	0.3	0.2
	free range, slatted floor	0.1	0.3
	deep litter	0.2	0
Young stock	tied, slatted floor	0.1	0.2
	slatted floor cubicle	–	0.3
	deep litter	0.1	0
Farrowers	pen, partly slatted floor	0.35	0.2
Pregnant sow	pen, partial slatted floor	0.1	0.1
	deep litter	0	0
Piglets	partly slatted floor	0.015	0.010
	fully slatted floor	0.005	0.015
Fatteners	fully slatted floor	0.005	0.010
	partly slatted floor	0.015	0.020
Fatteners	deep litter[a]	0	0

[a] For deep litter systems, the labour requirement for transport and field application includes removal from the house, so this system has no labour requirement in Steps 1–3.

accommodate 1 week of slurry production (capacity 10–30 m^3) and is emptied using a pump in the tank, which typically takes 1–2 h. For a deep litter system, there is no significant labour requirement for activities corresponding to Steps 1–3, as the deep litter is removed at monthly intervals. Cleaning may be required after the manure is removed. See Table 12.6.

12.8.2   Transport of Manure

Transport constitutes a major proportion of manure handling work, often 20–50% of the total (Sørensen, 2003). Different transport methods can be used.

12.8.2.1   *Transport of Slurry by Pumping*

When planning for slurry pumping, the pressure loss in pipes is related to total solids content of the effluent transported (Bashford *et al.*, 1977). Separation of the effluent to a fraction with particles of diameter below 0.5 mm reduces pressure head loss and improves its prediction, since the physical characteristics of the slurry are better known. Umbilical tubes can be used to transport the slurry long distances, up to 2 km from the store. These tubes are more flexible in use than pipes, but the laborious operation of moving the tubes from one field to the next must be considered. To avoid blockage of pipes and tubes, diameter should be above 150 mm, the flow should be above 150 m^3 h^{-1} and the pipe should be able to withstand a pressure of about 1000 kPa (Holjewilken and Zimmermann, 1982). The pumping capacity needed to transport slurry in tubes and pipes can be calculated using the algorithms given in Text Box – Advanced 12.1. An implementation of the calculations is shown in Example 12.1.

Text Box – Advanced 12.1   Calculation of pumping capacity needed for slurry transport in tubes (pipes)

The total pumping capacity (H, pressure height in m water column) needed to transport slurry in tubes (and pipes) and in some cases spread the slurry is:

$$H = H_1 + H_2 + H_3 + H_4 + H_5 \tag{12.1}$$

The resistance to transport is related to friction in the tubes (H_1), friction due to bends and valves (H_2), the height slurry is lifted (H_3), and a safety factor (H_4). If slurry is sprinkled on fields the pressure loss when sprinkling the slurry must be added (H_5).

The resistance to transport due to friction caused by tubes and slurry viscosity is:

$$H_1 = \lambda \frac{L \cdot v^2}{d \cdot 2 \cdot g} \tag{12.2}$$

where λ is the friction coefficient to transport (no units; often stated by the tube manufacturer), L is the length of transport (m), v is the velocity of the slurry (m s^{-1}), d is tube diameter (m) and g is the force of gravity (9.82 m s^{-1}). Standard friction constants (Table 12.7) generally include the effect of slurry dry matter content (viscosity) and tube material.

Table 12.7   *Resistance constant λ (no units) for transport of animal slurry in tubes (or pipes) as affected by dry matter content: at DM < 3%, the resistance constant is similar to that when transporting water (i.e. 0.015–0.025) depending on the tube (pipe) material (Rolfes et al., 1977).*

| | Cattle slurry | | | | Pig slurry | |
| | PVC tubes | | Aluminium tubes | | Galvanised steel tubes | |
Dry matter (g l^{-1})	Mean	Minimum–maximum	Mean	Minimum–maximum	Mean	Minimum–maximum
60–100	0.030	0.021–0.044	0.036	0.027–0.048	0.032	0.019–0.050
100–120	0.046	0.024–0.081	0.038	0.029–0.047	0.040	0.020–0.069
120–150	0.090	0.036–0.170	0.042	0.022–0.069	0.060	0.036–0.085

The sum of resistance in bends, valves and so on, ξ_n (no dimension), is summed to give H_2

$$H_2 = \sum_n \xi_n \frac{v^2}{2 \cdot g} \tag{12.3}$$

The value of ξ_n is given by the manufacturer or can be found in Table 12.8.

The resistance due to the height the slurry is pumped is equal to the height difference Z_H (m):

$$H_3 = Z_H \tag{12.4}$$

Table 12.8   *Sum of resistance of different components when pumping slurry in tubes (or pipes).*

$r > 4d$: $\xi \sim 0.2$	(uneven tubes)
$r = d$: $\xi \sim 0.1$	(smooth tubes)
$\xi \sim 0.5$	(uneven tubes)
$\xi \sim 0.2$	(smooth tubes)

$\xi \sim 0.5$: $d > 20$ mm

Two 90° bends
$\xi \sim 2 \times \xi_{single}$
$\xi \sim 3 \times \xi_{single}$
$\xi \sim 4 \times \xi_{single}$

Angles

β	0°	15°	30°	45°	60°
ξ_{smooth}	0	0.03	0.12	0.28	0.5
ξ_{uneven}	0	0.1	0.2	0.35	0.7

Slide valve
Without narrowing $\xi = 0.3$–0.1 (fully open)
With narrowing $\xi = 1.2$–1.3 (fully open)

Throttle valve
$\xi = 0.2$ (fully open)

Ball tap
$\xi = 0.2$–0.1 (fully open)

Flap valve
$\xi = 1.0$–0.4 (fully open)

Ball valve
$\xi = 2.0$–0.5 (fully open)

When designing the pumping capacity, a safety factor R_2 may be included if not already included as H_4 as in eq. 12.1.

$$H_4 = (H_1 + H_2 + H_3) \times R_4 \tag{12.5}$$

where R is a safety factor that ensures enough capacity to accommodate variations in slurry viscosity, often set between 0.25 and 0.4. The total resistance H can then be assessed (Equation 12.1). The pump must provide a pressure of ΔP (kPa) to transport and spread the slurry:

$$\Delta P = H \cdot \delta \cdot g \tag{12.6}$$

where δ is the density of the slurry (kg m^{-3}) and g is the force of gravity.

The pump effect needed can be calculated as:

$$N = \frac{\Delta P \cdot Q}{\eta \cdot 3600} \tag{12.7}$$

where N is the energy required to give this pressure (kWh), Q is the flow of liquid (m s^{-1}), η is the efficiency of the pump, which can be set to 0.4 for a centrifugal pump and 0.6 for a displacement pump, and 3600 is the conversion from seconds to hours.

Example 12.1   Estimation of pump capacity (Table 12.9)

Table 12.9   Calculation of the pump effect needed to transport pig slurry 800 m or cattle slurry 500 m from the animal house to a store 2 m higher (pig slurry dry matter is 40 g kg^{-1} and density 1.02 kg l^{-1}, while cattle slurry dry matter is 80 g kg^{-1} and density 1.02 kg l^{-1}).

	Units	Pig slurry	Cattle slurry
Dry matter content	g kg^{-1}	4	8
Slurry density (d) (Chapter 3)	kg l^{-1}	1.02	1.02
Distance of transport (x)	m	800	500
Height to which the slurry is pumped dZ$_H$	m	2	2
Flow velocity (Q)	m^3 h^{-1}	120	120
Tube material		PVC	PVC
Diameter of tube (d)	m	0.16	0.16
Resistance constant (λ) (Table 12.7)		0.03	0.04
Sum of resistance components (v)^a		3.5	3.5
Safety capacity factor (R)		0.25	0.25
Resistance to transport in tube H_1	m wc	21.0	17.5
Resistance of components H_2	m wc	0.5	0.5
Resistance to pumping level, H_3	m wc	2.00	2.00
Reserve pumping capacity H_4	m wc	5.9	5.0
Resistance to transport H_5	m wc	29.4	25.0
Pump efficiency		0.6	0.6
Pump pressure (ΔP)	kPa	294	250
Slurry flow rate (v)	m s^{-1}	1.66	1.66
Energy demand	kW	16.3	13.9

^{a}m wc = metres water column.

Following use, clean water (10 times pipe/tube volume) should be pumped through tubes or pipes to reduce the risk of clogging with dried manure.

12.8.2.2   Tanker Transport

Operating data and observed work methods form the basis for developing normative models for estimating labour input and machine capacity for specific manure handling systems (Text Box – Advanced 12.2). The work models used here for the operations of loading and transport are based on the principles outlined by Sørensen (1993), by which five types of labour requirements involved in the tanker transport system can be distinguished and modelled.

Text Box – Advanced 12.2   Calculation of labour requirement for manure transport systems

Labour requirements for loading at the store, L (min ha^{-1}):

$$L = \left(\frac{u}{r}\right) \times \left(\frac{60 \times r}{d} + m\right) \times (1+q) \tag{12.8}$$

where u (tonne ha^{-1}) is the application rate, r (tonne) is the tanker load, d (tonne h^{-1}) is the pump capacity, m (min load^{-1}) is the preparation time for loading and q (%) is the rest allowance.
   Labour requirements for transport from storage to field, T (min ha^{-1}):

$$T = \left(\frac{u}{r}\right) \times \left(\frac{0.12 \times t}{v} + m\right) \times (1+q) \tag{12.9}$$

where t (m) is the one-way transport distance and v (km h^{-1}) is the transport speed. Labour requirement for transport in the field is T_1 (min ha^{-1}):

$$T = \left(\frac{u}{r}\right) \times \left(\frac{0.12 \times t_1}{v_1} + m\right) \times (1+q) \tag{12.10}$$

where t_1 (m) is the one-way transport distance in the field and v_1 (km h^{-1}) is the transport speed in the field. The transport distance in the field can be estimated based on the assumption that each load is transported a distance that runs from the centroid or geometric centre of the field lengthwise down the field to the end of the field and from there to the corner of the field (Nielsen and Sørensen, 1993). Alternatively, a detailed estimation can be based on simulation models (e.g. the discrete event modelling approach; Bochtis *et al.*, 2009), which accounts for the field polygon coordinates, the driving direction and all the related machinery features (tanker capacity, machine kinematics, etc.). Such an approach can provide estimates of in-field transport when spreading slurry using tank spreaders with an accuracy of 1–3%.

12.8.3   Application in the Field

The handling and application of manure makes considerable demands on the capacity of technical systems, as significant quantities of manure must be applied within a short time frame (window of workability) in order to achieve the maximum use of the nutrients. Consequently, from an operational machine planning point of view, it is essential to understand the consequences in terms of labour input and system capacity of using various technologies, which might also involve different degrees of manure nutrient use and environmental impacts.

The labour requirement and capacity of the man–machine manure handling system depend on: type of machine system, work methods, type of crop, treated area, application rate, type of soil, terrain and weather, load weight, operating width, working speed, and transport distance (Text Box – Advanced 12.3). Decision support systems have been developed to help farmers choose methods for transport and application of slurry, and to assess the manpower needed (e.g. Sørensen, 2003). The decision support model predicts the capacity and labour requirement of the system in terms of field size, machinery size, travel distance to the field and application rate. Example 12.2 gives an example of estimating the labour requirement for specific handling chains.

Text Box – Advanced 12.3   Calculation of labour requirement for field application of manure

Transport within the field:

$$T_1 = \left(\frac{u}{r}\right) \times \left(\frac{t_1 \times 120}{v_1}\right) \times (1 + q) \tag{12.11}$$

where T_1 is labour requirement (min ha^{-1}), u is application rate (tonne ha^{-1}), r is net tanker load (tonne), t_1 is transport distance to field (m), v_1 is transport velocity in field (km h^{-1}) and q is rest allowance (normally 5% of the estimated labour requirement). t_1 is arbitrarily calculated by assuming that each load, on average, is transported a distance running from the centroid, lengthwise down to the end of the field and from there to the corner.

Labour requirements for preparation in field, P (min ha^{-1}):

$$P = \left(\frac{u}{r}\right) \times m_1 \times (1 + q) \tag{12.12}$$

where P is labour requirement (min ha^{-1}), u is application rate (tonne ha^{-1}), r is net tanker load (tonne) and m_1 is preparation time when starting unloading in field (min load^{-1}).

Application in the field:

$$S = \frac{\left(\left(\frac{h \times u}{d_1}\right) + \frac{p \times b}{e} + k + \left(\frac{u}{r} \times s\right)\right) \times (1 + q)}{h} \tag{12.13}$$

where S is labour requirement (min ha^{-1}), u is application rate (tonne ha^{-1}), h is area (ha), d_1 is unloading capacity (l min^{-1}), p is turning time (min turn^{-1}), b is width of field (m), e is effective working width (m), k is turnings on headlands (min field^{-1}), and s is crop and soil stops, adjustments, control, tending of machine and so on (min ha^{-1}).

Example 12.2   Model calculations for three different application techniques for slurry

Table 12.10 shows the work requirement/capacity for three application systems:

System 1: Self-propelled tanker with injector: working width 7.0 m, payload 16 tonne, loading capacity 500 m^3 h^{-1}, emptying capacity 220 m^3 h^{-1}.

System 2: Tractor-driven tanker with injector: working width 6.0 m, payload 16 tonne, loading capacity 300 m^3 h^{-1}, emptying capacity 160 m^3 h^{-1}.
System 3: Tractor-driven tanker, trailing hoses: working width 16.0 m, payload 16 tonne, loading capacity 300 m^3 h^{-1}, emptying capacity 360 m^3 h^{-1}.

Table 12.10   *Work requirement/capacity for three application systems.*

		Person-min per load for system		
Application system		1	2	3
Loading from storage	preparation/termination	1.3	1.8	1.8
	loading	1.9	3.2	3.2
	loading, total	3.2	5.0	5.0
Transport	transport, 500 m^a	2.4	2.4	2.4
Application on field	transport on field, etc.	3.4	4.5	2.6
Loading from storage	spreading	4.4	6.0	2.7
	spreading, total	7.8	10.5	5.3
	application, total	13.4	17.9	12.7
	min tonne^{-1}	0.84	1.12	0.79
	tonne h^{-1}	71.4	53.6	75.9

^{a}Driving speed on road, 25 km h^{-1}.

   This example shows the time used in operations with conventional and more innovative technologies. The time required for field operations depends on the circumstances (plot dimensions, working speed, machine width, distance to manure storage, etc.) and the work organisation. The model breaks down the time spent over the activities spreading, transport and loading for given values of specific parameters such as working width, payload and transport distance. Table 12.10 gives the labour requirement and system capacities for three different application systems. The calculations are made in minutes per load and then transformed into minutes per tonne and tonnes per hour. For the technical assumptions stated, the self-propelled tanker with injector (1) has 6% lower capacity than the system with trailing hoses (3), and the tractor-driven tanker injector (2) has 30% lower capacity. Shallow injectors have a reasonable capacity compared with band spreading.

Transport distances from farm to field that exceed 2–3 km often make it advantageous to include a separate transport unit in order to create a sufficient system capacity. Example 12.3 gives the estimated labour input and capacity for transport of slurry by truck from on-farm storage to intermediate storage at the edge of the field and spreading on the field by a tractor-driven tanker fitted with an injector.

Example 12.3   Manure handling chain with separate transport unit

Table 12.11 shows the work requirement/capacity for an application system with a separate transport unit:

System 1: Truck: payload 30 tonnes, loading capacity 300 m^3 h^{-1}, unloading capacity 300 m^3 h^{-1}.
System 2: Tractor-driven tanker with injector: working width 6.0 m, payload 16 tonnes, loading capacity 300 m^3 h^{-1}, emptying capacity 160 m^3 h^{-1}.

Table 12.11   *Work requirement/capacity for an application system with a separate transport unit.*

	Transport and application	Person-minutes per load for system	
		1	2
Loading from storage	preparation/completion of loading	4.3	1.8
	loading	6.0	3.2
	loading, total	10.3	5.0
Road	transport on road[a]	11.5	
	transport on field, etc.		4.8
	spreading		6.0
	prepare/terminate unloading	3.0	
	unloading at intermediate storage	6.0	
	application, total	30.8	15.8
	min tonne^{-1}	1.03	0.99
	tonne h^{-1}	58.3	60.6

[a]5000 m at a driving speed on the road of 52 km h^{-1}.

Compared with Table 12.10, the capacity of the spreader in the field has increased slightly, because no transport distance is involved. The overall system capacity at the defined transport distance of 5 km is approximately 60 tonne h^{-1}, but the overall labour input is 2.0 man-min tonne^{-1}, as one person is required to operate the transport unit and another the spreader.

Other approaches and simulation models for transport and application of manure in the field include MANURE$HAUL (Hadrich *et al.*, 2010), which estimates the resource input and costs for different animal production unit sizes, transport distances, types of machinery, etc. For example, simulations with this model clearly show the benefit of using separate transport units.

An alternative to manure transport with tankers or trucks is transport of slurry through tubes or pipelines to the field and application of the slurry with self-propelling machines (Figure 12.4).

One drawback of this system is the labour-intensive task of unrolling and rolling up the pipeline arrangement for each field, which is approximately 1.25 h km^{-1} of pipeline, plus an additional 0.2 h for connecting the pumps at the storage unit. Users report that around 40% of the total working time is used for preparing the system for operation. Table 12.12 gives an indication of the operational performance of an umbilical system under specific conditions. The results indicate that the gross labour input ranges from 0.52 to 0.99 person-min tonne^{-1}.

An alternative is to build a system of permanent pipelines (Iwars, 1992; Provolo *et al.*, 2000; Jacobsen *et al.*, 2002). The system configuration involves a network of underground pipelines, carrying the liquid manure from the storage location to an entry point (hydrant) comparable with irrigation systems. The drag hose for the application unit is coupled directly to a hydrant placed at the field boundary. The tractor used as the application unit pulls the drag-hose through the field during the application. A second tractor is used for unwinding and rolling up the drag hose at the beginning and end of the operation. This type of system requires a large investment and is only cost-effective if managing large amounts of manure (e.g. more than 40 000 tonne year^{-1}).

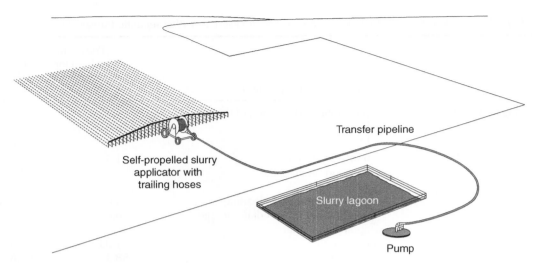

Figure 12.4   *Application system with pipelines for transport of slurry. (© University of Southern Denmark.)*

12.9   Farm Scenarios

It is possible to evaluate whole farm manure systems using the results from the individual operational studies of different application techniques and systems. Table 12.13 outlines two farm scenarios, one for cattle production and one for pig production. The dairy farm produces 3510 m^3 slurry year^{-1} and the pig farm 2688 tonne slurry year^{-1}. No specific crops are identified as the evaluation focuses on land area and the number of fields.

The capacity and labour requirement for different application techniques in Scenarios 1 and 2 were evaluated using the normative operational models described in previous sections. Table 12.14 presents results for tanker transport/trail hoses, tanker transport with injection and pipeline transport with hoses. Agitation in storage was not included in the analysis because it is similar for the different application techniques. Incorporation of the slurry was not included either, as this is assumed to be part of the normal field operations in sowing on bare soil.

Table 12.14 shows the differences in labour input and capacity for the different application methods. However, it should be noted that factors such as size of payload for the tanker systems might change the

Table 12.12   *System performance for pipeline system.*

Distance (m)	Dose (tonne ha^{-1})	Preparing system (h)	Net capacity (tonne h^{-1})	Gross capacity (tonne h^{-1})	Net labour input (min tonne^{-1})	Gross labour input (min tonne^{-1})
0	50	0.20	121.6	114.6	0.49	0.52
500	50	0.83	121.6	97.2	0.49	0.62
1000	50	1.45	121.6	84.4	0.49	0.71
1500	50	2.08	121.6	74.6	0.49	0.80
2000	50	2.70	121.6	66.8	0.49	0.90
2500	50	3.33	121.6	60.5	0.49	0.99

Table 12.13   *Farm scenarios: scenario 1: the farm has 100 cows, 105 young stock (heavy breed) and access to 94 ha of field; scenario 2: the farm has 150 sows, 822 fattening pigs and access to 150 ha of field.*

Scenario	Crop	Area (ha)	Dosage (tonne ha^{-1})	No. of fields	Distance (m)
1	1	18.5	25	2 fields of 9.2 ha	1000
	2	33.4	25	4 fields of 8.4 ha	500
	3	29.3	35	4 fields of 9.3 ha	700
	4	12.8	60	2 fields of 6.4 ha	500
2	1	40.5	15	4 fields of 10.1 ha	500
	2	55.0	20	5 fields of 11.0 ha	800
	3	30.5	20	3 fields of 10.2 ha	1000
	4	24.0	20	3 fields of 6.0 ha	500

results. The results for pipeline transport indicate that this system is the most time-saving method for high application rates, whereas a combination of lower application rate and many fields tends to increase the labour requirement for this system, because the time used for preparing and finishing the machine operation increases.

In the case of deep litter, handling and application are carried out using dedicated manure spreaders with rotating spreader tools that disintegrate and distribute the material across the working width of the spreader. The system is robust and can handle a range of organic manure types. In connection with the loading of the deep litter from within the animal house, a common technique is the tractor mounted front loader, self-propelled mini loader, etc. The load capacity of typical spreaders is up to 25 m^3 and the application rate is 15–40 tonne ha^{-1} (Table 12.15).

Table 12.14   *Capacity and labour requirement for different application methods.*

Farm category	Crop	Tanker transport[a] with trailing hoses			Tanker transport[a] with injection			Pipeline transport[b]		
		h ha^{-1}	min m^{-3}	Total h	h ha^{-1}	min m^{-3}	Total h	h ha^{-1}	min m^{-3}	Total h[e]
Dairy	1	0.41	0.99	7.6	0.55	1.33	10.3	0.33[d]	0.66	9.0
	2	0.35	0.83	11.6	0.49	1.17	16.3	0.33	0.66	14.3
	3	0.52	0.89	15.2	0.72	1.24	21.7	0.36	0.55	14.8
	4	0.83	0.83	10.6	1.17	1.17	15.0	0.48	0.44	7.8
	total			45.0			63.3			45.9
Pig	1	0.21	0.83	8.4	0.29	1.17	11.8	0.26	1.04	13.8
	2	0.31	0.93	17.1	0.42	1.27	23.3	0.28	0.88	21.4
	3	0.33	0.99	10.1	0.44	1.33	13.5	0.28	0.84	12.9
	4	0.28	0.83	6.6	0.39	1.17	9.4	0.28	0.84	8.1
	total			42.2			58.0			56.2

[a]Technical specifications as prescribed in Table 12.12.
[b]Technical specifications as prescribed in Table 12.13.
[c]Excludes time required for setting up the machine to operate in a specific field.
[e]Total time requirement including time for setting up the machine to operate in a specific field.

Table 12.15 *Systems for handling and applying deep litter.*

System	Distance	Loading equipment	Load capacity (m³)	Labour requirement[a] (h ha⁻¹)	System capacity[a] (tonne h⁻¹)	No. of units
1	animal house to field	front loader	9	2.5	39.0	2
2	animal house to field	dedicated loader	12	1.34	108.0	3
3	animal house to field	multi-purpose loader	14	1.64	73.8	2

[a]Transport distance 500 m, dose 30 tonne ha⁻¹.

In general, a tanker system is a flexible system, adaptable to varying constraints in terms of transport paths and field dimensions but with negative impacts in terms of heavy road transport, risk of soil compaction, compromising timeliness in the spring, etc.

Implementation of specific management strategies such as controlled traffic can increase in-field transport (Bochtis *et al.*, 2010). In comparison, the umbilical system uses a pipeline for transport and a lightweight in-field application unit, providing high capacity once the operation is set up, no road transport and reduced soil compaction, but is less flexible than the tanker system and operating costs can be higher for specific field configurations. Daugherty *et al.* (2001) compared the two systems in a study involving application of 60 tonne ha⁻¹ to a maize silage crop and found that the tanker system was most effective for small cattle farms with short transport distances, whereas the umbilical system had the lowest costs for larger operations with longer transport distances. Sørensen et al (2003) reported the same conclusions and stated that an economically feasible on-farm operation with an umbilical system requires an annual distribution exceeding 40 000 tonne.

12.10   Summary

Organising and managing the whole manure handling chain from animal house over transport to the point of use (e.g. in the field) is a challenging task requiring consideration of manure type and operating conditions. Solid and liquid manure must be handled differently, using very different technologies, and the environmental concerns are very different from country to the next. Decision support for farmers must advise on the method to use and on planning and execution of day-to-day manure handling operations.

From an operations management point of view, a slurry tanker is a flexible system, adaptable to varying constraints in terms of transport paths and field dimensions, but has negative impacts in terms of heavy road transport, risk of soil compaction, poor timeliness in the spring and so on. Alternative methods such as the umbilical system with pipeline transport of slurry have lower labour requirements, less soil compaction and so on, but require large annual amounts of manure to be economically feasible.

Simulation models that predict labour input, operations capacity and so on based on factors such as machine system, work method, crop type, treated area, application rate, soil type, terrain, weather, operating width, working speed, transport distance and so on are useful in evaluating the operational performance of different manure handling systems. In addition, the results may provide input to economic evaluations by the farmer.

References

Benjamin, J.G., Porter, L.K., Duke, H.R. and Ahuja, L.R. (1997) Corn growth and nitrogen uptake with furrow irrigation and fertilizer bands. *Agron. J.*, **89**, 609–612.

Bashford, L.L., Gilbertson, C.B., Nienaber, J.A. and Tietz, D. (1977) Effects of ration roughage content on viscosity and theoretical head loses in pipe flow for beef cattle slurry. *Trans. ASAE*, **20**, 1106–1109.

Bochtis, D., Sørensen, C.G., Jørgensen, R.N. and Green, O. (2009) Modelling of material handling operations using controlled traffic. *Biosyst. Eng.*, **103**, 397–408.

Bochtis, D.D., Sørensen, C.G., Green, O. and Olesen, J. (2010) Effect of controlled traffic in field efficiency. *Biosyst. Eng.*, **106**, 14–25.

Bochtis, D.D., Sørensen, C.G. and Green, O. (2012) A DSS for planning of soil-sensitive field operations. *Decis. Sup. Syst.*, **53**, 66–75.

Burton, C.H. and Turner, C. (2003) *Manure Management: Treatment Strategies for Sustainable Agriculture*, 2nd edn, Silsoe Research Institute, Silsoe.

Cumby, T.R. (1990) Slurry mixing with impellers. Part 1. Theory and previous research. *J. Agric. Eng. Res.*, **45**, 157–173.

Daugherty, A.S., Burns, R.T., Cross, T.L., Raman, D.R. and Grandle, G.F. (2001) Liquid dairy waste transport and land application cost comparisons considering herd size, transport distance, and nitrogen versus phosphorus application rates, presentation at the *ASAE Meeting*, St Joseph, MI, paper 01-2263.

Groenestein, C.M., Oosthoek, J. and van Faassen, H.G. (1993) Microbial processes in deep-litter-systems for fattening pigs and emission of ammonia, nitrous oxide and nitric oxide, in *Nitrogen Flow in Pig Production And Environmental Consequences: Proceedings of the First International Symposium on Nitrogen Flow in Pig Production* (eds M.W.A. Verstegen, L.A. den Hartog, G.J.M. van Kempen and J.H.M. Metz), *EAAP Publication 69*, PUDUC, Wageningen, pp. 307–312.

Groot Koerkamp, P.W.G. (1994) Review on emission of ammonia from housing systems for laying hens in relation to sources, processes, building design and manure handling. *J. Agric. Eng. Res.*, **59**, 73–87.

Hadrich, J.C., Harrigan, T.M. and Wolf, C.A. (2010) Economic comparison of liquid manure transport and land application. *Appl. Eng. Agric.*, **26**, 743–758.

Håkansson, I. and Danfors, B. (1998) The economic consequences of soil compaction by heavy vehicles when spreading manure and municipal waste, *Report 96:2*, Jordbruktekniska Institutet, Uppsala.

Hao, X., Benke, M., Larney, F.J. and McAllister, T.A. (2011) Greenhouse gas emissions when composting manure from cattle fed wheat dried distillers' grains with solubles. *Nutr. Cycl. Agroecosyst.*, **89**, 105–114.

Holjewilken, H. and Zimmermann, K.-H. (1982) Hinweise zum hydromechanishen gülletransport über druckrohrleitungen. *Agrartechnik*, **6**, 261–263.

Huijsmans, J.F.M. and de Mol, R.M. (1999) A model for ammonia volatilization after surface application and subsequent incorporation of manure on arable land. *J. Agric. Eng. Res.*, **74**, 73–82.

Jacobsen, B., Sørensen, C.G. and Hansen, J.F. (2002) *Handling of Animal Manure in Denmark – A Technical and Economical System Analysis*. Danish Research Institute of Food Economics, Copenhagen.

Iwars, U. (1992) *Transport av Flytgodsel i Rorledning [Transport of Slurry in Pipelines]*, Jordbrukstekniska Instituttet, Uppsala.

Larney, F.J. and Hao, X. (2007) A review of composting as a management alternative for beef cattle feedlot manure in southern Alberta, Cananda. *Bioresour. Technol.*, **98**, 3221–3227.

Misselbrook, T.H., Smith, K.A., Jackson, D.R. and Gilhespy, S.L. (2004) Ammonia emissions from irrigation of dilute pig slurries. *Biosyst. Eng.*, **89**, 473–484.

Muck, R.E., Guests, R.W. and Richards, B.K. (1984) Effect of manure storage design on nitrogen conservation. *Agric. Wast.*, **10**, 205–220.

Nielsen, V. and Sørensen, C.G. (1993) DRIFT: a program for calculation of work requirement, work capacity, work budget, work profile, *Bulletin 53*, National Institute of Agricultural Engineering, Horsens.

Nyord, T., Kristensen, E.F., Munkholm, L.J. and Jørgensen, M.H. (2010) Design of a slurry injector for use in a growing cereal crop. *Soil Till. Res.*, **1007**, 26–35.

Phillips, V.R., Pain, B.F. and Klarenbeek, J.V. (1991) Factors influencing the odour and ammonia emissions during and after the land spreading of animal slurries, in *Odour and Ammonia Emissions from Livestock Farming* (eds V.C. Nielsen, J.H. Voorburg and P. L'Hermite), Elsevier, London, pp. 989–106.

Poulsen, H.D., Børsting, C.F., Rom, H.B. and Sommer, S.G. (2001) Kvælstof, fosfor og kalium i husdyrgødning – normtal 2000 [Standard values of nitrogen, phosphorous and potassium in animal manure], *DJF Rapport 36*, Ministry of Food, Agriculture and Fisheries, Danish Institute of Agricultural Sciences, Tjele.

Provolo, G., Tangorra, F.M. and Bettati, T. (2000) A tool to support the design of pipeline slurry transport systems, presented at *9th International Workshop of the Network Recycling of Agricultural, Municipal and Industrial Residues in Agriculture*, Gargnano.

Raven, R.P.J.M. and Gregersen, K.H. (2007) Biogas plants in Denmark: successes and setbacks. *Renew. Sust. Ener. Rev.*, **11**, 116–132.

Rolfes, M.F., Gilbertson, C.B. and Nieaber, J.A. (1977) Head loss of beef manure flow in polyninylchloride pipe. *Trans. ASABE*, **20**, 530–533.

Smith, K.A. and Baldwin, D.J. (1999) A management perspective on improved precision manure and slurry application, in *RAMIRAN 98 Posters Presentations – FAO: 8th International Conference on Management Strategies for Organic Waste Use in Agriculture*, Rennes.

Sommer, S.G., Mathanpal, G. and Dass, T.T. (2006) A simple biofilter for treatment of pig slurry in Malaysia. *Environ. Technol.*, **26**, 303–312.

Sørensen, C.G. (1993) Slurry versus solid and liquid farmyard manure – primarily illustrated on the basis of operational technical and environmental consequences, *Bulletin 54*, National Institute of Agricultural Engineering, Tjele.

Sørensen, C.G. (2003) A model of field machinery capability and logistics: the case of manure application. *CIGR E-J.*, **5**, http://hdl.handle.net/1813/10358.

Sørensen, C.G. and Møller, H.B. (2006) Operational and economic modelling and optimization of mobile slurry separation. *Appl. Eng. Agric.*, **22**, 185–193.

Sørensen, C.A., Jacobsen, B.H. and Sommer, S.G. (2003) An assessment tool applied to manure management systems using innovative technologies. *Biosyst. Eng.*, **86**, 315–325.

Taiganides, E.P. (1992) *Pig Waste Managment and Recycling. The Singapore Experience*, International Development Research Centre, Ottawa.

Thorman, R.E., Hansen, M.N., Misselbrook, T.H. and Sommer, S.G. (2008) Algorithm for estimating the crop height effect on ammonia emission from slurry applied to cereal fields and grassland. *Agron. Sust. Dev.*, **28**, 373–378.

Tran, M.T., Vu, T.K.V., Sommer, S.G. and Jensen, L.S. (2011) Gaseous loss of nitrogen from pig manure during slurry storage and composting under typical Vietnamese farming conditions. *J. Agric. Sci.*, **149**, 285–296.

Vu, T.K.V., Vu, C.C., Médoc, J.M., Flindt, M.R. and Sommer, S.G. (2011) Management model for assessment of nitrogen flow from feed to pig manure after storage in Vietnam. *Environ. Technol.*, **33**, 725–731.

Webb, J., Sommer, S.G., Kuppert, T., Groenestein, K., Hutchings, N.J., Eurich-Menden, B., Rodhe, L., Misselbrook, T.H. and Amon, B. (2011) Gaseous emissions during the management of solid manures. A review. *Sust. Agric. Rev.*, **8**, 67–107.

13

Bioenergy Production

Sven G. Sommer[1], Alastair J. Ward[2] and James J. Leahy[3]

[1]*Institute of Chemical Engineering, Biotechnology and Environmental Technology,*
University of Southern Denmark, Denmark
[2]*Department of Engineering, Aarhus University, Denmark*
[3]*Department of Chemical and Environmental Sciences, University of Limerick, Ireland*

13.1   Introduction

Since the dawn of history, human beings have had a need for energy for cooking, for heating and more recently for transport. Since the late seventeenth century fossil energy has been contributing an increasing share of energy production. Coal was used as the main fuel of choice until World War II, after which it was supplanted with oil and more recently also with gas. A range of renewable technologies as well as nuclear power have contributed to energy production since the 1950s (Figure 13.1).

In 2008, fossil fuel supplied 80% of primary energy and it is expected that its share of energy production will remain at similar levels until 2030 (World Energy Council, 2010). Coal and gas are used primarily for power production, but also for heating and cooking, while oil is the principal source for transport. Energy consumption is crucial for all countries, and is known to be one of the main drivers of economic development and lifestyle choice. The impact of any society on the environment is a function of the quantities and types of energy resources it exploits, and the efficiency with which it converts potential energy into work and heat. Future long-term societal development is threatened by a growing energy demand and a limited total amount of energy resources. Known fossil resources of oil are expected only to be sufficient to support a growing oil energy demand for 41 years and natural gas demand for 54 years (World Energy Council, 2010). Furthermore, these are concentrated in a small number of countries (Figure 13.2), creating a dependency within the industrialised world, which has the greatest demand for imports of oil and gas.

Animal Manure Recycling: Treatment and Management, First Edition. Edited by Sven G. Sommer, Morten L. Christensen, Thomas Schmidt and Lars S. Jensen.
© 2013 John Wiley & Sons, Ltd. Published 2013 by John Wiley & Sons, Ltd.

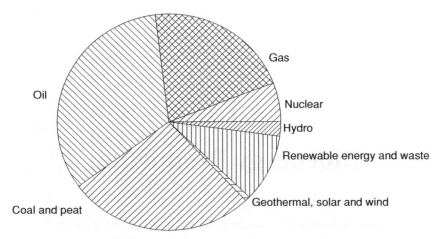

Figure 13.1 *Share of total primary energy supply in 2008 with total annual energy production in the world of 12 267 Mtoe (1 toe (ton oil equivalent) corresponds to 41.868 GJ). (Published with data from* World Energy Outlook *(IEA, 2010). © OECD/IEA 2010.)*

Energy-related activities are recognised as being responsible for the majority of global greenhouse gas (GHG) emissions. In order to meet the goals on carbon dioxide (CO_2) emissions to which countries have committed under the UN Framework Convention on Climate Change (UNFCCC), there has been an increased global search for sustainable energy sources.

Renewable bioenergy accounts for about 10% of global energy demand (Figure 13.1) and is thus larger than any other renewable energy source. Much of the biomass is used in wood-stoves and for cooking in developing countries. Biomass with low moisture content is also becoming an important renewable source for power and heat production. Wet biomass such as manure and organic waste (e.g. from food processing)

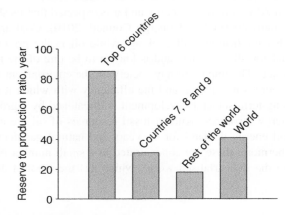

Figure 13.2 *Reserve to production (R/P) ratio of oil and natural gas, calculated by dividing the volume of proven recoverable reserves at the end of 2008 by volumetric production in that year. The resulting figure is the time in years that the proven recoverable reserves would last if production continues at the 2008 level. (Data from* Survey of Energy Resources *(World Energy Council, 2010). © 2010 World Energy Council.)*

is used for biogas production, which is an energy carrier that can be used for power production but also as a transport fuel. In addition, biomass is converted to liquid fuels such as bioethanol and biodiesel, which are the energy carriers most easily used as "drop-in" fuels for transport.

Bioenergy is considered one of the solutions to replace fossil fuels, which will become depleted in the foreseeable future (EU Council, 2007). To achieve this goal, there is a need to improve bioenergy production systems and increase the range of biomass feedstocks that can be used without harming the production of food or damaging natural ecosystems. When considering the use of plant material for energy production, one should consider whether the production competes with food and feed consumption (first-generation bioenergy production), which is marginally sustainable in a world with an increasing demand for food and animal feed. In contrast, second-generation energy production uses plant material that does not interfere with the food supply chain (e.g. plant residues or waste products) or is produced in areas where feed or food crops would not be produced (e.g. algae).

See Text Box – Basic 13.1 for definitions used in this chapter.

Text Box – Basic 13.1   Definitions

Adiabatic: Without transferring heat energy to the surroundings.

ar (as-received): The composition of fuels can be expressed as *ar* (i.e. as it is presented to a user), *db* where the composition is expressed on a moisture-free basis (e.g. ash content (wt% dry) = ash content (wt% ar) × 100/(100 − water content (wt%)) or *daf* (dry ash free) where the composition is normalised to account for the ash and moisture.

db: The units of a component given as a fraction of the dry biomass (e.g. g kg^{-1} (db)).

Digester: Fermentation unit where biomass is fermented to produce biogas.

District heating: Heat-producing plant providing warm water or steam that is distributed to users in the district.

First- and second-generation bioenergy production: In first-generation bioenergy production, feed, food and woody biomass are converted into energy carriers (petrol, gas); in second-generation bioenergy production, biowaste or plant residues are used.

Fluid bed: Boiler where the finely ground biomass is mixed with a hot bed of silica particles, for example, which are floating due to air addition from below.

Grate firing: Furnace where the biomass is moving on grates through which air is often added. The grates can move or the bottom with the grates is inclined so that the biomass slides from inlet to outlet.

Higher heating value (HHV) or gross calorific value: Potential energy of the fraction of the biomass that can be oxidised, given in MJ kg^{-1} wet or dry biomass (MJ kg^{-1} (db)).

Lower heating value (LHV) or net calorific value: Potential energy of the biomass with reduction of volatilisation of water in the biomass and condensation of water produced during combustion and also of other flue gases, given in MJ kg^{-1} dry biomass or wet biomass (MJ kg^{-1} (wb)).

Pyrolysis: Volatilisation of organic components in biomass heated without supplying oxygen.

Stoker: Boiler classified according to the method of feeding fuel to the furnace and the type of grate. The main classifications are spreader stoker and chain-gate or travelling-gate stoker.

STP: Standard temperature and pressure. The standard temperature is 273 K (0 °C) and the standard pressure is 1 atm pressure. At STP, 1 mol of gas occupies 22.4 l of volume (molar volume). STP is most commonly used when performing calculations on gases, such as gas density.

Thermal gasification: Volatilisation of gases at oxygen supply below stoichiometric addition relative to the biomass being heated and oxidised in a furnace.

wb: Units of a component given as a fraction of the wet biomass (e.g. g kg^{-1}(wb)).

13.2   Biomass and Energy

Biomass contains a high amount of partially oxidised organic components. The oxidation of these components releases energy that can be used either as heat or work. Chemical energy stored as bonding energy can be released either directly as heat (combustion) or as electrical work (fuel cells):

$$\Delta E = Q - W \tag{13.1}$$

where ΔE is the change in energy stored in the system (kJ kg^{-1}), Q is the heat added to the system (kJ kg^{-1}) and W is the work done by the system (kJ kg^{-1}). Work may be the movement of a component, such as electron transport in a power system. Energy released as work can be transformed totally into heat, while only a fraction of energy released as heat can be transformed into work. Work therefore constitutes a convenient basis for evaluating the usefulness or efficiency of an energy conversion process. From a theoretical basis, the highest possible useful work done by either a combustion process or a fuel cell can be estimated. For a fuel cell, in theory 100% of the chemical energy released can be converted into work (electricity) but in practice 70% has been achieved (Pilatowsky et al., 2011). For a combustion process, the maximum possible theoretical work obtained can be calculated from the Carnot cycle efficiency equation:

$$\eta_{\text{Carnot}} = 1 - \frac{T_L}{T_H} \tag{13.2}$$

where T_H is the highest temperature at which energy can be transported to the process (K), T_L is the lowest temperature at which surplus energy can be released from the process (K) and η_{Carnot} is the fraction of energy released that is transformed into work, the remainder being released as heat. As an example, a deep litter manure heap will self-heat to 50–60 °C due to microbial activity, while the ambient temperature in northern countries is around –5 to 20 °C. According to Equation (13.2), a maximum of only 10–20% of this heat energy can be converted into work and the remaining thermal energy will be of no practical value. Thus, the microbial activity in itself is not useful for the conversion of biowaste to bioenergy. Combustion of straw in an large efficient power plant produces a flue gas with a temperature of about 540 °C (713 K) and the flue gas emitted from the chimney has a temperature of 90 °C (363 K) or, in very efficient plants, 70 °C (343 K). This gives a maximum possible power production of 55–58% of the energy in the biomass (Thomsen and Larsen, 2012). In practice the power production will be much lower, partly because the process is not reversible. Heat is also lost through friction losses and through the use of energy for operating equipment within the power plant. Consequently, even the most efficient power plant will only convert about 30% of the biomass energy into power.

A more efficient use of the energy stored in biowaste would be to convert the chemical components in the biowaste into useful fuels that can be combusted to drive traditional power-generating engines. Two different types of engines that are relevant in this regard (i) internal combustion engines such as diesel engines, Otto engines (petrol engines) and gas turbines, and (ii) externally heated gas engines such as the Stirling engine or vapour engines.

Internal combustion engines have the advantage that they are fairly light compared with their power output and are therefore useful in cars, ships or aeroplanes. Their main drawback is the requirement for a high-quality fuel, either liquid fuels such as diesel, petrol or ethanol, or combustible gases such as methane (CH_4). Furthermore, the engine-specific theoretical and indeed practical work efficiency is far below the maximum theoretical work efficiency. Externally heated engines such as the steam engine and the Stirling engine are not sensitive to the fuel used to generate the heat. Power plants can be run on coal, any fossil fuel or indeed any combustible organic waste available. At the same time, their efficiency can at least in theory reach the

Table 13.1 *Maximum adiabatic flame temperatures for various biofuels, their Carnot cycle work potential, specific cycle work potential and actual work efficiency based on an ambient temperature of 25 °C.*

Fuel	Maximum adiabatic combustion temperature (°C)	Maximum Carnot cycle efficiency (%)	Typical use	Theoretical maximum efficiency of engine (%)	Typical work efficiency of engine (%)
Hydrogen	2307	88	as fuel for fuel cells; a theoretical energy-to-work efficiency of 100% is possible		30–70
Methane (biogas)	2027	87	as fuel for:		
			gas turbines	40–60	30–37
			power stations	40–42	35–38
			Stirling engines	87	15–30
Bio-oil	2170	88	as fuel in diesel engines	60–63	45
Bioethanol	2098	87	as mixed blend fuel for Otto engines	50–57	37
Biochar	2200	88	as fuel for:		
			power stations	40–42	35–38
			Stirling engines	88	15–30

Carnot cycle efficiency. Their main drawback is that they need a separate combustion chamber to generate the heat, making them heavier than internal combustion engines, and therefore less attractive as automobile and ship engines or indeed aeroplane engines, but ideal for power plants. The choice of fuel to produce from biowaste depends on the choice of power engine and so indirectly on the purpose of the engine.

Useful fuel types that might be produced from biowaste are hydrogen, CH₄ (biogas), bioethanol, biodiesel (bio-oil) and biochar. Their maximum adiabatic combustion temperatures, Carnot efficiency and practical use as fuels are shown in Table 13.1.

Fuel cells can currently only work with high purity hydrogen, CH₄ or methanol, whereas gas or methanol produced from biomass is not clean. Therefore, the use of biomass gases for fuel cell energy production is not considered a realistic application in the short-term.

Biodiesel and bioethanol can be produced from organic products that can also be used as food or feed, but also from plant residues and waste products (Demirbas, 2008). The raw materials for biodiesel production are vegetable oils and waste fat from animals; algal lipids may be used for second-generation production. As the fat content in manure is low, its use for biodiesel production is not economically competitive with either pyrolysis or biogas production.

First-generation bioethanol is typically produced from sugar cane and maize corn. Bioethanol can also be produced from straw and other cellulose plant products in second-generation ethanol production, where pre-treatment using steam and enzymes is needed to break down the lignocellulose before microbial degradation of carbohydrates to ethanol. Manure contains organic material that has been used by the livestock and thus it contains low amounts of easily degradable organic matter that can be used for bioethanol production. Deep litter may be an exception to this, as much of it is plant material that has not been digested by the animals and therefore has the potential to be used for bioethanol production in processes similar to those used when transforming straw to ethanol.

Liquefaction of biomass dissolved in water is carried out at high pressure (100–200 bar) and temperatures of 250–350 °C in the presence of a catalyst to enhance the process of transforming organic components into liquids. At high temperature and pressure the water is a supercritical fluid (i.e. neither a gas nor a liquid) and is very reactive in producing oil from organic matter, which after refining can be used as transport fuel. The process has been demonstrated at pilot scale, but is not used commercially to date.

The benefits resulting from the utilisation of biomass can be calculated using very different assumptions that are related to the objective of production of the energy source. Thus, the benefits may be calculated by comparing the use of fossil energy to produce the biofuel and relating this to the energy content of the biofuel. Alternatively, a life cycle analysis (LCA; Chapter 16) may be made considering the fossil energy used to produce the biofuel and relating this to the energy consumed by the substituted fossil fuel (Wenzel, 2006). In this calculation one must take into account the fact that the biomass resource is indeed limited. Thus, in the LCA analysis by Wenzel (2006), the energy surplus arising from the combustion of straw in a power plant is higher than when producing bioethanol, as bioethanol production is energy-intensive. Consequently, the reduction in GHG emissions is larger from straw combustion than from production of ethanol. The argument is that the oil is substituted by straw for power production, making it available for transport.

On the other hand, if the objective of the energy production is to produce transport fuel (i.e. ethanol) as well as power and heat, then the straw can be processed in an ethanol plant and the remaining undigestible organic matter combusted after separation. In this process salts are also removed during ethanol processing, which significantly reduces corrosion associated with combustion of the residues from fermentation of straw. In this scenario, transport fuel is produced at the same time as power and heat.

It seems that, in general, this assessment is accepted (i.e. biomass used for combined heat and power production is most beneficial for the environment). However, this assessment is only valid if the dry matter content of the biomass is high (above 55%), as otherwise the evaporation of water consumes the energy released during thermal conversion. At present wet biomass is ideally suited for anaerobic biogas production. Therefore, there is a need for discussing whether there is a valid argument that producing bioethanol, biodiesel or bio-oil from wet biomass is a sustainable option; the rationale being that there is a great need for the ethanol in the transport sector. At least in the transition phase from internal combustion engines to when sufficiently powerful electric-powered engine systems or the like become available, biofuels will provide reliable supplies and biofuel production will provide jobs.

For biowaste with a dry matter content higher than 55%, it has been estimated that direct combustion using the heat in co-generation power plants producing electricity and using the surplus heat either for district heating (in cold climates) or for potable water from seawater (distillation, in hot arid coastal areas) will produce more useful energy than fermentation of the biomass for liquid fuels or biogas. Combustion of biomass can sometimes cause problems due to the corrosive nature of the combustion gases, but solutions exist for these problems and the combustion of biomass feedstock, primarily straw, is currently a well-proven technology used in hundreds of small farm-scale and large power plants. The solutions require expensive maintenance of the plant equipment and a reduction in steam pressure and temperature, with the knock-on effect of a reduction in the overall efficiency of the power plant.

The combustion of biomass can be visualised as a sequence of events, including drying, pyrolysis, gasification and combustion. These steps occur in batch combustion processes found in small, manually managed furnaces or in continuous processes in medium- to large-scale boilers and power plants with forced aeration. Drying of the biomass is the initial step in the process, followed by pyrolysis or gasification releasing the most volatile components into the vapour phase, leaving a carbon-rich char, which is finally oxidised (Figure 13.3). Thus, oxidation of volatile gases provides the initial heat release and later in the process it is char oxidation that is the source of heat.

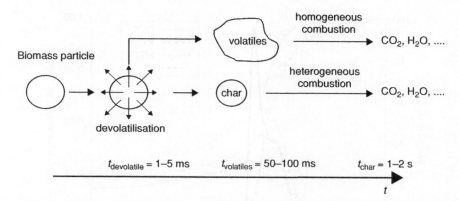

Figure 13.3   *Illustration of the volatilisation and oxidation of biomass components in a batch combustion process, which could take place in a simple oven. (© University of Southern Denmark.)*

The water evaporates from the feedstock at temperatures below 100 °C, consuming energy. Energy is used to transfer the water from the liquid to the gas phase (evaporation) and to heat the vapour to the temperature of the furnace, resulting in a lowering of the temperature in the combustion chamber. In biomass boilers or stoves, the combustion process will stop if the dry matter content of the biomass is below 40%. The energy balance is also affected by the mass fraction of oxygen (O_2) in the biomass; thus biomass with a high lignin content ($C_{40}H_{44}O_{14}$) and thus a low oxygen-to-carbon content has a higher enthalpy of oxidation than biomass containing cellulose ($C_6H_{10}O_5$) with a high oxygen-to-carbon content. When producing charcoal the oxygen is volatilised with light gases and the remaining charcoal consequently has a high enthalpy of oxidation.

In the absence of O_2, the dry biomass may be subject to pyrolysis, which is the volatilisation of organic components in the biomass at temperatures between 250 and 500 °C. The products of pyrolysis are tar (bio-oil), charcoal and low-molecular weight gases (CH_4, ethane, carbon monoxide (CO), etc.) and CO_2, which is produced in amounts proportional to the oxygen content of the biomass. The tar and charcoal contain very little oxygen. Decomposition of hemicellulose takes place at the lower temperatures and cellulose at the higher temperatures (Figure 13.4). At 300–400 °C, mass loss due to volatilisation of lignin is observed. The product composition depends on the process temperature and the biowaste composition. At low temperature, biochar is the primary product, while at higher temperatures the primary products are bio-oils and volatiles. A high content of hemicellulose and cellulose yields higher amounts of light gases, whereas high lignin content may yield more tar and char. The pyrolysis products have different uses. Biochar can be used for direct combustion in power plants for energy production or for domestic heating in rural areas where district heating might not be available. However, a more profitable use would be as activated carbon, used primarily to remove remnants of odorous or harmful compounds from wastewater or air. After further refining and upgrading, the bio-oils can be blended with conventional diesel or gasoline, while the volatile hydrocarbons can be used either as syngas in the chemical industry or as fuel for power plants or gas turbines. The benefit of using syngas in gas turbines or power plants is that the content of corrosive components in the gas is reduced and corrosion problems due to high contents of salts in biomass can be mitigated.

Gasification (Figure 13.5) is the volatilisation of low-molecular-weight organic components from the biomass in the presence of oxygen or other oxidising agents (e.g. steam or CO_2). The objective of gasification is to produce high yields of gas (CO, CO_2, H_2, H_2O, H_2, CH_4 and other hydrocarbons)

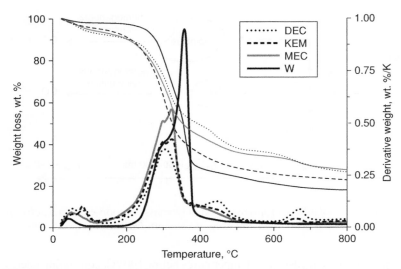

Figure 13.4   *Weight loss during heating of the fibre fraction from separated pig manure and wood at increasing temperature. The samples shown are raw slurry that is chemically pre-treated (polymers for flocculation) and mechanically separated (belt press + screw press) (KEM), decanter centrifuged anaerobically digested (DEC), mechanically separated by screw press or other simple separator only (MEC) and wood (W). Volatilisation of components is related to temperature as follows: 200–375 °C, a broad peak attributed to the conversion of hemicellulose at lower temperatures of the range and cellulose at higher temperatures of the range; 400–500 °C, a small shoulder due to devolatilisation of residual charcoal lignin; 600–700 °C, represents the aromatisation of the lignin; 0–900 °C, lignin volatilises slowly over the entire temperature range from ambient to approximately 900 °C. (© University of Southern Denmark.)*

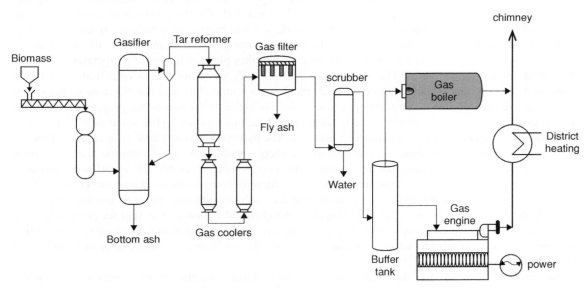

Figure 13.5   *Thermal gasification unit in which production of gas is the objective. (© University of Southern Denmark.)*

and the process is carried out at high temperatures (1073–1373 K). This is in contrast to pyrolysis, where the process is carried out at lower temperatures with the objective of producing tar and char. See Example 13.1.

Example 13.1

Calculation of the energy content of biomass combustion stoichiometry (it is the convention in the power engineering industry that calculations are made on a mass (kg or tonne) basis and that convention is followed here.)

Theoretical oxygen and air supply needed for combustion

The net amount of air or oxygen just sufficient to burn the net carbon, hydrogen and sulfur to CO_2, water vapour and sulfur dioxide (SO_2). The general expression for combustion of a fuel is:

$$C_mH_n + \left(\frac{4m+n}{4}\right)O_2 = mCO_2 + \left(\frac{n}{2}\right)H_2O \qquad (13.3)$$

Thus, oxidation of 1 mol CH_4 requires 2 mol O_2 to produce 1 mol CO_2 + 2 mol H_2O.

This calculation illustrates the theoretical oxygen demand for the combustion of biomass. More oxygen than the theoretical amount of air is required to achieve complete combustion. This excess air is expressed as a percentage of the theoretical air. The complete combustion of a fuel can be represented by a series of chemical equations:

$$C + O_2 \rightarrow CO_2 \qquad \Delta H = -393.7 \text{ MJ kg}^{-1} \text{ mol} \qquad (13.4)$$

$$H_2 + \frac{1}{2}O_2 \rightarrow H_2O \qquad \Delta H = -286 \text{ MJ kg}^{-1} \text{ mol} \qquad (13.5)$$

$$S + O_2 \rightarrow SO_2 \qquad \Delta H = -26.9 \text{ MJ kg}^{-1} \text{ mol} \qquad (13.6)$$

Using these equations, the theoretical oxygen use and energy outcome are calculated using kilograms instead of moles (i.e. 1 mol C is 12 kg C and 1 mole O_2 is 32 kg O_2):

$$12 \text{ kg C} + 32 \text{ kg } O_2 \rightarrow 44 \text{ kg } CO_2 \qquad \Delta H = -393.7 \text{ MJ}$$

$$1 \text{ kg } H_2 + 8 \text{ kg } O_2 \rightarrow 9 \text{ kg } H_2O \qquad \Delta H = -143 \text{ MJ}$$

In practice, only a part of the fuel is partially combusted. Partial combustion means that the elements are not oxidised to CO_2, H_2O and SO_2. For S and H, the unconsumed portion remains as S and H_2 and for C as CO. The energy loss due to incomplete C combustion is:

$$2C + O_2 \rightarrow 2CO \qquad \Delta H = 20.224 \text{ MJ kg}^{-1} \qquad (13.7)$$

Energy can be lost due to incomplete combustion caused by too little oxygen supply. For example, if the C is only oxidised to CO, 30% of energy is lost from the process.

When assessing the amount of ambient air needed (Table 13.2), one should bear in mind that air contains 79% nitrogen (N_2) by volume of nitrogen or 76.8% by weight. Fuels contain less than 2% by weight. Nitrogen is non-combustible and acts as a diluent to O_2 in air. The molecular ratio of nitrogen to oxygen is $79/21 = 3.76$ and the weight ratio of $N_2 : O_2$ in air is $76.8/23.2 = 3.31:1$, so Equation (13.3) can be rewritten as:

$$C + O_2 + 3.76\,N_2 \rightarrow CO_2 + 3.76\,N_2 \text{ and energy production } E = 393.7 \text{ MJ} \qquad (13.8)$$

Thus, to combust 1 kg of C, $\Delta H = 31.6$ MJ.

Table 13.2 *Amount of air needed to combust the C, H_2 and S in biomass.*

	Air combustible required for combustion (kg kg^{-1})			Flue products (kg kg^{-1} burned)			
	O_2	N_2	Air	CO_2	H_2O	N_2	SO_2
C	2.66	8.82	11.48	3.66		8.82	
H_2	8.0	26.40	34.40		9	26.40	
S	1.0	3.31	4.31			3.31	2

Example of air needed when combusting Miscanthus

From the proximate and ultimate analysis of *Miscanthus* given in Table 13.3 (below), the air required for combustion and the products of combustion (POC) can be calculated as follows.

First, the need for O_2 in air is calculated:

Component to be combusted (oxidised)	kg C, S or H in biomass	Amount of O_2 required for combustion (kg kg^{-1})	Total kg O_2
C	0.415	2.66	1.04
S	0.0009	1.0	0.0009
H	0.054	8.0	0.432
Total O_2			1.473

For combustion, ambient air is used and the weight of air required is: $(100/23.2) \times 1.473 = 6.35$ kg. At STP, the volume of 1 kg of ambient air is 0.776 m^3. The volume of ambient air needed is therefore 6.35 kg $\times 0.776$ m^3 kg$^{-1} = 4.93$ m^3 at STP.

The volume of gases in air can be calculated as follows: O_2 is $(1.473/32) \times 22.4 = 1.03$ m^3; N_2 is 4.88 kg $(6.35 - 1.473)$ or $(4.88/28) \times 22.4$ m$^3 = 1.03$ m^3.

Knowing the fuel composition, it is possible to calculate the theoretical composition of the products of combustion (POC).

Nitrogen in flue gas			3.65 kg	2.92 m^3
Weight of CO_2 produced	0.415×3.66		1.52 kg	0.773 m^3
Weight of H_2O produced	0.054×9	0.29 kg		
Weight of H_2O in willow		0.142 kg		
Total H_2O			0.432 kg	0.538 m^3
Weight of SO_2 produced	0.0009×2		0.0018 kg	0.0026 m^3
Total POC (wet)			5.60 kg	4.234 m^3
Total POC (dry)			5.17 kg	3.696 m^3

$$\% - CO_2 \text{ v/v} = (0.773/3.696) \times 100\% = 20.92\%.$$

The volume percentage of CO_2 in flue gas can be estimated as 20.92%.

The flue gas obtained from any fuel when burned with the amount of air required for complete combustion, but without excess, contains a fixed percentage of CO_2 (i.e. the theoretical %-CO_2). With pure dry carbon, the O_2 of the air is replaced quantitatively with CO_2 so that the theoretical percentage is 21% v/v. Any air supplied in excess of that theoretically required reduces the %-CO_2 in the flue gases proportionally. In practice, the %-CO_2 in flue gas is determined by volume on a dry basis. The theoretical %-CO_2 used in calculations is also the %-CO_2 on a dry basis. With fuels containing hydrogen or hydrocarbon gases, the theoretical %-CO_2 is lower than that for carbon because the air supplied to burn the hydrogen forms water, which condenses leaving nitrogen to dilute the flue gases from the carbon alone.

13.2.1   Operation of Combustion Plant

The energy production in a power plant using biomass is part of a chain of operations comprising collection of the biomass, drying the fuel, storing it until use, cutting the biomass into small particles and feeding the furnace or boiler. The furnace can be a grate furnace where the biomass is moved with travelling grates in power plants or by gravity on inclining grates in small furnaces. The grate furnaces can be used for material with high moisture content and varying particle size, and are cost-effective for plants smaller than 20 MW_{th}. Alternatively, in large power plants the biomass can be burned in fluidised beds where the fuel is pulverised and mixed with a hot, inert bed material that is suspended in an air stream, which also provides oxygen for the combustion. Good mixing reduces the excess air requirement and reduces heat losses from air pre-heating. These plants are economical at larger than 20MW_{th} (MW thermal energy).

In the boiler, heat exchangers are positioned so that the circulating feed water is heated and evaporated to steam in the lower "cooler" part, while in the upper part of the furnace the steam is heated to very high temperatures and energy content (Figure 13.6). The steam is transferred to the turbine, where it expands to provide steam velocities between 60 and 300 m s^{-1}, and the flow of steam makes the turbine blades move. After the turbine, the remaining heat in the steam can be recovered by heat exchange to be used partly for district heating, but also for pre-heating air entering the furnace. The flue gas leaving the furnace has to be cleaned of ash particles, NO_x and SO_2 before being released to the atmosphere.

The moisture content of any fuel affects the energy available for heat and power production because of the endothermic process of water evaporation, so power plant operators take account of this when expressing the energy content of biomass. The higher heating value (HHV) is defined as the heat released during combustion of a fuel due to oxidation of the organic and inorganic components (Text Box – Basic 13.1), including the heat released by condensation of water (evaporated or formed in the process) and dissolution in water of inorganic acids (from sulfur and nitrogen in the sample). The HHV can be measured experimentally by combustion of a small portion of sample in an adiabatic bomb calorimeter. The end point of the measurement is at room temperature, and thus condensation and dissolution processes are inevitably included in the outcome. The lower heating value (LHV) or net calorific value is the energy released during combustion assuming that water and acids are not condensed – this is typically the case when a fuel is burned in a boiler. Here the water vapour and the nitrogen and sulfur oxides are elutriated with the flue gas. The LHV is calculated from the HHV by subtraction of the energy used for evaporation of the incipient moisture, as well as the water produced by oxidation of hydrogen in the fuel. Dry cereal straw has a HHV of about 17 MJ kg^{-1} (ar) (Table 13.3) and at the normal dry matter content of 860–890 g kg^{-1} the LHV is about 15 MJ kg^{-1} (ar). If the dry matter content is low (i.e. 750 g kg^{-1}), the LHV is approximately 12 MJ kg^{-1}. Also see Text Box – Basic 13.2.

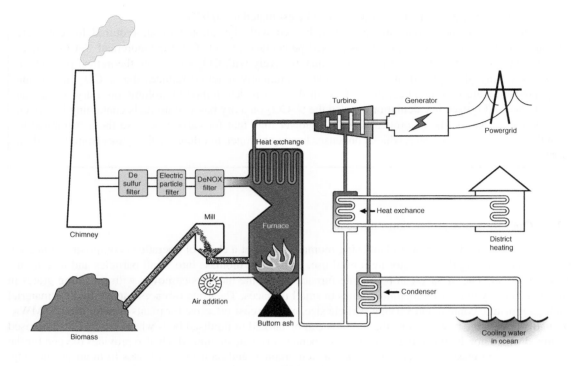

Figure 13.6   *Illustration of the operations in a power plant. (© University of Southern Denmark.)*

Table 13.3   *Gross calorific and net calorific value of biomass, including the composition of the biomass (PHYLLIS, 2012).*

| | Proximate analysis (ar) | | | | Ultimate analysis (wt-% dry basis) | | | | | |
| | Calorific value (MJ kg^{-1}) | | Moisture (wt%) | Ash (wt%) | C | H | O | N | S | Cl |
Feedstock	HHV	LHV								
Willow	16.32	14.95	15.0	1.7	41.5	5.1	36.8	0.36	0.03	0.013
Miscanthus	16.80	15.27	14.2	1.3	41.5	5.4	37.1	0.26	0.09	0.11
Forest residues	12.74	13.32	29.6	0.8	33.9	4.2	31.4	0.49	0.07	0.07
Bark eucalyptus	16.25	14.90	12.0	4.2	41.7	4.8	38.8	0.26	0.04	0.23
Barley straw	16.52	15.14	11.5	5.2	40.9	5.0	36.5	0.53	0.07	0.24
Wheat straw	17.00	15.60	10.3	4.2	42.5	5.3	37.1	0.52	0.06	0.15
Sunflower husks	17.87	6.45	10.5	3.8	43.8	5.4	35.6	0.77	0.07	–
Olive residues	20.05	18.70	5.5	4.3	47.9	5.6	35.0	1.29	0.28	0.17
Poultry litter	10.33	8.54	39.7	10.6	24.8	3.8	17.0	3.67	0.3	0.15
Cow manure	14.95	13.61	13.9	13.7	39.1	4.6	26.7	0.83	0.25	0.99
Pig manure	1.09	−1.24	92.1	2.8	2.8	4.38	21.3	2.79	0.04	0.09

Text Box – Basic 13.2 Assessment of LHV and HHV

The HHV can be measured by adiabatic (without heat exchange with the surroundings) calorimetric measurements. Alternatively, the HHV_{db} (MJ kg^{-1}) can be determined on the basis of the elemental composition (e.g. using the Milne equation):

$$HHV_{db} = 0.341C + 1.322H - 0.12O - 0.12N + 0.0686S - 0.0153ash \qquad (13.9)$$

where C, H and so on are the mass and the ash fractions in wt-% of dry material and HHV the heating value in MJ kg^{-1} (see Table 13.4). The ash content reduces the HHV as these mineral components cannot be oxidised. Thus, by using the hydrogen and ash fractions (wt-% dry) and moisture fraction w (wt-% ar), the different HHVs and LHVs can be calculated:

$$HHV_{ar} = HHV_{db} \times (1 - w/100) \qquad (13.10)$$
$$HHV_{db} = HHV_{daf} \times (1 - ash/100) \qquad (13.11)$$

where w is the wet fraction and ash is the ash fraction and HHV_{daf} is the HHV related to the dry ash fraction.

The LHV (MJ kg^{-1} (wb)) can be determined from the HHV taking into account the moisture and hydrogen content by:

$$LHV_{db} = HHV_{db} - 2.442 \times 8.936H/100 \qquad (13.12)$$
$$LHV_{ar} = LHV_{db} \times (1 - w/100) - 2.442 \times w/100 \qquad (13.13)$$
$$LHV_{ar} = HHV_{ar} - 2.442 \times (8.936H/100(1 - w/100) + w/100) \qquad (13.14)$$

where w is the moisture content (wt-% (ar)) and H is the concentration of hydrogen (wt-% (db)), see Table 13.3. Thus, increasing the water content reduces the LHV, because water has to be volatilised at an energy cost. The term with the hydrogen content signifies that the energy released by condensation of the vapour formed by combustion is not included in the LHV. This is a simplified assessment not taking into account all components that will affect the LHV.

	Proximate analysis				Ultimate analysis (wt-%)		
	db	daf	ar		dry	daf	ar
Ash	1.6		1.6	C	48.7	49.5	47
Water			3.5	H	5.84	5.9	5.6
Calorific value (MJ kg^{-1})				O	43.4	44.1	41.9
HHV	19600	19921	18924	N	0.41	0.42	0.4
LHV	18326	18625	17609	S	0.04	0.04	0.03

Example: European power stations pay a set price for fuel per GJ on a LHV basis

Using the information provided for willow, prepare a spreadsheet using Excel to allow calculation of the LHV for a range of moisture contents from 15–50%:

$$HHV_{db} = 19.60 \text{ MJ kg}^{-1}$$

$$LHV_{db} = HHV_{db} - 2.442(8.936H/100) = (H = 5.84\%) = 18.326 \text{ MJ kg}^{-1}$$

At 20% moisture, for instance, net heating value is: $18.326(1 - 0.2) - 2.442(0.2) = 14.172$ MJ kg^{-1}. Figure 13.7 depicts the effect of moisture content on net heating effect.

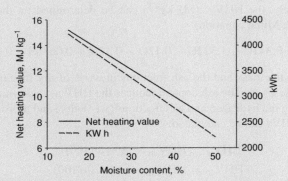

Figure 13.7   *Net heating effect when combusting willow at increasing moisture content. (© University of Southern Denmark.)*

13.2.2   Pre-Treatment of Biomass

Pre-treatment of manure prior to combustion is essential if automatic fuel feeding of the boiler is required. Ideally, any such feedstock should be homogeneous with a low moisture content, high energy density and with minimum impurities in the fuel.

Manures that offer the potential for combustion include solid chicken manure, the fibre fraction from separated animal slurry and deep litter. Deep litter has fuel properties similar to straw and may be used as a fuel without drying. Drying the biomass may be necessary to improve the overall energy balance arising from combustion. In order to reduce the risk of self-heating and ignition during storage, a moisture content below 18% is recommended. Drying manure can significantly increase the energy input (e.g. drying of deep litter can require an energy consumption of 20% of the LHV). However, storage of solid manures at higher moisture can give rise to composting where many organic components may be lost (up to 40%, see Chapter 4). The need for drying may be at the expense of losing some of the fibre fraction from the combustion process. As mentioned earlier, as a rule of thumb the dry matter content should be above 55% before the biomass can be used for power production in combustion plants. Chicken litter can be used as a fuel without additional treatments.

Torrefaction is a technology where the biomass is heated to moderate temperatures (200–300 °C), releasing moisture and some volatile organics, and significantly improving the energy density of the residual solid fuel. In some cases the volatiles released can be recovered for use as an energy carrier. The increased energy density of the torrefied biomass can significantly improve the supply chain logistics and economic viability of transporting biomass from remote regions (e.g. Russia or Canada). Thus, torrefaction technologies where energy consumption is minimised may also offer a pre-treatment option for manures if the fibre fraction is to be used for combustion.

Manure from livestock production is usually a heterogeneous, low-energy-density feedstock and may contain longer structures such as straw that can give rise to bridging in fuel feed systems. After drying, the manure may need to be chopped or milled to smaller particle size depending on the precise requirements of the furnace or boiler where the fuel is to be burned. Grate burners can manage biomass fuels of varying particle size, whereas in fluid bed boilers the particle size is restricted to below 40–80 mm depending on the technology. If combustion of manure is an option fluidised bed boilers have the advantage that they can handle material with variable moisture content (e.g. peat harvested in Ireland is burned at an average moisture content of 55%, but the boilers can handle between 25 and 70% moisture).

The transport and management of the dried biomass can be improved by pelletisation. This requires the dry matter content to be above 90%. Prior to pelletisation, the biomass is typically reduced in size with a hammer mill and steam-conditioned to improve adhesion before a flat die or ring pelletiser is used to produce the pellets. Such technology has been developed for farm-scale and central heating plants and may be used to upgrade the energy density of the manure before delivery to the power plant.

13.2.3 Energy Production Using Straw Residues

Straw residue combustion at the Danish power plant Fynsværket is a good example of how manure can be used for energy production. This plant burns 700 tonnes of straw daily, which is equivalent to 29 truck deliveries per day. As there are no deliveries during weekends, about 40 truckloads a day are received, with on-site storage capacity of 1400 tonnes.

Combustion of straw has a daily net calorific energy (LHV) production of in total 117.5 MW (Text Box – Basic 13.2). Of this, 109 MW are transferred to the water/steam circuit, and used to transfer the energy to the turbine and the district heating system, while the remainder stays in the flue gas and is emitted to the atmosphere (Figure 13.8). The energy of the water/steam system is transferred to electricity (work) in the turbine and generator, producing 38 MW power. Of this, 3 MW electric power are used internally at the power plant and 35 MW are exported to the power network (i.e. 29.8% of the biomass LHV). After the turbine, 71 MW of the energy of the water steam system are transferred to the district heating system through a large heat exchanger. The flue gas contains an energy reserve unaccounted for in LHV as water evaporated from

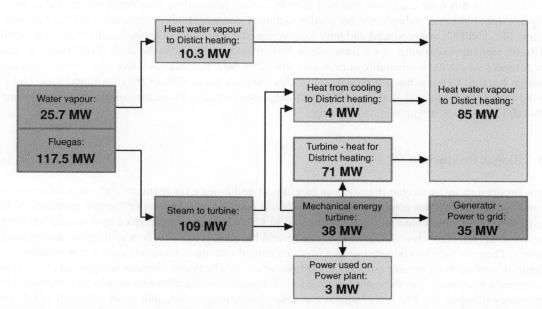

Figure 13.8 *Energy flow in a power plant (Fynsværket, Denmark) incinerating wheat straw. The flue gas energy corresponds to the LHV and flue gas + vapour energy are almost similar to the HHV. (© University of Southern Denmark; data from Thomsen and Larsen (2012).)*

the biomass and produced during combustion of the biomass. A part of this energy is through condensation transferred to the district heating system and delivers 10.3 MW to the heating potential of district heating. Furthermore, heat exchange from different cooling processes in the plant contributes 4 MW. In total, the power plant exports 85 MW heat to the district heating system.

13.2.4   Residues from Incineration

The amount of ash produced from manure depends on animal husbandry and feeding practices, and the amount of supplementary straw added as bedding, which can vary from about 14% to about 20%. This ash is collected as bottom ash in the furnace (80–90%) or retained in filters as fly ash (10–40%), as only a minute fraction of the fly ash is emitted to the atmosphere (van Loo and Koppejan, 2008). At Fynsværket, the relative distribution of bottom ash and fly ash is 90% and 10%, respectively.

The bottom ash contains Si, Al, Fe, Ca, Mg, Mn and P, which are non-volatile components, as well as some of the more volatile components such as K, Na, Cl and S.

From the fuel, a fraction of the volatile components (K, Na, Cl and S) and the heavy metals Zn, Cd, Hg and Cu are released to the flue gas and carried up in the furnace. These components condense on particle surfaces when the air cools down. In this process, CaO particles may act as nucleation sites to which other components can attach. Most of the particles formed through nucleation have a diameter below 1 μm. Some of the volatised components also adsorb to the larger fly ash particles.

The bottom ash can be used as a source of fertiliser P and K for crop production, because the content of heavy metals is low (Table 13.4). The fly ash particles are separated from the air before emission to the atmosphere. If this is carried out in two steps with filters retaining the coarse particles initially (cyclone) and thereafter filters retaining the fine particles (bag filters), then a coarse fly ash fraction with a relative low heavy metal concentration (10–35 wt-%) and a fine fraction with a high heavy metal concentration (2–10 wt-%) can be produced. Fly ash from coal combustion is used in cement production, road construction and so on, but the fly ash from biomass combustion is not used in Denmark due to its high concentration of heavy metals, In addition, SO_x and NO_x are produced and have to be retained or removed. SO_x can be adsorbed by scrubbing with basic reagents consisting of calcium, while NO_x may be reduced to N_2 with NH_3 (urea). In small straw furnaces where the temperature is relatively low (250–500 °C), there is a risk of dioxin dibenzofuran formation because biomass has a high content of C, Cl and O; about 80% of the dioxin and furans adsorb onto the fly ash (van Loo and Koppejan, 2008). In large power plants where the combustion temperature is 1200–1400 °C, there is no production of dioxin.

13.3   Biogas Production

Biogas production by anaerobic digestion is an efficient technology for manures and many other organic materials with a high water content that are not relevant fuels for combustion or thermal gasification. In a biogas plant the organic components are fermented to CH_4 and may also produce hydrogen (H_2). Biogas treatment of animal slurry is also claimed to have several beneficial side-effects in addition to the production of energy. These include less odour emission from the animal manure or waste material due to a reduction in concentration of odorous organic components (Masse *et al.*, 1997). Waste biomass is sterilised by reduction of pathogenic bacteria due to the retention time in the digester, the effectiveness of which decreases at low temperatures (Chapter 6). The use of biogas can also significantly reduce indoor air pollution in kitchens, and thus prevent poor health and even death of millions of women in developing countries, where firewood or coal is burned in open fires for cooking (Bruun et al., in preparation). Chemical oxygen demand (COD) is reduced (Masse *et al.*, 1997), which can be of importance when the digested material is discharged to

Table 13.4 Composition of ash from combustion of biomass: average values are shown, but these can vary by several milligrams for a component depending on combustion technology and biomass characteristics.

		Chloride (g kg^{-1})	Sulfate (g kg^{-1})	Ca (g kg^{-1})	Cd (mg kg^{-1})	Cr (mg kg^{-1})	Cu (mg kg^{-1})	Zn (mg kg^{-1})	Pb (mg kg^{-1})	Reference
Bottom ash	municipal solid waste Denmark in 2002		10.8	95	3.7	600	2700	3300	990	Hjelmar et al. (2009)
	municipal solid waste Europe and North America		9	90	36	1600	4100	4205	7000	Hjelmar et al. (2009)
	municipal solid waste	6	19	7.4	223		2456	1980	1200	Lisk (1988)
	wood chips, Sweden	1.0	9.1	13.4	2.8	167	312	1380	133	Hjelmar et al. (2009)
Fly ash	straw	160	170	27	10	97	120	620	24	Hjelmar et al. (2009)
	wood chips, Denmark	79	110	150	8	18	150	500	42	Hjelmar et al. (2009)
	wood, Sweden	5.1	40	18	12.6	95.4	231	3080	223	Hjelmar et al. (2009)

recipient waters. During fermentation, organic N is transformed to inorganic NH_4 which is readily available for plant uptake, but at the same time pH is increased and the NH_3 loss potential is high (Chapter 5). Biogas production reduces the use of wood and fossil fuels and thus this renewable and CO_2-neutral energy source reduces greenhouse gas emissions (Møller *et al.*, 2004a; Bruun *et al.*, 2013).

The biogas production process comprises anaerobic microbial fermentation of organic matter, which is transformed primarily to CH_4 and CO_2. Biogas plants can treat a wide variety of feedstocks, including animal manures and other agricultural residues, wastes from food and feed industries and purpose-grown energy crops (Ward *et al.*, 2008). In many cases the feedstock is a mixture of substrates in what is termed co-digestion. Biogas plants on livestock farms or plants receiving manure from livestock farms can use either solid or liquid manure. However, to our knowledge very few biogas production units use only solid manure to feed the digester (fermentation unit) and therefore the following presentation of the technology focuses on biogas plants processing slurries that can be transported by pumping. See Figure 13.9.

13.3.1 Biogas Process

Biogas production is an anaerobic process that is almost immediately stopped if oxygen is added. The process is usually divided into four steps in series: hydrolysis, acidogenesis, acetogenesis and methanogenesis (Figure 13.10), although the hydrolysis step may not be relevant if the substrate is already dissolved, as is often the case with wastewater.

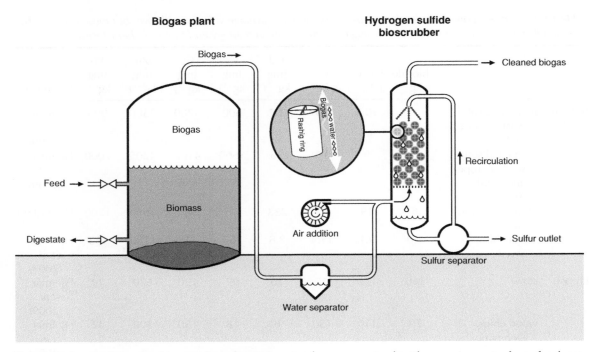

Figure 13.9   *Outline of a biogas plant digesting animal manure mixed with organic waste from food processing. The digester is the reactor seen to the left and to the right is a H_2S unit removing H_2S through oxidation by microorganisms. At the top of the tower the liquid is sprayed over filter blocks and percolates down over these, which provides a large surface on which the bacteria can adhere. (© University of Southern Denmark.)*

During hydrolysis, the complex organic polymers are hydrolysed in a chemical process by enzymes secreted from strictly anaerobic bacteria such as *Bacterioides* and *Clostridia* (Rivard *et al.*, 1990) but also facultative bacteria such as *Streptococci* (Hobson and Shaw, 1974). The enzymatic process is extracellular and transforms carbohydrates into sugars, proteins into soluble proteins or amino acids and lipids into long-chain fatty acids (LCFAs). The rate of this process depends on the substrate (i.e. complex fibrous material may take months to degrade whereas lipid degradation to LCFAs is very rapid), temperature, pH and the presence of inhibitory compounds. Hydrolysis is often the rate-limiting step in the process, especially if the biomass contains straw and other plant residues, which includes feed residues excreted by animals. This is due to the content of ligno-cellulose, which is the plant's defence system against degradation by microorganisms.

In the second step, acidogenesis, dissolved organic components are taken up by fermentative bacteria such as *Thermoanaerobacterium* and *Bacillus* and transformed to fatty acids, alcohols, H_2 and CO_2.

In the third step, acetogenesis, acetic acid is produced along with H_2 and CO_2 by the oxidation of reduced compounds. The production of acetate from propanoate and ethanol is shown in equations (13.15) and (13.16), respectively:

$$CH_3CH_2COO^- + 3H_2O \rightarrow CH_3COO^- + H_3O^+ + 3H_2 + HCO_3^- \qquad (13.15)$$
$$CH_3CH_2OH + H_2O \rightarrow CH_3COO^- + H_3O^+ + 2H_2 \qquad (13.16)$$

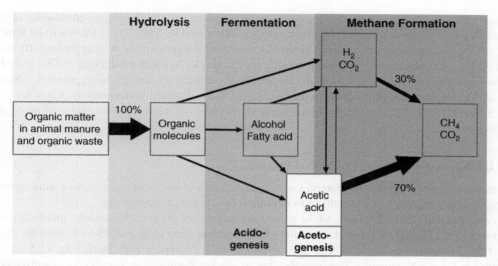

Figure 13.10 *Fermentation processes producing CH₄ in a biogas digester. The four main stages in the digestion process are: (1) hydrolysis, where a cocktail of exoenzymes degrades the biomass to smaller compounds (LCFAs, sugars, soluble proteins and amino acids); (2) acid formation (acidogenesis), where organic compounds are transformed to (a) acetic acid (CH₃COOH), (b) organic acids and alcohols, and (c) molecular hydrogen (H₂) and CO₂; (3) acetic acid formation (acetogenesis), where organic acids are transformed to (a) acetic acid and (b) H₂ and CO₂; and (4) methanogenesis, where (a) acetic acid is transformed to CH₄ and CO₂ (acetoclastic methanogenesis) and (b) H₂ and CO₂ are transformed to CH₄ (hydrogenotrophic methanogenic). (© University of Southern Denmark.)*

The fourth and final step is methanogenesis, where CH_4 is produced in two totally distinct processes. During acetoclastic methanogenesis, acetic acid is transformed to CH_4 and CO_2:

$$CH_3COOH \rightarrow CH_4 + CO_2 \tag{13.17}$$

while in the hydrogenotrophic methanogenic process the archaea transform H_2 and CO_2:

$$4H_2 + CO_2 \rightarrow CH_4 + 2H_2O \tag{13.18}$$

About 70% of the CH_4 is produced via the acetoclastic pathway (Equation 13.17) and 30% through the hydrogenotrophic pathway (Equation 13.18) (Batstone *et al.*, 2002). Recent research based on online isotope ratio measurements indicates that the acetoclastic methanogenic pathway may be less important and that especially at high organic load rates the hydrogenotrophic archaea are the main pathway for CH_4 production (Blume *et al.*, 2010).

Acidogenesis, acetogenesis and methanogenesis are intracellular processes resulting in direct bacterial growth. Thus, bacterial growth and loss due to wash-out from the reactor has to be accounted for when calculating mass balances of biogas production.

The individual steps of transformation of organic material to CH_4 and CO_2 are very much dependent on each other, not only because the product from one process is the substrate for a downstream process, but also because many of the biochemical processes are substrate- and product-inhibited. An example is the acetogenetic step producing H_2. This process requires syntrophy with hydrogen-consuming organisms such as *Methanospirillum hungatei* to be thermodynamically feasible. Thus, H_2 is known to be transferred between acidogenic and hydrogenotrophic syntrophic bacterial consortia present in aggregates (flocs). This interspecies transfer links the degradation of volatile fatty acids (VFAs) with production of CH_4 from H_2 and CO_2. Consequently, the concentrations of the intermediate products are low in the aggregates, often much lower than the concentration of products in the bulk liquid in the digester. Unfortunately, it has not yet been possible to measure the concentration of hydrogen in flocs, although microsensors capable of doing this are under development.

13.3.2 Inhibition of the Biogas Production Process

As the anaerobic digestion process is complex with many species of microorganisms, often working together in syntrophic relationships, it is also subject to inhibition by a range of mechanisms.

High concentrations of LCFAs and VFAs reduce the activity of the microorganisms involved in biogas production. LCFAs adsorb on the surface of the microorganisms and reduce uptake of nutrients (Batstone *et al.*, 2002). Furthermore, the activity of hydrolytic enzymes decreases with increasing VFA concentration, from 2 g l^{-1} VFA onwards. This may be due to feedback inhibition, (i.e. VFAs are produced and the hydrolytic bacteria respond by reducing excretion of enzymes, thus governing a process producing VFAs) (Siegert and Banks, 2005). Thus, VFA are toxic at high concentrations and thereby reduce the activity of the microorganisms. This also reduces the activity of the bacteria producing the enzymes needed for the hydrolysis and therefore stops production of the substrate for acidogenesis. The reduced production of acids and a steady, although slow, degradation of these reduces the acid concentration and the biogas process can start again. This recovery can last months, and therefore the digester may instead be partly emptied and fresh biomass with a low concentration of acids (possibly the stabilised digestate from another biogas reactor) added, thereby solving the problem by dilution. The concentration of propionate has been proposed as a good indicator of process stability (Nielsen *et al.* 2007), as have butyrate and isobutyrate (Ahring *et al.*, 1995). However, there are no clearly defined inhibition threshold levels of VFAs that are applicable in all cases. Instability of the process is seen over a wide range of concentrations of the acids due to experimental conditions and differences in the microbial populations, one example being that inhibition may take place at a propionate : acetate ratio above 1.4 and acetate above 0.8 g l^{-1} (Bitton, 2011).

Due to its small size and electrical neutrality, NH_3 can be transported over the cell membrane and after protonation of NH_3 inside the cell the component distorts cell homeostasis, including transport of, for example, potassium across the cell membrane. The inhibition due to NH_3 leads to an increase in VFAs, causing further inhibition (Angelidaki *et al.*, 1993). Ammonium concentrations over about 1500 mg N l^{-1} are in general not desirable due to the inhibitory nature of NH_3 to the archae (Zhang *et al.*, 2000), but in practice it has been shown that the microorganisms can acclimatise and biogas reactors perform well on pig slurry at higher NH_3 concentrations than this limit. It is not only the total ammoniacal concentration (TAN) in material fed to the biogas digester that should be considered, as about 80% of the organic N is transformed to TAN when fermenting the biomass. Thus, often it is the total N that is considered when assessing when N may be a problem for the biogas process. Furthermore, it is the NH_3 that inhibits methanogenesis and the NH_3 : NH_4^+ ratio increases at increasing temperature and pH (Chapter 3) (Figure 13.11). Thus, if N concentration is expected to be high, then one solution may be to choose a lower process temperature at the design stage. However, the microorganisms still need N for efficient growth and therefore if the N concentration in the feedstock is too low in relation to carbon, cellular growth and therefore biogas production will be reduced. Thus, microbial growth will be reduced if the C : N ratio exceeds 50 : 1 and at C : N ratios below 10 : 1. An optimal C : N ratio is considered to be 30 : 1.

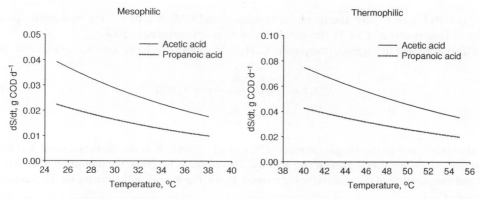

Figure 13.11 *Effect of NH$_3$ inhibition at increasing temperature: TAN = 0.2 M; pH 8, rate constants and inhibition constants given in Text Box – Advanced 13.1. (© University of Southern Denmark.)*

Text Box – Advanced 13.1 Degradation of substrate assessed using Monod kinetics including the effect of NH$_3$ inhibition

The degradation of the substrate is characterised by Michelis–Menten kinetics and can be depicted using the Monod equation:

$$\frac{dS}{dt} = \frac{K_{max}S}{(S + K_s)}X \tag{13.19}$$

where dS/dt is the rate of reduction of the component (g COD l^{-1} day^{-1}), K_{max} is the maximum specific substrate utilisation constant (g COD day^{-1} g^{-1} COD$_b$ bacterial biomass), S is the substrate concentration (g COD l^{-1}), K_S is the half-saturation concentration of the substrate (g COD l^{-1}) and X is the microbial biomass concentration (g COD$_b$ l^{-1}) (Table 13.5)

Table 13.5 *Parameters K$_{max}$ and K$_S$ used in the Monod equation for the degradation of acetic and propanoic acids.*

	K_{max} (g COD l^{-1})	K_S (g COD l^{-1})
Acetic acid	0.59	1.05
Propionic acid	0.22	0.34

X can be set at 20 g COD$_b$ l^{-1}.

The effect of inhibiting factors on the processes involved in biogas production can be assessed by including inhibition factors (I_n, dimensionless) in the equation There may be several inhibitors:

$$\frac{dS}{dt} = \frac{K_{max} \cdot S}{(S + K_s)}X \cdot I_1 \cdot I_2 \cdot I_3, \ldots, I_n \tag{13.20}$$

NH$_3$ inhibition of the transformation of propionic acid to acetic acid and H$_2$ can be assessed as:

$$I = \frac{1}{1 + \frac{[NH_3]}{K_I}} \tag{13.21}$$

where K_I is 0.011 mol l^{-1} for thermophilic processes and 0.0018 mol l^{-1} for mesophilic processes. According to Batstone *et al.* (2002), the variability of K_I is approximately 30%.

The COD (mg l^{-1}) of the organic component $C_nH_aO_bN_c$ (mol l^{-1}) can be assessed as (Speece, 2008):

$$COD = \frac{2n + \frac{a-3c}{2} - b}{2} \times 32000 \qquad (13.22)$$

H_2S is also inhibitory to the biogas process (Chen *et al.*, 2008). It is the un-dissociated and uncharged component that affects microbial growth and activity. Therefore, as for NH_3, the inhibition of H_2S is affected by the pH and temperature of the residue being treated. Reducing the pH will increase the H_2S concentration over HS^- and HS^{2-}.

There are upper and lower limits for optimal pH conditions for biogas production. This is partly due to the effect of H_2S and NH_3, and to combinations of environmental conditions on growth and activity of microorganisms. The effect of low pH is also due to interactions with the VFAs, of which a higher proportion will be uncharged at low pH, and the uncharged acids move more easily through the cell membrane of the bacteria, inhibiting the metabolic processes. Thus, neutral pH values are optimal for biogas production (Yadvika *et al.*, 2004).

As mentioned above, under stable digester conditions the main fraction of hydrogen used for methanogenesis is believed to be transferred directly from acetogenic to methanogenic organisms. A sudden increase in the organic load may cause H_2 to accumulate, and this causes the degradation of organic acids to be thermodynamically unfavourable (Text Box – Advanced 13.2), leading to the accumulation of VFAs.

The resulting decline in pH causes failure of biogas production because the acids inhibit the biogas process. Thus, it has been shown that VFA degradation rates are inversely correlated with the partial pressure of H_2. Dihydrogen concentration increases before VFAs start to accumulate, and therefore H_2 is proposed to be used as a parameter to control the biogas production and avoid digester failure (Rodriguez *et al.*, 2006). However, $H_2(g)$ in the gas phase has little correlation with the liquid phase $H_2(aq)$, which actually has an effect on microorganisms.

Text Box – Advanced 13.2 Thermodynamic constraints caused by hydrogen on the biogas process

A high concentration of $H_2(aq)$ renders the production of biogas thermodynamically unfavourable under standard temperature and pressure (STP) conditions. H_2 is an intermediate product in the reduction of organic matter to CH_4 and CO_2 (Figure 13.9). The microorganisms do not produce much energy in the oxidation processes, so if the H_2 in solution is high then the oxidation process will be unfavourable (Equations 13.12–13.15).

Reaction	ΔG (kJ mol^{-1})	Equation
$CH_3CH_2OH(aq) + 2H_2O \rightarrow CH_3COO^-(aq) + H_3O^+ + 2H_2(aq)$	+9.6	(13.23)
$CH_3CH_2COO^-(aq) + 3H_2O(l) \rightarrow CH_3COO^-(aq) + H_3O^+ + 3H_2(aq) + HCO_3^-$	+76.1	(13.24)
$CH_3CH_2CH_2COO^-(aq) + 2H_2O(l) \rightarrow 2CH_3COO^-(aq) + H_3O^+ + 2H_2(aq)$	+48.1	(13.25)
$C_7H_5O_2(aq) + 7H_2O \rightarrow 3CH_3COO^-(aq) + 3H_3O^+ + 3H_2(aq) + HCO_3^-(aq)$	+53	(13.26)

The process will only be thermodynamically favourable and take place if ΔG for the overall processes is negative. That is why collaboration between the syntrophic bacteria clusters is so important (i.e. hydrogenotrophic methanogens using the H_2).

Fortunately the acetoclastic and hydrogenotrophic methanogenic reactions have a negative ΔG (Equations 13.27 and 13.28), and these processes will run spontaneously.

Reaction	ΔG (kJ mol^{-1})	Equation
$CO_2(aq) + 4H_2(aq) \rightarrow CH_4(aq) + 2H_2O$	−130.4	(13.27)
$CH_3COOH(aq) \rightarrow CO_2(aq) + CH_4(g)$	−32.5	(13.28)

When combining the processes, the focus may be on the degradation pathway of propanoic acid and acetic acid (Equations 13.24 and 13.28). Combining these two processes gives a positive ΔG (Equation 13.18), but combining the propanoic acid and acetic acid degradation processes and the hydrogenotrophic process (Equation 13.27) gives a negative ΔG (Equation 13.28).

It can be assessed that at H_2 partial pressure below 10^{-4} atm, the propanoic acid combined degradation process is thermodynamically favourable with regard to propanoic acid degradation. At too low H_2 partial pressure (below 10^{-6} atm in the headspace), the CH_4-producing process is not favoured (i.e. H_2 is substrate for this process). Thus, H_2 should be in the interval 10^{-6} to 10^{-4} atm (headspace concentration) for the propanoic acid to CH_4 production pathway to be effective.

Reaction	ΔG (kJ mol^{-1})	Equation
$CH_3CH_2COO^-(aq) + 3H_2O(l) \rightarrow CH_4(aq) + CO_2(aq) + 3H_2(aq) + HCO_3^-(aq)$	43.7	(13.29)
$CH_3CH_2COO^-(aq) + 3H_2O \rightarrow 1.75CH_4(aq) + 0.5CO_2(aq) + 1.5H_2O(aq) + HCO_3^-(aq)$	−86.7	(13.30)

The upper limit of H_2 concentration, where the fermenting bacteria can transform propanoic acid to carbonate and CH_4 (Equation 13.21), can be estimated. At equilibrium, ΔG is zero:

$$\Delta G = RT\ln\frac{(C)^c \times (D)^d}{(A)^a \times (B)^b}, \ln K = -\frac{\Delta G^\theta}{RT} \tag{13.31}$$

where K is the reaction equilibrium value, R is the gas constant (8.3143 J mol^{-1} K^{-1}) and T is the temperature (K).

As mentioned earlier, it is not currently possible to measure the partial pressure of H_2 inside the flocs. Observations show that headspace hydrogen pressure far exceeds the concentration where the processes become unfavourable. However, inside the flocs it is assumed that the H_2 is consumed as soon as it is produced, and organisms in flocs are not subjected to the same hydrogen concentrations as in the bulk liquid.

The supply of biomass to the digester must be as homogeneous as possible, both in terms of the rate of addition of biomass to the digester and the composition of biomass. Mixing a new biomass to existing biomass used in the digester may result in failure of the process, because the microorganisms are not adapted to the biomass. Furthermore, the rate of addition must be fairly constant as sudden increases may cause an increase in organic acid production because the methanogenic microorganisms do not have the capacity to transform the acids or H_2 fast enough. The resulting high concentration of organic acids will inhibit most processes in the biogas digester and the biogas production will decline significantly.

Temperature variation in the digester also reduces the rate of biogas production. Therefore, the temperature must be maintained within a narrow interval to ensure a high production rate. Thus, in large-scale processes, particularly those in cold climates, digesters are insulated and heated to ensure a constant process. Small digesters in tropical and subtropical countries are buried in the soil to ensure that the temperature changes are slow.

The process may also be inhibited by high concentrations of antibiotic residues in excreta if these substances have been used to treat the animals, or by high concentrations of trace metals in the biomass.

13.3.3   Gas Production Rates

The rate and efficiency of the fermentation process is very temperature-dependent. The reaction rate increases exponentially with temperature. It is assumed that the fermentation producing CH_4 is carried out by microorganisms that are adapted to three temperature intervals: psychrophilic (below 20 °C), mesophilic (20–40 °C, optimum 35 °C) and thermophilic (above 40 °C, optimum around 55 °C). At each temperature interval, the CH_4 production rate is a first-order process (i.e. increases exponentially at increasing temperature to a maximum value). At temperatures higher than 55–60 °C many of the microorganisms die, below 20 °C the rate of the process is significantly reduced and at temperatures below about 10 °C biogas production is insignificant. The rate of methanogenic activity is reduced more than the rate of acetogenic activity at low temperatures and therefore the rate of biomass addition must be reduced at low temperatures to avoid accumulation of acetic acid and failure of biogas production.

The volume of biogas produced varies according to the concentration of easily digestible organic components. Digestible organic components may be estimated as volatile solids, BOD_5 or COD. The production of CH_4 varies in relation to the volatile solids, BOD and COD, because metabolism of different organic components gives different volumes of biogas with different concentrations of CH_4 and CO_2 (Equations 13.23–13.25; Table 13.5). BOD and COD are used as indicators of the need for aeration in the sewage water treatment industry, where the substrate is relatively dilute. Livestock slurry and solid organic materials contain much higher concentrations of suspended solids, and volatile solids content has been shown to be a more appropriate variable for estimating the organic matter. Furthermore, the volatile solids content is an expression of the total concentration of organic components, and can be distributed into its components, which in a mass balance give the weight of the volatile solids.

Symon and Buswell (1933) presented the following general equations that can be used to calculate the maximum amount of CH_4 and CO_2 that can be produced in anaerobic fermentation of organic material:

$$C_nH_aO_bN_c + xH_2O \rightarrow \left(\frac{n}{2} + \frac{a}{8} - \frac{b}{4} - \frac{3 \cdot c}{8}\right)CH_4 + \left(\frac{n}{2} - \frac{a}{8} + \frac{b}{4} - \frac{3 \cdot c}{8}\right)CO_2 + cNH_3 \quad (13.32)$$

On knowing the composition of volatile solids in animal manure, the amount of CH_4 that theoretically can be produced can be calculated. This theoretical maximum CH_4 production is termed B_u or theoretical yield. Lipids have a high C : O ratio than carbohydrates and therefore lipid fermentation produces a biogas with a much higher CH_4 concentration than carbohydrate fermentation.

However, the organic matter often contains lignified organic components that are only slowly degradable, if at all, under anaerobic conditions. Thus, a fraction of the organic matter may never be transformed to biogas in a digester managed at the economic optimum. Therefore, the specific biogas yield B_o (1 kg^{-1} VS) is determined by anaerobic batch incubation of the biomass. The accumulated gas volume production is

Table 13.6 *Organic components making up the volatile solids content in selected animal manures and their theoretical and specific biogas yields (after Møller et al., 2004b).*

	Volatile solids (VS) composition (%)						Theoretical yield (NL	
	VS_{lipid} $(C_{57}H_{104}O_6)$	$VS_{protein}$ $(C_5H_7O_2N)$	$VS_{D,carbohydrate}$ $(C_m(H_2O)_n)$	$VS_{SD,carbohydrate}$ $(C_m(H_2O)_n)$	VS_{lignin} $(C_{10}H_{13}O_3)$	VS_{VFA} $(C_2H_4O_2)$	CH_4 kg^{-1} VS)	B_o (NL CH$_4$ kg^{-1} VS)
Pig fatteners	13.7	22.9	34.7	16.6	4.9	7.2	516	356 ($\pm$28, $n = 7$)
Dairy cows	6.9	15.0	43.4	19.1	12.1	3.6	469	148 ($\pm$41, $n = 7$)

NL: normal litres (1 atm, 0 °C).

determined during incubations of the 1 : 1 biomass and inoculum ratio (adapted to 36 °C) until the biogas production rate is low or for 60 days (Pham *et al.*, 2013). It is also recommended that measurement be continued until there is no further gas production, for example for 90 days. Typical composition, B_u and B_o values for animal manures are shown in Table 13.6.

13.3.4 Biogas Digester Design

There are many designs for digesters used in biogas production, but a description of some of the most widely used designs is provided below. The gas from these digesters has to be treated in units removing the H_2S, as otherwise the biogas will corrode the burners or the engines using the biogas. If the biogas is transferred to the gas grid, then CO_2 and water in the gas also have to be removed.

Globally, animal manure is the most used biomass feed for biogas production. The animal manure on large farms is collected as a slurry and transferred to an intermediate storage unit at the farm or the biogas plant (Figure 13.8). Intermediate storage provides a buffer that facilitates a constant rate of slurry transfer to the digester and can also be used to mix organic waste into the slurry. Alternatively, the dilute slurry runs by gravity directly from the animal house to the biogas plant, a design often used in tropical and subtropical countries. In many biogas plants the feed is a mixture of organic waste and slurry, and this slurry is heated for about 1 h at 70 °C in a heat exchanger to reduce the content of pathogens. In industrialised countries power and heat is produced and the heat in the form of warm water from electricity generation is used to heat the influent to the digester. On small farms with unheated digesters the gas is used directly for cooking, heating and lighting. The digested slurry is transported and spread on crop fields

13.3.4.1 Covered Slurry Store or Lagoon

The lagoon is a low-cost biogas digester with a simple design (Figure 13.12). It consists of a cheap PVC cover over a pit or channel containing the biomass, which is not heated and very often not mixed.

Although simple and cheap, the covered lagoon is far from ideal; it has a low biogas production rate and can only be used in areas with low groundwater levels. As most covered lagoons are not heated, they operate in the psychrophilic or low mesophilic temperature ranges. The technique is therefore best suited to a warm climate. Biogas production rates vary based on the temperature of the lagoon, which in turn is affected by seasonal fluctuations in ground temperature, air temperature and feedstock temperature.

13.3.4.2 Simple Brick Design Underground Digester

An alternative to the covered lagoon digester is the Chinese brick digester with either concrete or brick walls and a dome. These digesters are always built below ground (Figure 13.13a). The reactor is operated through

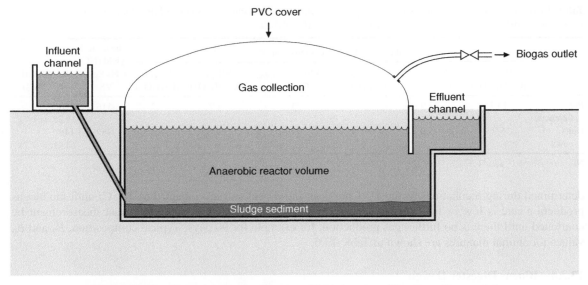

Figure 13.12   *Tunnel or lagoon digester. (© University of Southern Denmark.)*

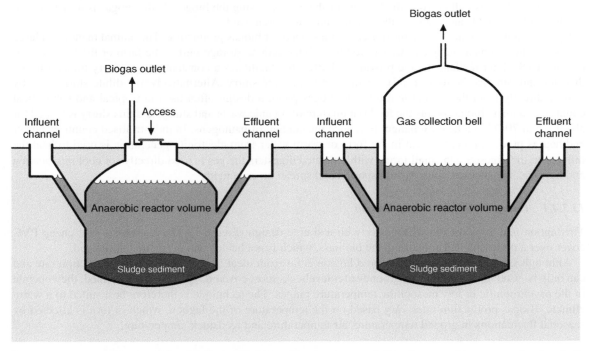

Figure 13.13   *(a) Drawing of a simple brick design biogas plant and (b) digester with a gas collection bell.*
(© University of Southern Denmark.)

biomass being transferred to the inlet when the animal house is cleaned, which is usually once per day. The reactor is not stirred, so the sedimented solids must be periodically removed. An alternative to the fixed dome digester is the similar Indian design, which differs in that the headspace is a floating gas bell which functions as a variable volume biogas reservoir (Figure 13.13b).

The dimensions of the digester depend on the expected biogas demand and available raw material, but the size is typically between 5 and 20 m^3. The digester (Figure 13.13) consists of an inlet with a mixing chamber (influent chamber), where animal manure, if necessary, can be mixed and diluted with water. From here a pipe leads directly to the digester chamber. Human excreta can be fed from a sewer either directly to the digester or to the mixing chamber, depending on the toilet construction. The digester chamber consists of a brick cylinder coated with hydraulic cement so as to form a water-tight vessel. The biogas is removed through the central pipe which can be closed off by a gas valve. The digested material is removed via the outlet chamber. In the digester shown in Figure 13.13a, the height difference between the material in the digester and that in the outlet controls the biogas pressure. If the biogas pressure becomes too high, biogas will bubble through the outlet chamber. The UN Food and Agriculture Organization (FAO, 1996) has uploaded a handbook on the internet that gives detailed information on the construction and cost of such a digester.

13.3.4.3 *Continuous Stirred Tank Reactor (CSTR)*

The CSTR is a more complex and expensive digester type to construct, but is the most common type for large-scale digestion. Typically, the feeding of effluent and the temperature are controlled within a narrow range (Figure 13.9). The retention time in a mesophilic digester (around 37 °C) is about 20–25 days. In a thermophilic reactor the process is faster and retention times are shorter, often in the range of about 10–20 days.

Stirring is needed to reduce stratification of particles in the slurry, and also to enhance contact between substrate and microorganisms. Stirring is also used to avoid layers of sediment building up at the bottom of the digester. The disadvantage of stirring is that the syntrophic clusters of microorganisms are disturbed and the process becomes more prone to H$_2$ inhibition. To overcome this, the development of gentle and yet effective stirring methods has been a consideration when developing the CSTR technology. Thus, failure of the process due to overloading can in some cases be alleviated by stopping mixing of the digester.

A CSTR design with two digesters in series has been proposed. The rationale is that the biomass is hydrolysed in the first digester, where the retention time is short and temperature may be higher than in the second digester, where CH$_4$ is produced. It has been shown that this design increases biogas production in small pilot digesters and at laboratory scale, but at full scale the effect has not been significant.

13.3.4.4 *Anaerobic Sequence Batch Reactor*

This "much-used" reactor type operates in a cyclic batch mode with four distinct phases per cycle: feed, react, settle and decant. During the feed and reaction phases the contents are mixed to ensure good contact between the microbial biomass and the substrate. In this system the solids in the substrate have a longer retention time than the dissolved substrate. The solid fraction of the substrate takes longer to degrade than the dissolved substrate, so the feeding technique ensures a high degree of digestion of the solid and the liquid phase without demanding a high reactor volume.

13.3.4.5 *Reactor with Immobilised Microorganisms*

This reactor type includes anaerobic filters, a fixed film reactor and a fluidised bed reactor, and uses inert medium with a high surface area in the reactor to promote growth, attachment and retention of bacterial cells.

These reactors facilitate short retention time and small reactor volume. The materials for fabricating the filter or film are an additional cost and are susceptible to clogging problems associated with the suspended solids. They are therefore not very suitable for anaerobic treatment of substrates with relatively high suspended solids such as animal slurries, but work effectively with wastewater. These reactors are described in detail by Hill and Bolte (1988).

13.3.4.6   Upflow Anaerobic Sludge Blanket (UASB) Reactor

The UASB reactor is a very efficient digester for fermenting waste with little or no suspended solids, such as wastewater. The UASB reactor contains no packing medium for retaining microbial cells. Instead, the organisms form dense clumps called granules, which can be several millimetres in diameter. The granules form a dense sludge bed which covers the lower part of the reactor. The influent wastewater is pumped into the bottom of the reactor, lifting the granules to form a sludge blanket, and thus ensuring good substrate to microorganism contact and efficient biogas production. The density of the granules allows them to sink back down towards the sludge bed, preventing them being washed out of the reactor with the effluent, which is pumped out from the top of the reactor. The reactor facilitates a short hydraulic retention time and thus the reactor volume can be small in relation to the volume of substrate treated. However, the continuous one-directional upflow of effluent may potentially cause problems with solids build-up in the reactor when treating effluents containing large amounts of suspended solids, such as animal slurry, which may also cause problems with granule formation. Furthermore, the UASB reactor is very sensitive to variations in the chemical and physical composition of the substrate and also changes in temperature. Thus, the technique places a great demand on the operator. For these reasons and because the UASB technology is costly, this digester cannot be recommended for biogas production on a farm.

13.3.5   Sizing Digesters

The optimal dimensions of a digester have to be modelled to ensure an economical and energy sustainable biogas production. The digester has to be dimensioned and designed in relation to the material available to feed the digesters, the end use of the biogas, climate, competence of the technicians managing the digester and so on. Below are a few algorithms that can be used for the dimensioning and design of biogas digesters.

The most important parameter determining construction costs is assessment of the correct retention time of the slurry (Table 13.7), because the volume of the biogas reactor is related to retention time. Retention time is related to temperature.

The biogas productivity as a function of temperature can be calculated from Equation (13.33), which is a modified Contois model (after Chen and Hashimoto, 1978):

$$\gamma \equiv \frac{B_o \cdot S_o}{\text{HRT}} \left(1 - \frac{K}{\text{HRT} \cdot \mu_m - 1 + K} \right) \text{ and } 20°C < T < 60°C \tag{13.33}$$

The variables and parameters are described in Table 13.7.

Knowing the amount of residues available and the specific gas yield, the reactor can be sized and designed, which also includes deciding the temperature at which the process will be run.

Heat losses (Q, W) from the digester and heating system can be estimated from first-order approximation as:

$$Q' = k_t \cdot \frac{A}{\Delta x}(T_d - T_a) \tag{13.34}$$

Table 13.7 *Definitions of retention times, parameters and variables used to assess biogas production yield.*

HRT or θ	hydraulic retention time	average time (days) the effluent is treated in the biogas digester: HRT $= V/Q$	V (m^3) is volume of the digester, and Q (m^3 day^{-1}) is the effluent flow to the digester
SRT	solid retention time	average time (days) effluent solids are retained in the digester: SRT $= (V \times C_{ds})/(Q_w \times C_w)$	C_{ds} and C_w (kg m^{-3}) are the solids concentration in the digester and in the effluent pumped into the digester, respectively
Nm^3	volume of CH$_4$ produced, calculated at 0 °C (273 K)	normal cubic metres of CH$_4$	m^3
γ	volumetric yield		Nm^3 CH$_4$ m^{-3} reactor volume day^{-1}
B_o	ultimate or specific CH$_4$ yield, measured with batch fermentation at more than 60 days and at 35 °C		Nm^3 CH$_4$ kg^{-1} (VS)
μ	maximal specific growth rate of the microorganisms, μ_m, a function of temperature and related to residues fed to the reactor	$\mu_m = 0.013 \cdot T - 0.129$	days^{-1}, T, is the temperature °C
K	a kinetic parameter depending on the rate of feed, feed composition and bacterial consortium	$K = 0.6 + 0.0206e^{(0.051 \times S_o)}$	No dimensions
S_o	concentration of organic components in feed to the reactor		kg VS m^{-3} feed

where k is the thermal conductivity (i.e. the property of a material describing its ability to conduct heat). It appears primarily in Fourier's law for heat conduction. Thermal conductivity is measured in W K^{-1} m^{-1}. Multiplied by a temperature difference $T_d - T_a$ (K) and an area (A, m^2), and divided by a thickness (Δx, m), the thermal conductivity predicts the rate of energy loss (in W) through a specific area of a material.

If the heat passes through more materials, the combined thermal conductivity is calculated as:

$$k_t = \frac{k_s \cdot k_b}{k_s + k_b} \tag{13.35}$$

where k_t is the overall heat transfer coefficient (W K^{-1} m^{-1}), k_s is the heat transfer coefficient for the insulation and k_b the heat transfer coefficient of the structural digester wall. T_d and T_a are the temperature inside the digester and the ambient temperature, respectively.

Furthermore, the efficiency and heat loss in the heat exchanger and in pumps has to be included in the calculation of the energy loss when designing a biogas plant. In heated biogas plants the energy efficiency can be improved by increasing insulation and also by decreasing the water content of the slurry, thereby reducing the volume that has to be heated.

13.3.6　Water Removal

When produced, biogas is saturated with water vapour, and this can cause corrosion and blocking of pipes and burners if not removed. The simplest way to remove water is by cooling the biogas to condense the water and collect it in a trap, although more complex methods such as activated carbon may be employed as part of a more complete biogas purification system, as described in Section 13.3.8.

13.3.7　Dihydrogen Sulfide Removal

Biogas contains small concentrations of H_2S, which is an odorous and toxic gas. This gas is oxidised to sulfite (SO_3^{2-}) upon combustion in gas burners or engines, which causes corrosion. Therefore, the H_2S has to be removed from the gas before being used as a fuel. The removal methods include devices that oxidise the sulfide either to elemental sulfur (S) or to sulfate by either bacterial or chemical means (Figure 13.9).

In the biological treatment of the gas, bacteria of the genus *Thiobacillus* transform the H_2S to S by addition of a small amount of oxygen, often as atmospheric air (2–6%) to the biogas. The process is carried out in an oxidation tower filled with packing medium, which has a large surface onto which the bacteria adhere (see Section 13.3.4.5). The bacteria are initially added to the tower by spraying slurry into the tower. During operation a slow stream of liquid circulates from the top to the bottom of the tower and S is transferred out of the system with a side stream of the liquid. The oxidation of S is a chemoautotrophic process that provides energy for microbial growth. The oxidation is a two-reaction process; in the first reaction S is the product and in the second sulfuric acid is produced. The process takes place between pH 6 and 8:

$$2H_3O^+ + 2HS^- + O_2 \rightarrow 4H_2O + 2S^0 \tag{13.36}$$

$$4H_2O + 3O_2 + S^0 \rightarrow 2HSO_4^- + 2H_3O^+ \tag{13.37}$$

The oxidation to S is a very fast process and the oxidation of S(s) to sulfuric acid is very slow. This gives the operator an opportunity to remove the S(s) so that sulfuric acid is not produced and the pH can consequently be maintained between 6 and 8. This method is mostly used in large-scale biogas plants.

Iron or copper filters are used to clean biogas in a chemical oxidation-reduction process where H_2S is removed. In these filters, iron or copper porous filter material is oxidised. The metal is reduced and the S^{2-} is oxidised. These filters can be used in large biogas plants and for treating biogas from small household digesters, where the gas is used in burners for cooking or for lighting.

13.3.8　Carbon Dioxide Removal and Upgrading Biogas for the Natural Gas Network and Transport Fuels

Removal of CO_2 is not necessary if the biogas is to be combusted for heat and/or electricity production. However, adding biogas to a natural gas network or using it as a vehicle fuel requires the application of strict criteria regarding the calorific value and the concentration of gases other than CH_4, and thus further upgrading processes are required. Typical processes for upgrading biogas are described below.

Absorption of CO_2 (and also H_2S) in a solvent such as water or poly(ethylene glycol) (PEG) is a simple yet effective upgrading technology. Using PEG has the added advantage of removing water and halogenated hydrocarbons, the latter sometimes being found in biogas from landfill sites. The process relies on the fact that CH_4 is much less soluble than the gases that require removal, and a high contact time between the solvent and the gas is ensured by bubbling gas up through a packed bed column (a reactor filled with plastic material with a high surface area), whilst the solvent is forced downwards to create a counter-flow. The solvent can be regenerated in a similar reactor, but with air bubbling through the solvent. This process can lead to some CH_4 emissions when the solvent is regenerated.

Pressure swing adsorption (PSA) uses pressure and porous molecular sieves to selectively remove gases. The binding of molecules in the pores is weak and they are released when the pressure is decreased, thus regenerating the sieve. The PSA process often employs a series of pressure vessels to reduce the energy requirement, as the release of pressure in one vessel assists the pressure increase in another.

Biogas can also be upgraded by the use of membranes. These can be gas–gas membranes operating at high pressure or gas–liquid membranes. The liquid absorbent is usually an alkaline solution such as NaOH.

13.4   Summary

Biomass can be used as feed for energy generation by incineration, thermic gasification or bio-oil, bioethanol and biogas production. A dry biomass with a dry matter content above 55% may be used for incineration and subsequent power generation or heating. A dry biomass may also be used for thermal gasification, producing syngas or bio-oils. The energy content is reduced at increasing water content of the biomass, but also at increasing content of components with high oxygen content. This is expressed by the net calorific value of a biomass being lower than the gross calorific value. Biomass is used for the production of bioethanol through fermentation with yeast. First-generation bioethanol production uses grain as feed and second-generation production uses plant residues.

For biomass with a high water content, biogas production may be the most efficient energy production method. The biomass fed to biogas digesters is very variable (e.g. animal slurries, crops, crop residues, waste from food and feed processing, and so on). Animal manures may be used for biogas production in combination with easily degradable waste from food processing (i.e. co-digestion). The addition of energy-rich materials such as food waste is needed, because animal slurry has a low organic matter content and much of the organic matter is slowly degradable. During the process, components that can inhibit biogas production need to be removed, while compounds that the microorganisms need in addition of carbon, e.g. N, P and micronutrients, must be present. Biogas plant construction depends on operating temperature and biomass characteristics. On large farms in colder climates biogas plants are often heated and power and heat are produced, whereas on small farms in the developing world the biogas may be used for cooking and heating.

References

Ahring, B.K., Sandberg, M. and Angelidaki, I. (1995) Volatile fatty acids as indicators of process imbalance in anaerobic digesters. *Appl. Microbiol. Biotechnol.*, **43**, 559–565.

Angelidaki, I. and Ahring, B.K. (1992) Effects of free long chain fatty-acids on thermophilic anaerobic digestion. *Appl. Microbiol. Biotechnol.*, **37**, 808–812.

Angelidaki, I., Ellegaard, L. and Ahring, B.K. (1993) A mathematical-model for dynamic simulation of anaerobic digestion of complex substrates – focusing on ammonia inhibition. *Biotechnol. Bioeng.*, **42**, 159–166.

Batstone, D.J., Keller, J., Angelidaki, I., Kalyuzhnyi, S.V., Pavlostathis, S.G., Rozzi, A., Sanders, W.T.M., Siegrist, H. and Vavilin, V.A. (2008) *Anaerobic Digestion Model: No. 1*, IWA Publishing, London.

Bitton, G. (2011) *Wastewater Microbiology*, 4th edn, John Wiley & Sons, Ltd, Chichester.

Blume, F., Bergmann, I., Nettmann, E., Schelle, H., Rehde, G., Mundt, K. and Klocke, M. (2010) Methanogenic population dynamics during semi-continuous biogas fermentation and acidification by overloading. *J. Appl. Microbiol.*, **109**, 441–450.

Chen, Y., Cheng, J.J. and Creamer, K.S. (2008) Inhibition of anaerobic digestion process: a review. *Bioresour. Technol.*, **99**, 4044–4064.

Chen, Y.R. and Hashimoto, A.G. (1978) Kinetics of methane fermentation, in *Proceedings of Symposium on Biotechnology in Energy Production and Conservation* (ed. C.D. Scott), John Wiley & Sons, Inc., New York, p. 269.

Demirbas, A. (2008) Biofuels sources, biofuel policy, biofuel economy and global biofuel projections. *Energy Conver. Manag.*, **49**, 2106–2116.

EU Council (2007) Brussels European Council 8/9 March 2007–Presidency Conclusions; http://www.consilium .europa.eu/ueDocs/cms_Data/docs/pressData/en/ec/93135.pdf.

FAO (1996) *Biogas Technology: A Training Manual for Extension.* Consolidated Management Services Nepal, Kathmandu, http://www.fao.org/docrep/008/ae897e/ae897e00.htm.

Gomez, X., Cuetos, M.J., Garcia, A.I. and Moran, A. (2007) An evaluation of stability by thermogravimetric analysis of digestate obtained from different biowastes. *J. Hazard. Mater.*, **149**, 97–105.

Hill, D.J. and Bolte, J.P. (1988) Synthetic fixed media reactor performance treating screened swine waste liquids. *Trans. ASAE*, **31**, 1525–1531.

Hjelmar, O., Wahlström, M., Andersson, M.T., Laine-YlijokI, J., Wadstein, E. and Rihm, T. (2009) Treatment methods for waste to be landfilled. *TemaNord 2009:583*, Nordic Council of Ministers, Copenhagen.

Hobson, P.N. and Shaw, B.G. (1974) Bacterial population of piggery-waste anaerobic digesters. *Water Res.*, **8**, 507–716.

IEA (2010) *World Energy Outlook*, IEA, Paris; http://data.iea.org.

Lisk, D.J. (1988) Environmental implications of incineration of municipal solid waste and ash disposal. *Sci. Total Environ.*, **74**, 39–66.

Masse, D.I., Droste, R.L., Kennedy, K.J., Patni, N.K. and Munroe, J.A. (1997) Potential for the psychrophilic anaerobic treatment of swine manure using a sequencing batch reactor. *Can. Agric. Eng.*, **39**, 25–33.

Møller, H.B., Sommer, S.G. and Ahring, B.K. (2004a) Biological degradation and greenhouse gas emissions during pre-storage of liquid animal manure. *J. Environ. Qual.*, **33**, 27–36.

Møller, H.B., Sommer, S.G. and Ahring, B.K. (2004b) Methane productivity of manure, straw and solid fractions of manure. *Biomass Bioenergy*, **26**, 485–495.

Nielsen, H.B., Uellendahl, H. and Ahring, B.K. (2007) Regulation and optimization of the biogas process: propionate as a key parameter. *Biomass Bioenergy*, **31**, 820–830.

Pham, C.H., Triolo, J.M., Cu, T.T.T., Pedersen, L. and Sommer, S.G. (2013) Validation and recommendation of methods to measure biogas production potential of animal manure. *Aust. J. Anim. Sci.*, in press.

PHYLLIS (2012) PHYLLIS – the composition of biomass and waste; downloaded 7 December 2012 from http://www.ecn.nl/phyllis/.

Pilatowsky, I., Romero, R.J., Isaza, C.A., Gamboa, S.A., Sebastian, P.J. and Rivera, W. (2011) *Cogeneration Fuel Cell-Sorption Air Conditioning Systems.* Springer, Berlin.

Rivard, C.J., Vinzant, T.B., Adney, W.S., Grohmann, K. and Himmel, M.E. (1990) Anaerobic digestibility of two processed municipal-solid-waste materials. *Biomass*, **23**, 201–214.

Rodriguez, J., Ruiz, G., Molina, F., Roca, E. and Lame, J.M. (2006) A hydrogen-based variable-gain controller for anaerobic digestion processes. *Water Sci. Technol.*, **54**, 57–62.

Sheppard, S. and Ridgen, C. (2005) Rapid non-equilibrium decompression (RND) 'Bug Buster' technology, presented at the *10th European Biosolids and Biowaste Conference*, Wakefield.

Siegert, I. and Banks, C. (2005) The effect of volatile fatty acid additions on the anaerobic digestion of cellulose and glucose in batch reactors. *Proc. Biochem.*, **40**, 3412–3418.

Speece, R.E. (2008) *Anaerobic Biotechnology and Odour/Corrosion Control – For Municipalities and Industries*, Archae Press, Nashville, TN.

Symons, G.E. and Bushwell, A.M. (1933) The methane fermentation of carbohydrate. *J. Am. Chem. Soc.*, **55**, 2028–39.

Thomsen, K. and Larsen, M.J. (2012) Biomasse til el- og varmeproduktion, in *Biotek 2 – Anvendt bioteknologi* (eds K. Overgaard and S.G. Sommer), L&R Uddannelse, Copenhagen, pp. 50–74.

Thygesen, A.M., Wernberg, O., Skou, E. and Sommer, S.G. (2011) Effect of incineration temperature on phosphorus availability in bio-ash from manure, *Environ. Technol.*, **32**, 633–638.

Van Loo, S. and Koppejan, J. (2008) *The Handbook of Biomass Combustion and Co-firing*, Earthscan, London.

Ward, A.J., Hobbs, P.J., Holliman, P.J. and Jones, D.L. (2008) Optimisation of the anaerobic digestion of agricultural resources. *Bioresour. Technol.*, **99**, 7928–7940.

Wenzel, H. (2006) Energi og mijø – hvordan får vi mest for pengene [Energy and environment – how to optimise the benefits from investments], presentation at *Landskongressen*, Herning.

World Energy Council (2010) *Survey of Energy Resources*, WEC, London; http://www.worldenergy.org/documents.

Yadvika, S., Sreekrishnan, T.R., Kohli, S. and Rana, V. (2004) Enhancement of biogas production from solid substrates using different techniques – a review. *Bioresour. Technol.*, **95**, 1–10.

Zhang, R.H., Tao, J. and Dugba, P.N. (2000) Evaluation of two-stage anaerobic sequencing batch reactor systems for animal wastewater treatment, *Trans. ASAE*, **43**, 1795–1801.

14

Animal Manure Residue Upgrading and Nutrient Recovery in Biofertilisers

Lars S. Jensen

Department of Plant and Environmental Sciences, University of Copenhagen, Denmark

14.1 Introduction

Livestock production is developing dramatically on a global scale, with trends towards increasing concentration on large specialist production units to improve profitability (Steinfeld *et al.*, 2006). These changes in production systems have resulted in increased pollution of air, aquifers, surface waters and soil. A major concern is also uncoupling of the sites of animal feed production and animal production, through the (economic) driving forces of specialisation, intensification and up-scaling. This leads to surplus amounts of animal manure in areas where animals are produced. As a consequence, an increasing number of livestock farms have insufficient land for efficient use of manure nutrients.

There is a strong relationship between livestock density and nitrogen (N) surplus at great risk of being lost to the environment (Velthof *et al.*, 2009). Regional variations in livestock density are large, with most livestock manures being produced in Europe, China, India, and particular regions of North America and Latin America (Steinfeld *et al.*, 2006). In the European Union, the maximum amount of manure to be applied to agricultural land is regulated, through the Nitrates Directive and the IPPC Directive, in order to avoid manure overloading to land and to force redistribution of livestock and/or livestock manure to regions with less intensity. As a consequence, there is an increasing need for manure processing in regions with intensive livestock production, and for the recovery and utilisation of valuable compounds from the manure.

Animal manures and slurries contribute to the fertility of the soil by adding organic matter (carbon (C)) and plant nutrients to the fields. The combined effect is hereafter defined as the biofertiliser effect. The total amounts of N, phosphorus (P) and potassium (K) in the livestock manures produced annually exceed the annual production of synthetic N, P and K fertilisers in the world. If used appropriately, manure can

Animal Manure Recycling: Treatment and Management, First Edition. Edited by Sven G. Sommer, Morten L. Christensen,
Thomas Schmidt and Lars S. Jensen.
© 2013 John Wiley & Sons, Ltd. Published 2013 by John Wiley & Sons, Ltd.

replace large amounts of mineral fertilisers (Chapter 15), indicating the high economic value of manure as a biofertiliser, while at the same time contributing to the maintenance of soil organic C stocks (Schröder, 2005).

However, at present, improper management and utilisation of manure results in waste of plant nutrients, which are a limited resource, and will therefore threaten the global feed and food supply. For example, P is a limited resource, with the mineable phosphate-rich rocks used for P fertiliser production projected to be exhausted within the next few hundred years. In addition, manure contains large amounts of organic material that, with the right technologies, can be used for energy production (Chapter 13). However, removing manure organic matter for energy recovery may jeopardise the maintenance of soil organic matter and this can potentially threaten soil fertility to the point of desertification, especially in sub-tropical and tropical regions. Hence, bioenergy technologies capable of recovering energy from manures, while at the same time supplying the more recalcitrant fraction of organic matter to the soil, should be a priority. Examples are anaerobic digestion, producing biogas and digestate with the more biologically recalcitrant organic C, or thermal gasification/pyrolysis producing syngas and more or less biologically inert biochar (for further details, see Section 14.6).

New and improved technologies for recycling nutrients and organic matter from animal manure to the soil (e.g. by utilising manures as a raw material for biofertiliser production) are therefore greatly needed (Martinez *et al.*, 2009). This is currently the focus of substantial public and private research and innovation efforts in countries with intensive livestock production around the globe (Oenema *et al.*, 2012).

The preceding chapters of this book present many of the processes for manure pollution prevention, management and energy recovery that are currently available or under development. This chapter examines the recovery and upgrading of organic matter and nutrients from manures in order to produce new valuable products, mainly biofertilisers and soil amendments, and describes their function and effectiveness.

14.2 Manure Upgrading Options

A number of different routes are available for recovering and upgrading nutrients and organic matter from different manure types (Figure 14.1). For slurries or other liquid manures such as digestate from anaerobic digestion of manure and other biowaste, basically all treatment steps start with mechanical separation into a relatively organic N- and P-rich solid fraction, and a liquid fraction, with low P, but relatively high mineral N and K contents.

Mechanical separation is a low-cost treatment strategy that increases the flexibility of manure disposal. Different separation techniques are available (Chapter 7), but these differ in efficiency of separating the solids and colloid fraction from the liquids. The higher the efficiency, the higher the costs, commonly in the range €2–4 m^{-3} (Schoumans *et al.*, 2010). However, when implementation of this treatment spreads more widely in practice, improvements in performance and economy of scale can be expected, lowering costs. Furthermore, biogas production (anaerobic digestion) will often serve as the "door-opener" for the introduction of other manure processing technologies. Many of the manure processing technologies are complementary to anaerobic digestion, either as a post-treatment, which helps to convert the digestate into products with properties more favourable for recovery, or as a pre-treatment to enhance biogas production by increasing the volatile solids concentration of digester inputs by adding separated slurry solids (Foged *et al.*, 2011a). Peters (2010) studied chemical and biochemical variations in solid fractions from animal slurry separation and developed the suggestions shown in Table 14.1 for solid fraction behaviour and improved utilisation.

The solid fraction typically has an initial dry matter content ranging from 16–36%, with most around 25–30% (Foged *et al.*, 2011b), and consists of a significant proportion of relatively degradable organic material making it unstable during storage, high concentrations of P and organic N, but also a significant

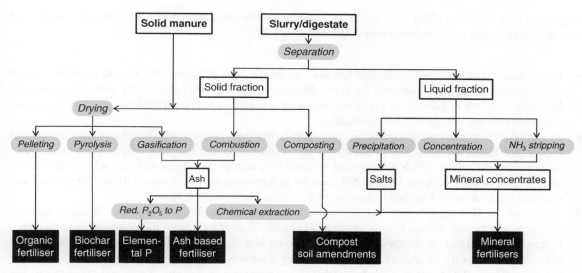

Figure 14.1   *Schematic overview of possible animal manure residue treatment steps in order to upgrade recovered nutrients in biofertilisers. (© University of Copenhagen.)*

Table 14.1   *Main characteristics and behaviour during various uses of the solid fraction from pig slurry separation, as affected by two main groups of separation technology (adapted from Peters, 2010).*

	Group	
	Separation techniques group A	Separation techniques group B
Solid fraction from	chemically pre-treated (flocculants) and mechanically separated slurries decanter centrifuged slurries and digestates from anaerobic digestion	mechanically separated slurries (screw-press based with or without additional sieving techniques)
Main characteristics	low (chemically pre-treated) to medium (centrifuge) dry matter content high ash content high N and P concentration large proportion of acid-soluble P high amount of small (less than 250 μm) particles large proportion of neutral detergent soluble C wet, dark, heavy physical structure	highest dry matter content low ash content low N and P concentration large proportion of water-soluble P high amount of large (greater than 250 μm) particles large proportion of hemicellulose- and cellulose-bound C dry, light brown and light physical structure
Behaviour during utilisation	potential for net N mineralisation in soil high mineral fertiliser equivalent value suitable for composting with high weight loss variable CH_4 production potential – tendency for the highest CH_4 potential	continuous N immobilisation in soil low mineral fertiliser equivalent value suitable for composting with low weight loss variable CH_4 production potential

proportion of ammonium (NH_4^+). It therefore resembles other solid manures (e.g. farmyard manure) and several different options for utilisation exist (Flotats *et al.*, 2011):

- *Direct land application* as an organic fertiliser on arable land (Chapter 15). Due to its high nutrient concentration it can be transported for some distance before the costs exceed the value as a fertiliser; however, it is often difficult to avoid emissions of ammonia (NH_3), greenhouse gases (GHGs) and malodours during storage and land spreading.
- *Composting* in order to create a stable and odourless biofertiliser product, with lower moisture content and most of the initial nutrients, free of pathogens and seeds. Composting significantly reduces mass (water evaporation and volatile solids decomposition) and hence transport costs; however, it is difficult to avoid some loss of manure N in the form of NH_3 and the process may also emit some GHGs. Composting and compost use is described further in Section 14.3.
- *Drying and pelletising* to create a more stable and odourless biofertiliser product. Drying is energy-intensive and thereby relatively expensive, unless excess energy (e.g. from the power plant engine on a biogas plant) is freely or cheaply available. Ammonia loss is inevitable in the process, unless exhaust filtering and recovery is applied. Drying is usually combined with a pelletising process to facilitate handling. The pelleted material can be marketed as an organic matter and P-rich soil amendment. For further details, see Section 14.4.
- *Combustion or thermal gasification* in a power plant to generate a net energy output for heat and/or electricity production (Chapter 13), but also producing ash residuals. These contain the non-volatile nutrients, concentrated relative to the solids, and can be used as an ash-based, P- and K-rich soil amendment or biofertiliser. However, nutrient availability in the ash is generally lower than for the raw manure. For further details, see Section 14.5.
- *Ash extraction or conversion* may be used in order to upgrade the nutrients further, using various acids in order to produce mineral fertilisers, or converting ash P into a pure elemental form using a thermochemical process, to be used in the food and feed industries (see Section 14.5.1).
- *Pyrolysis* (volatilisation of organics from the biomass heated without oxygen supply), producing a solid biochar, as well as bio-oil and syngas to be used as fuels. Like ash, the biochar contains the non-volatile nutrients and carbonised organic matter. Nutrient availability varies, but is often higher than for ash but lower than for raw manure. The biochar organics are very recalcitrant to biological decay and have a very large specific surface area, potentially charged; hence. biochar may be used for soil amendment, ameliorating soil fertility positively. For further details, see Section 14.6.

The liquid fraction can be utilised in various ways:

- *Land-applied locally* as a potentially very effective N and K fertiliser due to its high ratio of NH_4^+ to organic N and content of mineral K (Chapter 15).
- *Concentrated*, either by NH_3 stripping and absorption, ultrafiltration or reverse osmosis (Chapter 7) into mineral concentrates and water for discharge, eventually after further cleaning at a wastewater treatment plant.
- *Precipitation* of magnesium ammonium phosphate (struvite) or calcium phosphates (apatites) may also be applied (for theory, see Chapters 4 and 7).
- Both precipitates and mineral concentrates can be applied to land directly or used for mineral fertiliser production, although these are relatively costly processes (Velthof, 2011). For further details see Section 14.7.

Each of these treatment pathways and resulting products has certain advantages and disadvantages, and the net economic costs or profits differ greatly. A number of factors must be considered when prioritising the options (Foged *et al.*, 2011d):

- The primary aim should be nutrient recycling, mainly N and P; N is consumed in the largest quantities, is expensive and has impacts on energy consumption and GHG emissions, while P is a scarce and non-renewable resource, with a high price peak in 2008 and recent gradual price increases.
- Splitting N and P into different fractions is generally beneficial, as this enables balanced fertilisation in accordance with the needs of many crops.
- The technology or combination of technologies applied should preferably also produce energy or consume relatively little energy, so net energy production should be taken into account for both environmental and economic reasons.
- The quality of end-products and byproducts is assessed differently depending on the perspective of the user. For instance, an organic farmer will not appreciate combustion ash, where the majority of the N has been lost, while this is seen as a strong advantage for a poultry farmer in a livestock-dense region. Organic matter is often not valued by conventional crop production farmers, but highly so by private garden owners, vineyards and the like. Some fertiliser companies favour products with higher concentrations of P as interesting alternatives to conventional raw materials (rock minerals) with lower concentrations.
- Local solutions should be preferred, avoiding too high transport cost and impacts; regional or more central solutions are therefore only justified if the economy of scale via higher efficiency outweighs the negative impacts of transporting the manure to a common facility.

A considerable increase in manure processing has taken place in EU member states in the past 15 years, triggered by an increasing political focus on reduction of GHG emissions, recycling of waste and biosecurity aspects. This has resulted in tightened enforcement of environmental provisions and livestock producers are experiencing this as an additional driver for structural development towards increased livestock production unit sizes in most EU member states. Particularly in intensive livestock areas of the European Union, such as the Netherlands, Belgium (Flanders), France (Brittany), Denmark and Italy (northern part), manure processing is encouraged, often with economic incentives, in order to deal with surpluses (i.e. occurrence of more than 170 kg manure N ha^{-1} produced on livestock farms – the threshold for harmony between animal production and land base for manure spreading set in the EU Nitrates Directive; see Chapter 3).

The dominant types of manure processing in Europe are (in decreasing order of importance): (i) anaerobic digestion for biogas production, (ii) separation of slurry with mechanical, chemical and other technologies, and (iii) treatment of the liquid and solid fractions from separation. In total, around 7.8% of the total livestock manure production in the European Union is being processed, equal to 108 million tons, containing 556 000 tons of N and 139 000 tons of P. The largest proportion of livestock manure production is being processed in Greece, Germany and Italy, with 35%, 15% and 7%, respectively, of the national manure production being processed (Foged *et al.*, 2011a).

The end-product or byproduct produced from manure in by far the largest amounts in the European Union is "digestate" from anaerobic digestion in biogas plants, with more than 88 million tons of manure processed. The manure is co-digested with a substantial amount (around 60 million tons) of end-products and byproducts from other processes and non-livestock manure biomasses (Foged *et al.*, 2011b), which not only adds to the volume, but also increases the total amount of nutrients in the digestate significantly. Digestate is in principle relevant as a tradable product, but only in a local market as its rather voluminous and dilute properties mean that with longer distances, transport costs normally outweigh the fertiliser value.

Currently, only three manure end-product and byproduct categories attract the attention of the open market, namely "manure compost", "separation solids" and "dried manure pellets". This is due to their suitability for marketing and to the amounts currently produced (Foged *et al.*, 2011b). In the European Union, composting is the dominant market-orientated processing method for animal manure, with a very large number (more than 1000) of farm-scale and centralised medium-scale composting facilities (defined as processing less than 50 000 ton year^{-1}) in operation in Europe (Foged *et al.*, 2011a). A substantial section of this chapter is thus devoted to the composting process and use of compost.

Another two products, namely "ash and char" and "mineral concentrates", are in principle also relevant for marketing, but they are currently only produced in marginal amounts. However, with further development of manure processing they could become most interesting, as they can be used for the production of mineral fertiliser or mineral feedstuffs, or in some cases applied directly as fertilisers, as has been proposed for struvite, for example (Cabeza *et al.*, 2011). New disposal routes, especially for poultry litter, have been emerging in the past decade. In particular, combustion or fluidised bed gasification appears promising, with net energy output and ash as a valuable P fertiliser product (Kelleher *et al.*, 2002).

14.3   Composting of Manures

Composting involves thermophilic, aerobic degradation of the manure organic matter, followed by a curing phase where the temperature slowly decreases and complex organic macromolecules are produced (fulvic and humic acids). Aerobic composting of manure is most commonly practised to overcome some of the problems associated with management of fresh manure. It reduces mass (water evaporation and volatile solids decomposition) and yields a stabilised product that is cheaper to transport and easier to handle and apply. Controlled composting allows the safe storage and transport of the final product, adds value to the product because compost is a more concentrated and uniform product than the manure, and permits easy spreading and thus uniform distribution in the soil. Composting also eliminates animal and human pathogens and weed seeds, and reduces the risks of odour and gaseous emissions after application. The compost can also be used as a fertiliser for pots and as a basis for soil-less substrates (Bernal *et al.*, 2009). However, appropriate methods of composting manures with suitable amendments are crucial for reducing nutrient losses (e.g. in the form of NH_3), and for minimising environmental load and GHG emissions during production. Furthermore, high-quality compost with an improved economic value should be produced in order to balance the cost of composting. However, composting also has a number of drawbacks, as summarised in Table 14.2.

14.3.1   Basic Composting Concepts

Composting is the spontaneous biological decomposition that occurs with degradable organic materials in a predominantly aerobic environment. Compared with natural aerobic decay, composting is a process managed deliberately to speed up the decay, while retaining nutrients in mainly immobilised, organic form and stabilising the more recalcitrant organic compounds. The composting process leads to the production of carbon dioxide (CO_2), water, minerals and biologically stabilised (humified) organic matter. The latter, including part of the water and minerals, is called compost, and should be free of weed seeds, phytotoxicity and pathogenic microorganisms (see Chapter 6.3.3 on hygenisation by composting). The process is exothermic and energy is released, which generates a temperature increase in the mass, as long as heat is not lost to the surroundings at a higher rate than heat evolution.

Composting is typically characterised by a short initial mesophilic phase, followed by an intense thermophilic phase, a cooling phase and then a prolonged curing or maturation phase (see Text Box – Basic 14.1 for some definitions). More detailed reviews of manure and organic waste composting can also be found in Bernal *et al.* (2009) and Stentiford and de Bertoldi (2011).

Table 14.2   *Advantages and disadvantages of composting animal manures, compared with direct application.*

Advantages	Disadvantages

Easier handling:
    reduction in volume and moisture
    ease of storage, transport and use
	Cost of installation and management:
	    large areas for operation and storage required
	    need for a bulking agent
	    energy consumed for mixing/aeration
	    relatively little added value, costly compared
	        with direct application

Eliminates unwanted organisms in waste:
    animal, plant and human pathogens
    weed seeds killed off
	Few manures pose real risk from pathogens:
	    kill-off during solids storage and after land
	        application eliminates pathogen transfer
	    existing restrictions on manure application to
	        edible crops are sufficient

A stable and safe soil amendment:
    no odours – material is oxidised
    partial drying – low solubility, no leachate
    no phytotoxins and low soluble salts – proper
        germination of plants
    soil C sequestration and low soil emissions
	Gaseous losses occur during composting with
	impacts on the environment:
	    odours, CH_4 and N_2O may also be emitted,
	        especially from moist, easily decomposable manure
	    significant NH_3 losses may occur, a risk
	        especially from N-rich waste
	    low short-term N and P availability, reduces MFE[a]

Improves soil quality/fertility:
    active soil microbial life
    nutrients slowly but steadily available, more
        predictable MFE[a]
    nutrients less susceptible to loss
    improved aggregation
    higher water-holding capacity
    prevents erosion
    better root development
    easier tillage, improved soil tilth
    plant disease suppression
	No recovery of energy:
	    heat is normally lost, difficult to recover
	    better to use technologies which recover both
	        energy and nutrients (e.g. biogas)

[a]MFE: mineral fertiliser equivalent value (see Chapter 15 for definition).

Text Box – Basic 14.1   Some definitions of compost phases and terms

Composting: A bioxidative process, involving microbial transformation, mineralisation and stabilisation of heterogeneous organic matter under aerobic conditions and in the solid state. Composting consists of a bioxidative phase and a maturing phase.

Compost: The stabilised and sanitised product of composting, which has undergone an initial, rapid stage of decomposition and a stabilisation stage. It is beneficial to plant growth and has certain humic characteristics, beneficial for soil fertility.

Bioxidative phase: This phase develops in three steps, characterised by the temperature development, see also Figure 14.2:

  ○   *Initial mesophilic phase*: The first phase of the composting process is mesophilic (below 45 °C) and starts the aerobic decomposition of easily degradable organic matter, where mesophilic bacteria and fungi degrade simple compounds such as sugars, amino acids, proteins and so on. This rapid decay of material releases a great quantity of energy in the form of heat, which enhances the mass temperature and the degradation rate of the organic waste. Within a few days, this leads to the thermophilic phase.

○ *Thermophilic phase*: In this phase, thermophilic microorganisms (mainly bacteria) degrade fats, cellulose, hemicellulose and some lignin, and the maximum degradation of the organic matter occurs, together with the destruction of pathogens. Without control, the temperature can easily reach and exceed 70 °C. The main positive effect of operating at such high temperature is reduction of pathogenic agents present in the waste. In controlled composting processes this phase is limited in terms of temperature and exposure time (degrees and days) to obtain a balance between high stabilisation rates and good sanitisation, often to satisfy local legislation regarding sanitisation conditions.

○ *Cooling phase*: This phase is characterised by a decrease in temperature due to a reduction in microbial activity associated with the depletion of degradable organic substrates. The composting mass is re-colonised by mesophilic microorganisms, especially Actinomycetes, which are able to degrade the remaining sugars, cellulose and hemicellulose. Sometimes this causes a secondary temperature peak, which is usually short-lived (asterisk in Figure 14.2).

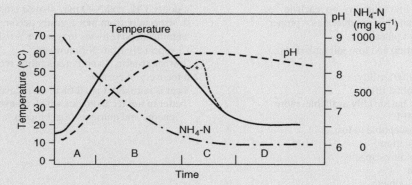

Figure 14.2 *Simplified temperature, pH and NH_4^+ changes during aerobic composting. Composting phases: (A) mesophilic, (B) thermophilic, (C) cooling and (D) maturation. *Secondary heating peak due to Actinomycete recolonisation and degradation of recalcitrant C can sometimes be observed. (© University of Copenhagen.)*

Maturation phase: The maturing phase, also called the curing phase, includes not only the mineralisation of slowly degradable molecules, but also the stabilisation and humification of lignocellulosic compounds. This phase can last some weeks to months, depending on the composition of the starting material. During microbial transformation intermediate metabolites are produced, which can make the composting material phytotoxic. These phytotoxic compounds are completely degraded at the end of the maturation process; thereafter, the final product becomes beneficial to plant growth. The composting process should ideally be stopped when the phytotoxicity is over. If the process goes on too long, there is an excessive loss of organic matter, reducing the beneficial impacts of the final product.

Composting is a discontinuous (batch) process resulting from sequential development of different microbial communities, mainly bacteria, Actinomycetes and fungi. These are normally present in the starting material; an inoculum is only needed when the starting material is deficient in microorganisms. Higher organisms, such as microarthropods and compost worms (*Eisenia foetida*), may also colonise and proliferate in maturing compost that has cooled to ambient temperature or may be used in specialised vermi-composting systems where they are added after the thermophilic phase. They mainly play a role in physically shredding and mixing the constituents, promoting microbial decay.

14.3.2 Control of the Composting Process

Composting optimisation involves the definition of adequate initial substrate conditions that must be controlled and maintained as composting progresses, since they determine the optimal conditions for microbial development and organic matter degradation. The factors affecting the composting process can be divided into two groups: those depending on the formulation of the composting mix, such as nutrient balance, C : N ratio, pH, particle size, porosity and moisture, and those depending on process management, such as oxygen concentration, temperature and water content (Bernal *et al.*, 2009).

Formulation of the composting mix includes:

- *Nutritional balance.* Mainly defined by the C and N content of the inputs. An adequate C : N ratio for composting is in the range 25–35, because microorganisms require around 30 units of C per unit of N. A high C : N ratio slows the process due to N limitation, but with a low C : N ratio there is an excess of N relative to degradable C and inorganic N is produced in excess, and can be lost by NH_3 volatilisation or by leaching from the composting mass. The low C : N ratio of manures can be balanced by adding a high C : N ratio bulking agent to provide degradable organic C. Similarly, optimal composting requires a C : P ratio of less than 120–240, but with the high P content of most manures, this is usually not limiting for the process.
- *pH.* A pH of 6.7–9.0 supports good microbial activity during composting and pH is usually not a key factor for composting, since most materials are within this pH range. However, pH is very relevant for controlling N losses by NH_3 volatilisation, which can be particularly high at pH > 7.5. Due to initial release of organic acids, pH may temporarily decrease, but subsequent oxidation and production of NH_3 will raise the pH, which may reach values of above 8.5 (Figure 14.2).
- *Microorganisms.* The microbial community involved in composting develops according to the temperature of the mass (Figure 14.2). Bacteria predominate early in composting, fungi are present throughout the process, but are not active at temperatures above 60 °C and only predominate at the lower moisture range (see below). Actinomycetes predominate during stabilisation and curing, and together with fungi are able to degrade resistant polymers.
- *Porosity.* Exerts a great influence on composting performance, since appropriate conditions of the physical environment for air distribution must be maintained during the process. Too high a porosity (above 50%) may cause too low a temperature due to the energy lost exceeding the heat produced. However, too low a porosity results in anaerobic conditions, GHGs and odour generation. The percentage air-filled pore space of composting piles should be in the range 35–50%.
- *Particle size and distribution.* Controls the porosity together with the shape, packing and moisture content of the composting mass. Balancing the surface area for growth of microorganisms and the maintenance of adequate porosity for aeration is critical. The larger the particle size, the lower the surface area-to-mass ratio, which impedes accessibility for microorganisms. However, too small particles can compact the mass, reducing the porosity too much. Usually a mixture of larger (bulking agent) and smaller particles is optimal.

Process management factors include:

- *Aeration.* Key factor; proper aeration controls the temperature, removes excess moisture and CO_2, and supplies O_2 for the biological processes. Aeration depends on porosity, but also on management factors such as ventilation and mechanical turning of the compost. Optimum O_2 concentration is between 10% and 20%, and should not decrease below 5%.
- *Moisture.* The optimum water content for composting varies with the waste to be composted, but generally the mixture should be in the range 45–60%. At higher moisture content, O_2 supply is inhibited and partly

anaerobic conditions prevail. Much water can evaporate, especially during the thermophilic phase, so rewetting may be required to maintain the optimum moisture for microbial activity, especially in warm, dry climates.

- *Temperature.* The temperature pattern shows the microbial activity and the occurrence of the composting process (Figure 14.2). The optimum temperature range for composting is 40–65 °C. If the temperature exceeds this (up to 70–72 °C) for an extended period, even the thermophilic decomposers will be killed off. Some regulation of the temperature is required for controlled composting, and may be achieved by controlling the size and shape of the composting mass, by promoting cooling and favourable temperature redistribution by turning operations or by applying systems with feedback-controlled ventilation, actively removing heat.

For the most commonly used windrow composting method, an illustration of the temperature distribution in a cross-section of the compost pile can be seen in Chapter 6 (Figure 6.2). There may be large temperature gradients within the pile, and hence regular turning of the pile or windrow is necessary to ensure that all material has been subjected to thermophilic conditions and to maintain aeration (Brito *et al.*, 2012). Alternatively, in a static, closed system, forced ventilation may be used to ensure aeration and insulation is then crucial to ensure exposure to thermophilic temperature for the entire mass. A third alternative is the in-vessel system, typically a slowly rotating horizontal drum that combines forced aeration with constant turning, and operates in a continuous flow-through mode.

Animal manure properties typically make it unsuitable for composting alone, owing to excess moisture, low porosity, high N concentration and thus a low C : N ratio, and in some cases a high pH value. In order to obtain quality compost with minimum emissions, careful formulation of the input mix and management of the composting conditions are required to reduce composting time and costs, and enhance the quality of the end-product. Addition of a bulking agent to the manure should optimise substrate properties such as air space, moisture content, C : N ratio, particle density, pH and mechanical structure, positively affecting the decomposition rate. A number of agricultural and forestry byproducts with high lignocellulosic content are commonly used as bulking agents: cereal straw, hay, other crop residues, tree leaves and twigs (green waste), bark, wood shavings, sawdust and paperwaste. All these have relatively low moisture, high organic C content and high C : N ratio (above 50–80) which can balance the properties of manures, see Example 14.1 on how to calculate optimal mixing ratios.

Example 14.1

Determine the ratio of straw to mix with separated pig slurry solid fraction in order to obtain parameters of the mixture (C : N and moisture) that lie within the optimum range for composting (FW = fresh weight, TS = total solids, VS = volatile solids).

- Solids properties: moisture 80% (TS 20%), VS 90% of TS, C 50% of VS, total-N 6% of VS.
- Straw properties: moisture 10% (TS 90%), VS 95% of TS, C 50% of VS, total-N 0.5% of VS.

Calculation of properties for 1 ton of compost mixture (kg ton^{-1} mix):

	FW	TS	Moisture	VS	Ash	C	N	C : N
Solids	**800**	240	560	*216*	24	108	13.0	8
Straw	**200**	180	20	*171*	9	86	0.9	100
Mix	1000	420	**580**	*387*	33	194	13.8	**14**

As can be seen from the calculation, a mixture with an 8 : 2 solids : straw ratio fulfils the criterion of lowering the moisture content of the manure from 80% to just below the optimum threshold of 60%. However, this yields a C : N ratio of 14 for the mix, which is too low (optimum C : N 25–35). This can only be raised by adding more straw, which increases the costs and lowers the moisture content below optimum, necessitating water addition. Thus, finding the right mixing ratio often requires balancing the different criteria or utilising several different bulking agents.

During the active phase of the composting process labile organic compounds (e.g. simple carbohydrates, fats and amino acids) are degraded quickly and the C : N ratio of the substrate mix decreases. As available C sources are depleted, the degradation rate decreases gradually as composting progresses with decomposition of more resistant organic substrates such as cellulose, hemicellulose and lignin. Synthesis reactions of new complex and polymerised organic compounds (humification) prevail over mineralisation during the maturation phase, leading to a stabilised compost end-product. Organic C losses during the entire composting process can reach 52% for poultry manure, 67% for cattle manure and 72% for pig manure (Bernal *et al.*, 2009), but the final amount depends greatly on the composting system and conditions, characteristics of the bulking agent added (recalcitrant materials such as wood chips resulting in lower C loss), and the environmental conditions of the season (winter or summer). Even with appropriate initial structure and porosity, partly anoxic zones in the compost may develop, particularly during the initial thermophilic phase, where the oxygen consumption is very high, the degradation of organic matter produces significant amounts of moisture and a gradual compaction of the composting mix occurs. In this phase some of the C degradation may result in emission of methane (CH_4), and although this represents a small fraction of total C lost – estimates range from 0.01% (Osada *et al.*, 2001) to 0.2% (Sommer and Møller, 2000) of compost mix total C emitted as CH_4 – it may still contribute significantly to the global warming impact of composting.

As mentioned earlier, composting of animal manure involves a high risk of N losses via NH_3 volatilisation, with losses of up to more than half of manure N (Tiquia and Tam, 2000). This is due to the fact that a significant proportion of manure N is already initially present as NH_4^+ or readily mineralised and during the thermophilic phase the pH increases (Figure 14.2, phase B), concomitant with a decrease in the pK_a value of the $NH_4^+ \leftrightarrow NH_3 + H^+$ chemical equilibrium due to the high temperature (for explanation, see Chapter 4). This results in a very high proportion of NH_3, which can potentially be lost with the convective flow of air through the composting mass. Several strategies can be applied to minimise this risk; elemental sulfur (S^0) has been used as an amendment for avoiding excessively high pH values, limiting NH_3 volatilisation (Mari *et al.*, 2005), while it also enhances the dissolution of P in compost and enriches the compost with S (Mahimairaja *et al.*, 1995). Peat, natural zeolite and alum addition may also reduce the N losses by adsorption of NH_4^+ (Kithome *et al.*, 1999; Delaune *et al.*, 2004; Venglovsky *et al.*, 2005; Guo and Song, 2009). Bulking agents with readily decomposable C (e.g. straw and other high C : N crop residues) can also be effective in reducing N losses (Bernal *et al.*, 2009), as they cause rapid immobilisation of NH_4^+-N during the mesophilic and initial thermophilic phase. Finally, the risk of NH_3 losses also depends on the transport of NH_3 to the atmosphere, and hence forced aeration and especially frequent mechanical turning of the compost promote NH_3 losses (Bernal *et al.*, 2009). Achieving a balance between sufficient aeration and minimising NH_3 losses is therefore a delicate matter. During the curing phase, small amounts of nitrous oxide (N_2O) may also be emitted as a byproduct from gradual nitrification of the NH_4^+ built up during the bio-oxidative phase. As for CH_4, although an insignificant proportion of N is lost as N_2O, this may contribute significantly to global warming potential, due to the high CO_2-equivalence value of N_2O.

14.3.3 Biofertiliser Value of Manure-Based Composts in Agriculture

As a result of the dry weight and C losses from the material during composting, the concentration of mineral elements increases, provided that losses of the volatile N compounds are minimal. However, the high proportion of NH_4^+-N in the manure has been immobilised in the composting process and needs to be mineralised again in order for the compost to act as an N fertiliser.

Chalk *et al.* (2013) showed that composts are generally inferior sources of N for crops compared with their raw materials due to lower NH_4^+ content and lower N mineralisation capacity in the year of application. Furthermore, with some immature compost, immobilisation of fertiliser N may increase in compost-amended soils and may reduce recovery by a crop, but fertiliser N losses are reduced overall. However, in subsequent years, mineralisation of residual N may contribute to the long-term fertiliser effect of the compost. When comparing fresh, dried and composted poultry manure, Munoz *et al.* (2008) found first-year N-MFE values of 57, 53 and 14%, respectively, and second-year values of 18, 19 and 12%; composted cow manure was also included but showed non-significant MFE values in both the first and second year. Therefore, manure composts, even if they have relatively low C : N ratios, must be considered slow-release biofertilisers with regard to the amount and pattern of plant-available N released, especially in the first year after application. Helgason *et al.* (2007) found that less than 5% of the organic N in composted cattle manures was mineralised over 425 days, suggesting that little of the organic N in compost becomes available in the year of application. They also showed that the initial inorganic N content of the compost, analysed prior to its application, can be used to predict plant availability of the N in compost. Loecke *et al.* (2012) compared application of composted versus fresh solid swine manure and found that maize N uptake efficiency was higher for composted than fresh manure, but found no consistent effect on *in situ* soil net N mineralisation and soil mineral N dynamics. They concluded that overall, composting of manures prior to soil application had no clear benefit for N synchrony in maize crops.

Composting has been shown to result in more or less the same or slightly lower P plant availability of the composted compared with fresh manure (Gagnon *et al.*, 2012), with fresh manures having an availability in the range of 90–100% of mineral fertiliser P. Similar results have been reported by Preusch *et al.* (2002) for poultry manure and by Jørgensen *et al.* (2010) for the manure solid fraction, where the content of water-soluble inorganic P in the composted manure remained unchanged or decreased slightly compared with that in untreated manure.

From a management perspective, composting of solid fractions should therefore not be viewed as a method for significantly increasing the potential plant availability of N or P. The beneficial effects of composting manure for agricultural use are therefore mainly linked to the reduction in volume, mass and water content, as well as odour and pathogens, enabling export of nutrients from areas of high nutrient loading to soils which may be deficient in nutrients. This was also the conclusion of Larney *et al.* (2006), who analysed options for beef cattle feedlot operations in southern Alberta, Canada. They indicated that composting solutions will only be prioritised if regulations on land application rates become stricter than they are today, as a larger land base would then be required, increasing the radius of manure haulage from source and making the composting option more attractive due to reductions in transportation costs.

However, for many small- and medium-scale farmers in Asia, composting of manure may earn significant additional income from sale of compost to other farmers (Vu *et al.*, 2007, 2012). The compost is used either for fertilising fish ponds, fertilising perennial crops grown on sloping land (e.g. coffee, fruits), where the risk of runoff from traditional soluble chemical fertiliser is too high, or for ameliorating soil fertility on depleted soils, where any fresh organic residues added decay very rapidly in a tropical climate.

14.3.4 Use of Compost Products in the Non-Agricultural Sector

However, there are actually many other uses for compost besides those in the agricultural sector and these other markets are often prepared to pay a higher price per cubic metre of compost. In the European Union, only about 40% of the compost produced (including biocompost from source-separated municipal solid waste and green waste from parks and gardens) is used in agriculture, while 60% is used for a variety of purposes in other sectors (Carlsbæk, 2011). Often these sectors require more from compost than merely a high level of plant nutrients, and the compost aimed at these markets must be more refined and have higher levels of product documentation. Non-agricultural applications for compost include soil improvement and mulching in professional landscaping, the private garden sector and specialist horticultural sectors such as fruit and vine growing, as well as use as a component in manufactured topsoils for landscaping or in growing media for greenhouses and nurseries. There is great potential for the use of compost in landscaping and private gardening because by applying compost to the soil, a number of problem areas common to these sectors can be effectively addressed, such as poor soil quality on construction sites and low microbiological activity caused by long-term storage of topsoil or by pesticide residue accumulation (Carlsbæk, 2011).

Growing media for greenhouses and nurseries are commonly produced from sphagnum peat, but in many parts of the world the harvesting of peat bogs exceeds the natural formation and has a negative impact on the environment; therefore, it is often necessary to find alternatives to peat. Compost may constitute a significant resource for replacing some of the peat-based products in horticulture (Favoino and Hogg, 2008) or as an organic fertiliser in forests, orchards and golf greens. Cáceres *et al.* (2006) studied the utilisation of composted solid fractions from separated cattle slurry as a growing medium and found these to be acceptable for use alone or in mixtures with peat. Although the salinity of the compost was at the high end of the range considered suitable in the horticulture industry, the authors concluded that it may be useful for outdoor ornamental plants or for mixing with other media. Similarly, MacConnell and Collins (2009) and Bustamante *et al.* (2012) utilised anaerobically digested and mechanically separated cattle slurry for composting and obtained promising results; the compost was equal or superior to peat moss when used as a major component of potting media for ornamentals, showing good physical stability and suppression of soil-borne plant diseases. However, complete replacement of peat-based growing media by manure-based compost is difficult to envisage for horticultural greenhouses, for example, mainly due to its high electrical conductivity and pH, and especially insufficient long-term physical stability, to fulfil quality requirements for all different horticultural production systems.

In private households, replacement of peat-based growing media with manure-based media will also be incomplete. Andersen *et al.* (2010) surveyed private customer behaviour, and found that their replacement of peat and fertiliser with compost was not 100%, but more likely around 50%, and thus less than often assumed when assigning environmental credits to compost. It was indicated that compost was used for many purposes in hobby gardening, including substituting for peat and fertiliser, to improve soil quality, and as a filling material (as a substitute for soil).

In conclusion, composting of manures in the agricultural context mainly has potential for reducing volume, mass, odour and pathogens, enabling more distant export and maintenance of soil fertility in regions of poor soil quality. However, there is a potential market for high-quality manure-based composts in the non-agricultural sector, with uses for landscaping, topsoil establishment and partial substitution of peat-based growing media.

14.4 Drying and Pelletising Solid Manures

The aim of manure drying is to obtain a dried, stable and odourless product, containing most of the nutrients, that is easier and cheaper to transport and spread as an organic fertiliser on land. Drying may be applied to solid manures (e.g. poultry litter) or to separated slurry solids. Thermal energy is usually recovered from combined heat and power (CHP) engines or other heat residual streams, since drying is energy-intensive and would otherwise be too expensive. For an industrial-scale facility, the thermal requirements are estimated to be around 15–18 kWh m^{-3} for slurry solids with 25–30% total solids (Flotats *et al.*, 2011). Various drying techniques can be applied: convection, radiation and conduction dryers. Conduction drying offers the best opportunities when combined with initial anaerobic digestion for biogas production. The gaseous emissions from the dryer must be recovered (by filtration or scrubbing) to avoid NH_3 or volatile organic compound (VOC) emissions (Maurer and Müller, 2012). If the product comes from an anaerobic digestion process, this reduces VOC emissions. Acidification of the input product with sulfuric acid, in order to control NH_3 emissions, can also be used, if scrubbing is not applied.

Drying is usually followed by a briquetting or pelletising process, which involves moulding the dried manure into pellets of a certain size range, to obtain a product easier to transport and to spread on land. Compression and temperature, as well as composition of the raw material, are the major factors affecting the process, in which the material is pressed through a die with holes of the desired shape/size and the material exits as a pellet-shaped product (Stelte *et al.*, 2012). The pelleted material can be marketed as an organic matter and P-rich biofertiliser or soil amendment. In practice, pelleted pig manure is less acceptable than, for example, poultry manure or ordinary P fertiliser (Oenema *et al.*, 2012). Mondini *et al.* (1996) compared composting and drying of poultry manure, and found that although both processes humified the organic matter, drying was much less effective in stabilising the organic matter, but lowering NH_3 losses with rapid drying caused a higher N fertiliser value. López-Mosquera *et al.* (2008) found that on average 4% of the C and 20% of the N in the manure was lost during drying and pelletising. The fertiliser N value in lowland rice was found to be similar for fresh and dried pelletised poultry litter, but generally only equivalent to 25% of the fertiliser value of urea (Golden *et al.*, 2006).

In addition to drying and pelletising, a process called torrefaction may be applied, to obtain a fuel of higher quality for combustion and gasification applications. It can be described as a thermal treatment process in which the biomass is heated to temperatures typically ranging between 200 and 300 °C in the absence of oxygen (Stelte *et al.*, 2012). Torrefaction leads to a dry product with changed biomass properties, condensation and reduced accessibility for biological decay, and hence higher stability during storage and handling. Torrefaction decreases the mass, but retains most of the energy content, and thus leads to a more energy-dense fuel. However, if applied to manure it would have to be adapted in order to recover NH_3 and VOCs emitted during the process.

Overall, most studies indicate that, provided N losses can be kept at a minimum, the N and P availability is not decreased significantly by drying and pelletising of solid manures. A number of larger-scale operations (more than 50 000 ton year^{-1}) are running on a commercial scale, especially in Spain (Foged *et al.*, 2011a).

14.5   Manure Combustion and Gasification Ash

Combustion of the fresh or dried solid fraction or solid manure can be carried out in a power plant to generate electricity (Chapter 13), thereby further decreasing the volume and weight of the manure into ash residuals. Thermal gasification can also be applied, a process that converts carbonaceous materials into CO, hydrogen, CO_2 and CH_4. Gasification relies on chemical processes at elevated temperatures above 700 °C, unlike biological processes such as in biogas plants. The resulting gas mixture is called syngas (synthetic gas) and may be more efficiently converted to high-quality energy such as electricity than would be possible by direct combustion of the fuel. Furthermore, corrosive ash elements such as chloride and K can be retained by the gasification process, allowing high-temperature combustion of the gas from otherwise problematic fuels, such as manures. The process is performed with a controlled amount of oxygen or steam (25–30% of the required O_2 for complete combustion; Flotats *et al.*, 2011).

The net energy yield depends on the moisture content of the solids. In practice, only relatively dry solids are feasible fuels, typically with above 50% dry matter (for details, see Section 13.2) in order to achieve net energy output and profitability. The manure N and S are mainly volatilised (as N_2, NO_x, SO_x and H_2S) in the combustion or gasification process (Chapter 13). However, the non-volatile nutrients (P, K, Mg, Ca, Na and microelements) are retained in the ash from both processes, concentrated relative to the solids, and these have the potential to be used as P- and K-rich soil amendments or fertilisers, with a whole range of secondary elements and micronutrients (Komiyama *et al.*, 2013).

Ash P is in the form of a mixture of carbonate and hydroxyapatite, which are relatively water-insoluble (less than 0.1% of total P). However, a significant proportion of the K and Na (greater than 50%) is water-soluble, and hence readily plant available. Availability of the more strongly bound nutrients in the ash (P, Ca, Mg and microelements) is therefore generally lower than in the raw manure solids, but this to some degree depends on incineration or gasification temperature and other process conditions. With increasing combustion temperature from 400 to 700 °C, the plant availability of P approximately halved (Thygesen *et al.*, 2011), due to the formation of crystalline hydroxyapatite. However, a low combustion temperature conflicts with energy production, which is optimal at temperatures above 800 °C, and thus thermal gasification, which can be conducted efficiently at lower temperatures, may be more attractive from a P recovery point of view. Thygesen and Johnsen (2012) studied a large number of manure solids and found a mean P concentration of 112, 123, 157 and 51 g P kg^{-1} in the combustion (850 °C) ashes from biogas digestate, pig slurry solids, mink solids and cattle slurry solids, respectively. They concluded that all the ash products except those from cattle slurry solids contained sufficiently high P concentrations to be suitable for fertiliser production. However, in the majority of manure ashes less than 1% of the total P is water-soluble and less than 7–20% extractable in citric acid, which is significantly less than in mineral P fertiliser (superphosphate) and pig slurry solids, which typically have more than 65% water-soluble and 85% citric acid-extractable P (Rubæk *et al.*, 2006; Kuligowski *et al.*, 2010).

As a consequence, ash from combusted or thermally gasified manure generally releases P at a slower rate than mineral fertilisers, providing lower immediate plant availability, but also a lower risk of leaching or runoff. Untreated manure ash can therefore not be recommended as a starter fertiliser (high solubility), but it may function as a slow-release fertiliser, having a more long-term effect (Pagliari *et al.*, 2010). The availability of P in ash from thermal gasification of animal manure products may actually be higher than indicated by traditional fertiliser P analyses (water and citric acid solubility). Rubæk *et al.* (2006) found that after soil application, the P in manure ash, which was water- or resin-extractable from the soil, was much higher than that indicated by the low solubility of P in the direct manure analyses. Low immediate plant availability may also be caused by the high pH of most ashes, decreasing the solubility of P, but soils have a large buffering capacity and, especially on acidic soils, some of the ash P may be solubilised. On acidic soil (pH 4.5) with ryegrass, Kuligowski *et al.* (2012) found the P fertiliser equivalent value of thermally gasified pig manure ash to be 6–12% relative to monocalcium phosphate, but if they neutralised the ash, this tripled to 19–33%.

14.5.1 Ash Extraction and Conversion

Due to the low solubility of P in combustion and thermal gasification ashes, they need to be treated if they are to be used as a readily available P fertiliser. Treatment of ash generally aims at (i) converting P into a plant available form for reuse as fertiliser or into a raw material for the P industry, and (ii) separating heavy metals and other unwanted elements from the valuable P. Two categories of P recovery technologies exist: wet chemical approaches and thermal approaches.

Among wet chemical approaches, a distinction is made first between acid and alkaline dissolution techniques. However as alkaline agents are usually more expensive than acids, these are not commonly applied. In the case of acidic dissolution, P and some of the metals are dissolved. Thus, after acidic leaching the dissolved P has to be separated from the (heavy) metals in order to create a P fertiliser product with high

P solubility and low metal content. The separation of P and dissolved metal ions requires further chemical input.

The first step is an almost complete acidic dissolution of P at pH < 2. This process is unavoidably accompanied by a partial dissolution of metals or their compounds. The amount of dissolved metals depends on the composition of the ash, as well as on the type and amount of the added acid (HCl or H_2SO_4). In manure ash the P component is predominantly present in the form of calcium phosphates, but may also be in aluminium and iron phosphates, and the chemical demand for complete acidic dissolution of P can be estimated from the reaction schemes (Petzet *et al.*, 2012):

$$Ca_9(Al)(PO_4)_7 + 21H^+ \rightarrow 9Ca^{2+} + Al^{3+} + 7H_3PO_4 \tag{14.1}$$

$$AlPO_4 + 3H^+ \rightarrow Al^{3+} + H_3PO_4 \tag{14.2}$$

$$Fe(PO_4)_2 + 6H^+ \rightarrow 3Fe^{2+} + 2H_3PO_4 \tag{14.3}$$

$$FePO_4 + 3H^+ \rightarrow Fe^{3+} + H_3PO_4 \tag{14.4}$$

Subsequent to dissolution, P and metals will be in solution, which cannot be used directly as a fertiliser, since the pH is too low, and metals and P will tend to precipitate again. Hence, a second step, with removal of cations from the acidic solution, is needed and different approaches are technically feasible, such as sequential precipitation, liquid–liquid extraction, sulfidic precipitation and a few others (Petzet *et al.*, 2012). The choice of method depends on the proportion of Ca, Fe and Al dissolved from the manure ash.

Equations (14.1)–(14.4) show that the theoretical chemical demand for a complete acidic dissolution of each of these P compounds is 3 mol H^+ mol^{-1} P. However, in practice more acid is needed since other acid-soluble components in the manure ash, such as CaO, $CaCO_3$, $Ca(OH)_2$, MgO, Fe_2O_3 and K_2O, are dissolved as well, typically 2–3 times as much, thus requiring around 0.39–0.78 kg H_2SO_4 kg^{-1} ash for more than 85% dissolution of ash P, depending on the total cation/phosphorus equivalent ratio (Cohen, 2009). In some biogas plants, lime ($CaCO_3$,) is used for pH control and this will add further to the acid requirement. For ash from thermally gasified pig manure, Kuligowski and Poulsen (2010) found an H_2SO_4 requirement of 0.98 kg kg^{-1} ash, corresponding to an acid demand of 19 kg H_2SO_4 kg^{-1} P recovered. This indicates that although some acid may be recovered and recycled in the process, it will be a relatively expensive process, and the problem is often that the number of secondary steps and the cost of processing become too high for the process to be profitable. In principle, the recovered P should be of similar availability to mineral fertiliser P (superphosphate), as also indicated in the study by Kuligowski *et al.* (2010), but the authors mention large variability. However, so far most research and development has focused on ash from wastewater treatment solids (sewage sludge) and less for manure ash – thus it is not commercially implemented (Flotats *et al.*, 2011), although increases in rock P world market prices may later prove this to be more profitable.

Thermal approaches for separation of P from ashes are carried out at very high temperatures of 1000–2000 °C, which enables transformation and condensation of ash P into elemental (white) P (Schipper *et al.*, 2001). The process is sensitive to impurities, especially Fe, but produces a very pure form of P. This is mainly used in the chemical, food and feed industries, as the costs are so far too high to utilise it for fertiliser production. The process is in commercial operation by the company Thermphos in the Netherlands (Schipper and Korving, 2009).

In conclusion, untreated manure ashes may function as a reasonably good K fertiliser and as more of a slow-release fertiliser for P, having a long-term, but rather limited starter effect. Manure combustion or gasification in Europe is in commercial operation in the United Kingdom and Ireland (Font-Palma, 2012), mainly with poultry litter as feedstock. High P availability can be achieved by chemical extraction or conversion of the ash to obtain soluble mineral P fertiliser, but these processes and products are not yet commercially available.

14.6   Biochar from Pyrolysis or Carbonisation of Solid Manures

Biochar is a carbonaceous material that contains polycyclic aromatic hydrocarbons with an array of other functional groups. Biochar is produced by carbonisation during pyrolysis, a process in which biomass is heated to 300–700 °C in the absence or at low levels of oxygen (Lehmann and Joseph, 2009). Upon pyrolysis a significant proportion of the original biomass C is retained within the biochar, but the recovery rates are highly dependent on the pyrolysis process. In addition to solid biochar, the process produces a bio-oil (tar) and a syngas that can be used as fuels. The process is endothermic, so requires energy input, and some or all of the energy may be covered by the bio-oil and syngas produced, depending on process conditions, but a positive net energy balance can only be achieved if the feedstock is relatively dry (Wnetrzak *et al.*, 2013).

Two types of pyrolysis may be applied:

- *Fast pyrolysis* involves finely ground biomass being rapidly (in the order of a few seconds) heated to 450–700 °C and a very short residence time. The process results in a high yield of bio-oil, up to 70% by mass of the feedstock, depending on material. Fast pyrolysis may yield a positive net energy outcome
- *Slow pyrolysis* involves slower heating to 300–550 °C with a longer residence time; the biomass is typically in the reactor for at least 30 min and possibly several hours. The process results in a higher yield of biochar, again depending on process conditions. This resembles charcoal production for energy purposes, but by definition biochar is produced as a soil amendment or biofertiliser, hence process conditions and feedstock may be modified.

The highly porous structure and very large surface area of biochars means that they are good adsorbents for a range of substances, and their surfaces can exhibit hydrophilic, hydrophobic, acidic and basic properties, all of which contribute to their ability to react with soil solution substances (Atkinson *et al.*, 2010). Biochar physical and chemical properties vary depending on the material used to produce it (feedstock), the availability of oxygen and the temperatures achieved during pyrolysis. Due to the aromatic structure of biochar, it is recalcitrant and has the potential for long-term carbon sequestration in soil; in the past decade a plethora of research evidence has been produced, indicating a range of soil quality enhancement effects from biochar application (Lehmann and Joseph, 2009), although many issues are still controversial and subject to scientific scrutiny.

Slow pyrolysis or carbonisation promotes the conversion of animal manures such as swine manure into nutrient-rich biochar (Ro *et al.*, 2009). Hence, manure biochars may be used for soil amendment, ameliorating soil fertility positively. Like manure ashes, manure biochar contains nutrients not volatilised and carbonised organic matter, so there is usually complete recovery of manure P, K, Ca and Mg, and a substantial part of the N and S (depending on pyrolysis temperature). Nutrient availability varies, but is often higher than for combustion ashes; biochars have been found to be as effective as mineral P fertilisers by some authors (e.g. Wang *et al.*, 2012), while others report somewhat lower fertiliser P-equivalent values of manure biochars (Sørensen and Rubæk, 2010). Biochar N is, of course, bound more strongly than in raw manure solids, but pig manure biochar still mineralises N even at a C : N ratio of 48, albeit at low rates (Ehlert and Oenema, 2010).

Biochar from manure may also have other uses, due its reactive nature. Streubel *et al.* (2012) demonstrated that biochar produced from anaerobic digester dairy solid fraction could be used to sequester P from dairy lagoons, through physical and weak chemical bonding, indicating that biochar could be a beneficial component to P reduction in dairy production systems. Biochar could also be blended with compost (Steiner *et al.*, 2010) in order to retain NH_4^+ and prevent NH_3 volatilisation, which could increase the value of the biochar once the compost is applied to soil.

In conclusion, the effect of manure biochars as biofertilisers has not yet been thoroughly studied, so there is no conclusive evidence. However, with the surge of interest in the implications of biochar application, more solid research evidence is likely to become available in the near future.

14.7 Precipitates and Mineral Concentrates from Liquid Manures

As described in Sections 7.4 and 7.5, a range of techniques can be applied to treat the manure liquid fraction, and two in particular are interesting as regards nutrient recovery.

14.7.1 Struvite

As already described in Sections 4.4.5 and 7.4.2, struvite ($MgNH_4PO_4 \cdot 6H_2O$) can be precipitated from liquid manures, provided that the appropriate conditions are present (pH $\sim$ 9, a molar ratio 1 : 1 : 1 of Mg^{2+} : NH_4^+ : PO_4^{3-}, conducive physical settling conditions), and as such constitutes a method for removal of both P and N from liquid manures. The method has been developed for wastewater treatment, where P removal can easily reach more than 70% (le Corre *et al.*, 2009) and is commercially available for treatment plants, although not widely applied as yet. For manures, the stuvite technique is particularly relevant for anaerobically digested slurries and the liquid fraction from digestate separation; hence, it has been the subject of massive research in the past decade and quite high P removal efficiencies have been achieved (56–93%; Zeng and Li 2006; Suzuki *et al.*, 2007; Liu *et al.*, 2011; Rahman *et al.*, 2011; Song *et al.*, 2011). However, further development is needed for application to liquid manures and only a few commercial-scale plants are in operation worldwide.

With respect to the use of the struvite product, not many studies of the fertiliser efficiency of the recovered struvite P are available yet. Latifiana *et al.* (2012) assessed the short-term release of N, P and Mg from struvite pellets, and found it to be considerably slower than from commercial mineral fertiliser. Similar results were found by Ganrot *et al.* (2007) in a study where human urine-derived struvite was tested on wheat and found to be effective as a slow release fertiliser, but not providing as high a starting effect (rapid plant P uptake) as mineral fertiliser P. However, Johnston and Richards (2003) found a range of struvites to have a similar effect to mineral fertiliser P on ryegrass. Similarly, wastewater treatment (Ponce and De Sa, 2007) or urine-derived (Antonini *et al.*, 2012) struvite was found to be as effective as a fertiliser P source for ryegrass biomass production; yields were comparable to or exceeding those achieved with commercial P fertilisers on deficient, low-fertility soils. Oenema *et al.* (2012) also reviewed a number of different struvite studies and found variable, but quite high, fertiliser equivalent values.

These results clearly indicate that further research is needed before the effectiveness of struvite-based fertiliser can be properly assessed. Furthermore, a significant proportion of manure N and P will still be bound organically, or in the case of P in other precipitated inorganic forms, in the manure solids, so the method will only be suitable for upgrading a fraction of the manure N and P.

14.7.2 Mineral Concentrates

Mineral concentrates are highly nutrient-rich solutions that may be obtained via ultrafiltration, evaporation or reverse osmosis of the liquid fraction from separation of slurry or digestate, as described in Section 7.5 and Masse *et al.* (2007). These mineral concentrates (the retentate) may be directly applied to agricultural land and the byproduct water low in nutrients (the permeate) may be directly discharged to surface waters or the sewage system (Schoumanns *et al.*, 2010).

The greatest experience in Europe with these technologies, and with the properties and utilisation of mineral concentrates, is in the Netherlands, where a number of centralised and large-scale farm manure processing

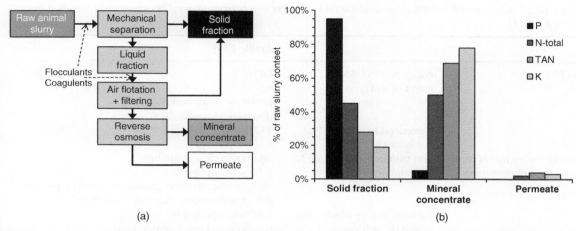

(a)   (b)

Figure 14.3 *Production process of mineral concentrate using chemical-mechanical separation and reverse osmosis (a) and relative mass distribution of nutrients (raw slurry = 100%) after treatment of pig slurry into the solid fraction, the mineral concentrate and the permeate from reverse osmosis (b). (Adapted with permission from Hoeksma et al. (2012). © 2012 CIGR.)*

plants utilising a range of technologies in combination have been operating for a number of years (Velthof, 2011). Hoeksma *et al.* (2012) monitored production of mineral concentrates from pig slurry using reverse osmosis for 2 years at 5 full-scale manure processing plants in the Netherlands. The treatment process at these plants was based on chemical/mechanical separation of the raw slurry, polishing of the liquid fraction by removing the remaining solids by coagulation and flotation filtering, and finally concentration of the dissolved nutrients by reverse osmosis to produce a mineral concentrate and a permeate (Figure 14.3a). The mineral concentrates could be characterised as an NK fertiliser containing 50% of the total N (majority as NH_4^+) and 78% of total K in the raw slurry (Figure 14.3b), and with a pH of 7.9, which implies a risk of N losses by volatilisation during and after spreading. From the raw slurry input, 95%, 45% and 19% of the P, N and K, respectively, ended up in the solid fraction (Hoeksma *et al.*, 2012).

Provided that the losses can be kept to a minimum, the mineral fertiliser equivalent value of the mineral concentrates theoretically should be relatively high, as they resemble commercial liquid fertilisers, with nearly all the nutrients in a mineral, plant-available form. This may require injection of the concentrate into the soil, to avoid gaseous losses; this can be carried out with a coulter, spoke-wheel or high-pressure injection tool, which have been shown to effectively limit potential NH_3 volatilisation from mineral concentrates with a high pH (Nyord *et al.*, 2008). Table 14.3 summarises the Dutch results on N-MFE value and the fate of N from mineral concentrates not taken up.

It was concluded that the N-MFE value of mineral concentrates compared to CAN fertiliser was on average 80–90% on arable land (basal dressing via injection) and around 58% on grassland. The variation in MFE value was large, but the N efficiency of the mineral concentrates was similar to that of liquid NH_4^+ nitrate in grassland and in arable land on clay.

Mineral concentrates thus have a similar N fertiliser value to liquid nitrogen fertilisers. In addition, K is important for many arable crops and maize. However, supplying K with mineral concentrate limits the applicable amount of mineral concentrate on dairy farms when the K status of the soil is sufficient or higher (Velthof, 2011). It has been estimated that if applied to all manure in the Netherlands, the mineral concentrates and the associated solid fraction from separation could decrease the need for imported mineral

Table 14.3   Summary of N-MFE value and the fate of ineffective N from mineral concentrates (modified from Velthof, 2011).

		Arable land	Grassland
MFE[a]	compared to CAN[c]	84%	58%
	effect of soil type	yes potatoes, sandy soil: 92% potatoes, clay soil: 80%	No
	compared to liquid NH_4^+ nitrate	117%	96%
Fate of ineffective N from mineral concentrates[b]	non-mineralised organic N	on average 5% of applied N	
	ammonia emission	<10% of applied N risk on sod injection grassland > deep injection risk on calcareous clay soil > sandy soil	
	gaseous loss by nitrification and denitrification	<10% of applied N risk on grassland > arable land	
	leaching	<5% of applied N risk on sandy soil > clay soil risk on arable land > grassland	
	immobilisation in soil	<10% of applied N risk on grassland > arable land	

[a] The MFE values in this table are based on field experiments in which mineral concentrates were tested at different N application rates.
[b] The fate of the inactive N is partly based on results from the experiments and partly on estimates.
[c] CAN: calcium NH_4^+ nitrate.

N and P fertilisers by up to 15% and 82%, respectively, with little change in total NH_3 losses, N_2O emissions and N leaching, due to increased N and P use efficiency on a national scale (Oenema *et al.*, 2012).

14.8   Summary

A number of different routes are possible for recovering and upgrading nutrients and organic matter from different manure types. In Europe, approximately 8% of all manure is currently being processed in one way or an other. For liquid manures all treatment steps start with mechanical separation into a solid and a liquid fraction, which may then be treated further. Composting of manures in the agricultural context mainly has the potential for a reduction in volume, mass, odour and pathogens, enabling more distant export and maintenance of soil fertility in regions of poor soil quality. However, there is a potential market for high-quality manure-based composts in the non-agricultural sector, with uses for landscaping, topsoil establishment and partial substitution of peat-based growing media. Manure combustion or gasification in Europe is in commercial operation in the UK and Ireland, mainly with poultry litter as feedstock. Manure ashes may function as a reasonably good K fertiliser, and as more of slow-release fertiliser for P, having a long-term, but rather limited, starter effect. High P availability can be achieved by chemical extraction or conversion of the ashes to obtain soluble mineral P fertiliser, but these processes and products are not yet commercially available. Manure biochars may also act as biofertilisers, but have not yet been thoroughly studied, with no conclusive evidence at present. Precipitated struvite containing Mg, P and N has been shown to yield very different fertiliser values, from similar to mineral P in some cases to much lower in other cases, and should probably be considered a slow release N and P fertiliser. Mineral concentrates from high-tech manure treatment, including reverse osmosis or ultrafiltration, potentially have a high MFE value, on average 80–90% on arable land and around 58% on grasslands, but only if incorporated into the soils, as their NH_3 volatilisation risk is high.

In conclusion, this chapter describes a few major manure upgrading pathways currently in operation, but also some of the promising new approaches currently under development. However, the manure upgrading arena is undergoing such rapid development these days, so in a few years the information provided is likely to need updating with more recent findings.

References

Andersen, J.K., Christensen, T.H. and Scheutz, C. (2010) Substitution of peat, fertiliser and manure by compost in hobby gardening: user surveys and case studies. *Waste Manag.*, **30**, 2483–2489.

Antonini, S., Arias, M.A., Eichert, T. and Clemens, J. (2012) Greenhouse evaluation and environmental impact assessment of different urine-derived struvite fertilizers as phosphorus sources for plants. *Chemosphere*, **89**, 1202–1210.

Atkinson, C.J., Fitzgerald, J.D. and Hipps, N.A. (2010) Potential mechanisms for achieving agricultural benefits from biochar application to temperate soils: a review. *Plant Soil*, **337**, 1–18.

Bernal, M.P. Alburquerque, J.A. and Moral, R. (2009) Composting of animal manures and chemical criteria for compost maturity assessment. A review. *Bioresour. Technol.*, **100**, 5444–5453.

Brito, L.M., Mourão, I., Coutinho, J. and Smith, S.R. (2012) Simple technologies for on-farm composting of cattle slurry solid fraction. *Waste Manag.*, **32**, 1332–1340.

Bustamante, M.A., Alburquerque, J.A., Restrepo, A.P., de la Fuente, C., Paredes, C., Moral, R. and Bernal, M.P. (2012) Co-composting of the solid fraction of anaerobic digestates, to obtain added-value materials for use in agriculture. *Biomembr. Bioenerget.*, **43**, 26–35.

Cabeza, R., Steingrobe, B., Römer, W. and Claassen, N. (2011) Effectiveness of recycled P products as P fertilizers, as evaluated in pot experiments. *Nutr. Cycl. Agroecosyst.*, **91**, 173–184.

Cáceres, R., Flotats, X. and Marfa, O. (2006) Changes in the chemical and physicochemical properties of the solid fraction of cattle slurry during composting using different aeration strategies. *Waste Manag.*, **26**, 1081–1091.

Carlsbæk, M. (2011) Use of compost in horticulture and landscaping, in *Solid Waste Technology and Management* (ed. T.H. Christensen), John Wiley & Sons, Ltd, Chichester, pp. 515–532.

Chalk, P.M., Magalhães, A.M.T. and Inácio, C.T. (2013) Towards an understanding of the dynamics of compost N in the soil–plant–atmosphere system using ^{15}N tracer. *Plant Soil*, **362**, 373–388.

Cohen, Y. (2009) Phosphorus dissolution from ash of incinerated sewage sludge and animal carcasses using sulphuric acid. *Environ. Technol.*, **30**, 1215–1226.

Delaune, P.B., Moore Jr, P.A., Daniel, T.C. and Lemunyon, J.L. (2004) Effect of chemical and microbial amendment on ammonia volatilisation from composting poultry litter. *J. Environ. Qual.*, **33**, 728–734.

Ehlert, P. and Oenema, O. (2010) Managing phosphorus cycling in agriculture-biochars from digested fattening pig slurry – phosphate availability of biochars, presented at the *Workshop Managing Livestock Manure for Sustainable Agriculture*, Wageningen.

Favoino, E. and Hogg, D. (2008) The potential role of compost in reducing greenhouse gases. *Waste Manag. Res.*, **26**, 61–69.

Flotats, X., Foged, H.L., Blasi, A.B., Palatsi, J., Magri, A. and Schelde, K.M. (2011) *Manure Processing Technologies. Technical Report No. II concerning "Manure Processing Activities in Europe" to the European Commission, Directorate-General Environment*, Agro Business Park, Tjele.

Foged, H.L., Flotats, X., Blasi, A.B., Palatsi, J., Magri, A. and Schelde, K.M. (2011a) *Inventory of Manure Processing Activities in Europe. Technical Report No. I concerning "Manure Processing Activities in Europe" to the European Commission, Directorate-General Environment*, Agro Business Park, Tjele.

Foged, H.L., Flotats, X., Blasi, A.B., Palatsi, J. and Magri, A. (2011b) *End and By-products from Livestock Manure Processing – General Types, Chemical Composition, Fertilising Quality and Feasibility for Marketing. Technical*

Report No. III concerning "Manure Processing Activities in Europe" to the European Commission, Directorate-General Environment, Agro Business Park, Tjele.

Foged, H.L., Flotats, X., Blasi, A.B., Schelde, K.M., Palatsi, J., Magri, A. and Juznik, Z. (2011c) *Assessment of Economic Feasibility and Environmental Performance of Manure Processing Technologies. Technical Report No. IV to the European Commission, Directorate-General Environment*, Agro Business Park, Tjele.

Foged, H.L., Flotats, X. and Blasi, A.B. (2011d) *Future Trends on Manure Processing Activities in Europe. Technical Report No. V concerning "Manure Processing Activities in Europe" to the European Commission, Directorate-General Environment*, Agro Business Park, Tjele.

Font-Palma, C. (2012) Characterisation, kinetics and modelling of gasification of poultry manure and litter: an overview. *Energy Conv. Manag.*, **53**, 92–98.

Gagnon, B., Demers, I., Ziadi, N., Chantigny, M.H., Parent, L.E., Forge, T.A., Larney, F.J. and Buckley, K.E. (2012) Forms of phosphorus in composts and in compost-amended soils following incubation. *Can. J. Soil Sci.*, **92**, 711–721.

Ganrot, Z., Dave, G., Nilsson, E. and Li, B. (2007) Plant availability of nutrients recovered as solids from human urine tested in climate chamber on *Triticum aestivum*, L. *Bioresour. Technol.*, **98**, 3122–3129.

Golden, B.R., Slaton, N.A., Norman, R.J., Gbur Jr, E.E., Brye, K.R. and DeLong, R.E. (2006) Recovery of nitrogen in fresh and pelletized poultry litter by rice. *Soil Sci. Soc. Am. J.*, **70**, 1359–1369.

Guo, M. and Song, W. (2009) Nutrient value of alum-treated poultry litter for land application. *Poultry Sci.*, **88**, 1782–1792.

Helgason, B.L., Larney, F.J., Janzen, H.H. and Olson, B.M. (2007) Nitrogen dynamics in soil amended with composted cattle manure. *Can. J. Soil Sci.*, **87**, 43–50.

Hoeksma, P., de Buisonjé, F.E. and Aarnink, A.A. (2012) Full-scale production of mineral concentrates from pig slurry using reverse osmosis, presented at the *International Conference of Agricultural Engineering CIGR-AgEng2012*, Valencia.

Johnston, A.E. and, I.R. Richards, I.R. (2003) Effectiveness of different precipitated phosphates as phosphorus sources for plants. *Soil Use Manag.*, **19**, 45–49.

Jørgensen, K., Magid, J., Luxhoi, J. and Jensen, L.S. (2010) Phosphorus distribution in untreated and composted solid fractions from slurry separation. *J. Environ. Qual.*, **39**, 393–401.

Kelleher, B.P., Leahy, J.J., Henihan, A.M., O'Dwyer, T.F., Sutton, D. and Leahy, M.J. (2002) Advances in poultry litter disposal technology – a review. *Bioresour. Technol.*, **83**, 27–36.

Kithome, M., Paul, J.W. and Bomke, A.A. (1999) Reducing nitrogen losses during simulated composting of poultry manure using adsorbents or chemical amendments. *J. Environ. Qual.*, **28**, 194–201.

Komiyama, T., Kobayashi, A. and Yahagi, M. (2013) The chemical characteristics of ashes from cattle, swine and poultry manure. *J. Mater. Cycl. Waste Manag.*, **15**, 106–110.

Kuligowski, K., Gilkes, R.J. Poulsen, T.G. and Yusiharni, B.E. (2012) Ash from the thermal gasification of pig manure – effects on ryegrass yield, element uptake, and soil properties. *Soil Res.*, **50**, 406–415.

Kuligowski, K. and Poulsen, T.G. (2010) Phosphorus and zinc dissolution from thermally gasified piggery waste ash using sulphuric acid. *Bioresour. Technol.*, **101**, 5123–5130.

Kuligowski, K., Poulsen, T.G., Rubæk, G. and Sørensen, P. (2010) Plant-availability to barley of phosphorus in ash from thermally treated animal manure in comparison to other manure based materials and commercial fertilizer. *Eur. J. Agron.*, **33**, 293–303.

Larney, F.J., Sullivan, D.M., Buckley, K.E. and Eghball, B. (2006) The role of composting in recycling manure nutrients. *Can. J. Soil Sci.*, **86**, 597–611.

Latifiana, M., Liua, J. and Mattiasson, B. (2012) Struvite-based fertilizer and its physical and chemical properties. *Environ. Technol.*, **33**, 2691–2697.

le Corre, K.S., Valsami-Jones, E., Hobbs, P. and Parsons, S.A. (2009) Phosphorus recovery from wastewater by struvite crystallization: a review. *Crit. Rev. Environ. Sci. Technol.*, **39**, 433–477.

Lehmann, J. and Joseph, S. (eds) (2009) *Biochar for Environmental Management: Science and Technology*, Earthscan, London.

Liu, Y.H., Kwag, J.H., Kim, J.H. and Ra, C.X. (2011) Recovery of nitrogen and phosphorus by struvite crystallization from swine wastewater. *Desalination*, **277**, 364–369.

Loecke, T.D., Cambardella, C.A. and Liebman, M. (2012) Synchrony of net nitrogen mineralization and maize nitrogen uptake following applications of composted and fresh swine manure in the Midwest, US. *Nutr. Cycl. Agroecosyst.*, **93**, 65–74.

López-Mosquera, M.E., Cabaleiro, F., Sainz, M.J., López-Fabal, A. and Carralet, E. (2008) Fertilizing value of broiler litter: effects of drying and pelletizing. *Bioresour. Technol.*, **99**, 5626–5633.

MacConnell, C.B. and Collins, H.P. (2009) Utilization of re-processed anaerobically digested fiber from dairy manure as a container media substrate. *Acta Hort.*, **819**, 279–286.

Mahimairaja, S., Bolan, N.S. and Hedley, M.J. (1995) Dissolution of phosphate rocks during the composting of poultry manure. *Fertil. Res.*, **40**, 93–104.

Mari, I., Ehaliotis, C., Kotsou, M., Chatzipavlidis, I. and Georgakakis, D. (2005) Use of sulfur to control pH in composts derived from olive processing by-products. *Compost Sci. Util.*, **13**, 281–287.

Martinez, J., Dabert, P., Barrington, S. and Burton, C. (2009) Livestock waste treatment systems for environmental quality, food safety, and sustainability. *Bioresour. Technol.*, **100**, 5527–5536.

Masse, L., Massé, D.I. and Pellerin, Y. (2007) The use of membranes for the treatment of manure: a critical literature review. *Biosyst. Eng.*, **98**, 371–380.

Maurer, C. and Müller, J. (2012) Ammonia (NH_3) emissions during drying of untreated and dewatered biogas digestate in a hybrid waste-heat/solar dryer. *Eng. Life Sci.*, **12**, 321–326.

Mondini, C., Chiumenti, R., da Borso, F., Leita, L. and De Nobili, M. (1996) Changes during processing in the organic matter of composted and air-dried poultry manure. *Bioresour. Technol.*, **55**, 243–249.

Munoz, G.R., Kelling, K.A., Rylant, K.E. and Zhu, J. (2008) Field evaluation of nitrogen availability from fresh and composted manure. *J. Environ. Qual.*, **37**, 944–955.

Nyord, T., Søgaard, H.T., Hansen, M.N. and Jensen, L.S. (2008) Injection methods to reduce ammonia emission from volatile liquid fertilisers applied to growing crops. *Biosyst. Eng.*, **100**, 235–244.

Oenema, O., Chardon, W.J., Ehlert, P.A.I., van Dijk, K., Schoumans, O.F. and Rulkens, W.H. (2012) Phosphorus fertilisers from by-products and wastes, *Proceedings 717*, International Fertiliser Society, Cambridge.

Osada, T., Sommer, S.G., Dahl, P. and Rom, H.B. (2001) Gaseous emission and changes in nutrient composition during deep litter composting. *Acta Agric. Scand. B*, **51**, 137–142.

Pagliari, P., Rosen, C., Strock, J. and Russelle, M. (2010) Phosphorus availability and early corn growth response in soil amended with turkey manure ash. *Commun. Soil Sci. Plant. Anal.*, **41**, 1369–1382.

Peters, K. (2010) Chemical and biochemical variation in solid fractions from animal slurry separation – improved opportunities for utilisation, *PhD Thesis*, University of Copenhagen.

Petzet, S., Peplinski, B. and Cornel, P. (2012) On wet chemical phosphorus recovery from sewage sludge ash by acidic or alkaline leaching and an optimized combination of both. *Water Res.*, **46**, 3769–3780.

Ponce, R.G. and De Sa, M.E.G.L. (2007) Evaluation of struvite as a fertilizer: a comparison with traditional P sources. *Agrochimica*, **51**, 301–308.

Preusch, P.L., Adler, P.R., Sikora, L.J. and Tworkoski, T.J. (2002) Nitrogen and phosphorus availability in composted and uncomposted poultry litter. *J. Environ. Qual.*, **31**, 2051–2057.

Rahman, M.M., Liu, Y., Kwag, J.H. and Raa, C.S. (2011) Recovery of struvite from animal wastewater and its nutrient leaching loss in soil. *J. Hazard Mater.*, **186**, 2026–2030.

Ro, K.S., Cantrell, K.B., Hunt, P.G., Ducey, T.F., Vanotti, M.B. and Szogi, A.A. (2009) Thermochemical conversion of livestock wastes: carbonization of swine solids. *Bioresour. Technol.*, **100**, 5466–5471.

Rubæk, G.H., Stoholm, P. and Sørensen, P. (2006) Availability of P and K in ash from thermal gasification of animal manure, in *Proceedings of the 12th RAMIRAN International Conference* (ed. S.O. Petersen), Danish Institute of Agricultural Sciences, Copenhagen, *Report 122*, Vol. **II**, pp., 177–180.

Schipper, W., Klapwijk, A., Potjer, B., Rulkens, W., Temmink, B., Kiestra, F. and Lijmbach, A. (2001) Phosphate recycling in the phosphorus industry. *Environ. Technol.*, **22**, 1337–1345.

Schipper, W.J. and Korving, L. (2009) Full-scale plant test using sewage sludge ash as raw material for phosphorus production, in *International Conference on Nutrient Recovery from Wastewater Streams* (eds K. Ashley, D. Mavinic and F. Koch), IWA Publishing, London, pp. 591–598.

Schoumanns, O.F., Rulkens, W.H., Oenema, O. and Ehlert, P.A.I. (2010) Phosphorus recovery from animal manure – technical opportunities and agro-economical perspective, *Alterra Report 2158*, Wageningen, Alterra.

Schröder, J.J. (2005) Revisiting the agronomic benefits of manure: a correct assessment and exploitation of its fertilizer value spares the environment. *Bioresour. Technol.*, **96**, 253–261.

Sommer, S.G. and Møller, H.B. (2000) Emission of greenhouse gases during composting of deep litter from pig production – effect of straw content. *J. Agric. Sci.*, **134**, 327–335.

Song, Y.H., Qiua, G.L., Yuan, P., Cui, X.Y., Peng, J.F., Zeng P., Duan, L., Xiang, L.C. and Qian, F. (2011) Nutrients removal and recovery from anaerobically digested swine wastewater by struvite crystallization without chemical additions. *J. Hazard. Mater.*, **190**, 140–149.

Sørensen, P. and Rubæk, G.H. (2010) Availability of P and K after application of ashes and biochars from thermally-treated solid manures to soil, presented at the *Workshop Managing Livestock Manure for Sustainable Agriculture*, Wageningen.

Steiner, C., Das, K.C., Melear, N. and Lakly, D. (2010) Reducing nitrogen loss during poultry litter composting using biochar. *J. Environ. Qual.*, **39**, 1236–1242.

Steinfeld, H., Gerber, P., Wassenaar, T., Castel, V., Rosales, M. and de Haan, C. (2006) *Livestock's Long Shadow – Environmental Issues and Options*, FAO, Rome.

Stelte, W., Sanadi, A.R., Shang, L. Holm, J.K., Ahrenfeldt, J. and Henriksen, U.B. (2012) Recent developments in biomass pelletization – a review. *Bioresources*, **7**, 4451–4490.

Stentiford, E. and de Bertoldi, M. (2011) Composting: process, in *Solid Waste Technology and Management* (ed. T.H. Christensen), John Wiley & Sons, Ltd, Chichester, pp. 515–532.

Streubel, J.D., Collins, H.P., Tarara, J.M. and Cochran, R.L. (2012) Biochar produced from anaerobically digested fiber reduces phosphorus in dairy lagoons. *J. Environ. Qual.*, **41**, 1166–1174.

Suzuki, K., Tanaka, Y., Kuroda, K., Hanajima, D., Fukumoto, Y., Yasuda, T. and Waki, M. (2007) Removal and recovery of phosphorous from swine wastewater by demonstration crystallization reactor and struvite accumulation device. *Bioresour. Technol.*, **98**, 1573–1578.

Tiquia, M. and Tam, N.F.Y. (1999) Fate of nitrogen during composting of chicken litter. *Environ. Pollut.*, **110**, 535–541.

Thygesen, A.M., Wernberg, O., Skou, E. and Sommer, S.G. (2011) Effect of incineration temperature on phosphorus availability in bio-ash from manure. *Environ. Technol.*, **32**, 633–638.

Thygesen, O. and Johnsen, T. (2012) Manure-based energy generation and fertiliser production: determination of calorific value and ash characteristics. *Bioresour. Technol.*, **113**, 176–172.

Velthof, G.L. (2011) Synthesis of the research within the framework of the Mineral Concentrates Pilot, *Alterra Report 2224*, Alterra, Wageningen.

Velthof, G.L., Oudendag, D., Witzke, H.P., Asman, W.A., Klimont Z. and Oenema, O. (2009) Integrated assessment of nitrogen emissions from agriculture in EU-27 using MITERRA-EUROPE. *J. Environ. Qual.*, **38**, 402–17.

Venglovsky, J. Sasakova, N., Vargova, M., Pacajova, Z., Placha, I., Petrovsky, M. and Harichova, D. (2005) Evolution of temperature and chemical parameters during composting of the pig slurry solid fraction amended with natural zeolite. *Bioresour. Technol.*, **96**, 181–189.

Vu, T.K.V., Tran, M.T. and Dang, T.T.S. (2007) A survey of manure management on pig farms in Northern Vietnam. *Livest. Sci.*, **112**, 288–297.

Vu, Q.D., Tran, T.M., Nguyen, P.D., Vu, C.C., Vu, V.K. and Jensen, L.S. (2012) Effect of biogas technology on nutrient flows for small- and medium-scale pig farms in Vietnam. *Nutr. Cycl. Agroecosyst.*, **94**, 1–13.

Wang, T. Camps-Arbestain, M., Hedley, M. and Bishop, P. (2012) Predicting phosphorus bioavailability from high-ash biochars. *Plant Soil*, **357**, 173–187.

Wnetrzak, R., Kwapinski, W., Peters, K., Sommer, S.G., Jensen, L.S. and Leahy, J.J. (2013) The influence of the pig manure separation system on the energy production potentials. *Bioresour. Technol.*, **136**, 502–508.

Zeng, L. and Li, X. (2006) Nutrient removal from anaerobically digested cattle manure by struvite precipitation. *J. Environ. Eng. Sci.*, **5**, 285–294.

15
Animal Manure Fertiliser Value, Crop Utilisation and Soil Quality Impacts

Lars S. Jensen

Department of Plant and Environmental Sciences, University of Copenhagen, Denmark

15.1 Introduction

From a historical perspective, land application of animal manure is as old as agriculture. The Chinese have used human and animal manure, and other household wastes, as a fertiliser for thousands of years. At the beginning of the twentieth century land application of animal manures in Europe and the United States was common. In those times the benefits of the manure were appreciated, and farmers were eager to use manure and other waste or residuals to improve crop yields. However, as industrially produced mineral fertilisers became available and affordable after World War II, manure and organic wastes were often seen as a disposal problem rather than a resource. For the EU member states, Oenema *et al.* (2007) have estimated that only around 50% of the nitrogen (N) excreted in livestock is recycled as a plant nutrient.

In sustainable agricultural production systems, "closing" the cycling of nutrients is a continuing goal, so that losses of nutrients from the system (to the atmosphere, to water bodies or to sediments) are minimised. As most of the world's soils have low contents of plant-available phosphorus (P) and P is a limited, non-renewable resource, recycling of manure and waste P as fertilisers is crucial for future food security and will increase in importance in the coming decades. Furthermore, fertiliser N production is highly energy demanding and currently reliant on fossil fuels, so this provides a further incentive for efficient recycling of N from manure to crop production.

The soil–plant–animal system offers an excellent opportunity for application and reuse of constituents present in manure and organic waste. Manure contains organic matter and essential plant nutrients including N, P, potassium (K), sulfur (S) and trace elements. It is therefore useful as a fertiliser and soil conditioner, and may replace the use of mineral fertilisers to a greater or lesser extent. In areas of intensive animal production, nutrients in the manure may contribute all or a substantial proportion of the local or regional fertiliser demand and some manure nutrients may even be in excess of local demand, requiring alternative utilisation to avoid environmental overload.

In order to fully understand how to optimise utilisation of animal manure fertiliser value, this chapter starts with a brief introduction to plant nutrition, crop fertilisation and nutrient use efficiency concepts. This is followed by a description of animal manure effects on soil fertility, including soil biological activity and nutrient turnover, as well as soil physical properties. The concept of mineral fertiliser equivalency (MFE) is introduced, as a key to quantifying and improving manure nutrient recycling efficiently in crops, and the effect of various application methods and manure treatment processes are described. Finally, manure impacts on long-term nutrient supply and strategies for optimal combination of animal manures and mineral fertilisers are described.

15.2   Fertilisation and Crop Nutrient Use Efficiency

The major purpose of fertilisation is to increase crop dry matter production and hence the yield of the harvested parts of the plants. However, the application of nutrients affects not only the size but also the chemical composition of the plant. In other words, the use of fertilisers or manures has consequences for both the quantity and quality of the harvested plant material. The effects of applied plant nutrients depend on climate conditions, interacting with the physical, chemical and biological properties of the soil.

Both macro- and micronutrients are of equal importance for plant growth (Text Box – Basic 15.1), but in most soils the availability of the macronutrients N, P and K is usually the most limiting factor for plant primary production. In intensive agriculture these macronutrients need to be added regularly in the form of fertilisers or organic manures to maintain high yields for most crops.

Text Box – Basic 15.1   Definitions of plant nutrition and agronomic terms used

Macronutrients: In addition to carbon (C), hydrogen (H) and oxygen (O), plants require 14 elements that are essential for their growth and reproduction. Macronutrients are defined as those required in concentrations of more than $1 \, g \, kg^{-1}$ plant dry matter, and comprise nitrogen (N), potassium (K), calcium (Ca), magnesium (Mg), phosphorus (P) and sulfur (S), listed in decreasing order of concentration.

Micronutrients: Essential plant nutrients required in concentrations of less than $0.1 \, g \, kg^{-1}$ plant dry matter are defined as micronutrients, and comprise chloride (Cl), boron (B), iron (Fe), manganese (Mn), zinc (Zn), copper (Cu), nickel (Ni) and molybdenum (Mo), listed in decreasing order of concentration.

Liebig's law of the minimum: Plant growth is controlled by the nutrient that is available in the smallest amount relative to plant requirements. Crop yield will increase with the supply of the limiting nutrient until another nutrient becomes limiting (shown by Justus von Liebig in 1840). Thus, both macro- and micronutrients can be limiting for crop production, although they are required in greatly differing quantities.

Soil fertility: Designates the ability of soil to provide plants with sufficient, balanced and non-toxic amounts of nutrients and water, and to act as a suitable medium for root development, to assure proper plant growth and maturity. Soil fertility is basically controlled by the inherent soil mineralogy and texture (the proportions of clay, silt and sand) given by its location and geology, and by the dynamic parameters of soil organic matter content (linked to soil biota and biogeochemical cycles of C, N, S and P; see Text Box – Basic 15.2 below), acidity, nutrient concentrations, porosity and water availability, all of which can be influenced by human activity and management.

Mineral (also termed chemical) fertiliser: Inorganic compounds of natural or synthetic origin added to supply one or more essential nutrients for the growth of crops. The discovery by Sir John Bennett Lawes in 1840 that sulfuric acid treatment of bone meal and later phosphate rocks produces a plant-available P fertiliser ("superphosphate") and the 1908 invention of the Haber–Bosch process, enabling industrial fixation of atmospheric N_2 into reactive NH_3, revolutionised mineral fertiliser supply. Fertiliser production and consumption have grown dramatically since World War II, and due to global growth in population and wealth continue to rise by 2–3% $year^{-1}$. According to conservative estimates, 50% of global crop yields can currently be attributed to the use of synthetic fertilisers. Fertiliser production is very energy-demanding, especially for N, consuming 40–50 MJ kg^{-1} N, but only 14 and 9 MJ kg^{-1} for P and K, respectively.

15.2.1   Source of Nitrogen Supply to Crops

Nitrogen is in most cases the plant nutrient that has the largest influence on crop dry matter production. Except for cropping systems with a significant proportion of N_2-fixing crops (leguminous species, such as beans, lentils, peas, clovers), most cropping systems therefore rely heavily on external N inputs from inorganic or organic fertilisers. These N inputs represent resources which should be utilised efficiently, both from an economic and an environmental viewpoint.

The dominant processes in N turnover through the soil–plant–atmosphere system are illustrated in Figure 15.1. The soil organic matter acts as a large source and sink for nutrients via mineralisation and immobilisation by the soil microbial biomass, which itself only constitutes a few per cent of the total C and N in soil. Once the substrate has become solubilised (i.e. by the action of extracellular enzymes), it may enter into the decomposer system. Depending on the properties of the material and the soil conditions, it may directly give rise to either net immobilisation or mineralisation. When the microorganisms proliferate, this will sooner or later activate their predators and following predation a release of some microbially bound nutrients will occur. The N thus released from the microbial biomass in the form of ammonium (NH_4^+) may be nitrified

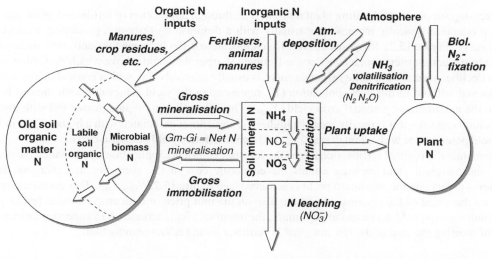

Figure 15.1   *The principal components of the N cycle in the soil–plant–atmosphere system. (© University of Copenhagen.)*

into nitrate (NO_3^-), taken up and utilised for plant growth, lost by leaching or lost in gaseous form due to denitrification or volatilisation.

It is important to understand that the two opposing processes in soil – mineralisation of organically bound nutrients to inorganic forms and immobilisation of inorganic to organic forms through assimilation by microorganisms – take place concurrently and continuously (i.e. even when no fresh inputs of organic matter to the soil have occurred). However, the rates of these two processes are usually affected quite differently by environmental factors. Net N mineralisation (the difference between gross mineralisation and gross immobilisation) therefore varies widely depending on total soil organic matter content, recent applications of organic N, soil structure, climate and tillage, which may affect the soil nutrient status dramatically. For agricultural soils in temperate climates, net N mineralisation rates in the growing season usually vary from less than 0.1 up to 1 or even 2 kg N ha^{-1} day^{-1}, the latter especially in organic soils or perennial pastures that have recently been ploughed under. This means that soil N supply for a crop can be quite substantial, providing that the soil is rich in labile organic N (Jarvis *et al.*, 1996).

Plants take up N from the soil mineral pool in the form of either NH_4^+ or NO_3^-. The sources of crop N uptake can thus be (see Figure 15.1):

- Net mineralisation of soil organic N, including both more recently added organic N and immobilised residual fertiliser N from previous fertiliser applications.
- Deposited N from the atmosphere, either as dry or wet deposition of mainly inorganic species.
- Fertiliser N and other direct sources of inorganic N (e.g. animal manure applied to present crop).

The balance between these sources depends on the soil type (soils high in organic matter having higher N mineralisation potential), crop type, fertiliser N application rates, farm type (intensity of manure application) and location (arable land in industrialised areas receiving much higher inputs of N through atmospheric deposition than in more remote areas).

15.2.2   Crop Yield Response to Fertiliser Nitrogen and the Economic Optimum

With increasing supply of the limiting plant nutrient (e.g. through application of fertilisers), plant growth and biomass production typically increase, but usually with a diminishing response, producing a convex yield response curve (Figure 15.2). Responses to fertiliser application differ between nutrients, with micronutrients and immobile macronutrients such as P demonstrating a sharper threshold, above which no further increase, or even a decline, occurs. For N, the response curve is usually relatively flat, with a gradual increase from the yield achieved without fertiliser N application (Y_0, representing the yield achievable with the soil N supply alone), to the maximum or potential crop yield (Y_{max}). On soils richer in organic matter and with high net N mineralisation capacity, this typically means a higher Y_0, sometimes associated with a higher Y_{max}, indicating higher yield potential, or with an unaltered Y_{max} and hence a flatter yield response curve.

The economically optimal fertilisation rate is the fertiliser (nutrient) application rate at which the economic value of the marginal yield increase equals the economic cost of the corresponding marginal fertiliser application – at this rate the maximum profit is achieved (see Figure 15.2, $N_{Econ.opt.}$). This economic optimum depends on the shape of the response curve, but also on the unit price of both crop products (yield, p_Y) and fertiliser nutrients (p_N). At the economic optimum, the marginal yield increase (ΔY) generates an income just capable of meeting the cost of the last marginal N fertiliser input (ΔN), meaning that:

$$\Delta Y \cdot p_Y = \Delta N \cdot p_N \Leftrightarrow \frac{\Delta Y}{\Delta N} = \frac{p_N}{p_Y} \qquad (15.1)$$

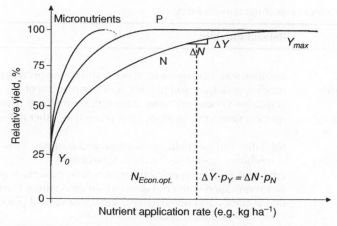

Figure 15.2 *Illustration of yield response curves for macronutrients (N, P) and micronutrients, and economically optimal N fertilisation rate (p_N, unit price of fertiliser N; p_Y, unit price of yield; Y_0, yield with soil N supply only; Y_{max}, maximum yield). (© University of Copenhagen.)*

or in other words, the economic optimum is where the marginal yield ratio equals the price ratio of fertiliser to crop (right part of Equation 15.1). The economic optimum may then be derived from experimental data by fitting a function $f(N)$ to the yield data (Y), differentiating $f(N)$, equalling it to the price ratio and then solving for N_{opt}:

$$Y = f(N) \Leftrightarrow \frac{dY}{dN} = f'(N) \Leftrightarrow f'(N_{Econ.opt}) = \frac{p_N}{p_Y} \qquad (15.2)$$

Typically f is either a quadratic or a cubic function. As long as the fertiliser has a price above zero, the economically optimal fertilisation rate ($N_{Econ.opt.}$) will always be below the fertilisation rate needed to achieve maximum yield (Y_{max}, Figure 15.2). The cost of spreading the fertiliser (usually a fixed cost) will influence the profit of fertilisation, but not the economically optimal rate.

15.2.3 Crop Nitrogen Uptake Efficiency

The efficiency with which applied fertiliser N is used by the crop for dry matter production is of major concern, both from an economic and an environmental viewpoint. For the farmer, fertilisation represents an economic cost, and hence an optimal efficiency is desirable, and losses to the environment will always be undesirable, albeit some losses are inevitable.

Use efficiency of applied N may be defined either as the *agronomic efficiency* (i.e. the amount of harvestable product (e.g. grain) gained kg^{-1} fertiliser N applied) or as the *fertiliser N uptake efficiency* (i.e. the amount of N absorbed kg^{-1} fertiliser N applied).

A range of factors influence the N uptake efficiency of arable crops (Table 15.1). Some of these affect crop uptake efficiency for both soil- and fertiliser-derived N, whereas some are related only to fertiliser N and are thus directly influenced by management strategies for applied mineral and organic fertilisers.

Crop N use efficiency can be determined from field experiments with increasing rates of fertiliser N application. The *agronomic efficiency* (AE) can be calculated from the yield (Y) increase between two N

Table 15.1   *Factors that affect crop N uptake efficiency.*

Factor	Related variables
Crop factors	
rate of N assimilation	weather, soil, management, species and genotype/variety
rooting depth and density	species, genotype, soil physics, fertiliser placement
crop growth stage	vegetative versus generative stage (i.e. grain filling), senescence and leaf drop
final yield of N in crop	species, genotype, weather, crop protection, other nutrients
Fertiliser factors	
formulation	solubility and reactivity, granulation and coatings, slow-release or with inhibitors against nitrification, for example
rate	crop response, economic optimum, within maximum yield
timing	synchronisation with crop demand, autumn/winter leaching
application method	rate accuracy and homogeneity, surface or subsoil placement, conventional or reduced tillage system
Availability of fertiliser N	
immobilisation	NH_4^+-fixing clay minerals, C : N ratio of crop residues
NH_3 volatilisation	inadequate application of anhydrous NH_3, $(NH_4)_2SO_4$, urea or manures (surface application/dry soil/too high rate)
denitrification	Soil aeration, temperature, pH, organic C, NO_3^-
leaching of NO_3^-	soil water balance (excess precipitation), autumn fertiliser application, crop type, plant uptake, rooting depth
competition	weeds, immobilisation by microorganisms
Availability of soil N	
organic N in root zone	amount and type of organic N added in previous years, previous crop(s), decomposition, soil aeration and temperature
inorganic N in root zone	residual fertiliser N, soil aeration, leaching and gaseous losses

fertiliser treatments (with fertiliser N rates of $N_2 > N_1$). AE has the units kg crop produced kg^{-1} fertiliser N applied:

$$AE = \frac{(Y_2 - Y_1)}{(N_2 - N_1)} \tag{15.3}$$

The *fertiliser N uptake efficiency* (NUE) – the amount of additional N absorbed by the crop from the fertiliser – can similarly be calculated from the increase in N uptake (N_{upt}) between two N fertiliser treatments and expressed as a percentage of fertiliser N applied:

$$NUE = \frac{(N_{upt2} - N_{upt1})}{(N_2 - N_1)} \cdot 100 \tag{15.4}$$

For practical reasons, usually only N content in the aboveground plant parts is determined or even just the N uptake in the harvested product (e.g. cereal grains). The NUE calculated in this way (Equation 15.4) is also sometimes termed the *apparent N recovery* (ANR) rate, as it builds on the assumption that the additional N uptake ($N_{upt2} - N_{upt1}$) is only derived from the applied additional fertiliser, implying that the uptake of soil N does not change with increasing fertiliser application. This is not always true, as increased growth of the crop may enhance root growth and uptake of additional soil mineral N; or the opposite, that increased N supply inhibits root development and decreases crop uptake of native soil N.

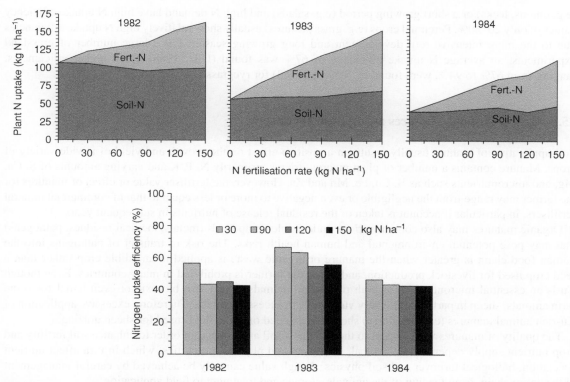

Figure 15.3 *Spring barley N uptake derived from fertiliser N and soil N, respectively (top graphs), and fertiliser N uptake efficiency (bottom graph) as affected by fertiliser N application rate in 3 consecutive years (same site all years) following a perennial white clover crop ploughed under before seeding of spring barley in 1982. (Modified with permission from Nielsen* et al. *(1988). © 1988 Elsevier.)*

Figure 15.3 illustrates a typical set of experimental data on N uptake efficiency of spring barley (Nielsen *et al.*, 1988). In this case the experiments were conducted with the use of isotopically labelled (^{15}N) fertiliser, so the plant N uptake can be differentiated into the contribution from soil N and fertiliser N, respectively. The uptake of soil N appears to be more or less unaltered by increasing fertilisation rate. The white clover, an N-fixing species, was grown for several years before the first crop in this experiment and had left a large pool of mineralisable organic N in the soil. This soil N pool contributed greatly to crop N uptake in the first year, but the contribution declined dramatically over the 3-year experimental period (i.e. from 107 kg N ha^{-1} in the first year to 40 kg N ha^{-1} in the third year). The N uptake efficiency (bottom graph in Figure 15.3) varied between 43% and 67%, and showed a slightly decreasing trend with increasing N fertilisation rate.

These N uptake efficiency values are typical for cereals grown in temperate regions. Pilbeam (1996) found that N uptake efficiency for winter wheat varied between 47% and 63% in a number of experiments in humid, temperate climates. For other types of crops, N uptake efficiency may be higher or lower. As an example, sugar beet has a long growing season and is known to be capable of very large N uptake; Dilz (1988) reviewed a large number of experiments yielding an average N uptake efficiency of 77%. Potatoes, on the other hand, have a highly variable capacity for uptake of N, partly due to varying length of growing season; Dilz (1988) cited results ranging from 20% to 70%, with an average of 51%. Similarly, oilseed rape usually has an N uptake efficiencies not exceeding 50% (Schjoerring *et al.*, 1995). Vegetable crops with shallow root systems (e.g. onions, leeks) or a short growing period (e.g. salads) and high N demand have high N uptake efficiency values of only 20–30%. Perennial crops (e.g. grass pastures) usually show relatively high N uptake efficiencies due to the more intensive root development and long growing season. For a large number of grassland experiments, an average N uptake efficiency of 67% was found (Dilz, 1988). Even higher efficiencies, ranging from 67% to 94%, were found by Stevens (1988) for ryegrass.

15.3 Use of Animal Manures as Organic Fertilisers

Soil application of manure usually results in a positive effect on the growth and yield of a wide variety of crops. Manure contains a number of plant macronutrients, primarily N, P, K and varying amounts of S, Ca, Mg, and micronutrients such as B, Cu, Fe, Mn and Zn. However, the fertiliser value or effect of manures for the farmer may range from the negligible or even negative to more or less equal to that of commercial mineral fertilisers, in particular if account is taken of the residual release of nutrients in subsequent years.

Organic manures may also contain various undesirable compounds (metals, medical residues, pathogens) that may pose potential environmental and human health risks. The risk of transfer of pathogens into the human food chain is greater when the manure or organic waste is applied to an edible crop rather than a feed crop used for livestock production, and hence the former is prohibited in many countries. Even though Cu is an essential micronutrient for both plants and animals, it can also be toxic or even lethal for some farm animals; sheep in particular are very vulnerable to excess Cu intake. Therefore, excessive application of Cu-rich animal manures (e.g. pig slurry) should be avoided on grass fields used for sheep grazing.

The quality of manures with respect to their use as a soil amendment in order to enhance soil fertility and crop nutrient supply depends on chemical, biological and physical properties, which in turn affect nutrient speciation, biological turnover and soil physics. A high value can only be achieved by careful management in the entire chain, from feeding of the animals, storage and treatment to field application.

15.3.1 Manure Chemical Properties and Effects on Soil Fertility

The fertiliser value of manure is mainly related to its chemical composition, and especially its contents of the macronutrients N, P and K, but also several micronutrients. The chemical composition of manures depends on the entire chain from which it is produced (i.e. the crop–feed–animal–excreta–storage–field chain). The speciation of each nutrient affects its plant availability and this is especially relevant for N, where the NH_4^+ form is highly plant-available, whereas the organic species are unavailable and require mineralisation to become available. The N speciation varies greatly between different manure types (Figure 15.4), from a very high proportion of NH_4^+ (70–90%) in many of the liquid manure types, to a very low proportion in deep litters and composts. For poultry manure, the proportion of uric acid is significant, and although organic, this is relatively readily mineralised after application to soil and hence may become available to plants within a growing season, even if not present as NH_4^+ initially.

After soil application, the NH_4^+ may take part in many reactions, of which the most important are (Figure 15.1): (i) uptake into growing plants, (ii) assimilation into active metabolising bacteria or fungi (immobilisation), (iii) oxidation by chemo-autotrophic bacteria (nitrification), and (iv) fixation into soil minerals (minimal in soils with low clay content). Each of these processes may affect a considerable amount of the added NH_4^+, depending on the actual conditions in the soil when the manure is added.

For each animal type, the balance between the microbial fraction (including microbial residues) and undigested feed is to a large extent a function of the digestibility of the feed, and this in turn has considerable implications for the fertiliser value of the excreta. This has been shown for ruminant feeding, where increasing digestibility of organic N content in feed increases the degradability of organic N in the manure produced.

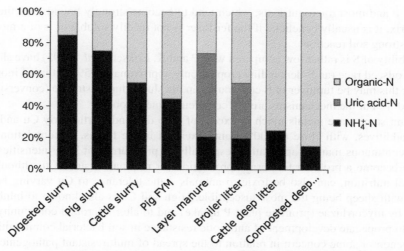

Figure 15.4 *Typical proportion of NH_4^+-N, organic-N and uric acid-N in various manure types. (© University of Copenhagen.)*

This in turn is significantly correlated with the proportion of organic N being mineralised during anaerobic storage as slurry. Consequently, more mineralised N is available for the crop after field application of the slurry, because net N mineralisation is faster (i.e. immobilisation occurs to a lesser degree) (Kyvsgaard *et al.*, 2000). A similar relationship has not been demonstrated for pig feeding, however (Sørensen and Fernandez, 2003). Decomposition and mineralisation during storage also increase the proportion of NH_4^+ (e.g. it has been estimated that 30–40% of the organic N in faeces and 90% of urine N is mineralised during a typical 6- to 9-month storage period) (Sørensen *et al.*, 2003). However, there is generally limited knowledge on interactions between environmental storage conditions (temperature, oxygen, etc.) and manure type (animal type, bedding material, etc.).

The P in manure is predominantly in inorganic form (80–90%; Huang *et al.*, 2011) and becomes more or less plant-available in neutral or slightly acidic soils, with the remaining organic forms taking part in the turnover of added organic matter as described for N. Manure K is exclusively in inorganic form, mostly as soluble compounds.

Manure contains micronutrients such as Fe, Mn, B, Zn and Cu, which are essential for plant growth. Consequently, it can also serve directly as a valuable source of micronutrients for plants when used as a macronutrient source. Most of the micronutrients in manure are chelated or complexed by organic compounds in the manure, which again may be either soluble or particle-bound. Manure application to soil may also indirectly influence the availability of native micronutrients through the action of dissolved manure organic matter, which can enhance solubilisation of metal micronutrients, sorbed or precipitated on the soil particles, through complexation and release into the soil solution and subsequent absorption by plants (Nikoli and Matsi, 2011). However, manure application may also alter soil pH, which has a marked influence on the availability of many trace elements, and hence the effect of manure on crop micronutrient uptake is often variable and inconsistent, depending on the interactions between these factors.

The nutritional value of P, K and micronutrients in manure is generally comparable or superior to that of mineral fertilisers, applied at a similar time. However, some caution is required, as the immediate availability may be somewhat lower (e.g. of P and micronutrients), and thus if the crop has an obvious deficiency, mineral fertilisers with higher solubility and availability may be more adequate. On the other hand, with reactive nutrients such as P and most micronutrients, which tend to become strongly bound once they start to react with the soil matrix, it is usually beneficial if the fertiliser is not readily soluble or is of a more slow release type, preventing strong soil reactions.

Manure availability of S is rather low compared with P and K. Eriksen *et al.* (1995) have shown that slurry S availability for oilseed rape (an S-demanding crop) is only approximately 5% of that of inorganic fertiliser S. The reason for this may be turnover of S compounds in the slurry during storage (conversion of sulfate to organically bound S), but the mechanisms are not completely understood.

Manures contain some trace metals much in excess of crop demand, particularly Cu and Zn, which are applied as feed additives, with close to 100% being excreted in the faeces. Accumulation of Cu and Zn may occur with continuous manure application, especially of pig slurries at high intensities (Benke *et al.*, 2008). This may become a problem for forage production for ruminants, since Cu, although an essential element in animal nutrition, can also be toxic to animals – the tolerance to Cu varying between species and breeds, and with sheep being the most susceptible. High soil Cu is also known to inhibit colonisation of soil and roots by mycorrhizae (promote plant P uptake) and to alter microbial community structure, and has been shown to propagate development of antibiotic resistance in soil bacterial communities (Berg *et al.*, 2010), a phenomenon of some concern in relation to the spread of multiresistant pathogenic bacteria in the environment.

Manure also contains unwanted trace metals, such as Cd, which are considered serious environmental contaminants. However, Benke *et al.* (2008) showed that total contents of Cd were not significantly increased after 25 years of manure application, even at excessive rates (up to 180 ton ha^{-1}), but that Ethylenedi-aminetetraacetic acid (EDTA)-extractable Cd had more than doubled, due to the increased solubilisation from organic matter. It should be noted that commercial mineral fertilisers also contain small amounts of metal contaminants. Phosphorus fertilisers in particular potentially contain contaminants such as Cd, F, Sr, As, Pb and Hg, depending on the origin of the ingredients, phosphate rock and sulfuric acid, used in the production of the fertiliser (McLaughlin *et al.*, 1996).

For a further description of manure chemical properties, please refer to Chapter 4.

15.3.2 Manure Effects on Soil Biological Activity and Nutrient Turnover

Addition of manure or other processed organic waste products affects soil fertility (Text Box – Basic 15.1) and biology (Text Box – Basic 15.2) in mostly positive ways. Organic waste application stimulates soil microbial and faunal (especially earthworm) activity, and hence improves soil fertility, as these soil organisms play a central role in mineralisation and turnover of organically bound nutrients. A study by Saison *et al.* (2006) found that compost addition affected the activity, size and composition of the soil microbial community, but the effect was mainly due to the physico-chemical properties of the compost matrix and the stimulation of native soil microorganisms by available substrate, rather than to compost-borne microorganisms. However, organic waste materials may also contain unwanted biota, including parasites and pathogens of man and animals such as bacteria (e.g. *Salmonella*, *Campylobacter*, *Enterococcus*), viruses and parasites (e.g. *Ascaris*); for further details, see Chapter 6. Biological treatment (e.g. by anaerobic digestion or efficient composting) prior to application to land most often reduces or eliminates these pathogens, depending on the pH, temperature and retention time of the biological treatment.

Text Box – Basic 15.2 Soil biology definitions

Soil organic matter (SOM): The organic matter in soil, except for materials identifiable as more or less undecomposed biomass, is called humus and is the solid, dark-coloured component of soil. It plays a significant role in soil fertility (Text Box – Basic 15.1), and is formed by microbial decay of added organic matter (e.g. plant residues, manures and other organic waste) and polymerisation of the cycled organic compounds – a process termed humification. This produces a complex mixture of highly aromatic and phenolic compounds, which are very stable and recalcitrant against decay. Humus contributes to the cation exchange capacity of soils (Chapter 4; i.e. the ability to retain cationic nutrients by adsorption, in a form accessible to plants, but in a non-dissolved form prevented from leaching). The biochemical structure of humus also enables it to moderate – or buffer – excessive acid or alkaline soil conditions. Toxic substances such as heavy metals, as well as micronutrients, can be chelated (i.e. bound to the complex organic molecules of humus) and so prevented from entering the wider ecosystem. Finally, the biogeochemical cycles of C, N, S and P are intrinsically linked to the turnover of soil organic matter brought about by the activity of the soil biota.

Soil biology and biota: The soil biota comprises the following types of organisms and functions:

- Microorganisms, including fungi and bacteria, dominate the soil biota in terms of numbers, mass and metabolic capacity; as primary decomposers they have a major influence on soil biological fertility.
- Fauna (animals), ranging in size from earthworms to single-celled protozoa, also play a role in soil fertility, especially by physically fragmenting and mixing organic residues, creating biopores and regulating bacterial and fungal populations through predation.
- Flora (plants) includes the roots of higher plants as well as microscopic algae, and these serve as major, continuous inputs of organic substrate for the soil microorganisms and fauna.

Fertiliser practices affecting soil fertility and biology

- Adding inorganic lime and fertilisers to an infertile soil generally increases chemical soil fertility, but only affects microbial and faunal activity indirectly, through increasing plant biomass that is likely to be returned to the soil as roots, root exudates and shoot residues.
- Adding organic fertilisers, such as animal manures and biosolids, to an infertile soil influences microbial and faunal activity directly, as they provide organic components serving as substrate for the microorganisms and fauna, and thus enhance soil fertility.

Immediately after manure is applied to the soil, microbial decomposition of the manure continues in the specialist soil environment. Urea (if present) is hydrolysed rapidly to NH_4^+ by the extracellular urease enzyme that is ever-present in the soil. The organic matter in different manures is not equally degradable (e.g. slurry contains a relatively large fraction of easily degradable C- and N-containing compounds, while for composted manure the fraction of easily degradable compounds is small, since they have already been metabolised in the composting process). If the manure is not fully metabolised prior to application to the soil, the availability of organic N corresponds initially to a biological half-life of approximately 4–8 years (approximately 10–20% of remaining organic N being mineralised each year), but a longer half-life later on (see Section 15.4.4). This is substantially slower than for decomposition of fresh plant litters, which is reflected in the finding in one study that soil C had only increased by 11–20% of the total straw C input during 30 years of continuous input of straw to soil, while for less decomposable solid manure the increase in soil C was 36% of total manure C inputs over the same period, indicating the much more recalcitrant nature of manure organic matter (Christensen and Johnston, 1997).

For those plant nutrients that are to any extent bound in soil organic matter, biological assimilation and mineralisation by soil microorganisms play a major role in regulating their concentration in the soil solution. The addition of any sort of organic material, whether plant root exudates, dead roots, plant litter or animal and

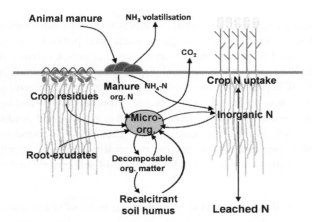

Figure 15.5   *Turnover of organically bound N in soil is affected by inputs of organic matter, including crop residues and manures, decay and competition between plant and microbial uptake, as well as water percolation. (© University of Copenhagen.)*

human organic waste materials, may therefore significantly alter the turnover of these nutrients in soil, and hence their availability for plant uptake and risk of losses to the environment (Figure 15.5). This is clearly the case for N, but also for S and to some extent for P, whereas for K biological activity does not have any direct effect. Indirect effects could be that the increased biological activity brought about by addition of the organic material may affect other parameters influencing the availability of the nutrient, such as by altering the redox potential, pH, aeration or amounts of organic chelates (e.g. affecting micronutrient solubility) in the soil solution.

Nearly irrespective of the $C : N_{org}$ ratio of fresh or stored manures, net immobilisation of soil inorganic N occurs temporarily after incorporation of the manure into the soil (Paul and Beauchamp, 1994) and reaches a maximum within approximately 1 week (Sørensen and Jensen, 1995; Sørensen and Amato, 2002). This net immobilisation conceals the fact that both gross mineralisation and gross immobilisation rates increase drastically due to manure incorporation, and that gross immobilisation only exceeds gross mineralisation for a relatively short period. The degree and duration of the immobilisation vary with the amount of easily degradable substances, but the degree is typically in the order of 20–30% of the initial manure NH_4^+ content.

An illustration of this mineralisation–immobilisation turnover (MIT) can be seen in Figure 15.6a, where inorganic N is rapidly immobilised over the first few days. A close relationship between the volatile fatty acid (VFA) content of the four different manures and the maximum immobilisation of soil inorganic N can be seen (Figure 15.6b), indicating that the initial immobilisation is mainly driven by easily degradable organic matter. As a consequence of this relationship, urine and other manures with a low content of VFAs do not seem to induce temporary N immobilisation to nearly the same degree when applied to soil (Sørensen, 1996). It has also been shown that the degree of N immobilisation after slurry incorporation increases with soil clay content (Sørensen and Jensen, 1996) and with the degree of mixing with the soil, with higher crop recoveries of injected slurry N (i.e. little contact between slurry and soil; Sørensen and Jensen, 1995). This implies that although losses through ammonia volatilisation and denitrification can be avoided through incorporation and mixing into the soil, crop utilisation of the manure N may be hampered in loamy and clayey soils.

From Figure 15.6a it can also be seen that a large remineralisation of immobilised N occurs over the following approximately 2 weeks (this experiment was conducted at 25 °C, so this is a relatively rapid remineralisation pattern compared with common natural conditions). Although remineralisation occurs to a great extent, this temporary immobilisation and delayed release of mineral N may well cause substantial

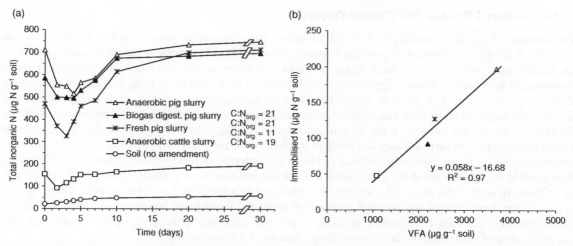

Figure 15.6 *Relationship between N immobilisation and initial VFA content of the manure. (a) Time course of soil inorganic N content (NH_4-N + NO_3-N) after application of various animal slurries (2.5 g slurry kg^{-1} soil) and (b) correlation between N immobilisation and initial VFAs in the slurries added to soil. (Modified with permission from Kirchmann and Lundvall (1993). © 1993 Springer-Verlag.)*

failure in the synchronisation between soil N supply and crop N demand, if the manure is applied immediately before sowing or in late spring in the growing crop. This is a major barrier to high and predictable N use efficiency when using manures as a source of N for crops. After the initial period of N immobilisation and remineralisation the organic N in the manure continues its turnover, albeit at a slower rate than initially, as the remaining organic matter becomes more and more recalcitrant. The overall net mineralisation trend may be quite variable. Van Kessel *et al.* (2002) found values for a wide range of manures to vary from a net mineralization of 55% to a net immobilization of 29% of the manure organic-N, and with single manure parameters capable of explaining less than 16% of variation or less than 50% when combined in a multiple regression. Although difficult to predict, a rule-of-thumb is that manures with an organic-C : organic-N ratio above 15 will immobilize N, and those with ratios below 15 will release N in the short to medium term (Webb *et al.*, 2013). The prediction of manure N mineralisation is further complicated as some studies indicate that mineralisation and immobilisation processes do not respond similarly to temperatures (Andersen and Jensen, 2001).

Application of manure to soil may also contribute positively to long-term stabilisation of C in the soil and hence to decreasing atmospheric CO_2 levels (Smith *et al.*, 2001). Carbon dioxide from degraded fresh organic matter such as manure is considered neutral with respect to global warming impact, because the plants used in the animal feed have recently removed an equal amount of CO_2 from the atmosphere during growth. Thus, organic matter from manure stored in the soil represents "saved" emissions of CO_2. This mechanism is referred to as carbon sequestration and carbon is sequestered (protected) in soils as soil organic matter. However, C is not sequestered indefinitely and soil C eventually reaches a new equilibrium with decomposition, where no further increase is achieved. Nevertheless, even with the low net C storage simulated after 100 years (Bruun *et al.*, 2006), this corresponds to avoided CO_2 emissions of approximately 180 kg CO_2 ton^{-1} dry matter of manure applied to the soil.

For a further description of manure organic matter characteristics and microbial transformations, please refer to Chapter 5.

15.3.3 Manure Effects on Soil Physical Properties

The addition of organic materials such as solid manure, deep litter or compost to a fine-textured soil can assist the development of structure (aggregation) and can stabilise soil aggregates against dispersion. During the humification process of the added organic material, microbes secrete sticky gum-like mucilages; these contribute to the crumb structure (aggregates of soil particles) of the soil by holding particles together, thus allowing greater aeration of the soil. Improvement of the soil structure makes it more friable (lower soil strength, desirable for cultivation and seedbed preparation), and increases the amount of pore space available for plant roots and the entry of water and air into the soil. Long-term (90 years; the Askov Long-Term Trials) manure application to a loamy soil resulted in a large increase (50–100%) in stable large aggregates in animal-manured soil compared with mineral fertilised and unfertilised treatments (Schjønning *et al.*, 1994). This resulted in a significant shift in the plastic limit of the soil (the water content below which the soil can be cultivated), so that the manured soil could be cultivated at a much higher water content. This effect is important for timely tillage and seedbed preparation under the usually wet spring conditions in temperate humid climates. The long-term manure application also increased total pore space by 2–4%, although this had relatively little influence on the water-holding capacity of this loamy soil.

In coarse-textured sandy soils, organic matter can increase the water-holding capacity of the soil significantly, and provide exchange and adsorption sites for nutrients. Humus can hold the equivalent of its weight in moisture and therefore increases the soil's capacity to withstand drought conditions. In more arid regions with limited water supplies, solid organic manures may be used as mulches (surface applications of 5–15 cm thickness) for water conservation, which increase water-holding capacity and reduce evaporation from the soil surface, thus reducing water requirements. Addition of manure and other organic materials to these soils is particularly beneficial.

Furthermore, diurnal variations in surface soil temperatures are reduced by mulches of solid manure (Dahiya *et al.*, 2001) and they may therefore serve to modulate variations in soil temperatures, buffering from temperature extremes at the soil surface, which is beneficial for seed germination and early seedling growth. In cooler climates, on the other hand, the dark colour of humus (usually black or dark brown) helps to warm up cold soils in the spring.

Finally, weed infestation of crops may be reduced by mulching with organic materials, e.g. manures (Preusch and Twokoski, 2003) or by injection of slurry, placing the available nutrients below the soil surface where most of the weed seeds germinate. However, slurry with a high content of VFAs may also be lethal to crop seedlings and hence care should be taken to separate the germinating or small plants from the manure application in space or time.

15.4 Manure Fertiliser Value as Affected by Application Method, Manure Type and Treatment

Although manures have many potentially beneficial properties when used as fertiliser, they are usually bulky, of variable composition and have high water contents. Accordingly, they cannot be transported very far before the transportation costs exceed the fertiliser value. In addition, application scheduling compatible with agricultural planting, weather conditions and harvesting requires careful management. This often limits the best application time for efficient plant nutrient use to certain months (e.g. in a temperate, humid climate on arable land to early spring) and may introduce significant costs for storage facilities. However, farmers may prioritise applying manures in autumn or winter, simply because they are less busy in the field during those periods than in the spring. Furthermore, some of the nutrients in organic manures are not readily available for plant uptake, but must be mineralised first. Thus, manures are much more difficult fertilisers to manage and utilise than the chemical counterparts they have the potential to replace. Maximum efficiency can only be achieved by careful management of the field application techniques in accordance with the soil, weather and manure conditions. In order to compare different manure treatments and application techniques, it is also necessary to establish a meaningful method for measuring fertiliser efficiency.

15.4.1 MFE Value

In agronomic practice, the term "N mineral fertiliser equivalent" (N-MFE) value may be used to describe how efficient an organic fertiliser or manure is in providing available N for the crop compared with a mineral fertiliser source of N. The term N fertiliser replacement value (NFRV) is also used and is a synonym for N-MFE value. The MFE value thus designates how much mineral N fertiliser manure N may replace during the first year after application, under a given set of conditions (e.g. time of application, crop type, climate). For example, if the MFE value is 40%, then 100 kg of manure N will be able to replace 40 kg of mineral fertiliser N. The residual 60 kg N is either lost or enters the soil organic N pools, where it may become plant-available in subsequent years, or may be lost through leaching or denitrification.

An experimental estimation of MFE value may be carried out by different methods, but it always includes comparison of the effect of the manure with that of a well-known and highly efficient reference fertiliser, typically an NH_4^+ nitrate-based mineral fertiliser. A standard method is to perform a traditional fertiliser N response experiment (as described in Section 15.2) with the crop in question, in which one or more additional treatments with animal manure are included (Schröder, 2005). From such an experiment, data on yield and possibly also N uptake in the crop in the various treatments may be utilised to calculate the MFE value of the manures as described in Figure 15.7.

The MFE value can be estimated based on the crop yield response (e.g. grain yield) to fertiliser and manure, which seems like the most logical basis, as farmers are usually paid for the grain/produce. The calculation is based on the experimental determination of agronomic efficiency (AE) of fertiliser N and manure total-N,

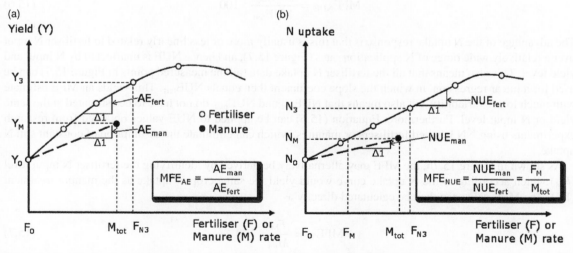

Figure 15.7 *Illustration of the principles for MFE value estimation, based on either (a) grain yield response data or (b) crop N uptake data. The crop experiment must contain several different rates of mineral N fertiliser (N application rates of F_0 to F_X) and at least one level of organic manure with a total-N content of M_{tot}. (© University of Copenhagen.)*

respectively, calculated according to Equation 15.3 (Section 15.2). The MFE value is then defined as the ratio between the AE of manure and fertiliser N as (Figure 15.7a):

$$\text{MFE}_{\text{AE}} = \frac{\text{AE}_{\text{man}}}{\text{AE}_{\text{fert}}} \cdot 100 \tag{15.5}$$

This approach is not entirely unproblematic, however, as the yield response to fertiliser N input is typically convex and hence the AE will vary depending on the N input and yield level. The question then arises as to whether the AE of the fertiliser N should be determined at the same yield level or the same N input level as for the manure treatment. Furthermore, the slope of the response curve is very low near the maximum yield and if estimated from AE near this maximum, the MFE value is subject to greater error. At high manure application rates a yield in excess of the fertiliser response curve maximum may also sometimes be achieved, producing MFE > 100% – an artefact caused by the manure having non-N effects (e.g. by improving soil structure, root growth, water retention, etc.) or by providing the other nutrients (P, K) in a more efficient way than the P and K in the mineral fertiliser.

Due to these problems, an alternative method of MFE estimation is often based on the plant N uptake in the same experiment. However, this requires chemical analyses of plant samples and representative sampling of the entire plant (including roots). The latter is very expensive, and usually only the above-ground biomass is sampled and analysed or, even simpler, the chemical N analyses are just conducted on the harvested part (e.g. the grain). In doing so, it is assumed that the sampled part can be used as a proxy for the entire plant, and that the ratio of N uptake between the entire plant and the sampled part does not differ between fertiliser and manure treatments.

Calculation of MFE from N uptake data is similar to Equation (15.5) based on the experimental determination of NUE of fertiliser N and manure total-N, respectively, calculated according to Equation (15.4) (Section 15.2). The MFE value is then defined as the ratio between the NUE of manure and fertiliser N (Figure 15.7b):

$$\text{MFE}_{\text{NUE}} = \frac{\text{NUE}_{\text{man}}}{\text{NUE}_{\text{fert}}} \cdot 100 \tag{15.6}$$

The advantage of the N uptake response is that this is usually more or less linearly related to fertiliser N input over a relatively wide range of N application rates (Figure 15.7), and hence NUE is unaffected by N input and yield level. This also means that all the fertiliser N uptake data (all the measured points in Figure 15.7) can be used for a linear regression, in which the slope coefficient then equals NUE_{fert}. This yields an MFE estimate with much lower variability, but also means that NUE_{fert} and NUE_{man} do not have to be estimated at the same yield or N input level. Furthermore, Equation (15.6) can be applied to NUE values obtained from research experiments using ^{15}N labelled fertilisers or manures, which can eliminate the assumption of constant soil N uptake.

As shown in Figure 15.7b, the MFE may alternatively be derived by identifying the fertiliser N input level (F_{M}) which according to the N uptake curve would yield the same crop N uptake as the manure treatment (Y_{M}). The MFE value can then be calculated directly as:

$$\text{MFE}_{\text{N}} = \frac{F_{\text{M}}}{M_{\text{tot}}} \cdot 100 \tag{15.7}$$

MFE values estimated from the N uptake curve are often slightly lower than those estimated from the yield response curve, due to differences in the protein content of plants fertilised with manure and mineral fertiliser N.

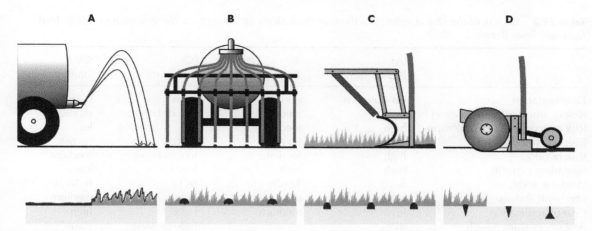

Figure 15.8 *Illustration of principles for the four major types of slurry application methods and the associated slurry distribution within the crop-soil after application: (A) splash plate spreader, (B) band spreader, (C) trailing shoe and (D) shallow injector. (© Southern Denmark University.)*

It is important to remember that the MFE value is specific for the experimental conditions in which it was determined (i.e. the particular soil type, weather conditions, crop, application rate and method). Therefore, a universal MFE value for a particular manure type does not exist, since although the NH_4^+ content is the single most important factor for the potential MFE value, as stated above, the actual MFE may be less, depending on the above factors.

15.4.2 Manure Application Methods

Owing to the bulkiness, high water content and high volume of livestock slurries, suitable technologies for field application have for many years been lacking and this has been a major barrier to optimal utilisation. However, today a range of efficient slurry application methods have been developed that have high capacity and provide low emissions of ammonia, potentially ensuring more optimal utilisation of the slurry.

In general, four principal methods for the application of slurry are available: broadcast spreading (splash plate), trailing hoses (band spreading), trailing shoes and injection. Figure 15.8 illustrates the principles of the four methods and Table 15.2 presents an overview and comparison of their characteristics. Some major aspects that must be considered are: (i) homogeneity of application rate over the entire field, (ii) low risk of ammonia volatilisation, (iii) high capacity and low cost, and (iv) low risk of soil compaction. Some of these may have conflicting effects, whilst others are synergistic. From a fertilisation and an environmental point of view, the most attractive methods are trailing hoses/shoes or injection. The higher costs are somewhat compensated for by higher utilisation of the nutrients in the slurry, but these methods do not necessarily result in higher yield or higher net profits.

In addition to these application methods, irrigation sprinklers are used to apply diluted slurries in some countries. This is a very cheap application method, but due to the formation of many small droplets or aerosols, it increases the risk of disease spread, odour problems and large losses of ammonia, and hence is either banned or not recommended for environmental reasons.

- *Broadcast spreading (splash plate)*. The slurry is pumped at high pressure onto a plate mounted at the back of the trailed tanker. A "fan" of slurry is sprayed into the air and the slurry drops in a 6- to 10-m wide band behind the trailer. Most of the soil surface and the crop is covered with a thin film of

Table 15.2   *Review of the characteristics of the four main slurry application methods shown in Figure 15.8 (adapted from Birkmose, 2009).*

	(A) Broadcast	(B) Trailing hoses	(C) Injection	(D) Trailing shoe
Distribution of slurry	very uneven	even	even	even
Risk of ammonia volatilisation	high	medium	low or none	medium
Risk of contamination of crop	high	medium	low	low
Risk of wind drift	high risk	no risk	no risk	no risk
Risk of odour	high	medium	low or none	medium
Spreading capacity	high	high	low	low
Working width (m)	6–10	12–28	6–12	6–16
Mechanical damage of crop	none	none	high	medium
Cost of application	low	medium	high	high
Amount of slurry visible	most	some	little or none	some
Especially suited for	all	winter crops	grass, bare soil	grass

slurry, with a high specific surface area. A major disadvantage is that the slurry is spread unevenly on the field. In particular, in windy conditions a satisfactory spreading pattern cannot be achieved, with the consequence that crop yield and quality decrease, unless supplementary fertiliser N is applied. Another disadvantage of broadcast spreading is the relatively limited effective working width, since overlapping is needed to approach an overall even application rate. A final disadvantage is that the risk of ammonia losses (Chapters 8 and 9) is greater due to the larger surface area of the applied slurry than with other methods, especially under warmer weather conditions. However, an advantage of the technology is that it is very simple, with low costs for investment, running and maintenance. Therefore, due to its low costs and simplicity, broadcast spreading is still the dominant slurry application method in many countries. However, in intensive livestock producing countries such as the Netherlands and Denmark, the method has been banned by the environmental authorities to reduce ammonia and odour emissions, forcing farmers to apply one of the following methods.

- *Trailing hoses (band spreading).* The slurry is pumped through numerous hoses (40–50 mm diameter) positioned along a boom at a spacing of around 30 cm. The total width of the boom can be up to 28 m (typically 18–24 m). In winter crops the common tramlines can therefore be used for slurry application and the damage to the crop is limited. The slurry is placed directly onto the soil surface (30–50% ground cover), even in a growing crop, and left in 10- to 15-cm wide bands and as the crop is not covered with slurry, the specific surface area is lower than for broadcast spread and emissions consequently reduced. The trailing hose technique has a great advantage due to its capability to reduce ammonia losses from slurry applications in a standing crop. However, as Figure 15.9a illustrates, in early spring when the crop height is limited, no significant differences can be found between the splash plate and trailing hose techniques. Sommer *et al.* (1997) showed that soil water content, solar radiation and leaf area index significantly influenced the potential reduction efficiency of the trailing hose technique, with the greatest effect at low soil water content, high solar radiation and high leaf area index. Thorman *et al.* (2008) analysed these and additional data, and found that the reduction in efficiency of the trailing hose technique relative to bare soil without a crop increases by slightly less than 1% for every 1 cm increase in crop height (Figure 15.9b). A most important effect of trailing hose application is that a very even distribution can be achieved across a width of up to 28 m, regardless of wind. Thus, the coefficient of variation for trailing hose applied slurry may be as low as that of mineral fertiliser spread with a conventional centrifugal spreader (10–15%). This effect often predominates because a large share of slurry application takes place when crops are still small,

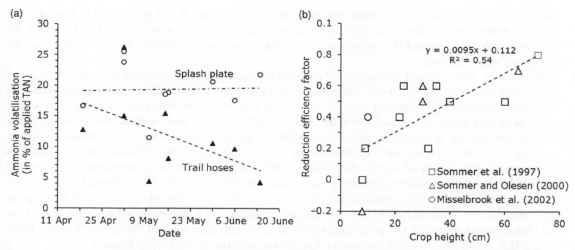

Figure 15.9 *(a) Accumulated ammonia volatilisation (0–7 days after application) as a per cent of total ammoniacal N (TAN) applied in pig slurry (rates varying from 19 to 31 ton ha^{-1}) to winter wheat at various dates during spring with either splash plate or trailing hose techniques. (Reprinted with permission from Sommer et al. (1997). © 1997 American Society of Agronomy.) (b) Reduction efficiency factor (relative to slurry applied to bare soil) for accumulated ammonia emissions by trailing hose spreading with increasing crop height. (Reprinted with permission from Thorman et al. (2008). © 2008 EDP Sciences.)*

and the main advantage of the trailing hose technique is then the higher precision and homogeneity of application. Furthermore, trailing hoses also have many practical advantages and lower odour emissions compared with the broadcast method, so in many European countries this is becoming the standard for spring application of slurry to growing crops early or later in the season, especially for larger farms and contractors.

- *Injection*. As with trailing hoses, the slurry is pumped though hoses, each connected to a tine or a disc, creating a furrow into the ground, which is filled with slurry. In grassland, injectors with one or two slightly angled discs are usually used to create a 4- to 5-cm deep V-shaped furrow and the slurry is not covered with soil after application as the furrows are left fully or partly open. In bare soil injectors, a strong tine (typically an S-shaped harrow tine) is used to place the slurry at approximately 10 cm depth and the slurry is completely covered by soil after application. Injectors have the same advantages as trailing hoses regarding evenness of slurry application rate, but the main advantage is the significantly lower risk of ammonia losses, with reduction efficiencies of 49–97% compared with 41–48% for trailing hoses, both relative to broadcasting (Webb *et al.*, 2010). Injectors have some great disadvantages, however, the working width is limited to maximum 12–14 m (typically 6–9 m), while injection is much more energy-demanding than any of the surface spreading techniques due to the high requirement for draught power. Effective injection is more or less impossible on stony soils or very heavy clay soils. Grassland injection may fail when either dry soil conditions or a dense layer of grass roots impedes the creation of a sufficiently deep furrow. Rodhe and Etana (2005) found no positive MFE effect from injection on grassland, even though ammonia losses were reduced by 50%, whereas Birkmose (2009) found that injection increased the MFE of cattle slurry on grassland to 40%, compared with 27% for trailing hoses. Finally, in a growing crop the damage from the wheels is much higher than when using trailing hoses and the cutting of roots by the injection discs also causes some crop damage. For winter wheat, crop damage by injection in most cases counteracted any potential beneficial MFE effect (Birkmose, 2009).

For this reason, injection is hardly ever used in winter cereals (although this application is developing in the Netherlands), but for bare soil, and to some extent grassland, injection is more or less becoming the standard application method in countries with strict environmental regulations on manure use (e.g. the Netherlands and Denmark).

- *Trailing shoes*. The so-called "trailing shoe" is a kind of "semi-injection". Shallow furrows are cut in the ground and the furrows are filled with slurry, but are too shallow to contain all the applied slurry. Although the furrow increases the infiltration rate, 20–40% of the ground may therefore still be covered by slurry, which is equally prone to ammonia volatilisation as with the trailing hose method. An advantage of the trailing shoe method is a lower draught requirement than proper injection, and hence potentially higher working width and capacity, but the disadvantage is the somewhat lower efficiency than true injection in reducing losses compared with trailing hoses or broadcasting under some conditions. On grassland the method has shown relatively high emissions reduction efficiencies of 65%, compared with 41% for trailing hoses, both relative to broadcasting (Webb *et al.*, 2010), and Hoekstra *et al.* (2010) found grass recovery of slurry ^{15}N-labelled NH_4^+ to increase by on average 13% compared to broadcast. Farmers in the United Kingdom, which has a high proportion of grassland, have adopted the method widely (Webb *et al.* 2013), but in other parts of the world it has not yet been taken on by many farmers.

For solid manures, much less technological development of application methods has been carried out and this basically means that solid manure spreaders are fairly standard, although some improvements to homogeneity of spreading have been achieved (e.g. modern solid manure spreaders are capable of spreading with a coefficient of variation of less than 20%). They usually comprise a carrier wagon, with a bottom conveyer belt that moves the manure continuously towards the rear-mounted spreading augers, which may be mounted either vertically (producing a broader spreading pattern) or horizontally (producing a narrow, but very homogeneous application strip).

15.4.3 First-Year N-MFE Value for Various Manure Types

A high fertiliser value of nutrients in manures requires compatibility (amount, timing, placement in the root zone) between availability of the manure nutrients and nutrient demand by the crop. Thus, when planning manure application one should consider the risks of nutrient losses, but also that of luxury uptake, which may have adverse effects on the crop, such as increased susceptibility to pests, lodging of the standing crop, decreases in nutritional value and in crop quality in general. This is difficult to achieve under field conditions and is further complicated by the fact that organic N in manures may be considered as a kind of "slow-release" fertiliser, with a release pattern that is very difficult to predict. The chemical composition and biological properties vary greatly between manures of different animal origin and type (liquid, solid, litters) (Table 15.3).

The actual N-MFE value of an animal manure or slurry application depends on a number of factors, and some generalisations may be made about the MFE value:

- The proportion of NH_4^+-N in the total manure N is the single most important factor for the magnitude of the MFE value. The effect of the NH_4^+ fraction depends on both the possible ammonia losses and on immobilisation or leaching losses of nitrified NH_4^+.
- The effect of the organic fraction, on the other hand, depends both on the decomposition dynamics (as affected by degradability, moisture and temperature) but also the C : N ratio of the decomposable fraction (Webb *et al.*, 2013).
- The time of application and thus the synchronisation of manure nutrient availability with crop nutrient demand is important. Thus, under temperate, humid climate conditions, very low MFE values are generally

Table 15.3 *Typical properties for a range of manure types: dry matter, total P, K and N content, NH$_4^+$ share of total-N, C : N ratio as well as a characterisation of the biodegradability of the organic material and typical potential first-year N-MFE.*

Manure type	Dry matter (%)	P (g kg^{-1} fw)	K (g kg^{-1} fw)	Total-N (g kg^{-1} fw)	NH$_4$-N (% of total-N)	C : N	Biodegradability	Potential N-MFE (%)
Deep litters	25–30	1.5	10–12	7–10	10–25	20–30	medium	20–30
FYM (pig)	20–25	4–5	8–9	9	30–45	12–15	medium	20–50
FYM (cattle)	18–20	1.7	3	6	20–30	15–20	low	15–30
Broiler manure	45–50	7–9	13–16	20	10–25[a]	5	high	50–65
Layer manure	50–60	7–12	9–16	20–30	5–35[a]	10	medium	40–50
Slurry (pig)	4–7	1,0	2–3	3–5	70–75	5–8	medium	40–70
Slurry (cattle)	7–10	0.9	4–6	4–5	50–60	8–10	low	35–50
Slurry (poultry)	10–15	1–2	2–3	6–10	60–70	4	medium	70–85
Biogas digestate[b]	1–5	1.0	2–3	3–10	60–85	3–5	low	60–90
Sep. slurry liquid	1–2	0.1–1	2–6	3–5	80–95	1–2	medium	80–100
Urine	2–3	0.3	2–6	3–4.5	90	1–2	high	90–100

[a] Depends on storage time and degree of uric acid hydrolysis.
[b] Mixed animal slurry with biowaste co-fermentation.

found if the manure is applied in the autumn, due to leaching losses over winter. Solid manures are somewhat less sensitive to the time of application, since they have comparatively lower inorganic N content and an equally larger fraction of organic N, but many solid manures still contain a substantial fraction of NH$_4^+$-N and hence MFE values may be improved with spring application in humid climates.

- The crop species, and especially the length of its growing season, is important for the MFE value. Notably, beets and maize, which have a long growing period and which take up nutrients right up until harvesting late in the growing period, may utilise the total-N in liquid manures almost as well as mineral fertiliser N, if the manure is applied directly into the soil without losses.
- The method of application is also important. Thus, broadcasting generally leads to the least efficient use of nutrients, because of increased losses of ammonia and uneven distribution of the manure. Injection of the manure gives a higher utilisation because of low losses, and placement on the ground with a system by which the slurry or urine is pumped through trailing hoses gives a rather even distribution and thus typically higher MFE values.
- The method to determine MFE value reflects the amount of mineral N fertiliser that may be replaced by manure total-N in the first crop to which the manure is applied, and thus normally disregards the residual effects to crops in subsequent years (see Section 15.4.4).

The actual first-year N-MFE value of a particular manure application therefore depends not only on the manure type and the application method, but also on the crop and the application timing. In the following, some aspects of the MFE values that can be obtained for N are described.

15.4.3.1 *Pig and Cattle Slurries*

Pig slurry is usually characterised by a relatively high proportion of NH$_4^+$, a neutral pH and a relatively low content of dry matter, with organic matter of medium degradability (Table 15.3). This means that pig slurry has a relatively high potential MFE value, but the actual MFE value depends greatly on the soil and weather conditions. After surface application, infiltration is usually relatively rapid due to the low dry matter content and hence low viscosity, but this may be inhibited under wet soil conditions, resulting in higher ammonia losses and consequently lower MFE value. However, if the slurry infiltrates rapidly, is incorporated or even better is injected, a high actual MFE value may be obtained. Conversely, high initial immobilisation and failure of rapid remineralisation may impede a high MFE.

Cattle slurry, on the other hand, is characterised by a slightly lower proportion of NH_4^+ and higher content of dry matter than pig slurry, but with an organic matter content of low degradability. This means that for cattle slurry the application method is of major importance for the actual MFE value, since following surface application infiltration is usually relatively slow due to the higher dry matter content and viscosity. However, if cattle slurry is incorporated rapidly or injected, ammonia losses may be avoided and reasonably high actual MFE values may be obtained, but seldom higher in the first year than the NH_4^+ proportion. In one study, application of cattle slurry to grassland using the trailing shoe method increased the NUE and MFE values by approx. 10% compared with the broadcasting method in a 6-week period following slurry application (Lalor *et al.*, 2011). However, the relatively higher content of dry matter also means that the N immobilisation potential of cattle slurry is substantial and only crops or pastures with a long growing season will be able to utilise the slowly remineralised slurry N (Gutser *et al.*, 2005).

15.4.3.2 Solid Manures

For the solid manure types, with a relatively high dry matter content (above 12%), the proportion of NH_4^+ is generally below 50% and hence the proportion of organic N is accordingly higher. While the proportion of NH_4^+ is still important for the actual MFE value, the degradability and C : N ratio of the manure organic matter gain in importance. The storage conditions of solid manure are of particular importance for the NH_4^+ content, since manure typically starts to partially compost with access to oxygen. Covering the manure with an airtight material may reduce air access and may give a higher crop NUE (23%) than for manure stored uncovered (14%), where composting will have caused high ammonia losses, while fresh manure application has an NUE of 17% (Shah *et al.*, 2012).

The relatively higher content of organic N means that solid manures are especially suitable for arable crops with a long growing season (e.g. maize or beets) since the manure can be applied prior to drilling, with rapid incorporation to minimise ammonia losses and maximise immediate uptake of manure NH_4^+, and the extended growing period enables crop uptake of slowly mineralised manure N.

For solid manure types such as deep litter, their relatively high C : N ratio (due to the litter material, typically straw) often causes the N immobilisation period to extend for so long that hardly any of the immobilised N is remineralised within the first growing season, leading to a negligible or even negative MFE value. Unfortunately this effect is very unpredictable, making deep litter particularly unfavourable as a fertiliser, and increases the risk of farmers disregarding its nutrient value and applying regular mineral fertiliser rates, increasing the risk of N losses when the deep litter releases more N than anticipated.

Poultry manures are a special case, due to their relatively high dry matter content and the significant proportion of N which may be present in the form of uric acid. Therefore, their NH_4^+ N content is not necessarily a good estimator of the N-MFE value of poultry manure (Thomsen, 2004). Instead, the sum of NH_4^+-N plus uric acid-N or a percentage of the total-N could be used. Delin (2011) found MFE values of 30–40% for hen and broiler manure when applied in early spring or before sowing of spring barley, whereas Thomsen (2004) found MFE values of 82% for chicken manure when applied in spring, but only around 30% when applied in the previous autumn for spring barley.

15.4.3.3 Acidified Slurry

Acidification of manure is an obvious treatment for reducing ammonia emissions from livestock production and field application of manure (Husted *et al.*, 1991), but there are risks of foaming (from degassing of CO_2) and potential work hazards associated with addition of strong acids to slurries. Until recently, technologies have not been available to address these problems, but systems for acidification of slurry both within animal houses and storage tanks and during field application have now been developed (Birkmose, 2009; Sørensen and Eriksen, 2009), and are being installed on farms in Denmark. Acidification of pig slurry to pH 5.5 with sulfuric acid reduced ammonia emissions from slurry storage by 90% and from field-applied slurry by 67%, producing an overall increase of 75% in the plant-available N in slurry, on the farm studied by Kai *et al.* (2008).

In the study by Sørensen and Eriksen (2009), the MFE value for the acidified slurry was 101–103%, compared with 74% for untreated pig slurry applied by trailing hoses to winter wheat, while for cattle slurry the corresponding MFE values were 63–66% and 39%, respectively. Thus, acidification in that study appeared capable of increasing the MFE value by about 25%. Similar results were found by Kai *et al.* (2008). This means that slurry acidification has the potential not only to reduce ammonia losses significantly during housing, storage and application, but also to improve the quantity of plant-available N on the farm and the fertiliser efficiency of slurry when surface-applied to crops. However, in spring-sown crops such as spring barley, where the slurry is typically incorporated before sowing, resulting in relatively low ammonia losses even for untreated slurries, lower or non-significant increases in MFE value by slurry acidification have been observed (Sørensen and Eriksen, 2009). For the field situation, slurry acidification should therefore mainly be seen as an alternative to injection or rapid incorporation, and will be particularly interesting where farmers want to avoid the crop damage and soil disturbance caused by injection, i.e. in winter cereals and grass leys or pastures.

15.4.3.4 Digested Manure from Biogas Plants

With the proliferation of biogas plants in many countries, an increasing proportion of manures applied to land will be in the form of biogas effluent or digestate. These have potential benefits regarding N availability and potential MFE value; the proportion of NH_4^+ increases due to mineralisation during the digestion and easily decomposable C in VFAs is converted to biogas and hence does not contribute to immobilisation upon soil application. However, the increased proportion of NH_4^+ in digested slurries does not guarantee improved utilisation efficiencies of slurry N, since digestate pH is usually 0.5–1.5 units higher than that of the raw slurry. Therefore, digestate needs to be injected or incorporated rapidly into the soil if higher MFE values are to be obtained (Nyord *et al.*, 2012). The available field data on actual N and P MFE of digestate from animal slurries show variable results, with small or inconsistent benefits compared with undigested slurries (Möller and Müller, 2012). However, in a review of 11 field trials with various types of digested animal slurries, Birkmose (2009) found that the MFE value increased on average by approximately 10%, compared with that of the raw slurry. A similar magnitude has been found by Schröder *et al.* (2007) and de Boer (2008).

There may be concerns that applying digestate with a reduced dry matter and C content compared to raw slurry will have an adverse effect on soil structure and fertility in the long term due to the lesser organic matter input. However, the organic matter left after anaerobic digestion is in more stable compounds and will contributes more to long-term carbon retention in soil; Thomsen *et al.* (2013) found that 48% and 76% of the C applied in faeces and digested faeces, respectively, was retained in the soil after 1–2 years, but in the long-term C retention was found to be similar, 12–14% of the C initially present in the animal feed, regardless of anaerobic digestion.

15.4.3.5   Separated Solid and Liquid Fraction

Slurry separation (Chapter 7) produces a liquid and a solid fraction, both of which could be land-applied, but typically not to the same crop or geographical location. Due to their differing properties, strategies for their optimal application method and crop use should be different.

The liquid fraction normally comprises 85–90% of the mass of the raw slurry, and the N pool is characterised by a very high proportion of NH_4^+, a pH above neutral, and a very low content of dry matter and P, but high content of K (Table 15.3). This means that the liquid fraction has a high potential MFE value. By surface application, infiltration is usually relatively rapid due to the low dry matter content and viscosity, but the relatively high pH makes it prone to ammonia losses if infiltration or incorporation does not occur rapidly. For a separated pig slurry liquid fraction incorporated into soil before sowing of spring barley, Sørensen and Thomsen (2005) found MFE values of 91–98%, compared with 75–81% for the raw slurry. In a similar field study with surface application to winter wheat, the corresponding MFE values were 76–83% and 59–69%, respectively. In a study by Meade *et al.* (2011) with liquid fraction application to winter wheat by surface application in Ireland, somewhat lower MFE values of around 50% were observed. After separating solids from cattle slurry and only applying the liquid fraction to a grass ley, Bittman *et al.* (2011) found that MFE values increased from 45–50% for whole cattle slurry to 61–76% for the liquid fraction, when surface-applied.

The solid fraction normally comprises 10–15% of the mass of the raw slurry, and is characterised by a high content of dry matter, a large proportion of organic N (but still 20–40% as NH_4^+), a pH above neutral and a variable content of P (depending on separation technology, Chapter 7). The solid fraction therefore resembles other solid manures in terms of fertiliser properties and strategy for management and utilisation (see above), including proper storage under cover, to avoid significant losses of plant-available and residual organic N (Petersen and Sørensen, 2008a). Using spring application of various solid fractions to winter wheat, Sørensen and Rubæk (2012) found MFE values of 23–39%. This is similar to the MFE values of 27–29% found by Sørensen and Thomsen (2005), who also demonstrated that the overall weighted efficiency of liquid and solid fraction application was not significantly higher than the MFE that could be obtained by direct application of the unseparated slurry. The main advantage of separation is that it facilitates a better distribution of nutrients (a better balance between N, P and K, with less surplus P).

In temperate, humid climates, autumn application of the solid fraction may result in leaching losses of N (Sørensen and Rubæk, 2012; Chapter 11), but only slightly lower MFE values (17–32%) for autumn than spring application (23–39%). Thomsen (2005) found similar MFE and leaching results for FYM application in autumn, winter or spring. These results can be explained by the fact that both mineralisation and nitrification occur at low temperatures around and even below 0 °C (Andersen and Jensen, 2001; Clark *et al.*, 2009), although of course at lower rates. So even though solid manures contain mostly organic N and NH_4^+, which should be bound in the soil, application in the autumn leaves enough time for these processes to convert a significant fraction of N into the leachable NO_3^- form. From an environmental perspective, solid manure fractions should therefore not be applied in autumn to a winter crop under humid conditions, as also concluded by Birkmose (2009). However, by applying the solid fraction to a spring barley crop (incorporation before sowing in early spring) Sørensen and Thomsen (2005) found much higher MFE values (42–50%) and therefore this is the optimal direct field utilisation of this product from slurry separation.

For the effect of further upgrading of the liquid fraction into mineral concentrates (e.g. by reverse osmosis or ultrafiltration) and the solids into compost or biofertilisers (e.g. biochar), please refer to Chapter 14.

15.4.3.6   Summary Recommendations for First-Year MFE Values

To summarise the preceding sections, manure type, crop type, application method and timing influence the actual MFE value obtained. In the literature, there is a predominance of studies from Scandinavia and

Table 15.4 Danish recommendations for first-year MFE (%) of total-N in liquid manure types depending on time of the year, application method and crop type (Birkmose, 2009).

	MFE (%)					
	Spring		Summer		Autumn	
	Injected	Trailing hoses	Injected	Trailing hoses	Before sowing	Growing crop
Pig slurry						
spring seed	75	70	–	45	–	–
beet or maize	75	70	70	40	–	–
winter cereal	70	65	–	65	–	–
winter rape	–	65	–	–	65	55
grass	60	60	55	45	–	55
Cattle slurry						
spring seed	70	50	–	35	–	–
beet or maize	70	55	60	35	–	–
winter cereal	55	45	–	40	–	–
winter rape	–	45	–	–	50	35
grass	50	45	45	35	–	40
Digested slurry						
spring seed	75	70	–	50	–	–
beet or maize	75	70	70	45	–	–
winter cereal	75	75	–	65	–	–
winter rape	–	75	–	–	65	55
grass	70	65	60	45	–	60
Liquid fraction, separation						
spring seed	90	90	–	70	–	–
beet or maize	90	90	90	70	–	–
winter cereal	90	85	–	85	–	–
winter rape	–	85	–	–	85	70
grass	80	75	75	65	–	70

the Netherlands – regions with intensive livestock production, but also relative cool, humid climates, with excess precipitation during winter. In Denmark, which has some of the strictest environmental regulations on agricultural manure management in Europe (Chapter 3), a large number of field experiments with animal manure (several hundred) were conducted during the 1990s (Birkmose, 2009) in order to provide the farm advisory system with experiment-based, well-informed and up-to-date knowledge. This large experimental database was the foundation for the recommendation guidelines for first-year MFE values listed in Table 15.4 for various liquid manure types, application times and methods to crops with different growing seasons. These recommendations have been implemented into fertilisation planning schemes and software to assist Danish farmers in decision making regarding optimal manure utilisation.

Recommendations for MFE values at the national scale may also differ due to differences in farming practices, crop rotation, climatic conditions, knowledge base, and degree of independent local research and development. However, as application technologies spread to different countries, recommended MFE values across these will converge (e.g. for cattle slurry they are quite similar among some European countries; Table 15.5). In the Netherlands and Denmark, MFE values for pig slurry are somewhat higher than values used in other countries, mainly because better application methods are more widespread in these two countries.

Table 15.5   *Recommendations from different European countries for first-year MFE values (%) of total-N in various manure types (adapted from ten Berge and van Dijk, 2009).*

Manure type	System, time, crop	Netherlands	Belgium (Flanders)	Germany	Denmark	France
Slurries						
cattle	arable, spring, maize/potato/beet	50–55	55	70	55–70	55
	arable, spring, winter wheat	40	55	70	45–55	
	grass, before first cut	45–50	55	70	45–50	50–60
	excreted on pasture			25		
pig	arable, spring, maize/potato/beet	70–75	65	60	70–75	60–75
	arable, spring, winter wheat	55	65	60	65–70	60–70
	grass, before first cut	45–55	65	60	60	50–65
Solid manures						
cattle	arable, spring, maize/potato/beet	30	30	60	45	15–30
chicken	arable, spring, maize/potato/beet	50–55	55	50	65	45–65
Separated slurries						
liquid fraction	arable, spring, maize/potato/beet	85–90	80–90		90	
	arable, spring, winter wheat	70	80–90		85–90	
	grass, before first cut	65–75	80–90		75–80	
solids, cattle	arable, spring, maize/potato/beet	25	25		55	
solids, pig	arable, spring, maize/potato/beet	50	35		55	
Compost						
all manures	arable, spring, maize/potato/beet	10	10			10–15

For Denmark and the Netherlands, the ranges in MFE given refer to different application methods, with the highest values for slurry injection and the lowest for band application.

MFE will also depend on application timing, with much lower values for autumn application in humid climates; but in many countries, environmental regulations will include closed periods for the application of those manures that contain large proportions of mineral N (slurries and other liquid manures; Webb *et al.*, 2013) in order to minimise N leaching risks, and hence MFE values for such application timing will not be relevant.

15.4.4   Long-Term Manure Nitrogen Turnover and Residual MFE Value

Although manures yield N fertiliser values below that of mineral N, a major fraction of the manure N not taken up by the crop remains in the soil in organic form. This residual soil manure N may partly be original manure organic N, but a significant proportion is actually inorganic manure N assimilated and immobilised by soil microorganisms (Sørensen and Amato, 2002), as described earlier in this chapter. The residual organic N in soil will contribute to the slow, continuous mineralisation into plant-available, but also leachable N, and the fate in subsequent years following application of the manure will depend on the synchronisation between release, crop demand and soil water percolation (see also Figure 15.5). Quantifying the magnitude of the residual MFE value of animal manures in subsequent years is not straightforward; Cusick *et al.* (2006) gives a good overview of available methods and examples of results.

One approach is to apply manure to cropped plots over a substantial number of years to approach some degree of steady-state conditions between input and output, and then estimate the residual value from the

Table 15.6 *Crop uptake of ^{15}N-labelled mineral fertiliser and animal manure components during 2–3 growing seasons measured in Danish experiments under field conditions (compilation of data from Sørensen et al., 2002; Petersen and Sørensen, 2008b).*

^{15}N-labelled component	^{15}N crop uptake (% of applied)		
	Application year	First res. year	Second res. year
Mineral fertiliser N	36–57	3–5	1.2–1.5
Manure applied in spring before barley			
Faecal-N in slurry	12–17	3–6	–
Faecal-N in solid manure	9	4	1.1–2.0
Urinary-N in slurry	32–36	3	–
Urinary-N in solid manure	25–27	3–4	1.3
NH$_4$-N in pig and cattle slurry	27–41	3–4	1.8–2.5
Straw-N in solid manure	9–10	3	1.1–1.3
Manure applied in autumn before winter wheat			
Total-N in solid manure	8–10	2.6	–

First res. year = the year after application, first year with residual N effect.

increase in MFE value of the manure over the years, but this is evidently very time-consuming. Using this approach, Nevens and Reheul (2005) evaluated long-term cattle slurry fertilisation of silage maize, and found that MFE increased from 20% initially to around 60% over the first 10 years of slurry application and then remained at this level for the following 12 years, indicating that the accumulated residual MFE value is much in excess of the first-year MFE. Similarly, Hernández *et al.* (2013) reported increasing MFE values with repeated application of pig slurry to winter barley under Mediterranean conditions. Generally, experiments with this approach often show the largest residual effect for manures with a high proportion of organic N (solid manures), with lower residual values for manures with a high proportion of NH$_4^+$ N (slurries, urine).

Another approach is to label the applied manure with the stable isotope ^{15}N and then study the uptake of the labelled residual N in subsequent years. Table 15.6 presents a summary of such experiments, where different manure fractions were originally labelled with ^{15}N. Surprisingly, the fraction of ^{15}N taken up in the crops in the first and second residual year is comparable between the manures and the parallel plots fertilised with ^{15}N-labelled mineral fertiliser in the first year. The residual fertiliser N value in the second year is due to the fact that the fraction (20–50%) of the fertiliser N not taken up by the crop in the first year is incorporated into the soil organic N through the continuous MIT in soil and this immobilised N may naturally be mineralised in the following year, irrespective of its origin. This means that once the initial turnover of manure N is over in the first year, the residual N value may not be larger from manures than from mineral fertiliser, when calculated as a fraction of applied N. However, as the total-N application with manures is usually somewhat larger than with fertiliser, the absolute residual value (kg N ha^{-1}) will still be larger after manures than after fertilisers. This second-year effect may for ruminant slurry be equal to around 2–5% of total-N applied. For manures from pigs and other monogastric animals, the residual effect appears to be of the same magnitude, although the effect in theory should be greatest for manures containing a large proportion of organic N (Webb *et al.*, 2013). Climate and length of growing season naturally also play a role; Cela *et al.* (2011) and Yagüe and Quílez (2013) compared the residual effects of pig slurry applied for 4–6 years to continuous maize under Mediterranean conditions and found somewhat higher residual effects than the results from humid temperate conditions in northern Europe (e.g. residual MFE of 15–45% (of the earlier annual slurry N application rate) in the first residual year and 7–15% in the second residual year.

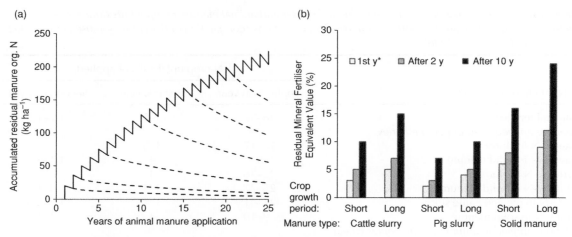

Figure 15.10   *Modelled (a) residual manure organic N accumulated (full line) in soil over 30 years with annual application of 100 kg total-N ha⁻¹ in pig slurry (25% organic N, dashed line indicating decay of organic N after first, second, fifth, 10th, 15th and 20th application) and (b) the corresponding residual MFE value of 1, 2 or 10 years of repeated applications of different types of animal manure, compared with soil only receiving mineral fertilisers. MFE given in per cent of annual manure total-N application for crops with a short (e.g. spring cereals, 50% of mineralised N available for crop) or long growing period (e.g. maize, beet, 75% of mineralised N available for crop). Solid manure: cattle FYM and deep litter. *Effect in the year after application, the first year with residual N effect. (© University of Copenhagen; data based on principles in Sørensen et al. (2002).)*

The residual effect of continued application of animal manures can now be estimated based on the above data on residual N-MFE of various manure products. Sørensen *et al.* (2002) and Petersen and Sørensen (2008b) made an estimate using a simple model, assuming that a certain percentage of the remaining organically bound N will be mineralised each year. Based on calibration on the above data, the yearly mineralisation rate was 20% in the year after application, gradually declining over the next few years to just 5% per year in all subsequent years. Mineralisation is a continuous process over the year, so crops with a longer growing period will achieve a higher MFE value. The estimated accumulated residual MFE values (Figure 15.10) vary from rather moderate residual effects of only 7–8% with a short season crop even after 10 years of pig slurry application, whereas for the solid manures and a crop with a long growing season the accumulated residual MFE after 10 years can be as much as 24% (Figure 15.10b). However, the residual effects continue even after 10 years (Figure 15.10a) and for a very long time horizon (e.g. after 50 years it may amount to 17–25% for cattle slurry, 9–14% for pig slurry and 20–43% for solid manures). These predictions with a very simplified model have been compared to more advanced mechanistic modelling, using the FASSET model, where Berntsen *et al.* (2007) found more or less the same magnitude of residual effects as with the simplified model by Sørensen *et al.* (2002). The estimated residual values are also within the range found by Schröder (2005) and further confirmed by modelling in Schröder *et al.* (2007).

As mentioned in Section 15.3.1, the majority of manure P is in inorganic form and has an MFE value of around 100% when compared with superphosphate P fertiliser, for example. However, residual manure P not taken up by the crop is adsorbed more or less strongly in the soil, and contributes to the pool of exchangeable and plant available P, which can be estimated with extraction by, for example, a bicarbonate solution (Olsen-P method). Huang *et al.* (2011) demonstrated that for a loamy soil in a temperate climate, Olsen-P levels in the soil (ranging from 1 to 80 mg P kg⁻¹) increased proportionally to the accumulated surplus of P (fertiliser and/or manure P input minus the P removal with crops during 14 years) and that the application of manure or fertiliser P had more or less identical effects on Olsen-P levels, with ΔOlsen-P (mg P kg^{-1} in 0–20 cm) = 0.033 × accumulated P surplus (kg P ha^{-1}). However, since manure P is often added in greater surplus to crop demands than mineral fertiliser P (see Section 15.3.5), this means that substantial residual manure P fertiliser will accumulate over time. This will positively affect soil fertility, but also potentially poses a risk of increased losses to the aquatic environment via runoff and leaching (see Chapter 11 for further details).

15.4.5 Strategies for Combined Manure and Mineral Fertiliser Use

When the farmer needs to plan optimal field utilisation of manure available on the farm, all of the above factors (manure type, crop type, soil type, application method, timing, expected MFE value) should be taken into account. However, in order to design a complete fertilisation strategy at the whole-farm level, several other factors add to the complexity of this task.

The supply of nutrients can be from three principal sources: soil, manure and mineral fertiliser (Figure 15.11).

In order to estimate the rate of manure and fertiliser to apply, the capability of the soil to supply at least the three major macronutrients N, P and K has to be assessed. This can be done by soil analyses, for N by analysis of N_{min} (i.e. the amount of N present as $NH_4^+ + NO_3^-$ in the rootzone just before fertilisation), and for P and K by various soil extraction methods (e.g. Olsen-P, Bray-P, etc.) that provide an empirically derived index for P or K availability. Alternatively, simple balance or simulation models may be used to estimate soil supply.

The amounts and ratios of nutrients in manure will typically not match a well-formulated chemical fertiliser and will rarely match all the nutrient needs of individual crops. Typically, the available N : P and available N : K ratios in manure are lower than that of crop demand. Therefore, if manures are applied to supply the crop fully with each plant-available nutrient, then one or two of the components N, P and K will usually be applied in excess of crop demand, as illustrated for K in Figure 15.11. This means they will gradually accumulate in the soil, where they are usually adsorbed and bound in non-soluble, but still plant-available, forms. This will be beneficial for soils of low P or K fertility status, but over the years the surplus will accumulate and the soil will gradually become saturated with P and K, potentially increasing the risk of leaching and surface runoff losses (see Chapter 11).

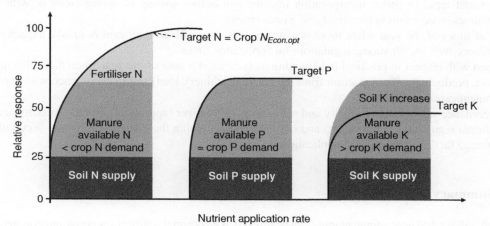

Figure 15.11 *Supplying crop N, P and K requirements from soil, manure and fertiliser. As an example N and P reach the target additions through application of manure for P and fertiliser and manure for N. Potassium is applied above target by the manure application. (© University of Copenhagen.)*

Accumulation of excess available nutrients may also cause antagonistic effects on uptake of other nutrients (e.g. as known for the depression of Mg and Ca uptake by excessive K supply and uptake in grasses). If cattle are fed such grass forages with an unduly high K : (Ca + Mg) ratio, this may cause them to contract the disease grass tetany – a mineral imbalance that results in involuntary contraction of muscles and potentially death of the animal.

Therefore, when deciding the actual rates of manure application to individual crops, these should:

- In principle be at rates to fulfil the nutrient with the lowest requirement compared with the manure content and then the remaining demand for the other nutrients can be supplied in mineral fertiliser – in the case shown in Figure 15.11, this means some additional mineral N fertiliser is supplemented up to the target N level. In this way, any excess of manure nutrients can be avoided.
- In practice, be adjusted to provide the most expensive of the macronutrients, which is currently P. This usually means that the deviation in K is minimal, but that additional N has to be added. Some deviation from P and K demand is unproblematic for most crops, as long as the soil is at a reasonable fertility level (to be confirmed by soil analyses).
- For logistics reasons, be coordinated with fertiliser application, so that the remaining nutrient demand is not too small, necessitating small amounts of extra mineral fertiliser to be applied (min. limit for precise spreading is 75–100 kg fertiliser ha^{-1}, corresponding to 15–25 kg N ha^{-1}).

When prioritising the available animal manure between crops, application of manure should be:

- Preferentially to crops with a long growing season (i.e. beet, maize, pasture grass and possibly winter oilseed rape), which can utilise the N mineralised from the organic N in the manure.
- Avoided on crops where the slurry may pose a hygiene risk for animals or humans (i.e. pastures to be grazed shortly afterwards by cattle; fresh edible crops such as vegetables, potatoes for human consumption) or crops with special quality requirements (i.e. low protein in sugar beet or malting barley).

When planning the timing of application, the manure application should be:

- At times and with methods that give the highest N utilisation (i.e. immediately before the crop growing season and rapid or direct incorporation into the soil before sowing of spring crops or with trailing hoses/shoes or injection in perennial and winter crops).
- Only at times of the year when there is a subsequent crop with a significant N uptake capacity and in accordance with local/national regulations on application times.
- Planned with respect to practical field conditions (i.e. not at a time of the year when the soil is normally too wet, producing soil compaction from the heavy machinery load) and in accordance with application equipment capacity.
- In accordance with storage capacity and status to avoid runover (appropriate storage capacity according to national regulations is important) and only in amounts within the quantity of manure accumulated in the storage facility at the time of application.

15.5 Summary

This chapter described how animal manure can be utilised as an optimal fertiliser for crops and grasslands. The main factors for crop productivity are soil fertility, water and nutrient supply, and crops respond differently to mobile (N) and immobile (P, micro-) nutrients, with N yielding the most gradual response. Crop nutrient uptake efficiency is affected by a range of soil, crop and fertiliser factors, and the latter can be optimised by assuring high nutrient availability, proper synchronisation with crop demand and minimisation of losses.

Animal manure application affects soil chemical, biological and physical properties, all of which are important for soil fertility and quality. Microbial MIT of organically bound nutrients (N, S and to some extent P) in the soil is greatly affected by manure application and has a marked influence on the different modes of action of animal manures compared with inorganic fertilisers. The concept of MFE, estimated from the ratio of manure to inorganic fertiliser nutrient uptake, is useful for quantifying how efficient manure is in supplying nutrients to crops. The main determinant for first-year MFE for N in animal manure is the proportion of NH_4^+ in total-N, but it is also affected by different application methods and manure treatment processes (separation, acidification and anaerobic digestion for biogas). Although first year N-MFE of the latter may approach 100%, most manures have lower first-year N-MFE values, but the residual MFE of these manure affects the long-term nutrient supply. Strategies for optimal utilisation of animal manures will often include supplementation with mineral fertilisers, and must therefore include estimation of soil supply and first-year and residual manure supply, followed by balancing crop N, P, K, S and micronutrient demand with supplementary inorganic fertiliser, in order to avoid significant excess application of one or several manure elements.

References

Andersen, M.K. and Jensen, L.S. (2001) Low soil temperature effects on short-term gross N mineralisation-immobilisation turnover after incorporation of a green manure. *Soil Biol. Biochem.*, **33** 511–521.

Benke, M.B., Indraratne, S.P., Hao, X., Chang, C. and Goh, T.B. (2008) Trace element changes in soil aft er long-term cattle manure applications. *J. Environ. Qual.*, **37**, 798–807.

Berg, J., Thorsen, M.K., Holm, P.E., Jensen, J., Nybroe, O. and Brandt, K.K. (2010) Cu exposure under field conditions coselects for antibiotic resistance as determined by a novel cultivation-independent bacterial community tolerance assay. *Environ. Sci. Technol.*, **44**, 8724–8728.

Berntsen, J., Petersen, B.M., Sorensen, P. and Olesen, J.E. (2007) Simulating residual effects of animal manures using ^{15}N isotopes. *Plant Soil*, **290**, 173–187.

Birkmose, T.S. (2009) Nitrogen recovery from organic manures: improved slurry application techniques and treatment – the Danish scenario, *Proceedings 656*, International Fertiliser Society, York.

Bittman, S., Hunt, D.E., Kowalenko, C.G., Chantigny, M., Buckley, K. and Bounaix, F. (2011) Removing solids improves response of grass to surface-banded dairy manure slurry: a multiyear study. *J. Environ. Qual.*, **40**, 393–401.

Bruun, S., Hansen, T.L., Christensen, T.H., Magid, J. and Jensen, L.S. (2006) Application of processed organic municipal solid waste on agricultural land – a scenario analysis. *Environ. Model. Assoc.*, **11**, 251–265.

Cela, S., Santiveri, F. and Lloveras, J. (2011) Residual effects of pig slurry and mineral nitrogen fertilizer on irrigated wheat. *Eur. J. Agron.*, **34**, 257–262.

Christensen, B.T. and Johnston, A.E. (1997) Soil organic matter and soil quality – lessons learned from long-term experiments at Askov and Rothamsted, in *Soil Quality for Crop Production and Ecosystem Health* (eds E.G. Gregorich and M.R. Carter), Elsevier, Amsterdam, pp. 399–430.

Clark, K., Chantigny, M.H., Angers, D.A., Rochette, P. and Parent, L.-E. (2009) Nitrogen transformations in cold and frozen agricultural soils following organic amendments. *Soil Biol. Biochem.*, **41**, 348–356.

Cusick, P.R., Kelling, K.A., Powell, J.M. and Muñoz, G.R. (2006) Estimates of residual dairy manure nitrogen availability using various techniques. *J. Environ. Qual.*, **35**, 2170–2177.

Dahiya, R., Malik, R.S. and Jhorar, B.S. (2001) Organic mulch decomposition kinetics in semiarid environment at bare and crop field conditions. *Arid Land Res. Manag.*, **15**, 49–60.

De Boer, H.C. (2008) Co-digestion of animal slurry can increase short-term nitrogen recovery by crops. *J. Environ. Qual.*, **37**, 1968–1973.

Delin, S. (2011) Fertilizer value of nitrogen in hen and broiler manure after application to spring barley using different application timing. *Soil Use Manag.*, **27**, 415–426.

Dilz, K. (1988) Efficiency of uptake and utilization of fertiliser nitrogen by plants, in *Nitrogen Efficiency in Agricultural Soils* (eds D.S. Jenkinson and K.A. Smith), Elsevier, London, pp. 1–26.

Eriksen, J., Mortensen, J.V., Kjellerup, V.K. and Kristjansen, O. (1995) Forms and availability of sulfur in cattle and pig slurry. *Z. Pflanzenern. Bodenk.*, **158**, 113–116.

Gutser, R., Ebertseder, T., Weber, A., Schraml, M. and Schmidhalter, U. (2005) Short-term and residual availability of nitrogen after long-term application of organic fertilizers on arable land. *J. Plant Nutr. Soil Sci.*, **168**, 439–446.

Hernández, D., Polo, A. and Plaza, C. (2013) Long-term effects of pig slurry on barley yield and N use efficiency under semiarid Mediterranean conditions. *Eur. J. Agron.*, **44**, 78–86.

Hoekstra, N.J., Lalor, S.T.J., Richards, K.G., O'Hea, N., Lanigan, G.J., Dyckmans, J., Schulte, R.P.O. and Schmidt, O. (2010) Slurry $^{15}NH_4$-N recovery in herbage and soil: effects of application method and timing. *Plant Soil*, **330**, 357–368.

Huang, S., Ma, Y., Bao, D., Guo, D. and Zhang, S. (2011) Manures behave similar to superphosphate in phosphorus accumulation in long-term field soils. *Int. J. Plant Prod.*, **5**, 135–146.

Husted, S., Jensen, L.S. and Jorgensen, S.S. (1991) Reducing ammonia loss from cattle slurry by the use of acidifying additives: the role of the buffer system. *J. Sci. Food Agric.*, **57**, 335 349.

Jarvis, S.C., Stockdale, E.A., Shepherd. M.A. and Powlson, D.S. (1996) Nitrogen mineralization in temperate agricultural soils: processes and measurement. *Adv. Agron.*, **57**, 187–235.

Kai, P., Pedersen, P., Jensen, J.E. Hansen, M.N. and Sommer, S.G. (2008) A whole-farm assessment of the efficacy of slurry acidification in reducing ammonia emissions. *Eur. J. Agron.*, **28**, 148–154.

Kirchmann, H. and Lundvall, A. (1993) Relationship between N immobilization and volatile fatty acids in soil after application of pig and cattle slurry. *Biol. Fertil. Soils*, **15**, 161–164.

Kyvsgaard, P., Sørensen, P. Møller, E. and Magid, J. (2000) Nitrogen mineralization from sheep faeces can be predicted from the apparent digestibility of the feed. *Nutr. Cycl. Agroecosyst.*, **57**, 207–214.

Lalor, S.T.J., Schröder, J.J., Lantinga, E.A., Oenema, O., Kirwan, L. and Schulte, R.P.O. (2011) Nitrogen fertilizer replacement value of cattle slurry in grassland as affected by method and timing of application. *J. Environ. Qual.*, **40**, 362–373.

McLaughlin, M.J., Tiller, K.G., Naidu, R and Stevens, D.G. (1996) Review: the behaviour and environmental impact of contaminants in fertilizers. *Austr. J. Soil Res.*, **34**, 1–54.

Meade, G., Lalor, S.T.J. and McCabea, T. (2011) An evaluation of the combined usage of separated liquid pig manure and inorganic fertiliser in nutrient programmes for winter wheat production. *Eur. J. Agron.*, **34**, 62–70.

Misselbrook, T.H., Smith, K.A., Johnson, R.A. and Pain, B.F. (2002) Slurry application techniques to reduce ammonia emissions: results of some UK field-scale experiments. *Biosyst. Eng.*, **81**, 313–321.

Möller, K. and Müller, T. (2012) Effects of anaerobic digestion on digestate nutrient availability and crop growth: a review. *Eng. Life Sci.*, **12**, 242–257.

Nevens, F. and Reheul, D. (2005) Agronomical and environmental evaluation of a long-term experiment with cattle slurry and supplemental inorganic N applications in silage maize. *Eur. J. Agron.*, **22**, 349–361.

Nielsen, N.E., Schjorring, J.K. and Jensen, H.E. (1988) Efficiency of fertilizer nitrogen uptake by spring barley, in *Nitrogen Efficiency in Agricultural Soils* (eds D.S. Jenkinson and K.A. Smith), Elsevier, London, pp. 62–72.

Nikoli, T. and Matsi, T. (2011) Influence of liquid cattle manure on micronutrients content and uptake by corn and their availability in a calcareous soil. *Agron. J.*, **103**, 113–118.

Nyord, T., Hansen, M.N. and Birkmose, T.S. (2012) Ammonia volatilisation and crop yield following land application of solid–liquid separated, anaerobically digested, and soil injected animal slurry to winter wheat. *Agric. Ecosyst. Environ.*, **160**, 75– 81.

Oenema, O., Oudendag, D. and Velthof, G.L. (2007) Nutrient losses from manure management in the European Union. *Livest. Sci.*, **112**, 261–272.

Paul, J.W. and Beauchamp, E.G. (1995) Availability of manure slurry ammonium for corn using ^{15}N-labelled $(NH_4)_2SO_4$. *Can. J. Soil Sci.*, **75**, 35–42.

Petersen, J. and Sørensen, P. (2008a) Loss of nitrogen and carbon during storage of the fibrous fraction of separated pig slurry and influence on nitrogen availability. *J. Agric. Sci.*, **146**, 403–413.

Petersen, J. and Sørensen, P. (2008b) Fertilizer value of nitrogen in animal manures – basis for determination of a legal substitution rate (in Danish with English summary), *DJF Rapport Markbrug 138*, Aarhus University, Aarhus.

Pilbeam, C.J. (1996) Effect of climate on the recovery in crop and soil of N-15-labelled fertiliser applied to wheat. *Fert. Res.*, **45**, 209–215.

Preusch, P.L. and Tworkoski, T.J. (2003) Nitrogen and phosphorus availability and weed suppression from composted poultry litter applied as mulch in a peach orchard. *Hortscience*, **38**, 1108–1111.

Rodhe, L. and Etana, A. (2005) Performance of slurry injectors compared with band spreading on three Swedish soils with ley. *Biosyst. Eng.*, **92**, 107–118.

Saison, C., Degrange, V., Oliver, R., Millard P, Commeaux, C., Montagne, D. and Roux, X.L. (2006) Alteration and resilience of the soil microbial community following compost amendment: effects of compost level and compost-borne microbial community. *Environ. Microbiol.*, **8**, 247–257.

Schjoerring, J.K., Bock, J.G.H., Gammelvind, L., Jensen, C.R. and Mogensen, V.O. (1995) Nitrogen incorporation and remobilization in different shoot components of field-grown winter oilseed rape (*Brassica napus*, L.) as affected by rate of nitrogen application and irrigation. *Plant Soil*, **177**, 255–264.

Schjønning, P., Christensen, B.T. and Carstensen, B. (1994) Physical and chemical properties of a sandy loam receiving animal manure, mineral fertilizer or no fertilizer for 90 years. *Eur. J. Soil Sci.*, **45**, 257–268.

Schroder, J. (2005) Revisiting the agronomic benefits of manure: a correct assessment and exploitation of its fertilizer value spare the environment. *Bioresour. Technol.*, **96**, 253–261.

Schroder, J.J., Uenk, D. and Hilhorst, G.J. (2007) Long-term nitrogen fertilizer replacement value of cattle manures applied to cut grassland. *Plant Soil*, **299**, 83–99.

Shah, G.M., Groot, J.C.J., Oenema, O. and Lantinga, E.A. (2012) Covered storage reduces losses and improves crop utilization of nitrogen from solid cattle manure. *Nutr. Cycl. Agroecosyst.*, **94**, 299–312.

Smith, P., Goulding, K.W., Smith, K.A., Powlson, D.S., Smith, J.U., Falloon, P. and Coleman, K. (2001) Enhancing the carbon sink European agricultural soils: including trace gas fluxes in estimates of carbon mitigation potential. *Nutr. Cycl. Agroecosyst.*, **60**, 237–252.

Sommer, S.G. and Olesen, J.E. (2000) Modelling ammonia volatilization from livestock slurry trailing hose applied to cereals. *Atmos. Environ.*, **34**, 2361–2372.

Sommer, S.G., Friis, E., Bach, A. and Schjorring, J.K. (1997) Ammonia volatilization from pig slurry applied with trail hoses or broadspread to winter wheat: effects of crop developmental stage, microclimate, and leaf ammonia absorption. *J. Environ. Qual.*, **26**, 1153–1160.

Stevens, R.J. and Laughlin, R.J. (1988) The effects of times of application and chemical forms on the efficiencies of N-15-labelled fertilizers for ryegrass at 2 contrasting field sites. *J. Sci. Food Agric.*, **43**, 9–16.

Sørensen, P. (1996) Short-term anaerobic storage of ^{15}N-labelled sheep urine does not influence the mineralization of nitrogen in soil, in *Progress in Nitrogen Cycling Studies* (ed. O. Van Cleemput), Kluwer, Dordrecht, pp. 141–145.

Sørensen, P. and Amato, M. (2002) Remineralisation and residual effects of N after application of pig slurry to soil. *Eur. J. Agron.*, **16**, 81–95.

Sørensen, P. and Eriksen, J. (2009) Effects of slurry acidification with sulphuric acid combined with aeration on the turnover and plant availability of nitrogen. *Agric. Ecosyst. Environ.*, **131**, 240–246.

Sørensen, P. and Fernandez, J.A. (2003) Dietary effects on the composition of pig slurry and on the plant utilization of pig slurry nitrogen. *J. Agric. Sci.*, **140**, 343–355.

Sørensen, P. and Jensen, E.S. (1995) Mineralization–immobilization and plant uptake of nitrogen as influenced by the spatial distribution of cattle slurry in soils of different texture. *Plant Soil*, **173**, 283–291.

Sørensen, P. and Jensen, E.S. (1996) The fate of fresh and stored N-15-labelled sheep urine and urea ap-plied to a sandy and a sandy loam soil using different application strategies. *Plant Soil*, **183**, 213–220.

Sørensen, P. and Rubæk, G.H. (2012) Leaching of nitrate and phosphorus after autumn and spring application of separated solid animal manures to winter wheat. *Soil Use Manag.*, **28**, 1–11.

Sørensen, P. and Thomsen, I.K. (2005) Separation of pig slurry and plant utilization and loss of nitrogen-15-labeled slurry nitrogen. *Soil Sci. Soc. Am. J.*, **69**, 1644–1651.

Sørensen, P., Jensen, E.S. and Nielsen, N.E. (1994) The fate of ^{15}N labelled organic nitrogen in sheep manure applied to soils of different texture under field conditions. *Plant Soil*, **162**, 39–47.

Sørensen, P., Thomsen, I.K., Jensen, B. and Christensen, B.T. (2002) Residual nitrogen effects of animal manure measured by ^{15}N, in *Optimal Nitrogen Fertilization Tools for Recommendation: Proceedings from NJF Seminar 322* (eds H.S. Østergård, G. Fystro and I.K. Thomsen), Danish Institute of Agricultural Sciences, Copenhagen, *DIAS Report Plant Production 84*, pp. 37–41.

Sørensen, P., Weisbjerg, M.R. and Lund, P. (2003) Dietary effects on the composition and plant utilization of nitrogen in dairy cattle manure. *J. Agric. Sci.*, **141**, 79–91.

ten Berge, H.F.M. and van Dijk, W. (2009) Management of nitrogen inputs on farm within the EU regulatory framework, *Proceedings 654*, International Fertiliser Society, York.

Thomsen, I.K. (2004) Nitrogen use efficiency of ^{15}N-labeled poultry manure. *Soil Sci. Soc. Am. J.*, **68**, 538–544.

Thomsen, I.K. (2005) Crop N utilization and leaching losses as affected by time and method of application of farmyard manure. *Eur. J. Agron.*, **22**, 1–9.

Thomsen, I.K., Kjellerup, V. and Jensen, B. (1997) Crop uptake and leaching of ^{15}N applied in ruminant slurry with selectively labelled faeces and urine fractions. *Plant Soil*, **197**, 233–239.

Thomsen, I.K., Olesen, J.E., Møller, H.B., Sørensen, P. and Christensen, B.T. (2013) Carbon dynamics and stabilization in soil after anaerobic digestion of dairy cattle feed and faeces. *Soil Biol. Biochem.*, **58**, 82–87.

Thorman, R.E., Hansen, M.N., Misselbrook, T.H. and Sommer, S.G. (2008) Algorithm for estimating the crop height effect on ammonia emission from slurry applied to cereal fields and grassland. *Agron. Sustain. Dev.*, **28**, 373–378.

Van Kessel, J.S. and Reeves III, J.B. (2002) Nitrogen mineralization potential of dairy manures and its relationship to composition. *Biol. Fertil. Soils*, **36**, 118–123.

Webb, J., Pain, B., Bittman, S. and Morgan, J. (2010) The impacts of manure application methods on emissions of ammonia, nitrous oxide and on crop response – a review. *Agric. Ecosyst. Environ.*, **137**, 39–46.

Webb, J., Sørensen, P., Velthof, G., Amon, B., Pinto, M., Rodhe, L., Salomon, E., Hutchings, N., Burczyk, P. and Reid, J. (2013) An assessment of the variation of manure nitrogen efficiency throughout europe and an appraisal of means to increase manure-N efficiency. *Adv. Agron.*, **119**, 371–442.

Yagüe, M.R. and Quílez, D. (2013) Residual effects of fertilization with pig slurry: double cropping and soil. *Agron. J.*, **105**, 70–78.

16

Life Cycle Assessment of Manure Management Systems

Sander Bruun[1], Marieke T. Hoeve[1] and Morten Birkved[2]

[1]Department of Plant and Environmental Sciences, University of Copenhagen, Denmark
[2]Department of Management Engineering, Technical University of Denmark, Denmark

16.1 Introduction

As the preceding chapters have described, new and advanced technologies for manure management are increasingly being developed and applied in order to reduce the environmental impacts or burdens associated with manure in animal housing, during storage or during or after field application. Environmental impacts may include global warming, eutrophication and human toxicity. However, the new technologies are often associated with other environmental impacts. For example, anaerobic digestion results in biogas, which can be combusted to produce electricity and heat, and thereby avoid the environmental impacts associated with the combustion of coal and/or natural gas. However, it also requires additional transportation of the manure to the biogas plant and may thus lead to impacts on the environment associated with this. The environmental consequences of the new technology may thus be of a completely different nature than those of the former technology and it is not easy to determine which technology option is most environmentally benign. Therefore, with the introduction of new technology it is most often relevant to ask: "Is the new technology that I wish to apply better than the former from an environmental point of view?". This seems like a very straightforward and legitimate question, but as it turns out, it is often very difficult to answer. One methodology that has been developed to answer such questions is life cycle assessment (LCA). LCA has been applied in a number of studies to compare alternative manure management technologies with existing technologies (Basset-Mens and van der Werf, 2005; Hamelin *et al.*, 2011; Prapaspongsa *et al.*, 2010; de Vries *et al.*, 2010). In this chapter, we provide a general introduction to LCA and give some examples of how it can be applied to

Animal Manure Recycling: Treatment and Management, First Edition. Edited by Sven G. Sommer, Morten L. Christensen, Thomas Schmidt and Lars S. Jensen.

manure management systems. The aim is to give readers a general understanding of the LCA methodology and how it can be used in connection with manure management systems, thereby enabling them to study and critically evaluate LCA papers and reports.

16.2 Introduction to the Life Cycle Assessment Methodology

LCA is a methodology that can be used to compare the environmental impacts of products or services providing the same function. The LCA framework and methodology are standardised in International Standardization Organization (ISO) standards ISO 14040 (International Standardization Organization, 2006a) and ISO 14044 (International Standardization Organization, 2006b). These standards provide very general guidelines for LCA application, research and reporting. In addition, numerous detailed guidelines on how to conduct LCAs have been developed. One example of such guidelines is the *International Reference Life Cycle Data System Handbook* (European Commission – Joint Research Centre, 2010).

In LCA, the entire life cycle of the product is considered through all life stages of the product from raw material extraction through manufacturing, transportation and use to final disposal. Secondary processes associated with the production system for the product also need to be included. Examples of such secondary processes are diesel production for machinery used for raw material extraction, electricity production and vehicle production for transportation. A holistic system perspective considering the complete life cycle is clearly necessary. If decisions are based on an analysis only focusing on one life stage, the option chosen may have the lowest impact in this life stage, but higher impacts in other life stages.

16.3 Four Phases of a Life Cycle Assessment

According to the ISO standards, there are four phases in an LCA: *goal and scope definition, inventory analysis, impact assessment*, and *interpretation*. These four phases are discussed in more detail in the following sections. The LCA practitioner has to go through these four phases. However, performing an LCA is an iterative process. This means that the LCA practitioner is constantly learning and updating the output of the different phases. For example, if it is found while working on the impact assessment that one of the processes has a major influence on the results, the LCA practitioner goes back to the inventory of this particular process, to find more numerous and ideally more representative (i.e. temporal or spatial) information about the process.

16.4 Goal and Scope

In the goal and scope definition, all major decisions have to be made. The goal of the LCA should be defined as clearly as possible. In order to do this, the intended application has to be defined (i.e. the products or services to be compared). If a study is looking at manure management systems, it could compare different storage options or a system with biogas production with a system without, or it might compare different methods for field application of manure. The stakeholders that may have an interest in the results also have to be defined. Who may be interested in the results and why? Who has commissioned the study?

To define the scope of the study, a functional unit has to be chosen. The functional unit represents the function that the product provides. In manure management systems the functional unit could be "treatment of 1 ton of manure" or when comparing different ways of producing biogas it could be "production of 1 m^3 biogas at the digester". The definition of the functional unit depends entirely on the application, but should always serve to make sure that the function or service of the products provided by the systems being compared is the same and comparable.

16.4.1 System Boundaries

An important step of the scope definition is to define the system boundaries. There are two different kinds of system boundaries that need to be defined: boundaries between the technical system and the natural system, and boundaries between processes that are included in the assessment and processes that are not. First, an explanation of what is meant by the boundaries between the technical (i.e. production) and natural system. When a substance leaves the technical system and enters the natural system, it is considered to be an emission, and thus may have an effect on the environment. In a similar way, when a substance leaves the natural system and enters the technical system, it is considered consumption of a resource. For example, ammonia (NH_3) that volatilises from a slurry storage tank is an emission because it is leaving the technical system. In this case the system boundary is the surface of the slurry tank. When nitrate leaches from a soil, it is also an emission, but how far down the soil does it have to leach before it is considered to be lost? Here a logical system boundary could be somewhere beneath the root zone where the chance of plant uptake is minimal. In an LCA, it is rarely possible to include all processes that are associated with a product. This can be seen for the example of a new technology which uses a pump in the transfer of animal slurry. To produce that pump, steel and rubber are needed, but a factory is also needed. Imagine for a second what it takes to make a factory. In practice, the LCA practitioner has to decide which processes to include and which not. The general rule is that all processes that make a significant contribution from an environmental perspective need to be included. It may also be impossible to find representative information for certain processes (e.g. confidential information that companies do not want to provide). In that case, estimates (i.e. extrapolations, interpolations, model estimates, forecasting and even predictions, etc.) have to be made for the processes involved. An LCA paper or report needs to clearly state which processes are included in the product system model and, even more importantly, which processes are omitted.

16.4.2 Allocation Problems

Another decision that has to be made in the scope definition is how to handle co-products resulting from processes with multiple outputs (e.g. crop production producing both grain and straw, both valuable products serving quite different purposes). Consider two products that provide a similar function, but there is a co-product associated with the production of one and not the other. Thus, the two systems are not providing the same service and cannot be compared because one of the product systems is also producing the co-product. For example, when conventional manure treatment is compared with anaerobic digestion of manure, there is a difference in the service provided by these two systems. Both systems provide disposal of the manure, but anaerobic digestion also produces biogas. The preferred way of solving this problem is to use system expansion. This is done by identifying which product the co-product would replace and then modelling the consequences of avoiding the production of these products. In the example above, system expansion would mean that the same amount of gas (i.e. gas energy) needs to be produced in the conventional scenario; this gas would most likely be natural gas. The problem can also be solved by allocation, meaning that the emissions are allocated to the primary product and the co-products according to either a physical relationship such as weight or volume, or something else, most often economic value. An example of an allocation problem is given in Example 16.1.

Example 16.1

In this example, conventional pig slurry management is compared with separation of pig slurry. In the conventional management scenario, slurry is applied to a field with a high phosphorus status. The

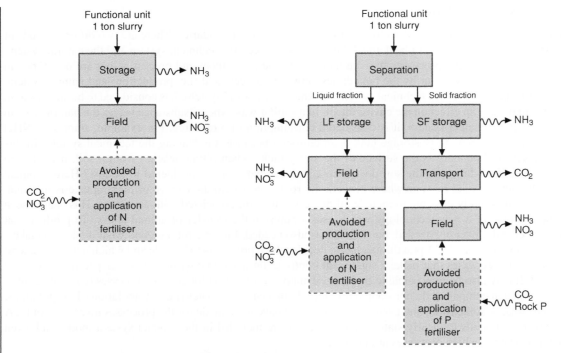

Figure 16.1 *Two scenarios for the handling of one functional unit (1 ton of manure), one with traditional land application and one with slurry separation, land application of the liquid fraction (LF) and transport of the solid fraction (SF) to a place where P fertiliser is needed where it is applied. Avoided processes are indicated with a broken frame. (© University of Southern Denmark.)*

liquid fraction after separation is applied to the same field as slurry in the conventional management case. The solid fraction after separation is applied to a field with a low phosphorus status. Figure 16.1 provides a graphical overview of the two scenarios.

In both scenarios the functional unit is defined as treatment of 1 ton of animal slurry. In the conventional management scenario, slurry is stored and then applied to the field. Ammonia is emitted during storage and after field application, and nitrate is emitted to the aquatic environment after field application. However, the application of slurry is useful in the sense that it acts as a fertiliser supplying nitrogen (N) to the plant. It also supplies phosphorus (P), but since there is already plenty of P this does not reduce the amount of P fertiliser that will be applied. In contrast, the N applied means that the farmer can reduce the amount of mineral N fertiliser used. This can be modelled by using system expansion. This is done by identifying the N fertiliser that is not used, and then running the production and application processes in reverse. Reversing these processes means that carbon dioxide (CO_2) is taken out of the atmosphere during reverse fertiliser production and nitrate is taken out of the groundwater and aquatic environment after reverse field application. The uptake from the environment or reverse leaching is likely to be smaller than the amount lost because of the application of manure. So the net nitrogen leaching will still be positive. In the separation scenario, NH_3 is emitted during storage and after field application of both fractions and nitrate is emitted to the aquatic environment after field application for both fractions. The liquid fraction is applied to the same field type as slurry and thus the N contained in it can replace mineral fertiliser in the same way as in the conventional scenario. The solid fraction is assumed to be transported to a field further away from the farm. Therefore, CO_2 emissions from transportation are higher than for the slurry and liquid fraction. Since the field is deficient in P, the solid fraction can replace mineral P fertiliser. This means that the production and application of P fertiliser is avoided. Again, system expansion can be used and the production and application process run in reverse; taking CO_2 out of the atmosphere and putting rock phosphate back in the mines. In that case, the service provided in both scenarios corresponds to the functional unit and is equal. The only difference between the scenarios is the environmental emissions and the resource consumption.

16.4.3 Consequential versus Attributional Life Cycle Assessments

There are two different approaches to LCAs that are usually distinguished – the *consequential* and *attributional* approach. In the attributional approach, the consequences are modelled as averages of current sources, whereas in the consequential approach the actual consequences of the actions are modelled. For an explanation of these modelling approaches see Text Box – Advanced 16.1.

Text Box – Advanced 16.1   Consequential versus attributional LCAs

There are two different approaches to LCAs that are usually distinguished – the *consequential* and *attributional* approach. In the attributional approach, the consequences are modelled as averages of current sources, whereas in the consequential approach the actual consequences of the actions are modelled. For example, if electricity is used for manure separation by a screw press, the question arises where this electricity comes from. In the attributional approach a weighted average would be used so that it represents a mix of electricity produced in coal-fired plants and by nuclear power, natural gas, hydro power and windmills used in the region where the electricity is used. In the consequential approach, on the other hand, the consequences of the additional use of electricity are modelled. As the wind is not likely to blow any stronger because of additional consumption of electricity, the additional electricity is not likely to come from windmills. It is not likely to come from nuclear power either, because nuclear plants run with constant effect all the time. Rather, the additional consumption of electricity is most likely to be produced from coal or natural gas. In the longer term, investments in windmills may be stimulated if there are some political targets for production of renewable energy. The source which is affected by the changed demand needs to be identified in order to be able to model the effects. As the above example illustrates, small, short-term changes in demand affect only capacity utilisation, whereas large or long-term changes may affect investments and capacity building. Identifying the source actually affected relies on having much information about the market for the product. This is very difficult, but there are some general guidelines that can be used (Weidema, 2003).

In the product systems shown in Figure 16.1, it can be seen that the production of P fertiliser is avoided for the solid fraction. This is due to the fact that the P level is assumed to be low in the field where the solid fraction is applied. In the attributional approach, the avoided production would be modelled as a weighted average of the P fertilisers being used in the country or region. This would mean a mixture of fertilisers, such as superphosphate and triple superphosphate. The phosphate rock that is used for the production of these fertilisers originates amongst others from China, Morocco and Florida, reflecting the current state of production. In the consequential approach, the fertiliser actually affected by the decreased demand for fertiliser would have to be identified. So according to the consequential assessment approach, the question is where in the world would extraction of P be reduced because of decreased demand? In the short term, this would tend to be the least competitive place.

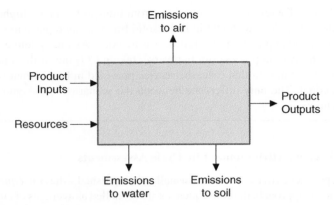

Figure 16.2   *General representation of a process receiving inputs in the form of natural resources and products from other processes and delivering other products and resulting in emissions to different compartments. (© University of Southern Denmark.)*

16.5   Inventory Analysis

The purpose of inventory analysis is to calculate all the environmental exchanges (i.e. emissions and resource consumption) that pass over the system boundaries in relation to the production of one functional unit. This is done by collecting input and output flows of each of the processes in the system and finding data for the emissions and resource consumption for each of them. This is illustrated in Figure 16.2. For example, how much manure goes into the digester per day (product input), how much biogas comes out (product output), how much digestate comes out per day (product output), what are the losses of methane from the digester per day (emissions to air).

Data for different processes have to be compiled from available sources such as measurements, experts, scientific papers, reports, databases and so on. Some emissions may also be calculated from principles used in systems analysis, such as conservation of energy, mass and elements.

When these data have been collected they are related to the functional unit so that a total inventory of environmental exchanges is calculated per functional unit. This is illustrated in Example 16.2.

Example 16.2

A simple example with two processes is illustrated in Figure 16.3. Data have been collected for both processes and they show that process A receives 0.5 kg of product a and produces 0.2 kg of product b. In this process 0.1 kg of e is emitted. Process B receives 8 tons of product b and produces 4 tons of product c. In this process, 2 tons of e is emitted. The LCA modeller investigating the resource need and emissions associated with the production of c has chosen a functional unit of 1 kg c and therefore first scales process B so that 1 kg of c is produced in the process. This means that 0.5 kg e is emitted and 2 kg b is used. Process is A is then scaled to produce the 2 kg b needed for process B. This means that 1 kg e is emitted and 5 kg a is consumed. The result of the inventory analysis is therefore that 5 kg of resource a is consumed and 1.5 kg e is emitted.

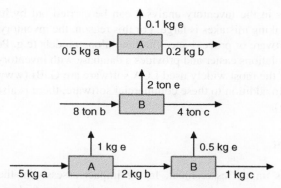

Figure 16.3   *A simple example illustrating how two processes are connected and scaled to the functional unit of 1 kg of product c. (© University of Southern Denmark.)*

The result of an inventory is a long list of resource consumptions and emissions of environmentally problematic substances to different compartments such as soil, ground water, surface fresh water, sea water and air. An example of inventory results can be found in Example 16.3.

Example 16.3

The resource use and emissions to the environment of the processes exemplified in Figure 16.1 are listed in detail in Table 16.1. In reality there will be much more emissions, but these have been omitted here for reasons of simplicity.

Table 16.1   *Resource use and emissions during life cycle stages.*

	Resources					Environmental emissions				
	Crude oil	Hard coal	Lignite	Natural gas	P rock	NH_3	N_2O	NO_3	CH_4	CO_2
Storage										
slurry						0.05			2.11	5.80
liquid						0.05			1.27	3.48
solid						0.00	0.00		0.41	1.50
Transportation										
slurry										
liquid										
solid	0.69	0.08	0.04	0.06					0.00	2.56
Field application										
slurry	0.64	0.35	0.11	0.11		0.71	0.15	8.34	0.00	49.61
liquid	0.57	0.30	0.10	0.10		0.48	0.13	7.14	0.00	34.60
solid	0.09	0.04	0.01	0.01		0.19	0.02	1.03		17.48
Avoided N + P fertiliser										
slurry	−0.12	−0.40	−0.95	−3.69		−0.12	−0.14	−6.45	−0.06	−19.54
liquid	−0.11	−0.39	−0.92	−3.58		−0.11	−0.14	−6.26	−0.06	−18.96
solid	−0.18	−0.23	−0.35	−0.54	−0.45	−0.01	−0.02	−0.82	−0.01	−3.65

Of course the calculations in the inventory analyses can be carried out by hand, but this is very time-consuming and the risk of making mistakes is large. For this reason, the inventory analysis is usually carried out using dedicated LCA software or process-specific stand-alone models (e.g. PestLCI 2.0; Dijkman *et al.*, 2012), which makes the calculations easier and provides a database with inventory data for some of the most common processes. Some of the most widely used LCA software are GaBi (www.gabi-software.com/LCA) and SimaPro (www.pre.nl). In addition to these commercial software, there is also the open-source program openLCA (www.openlca.org).

16.6   Impact Assessment

The third phase of LCAs is impact assessment. In the impact assessment the potential burdens of the environmental exchanges (emissions and resource consumptions) quantified in the inventory analysis are assessed. According to the ISO standards, the impact assessment consists of some mandatory steps and some optional steps. The mandatory steps include impact category definition, classification and characterisation, while the optional steps include normalisation and weighting.

16.6.1   Impact Category Definition, Classification and Characterisation

In the impact category definition, impact categories such as global warming, eutrophication, acidification, human toxicity and resource consumption are defined. In the classification, the categories to which the different environmental exchanges contribute are decided. For example, an emission of NH_3 will add to eutrophication because it contains bioavailable nitrogen which potentially ends up in sensitive terrestrial and aquatic ecosystems. Ammonia further contributes to acidification since once it ends up in an ecosystem it is oxidised to nitrate, releasing H^+ in the process. The final contribution of the different environmental exchanges (as calculated in the inventory analysis) is thus determined. This step is defined as characterisation. The factors used to calculate the contribution to each impact category are known as characterisation factors or equivalency factors. These factors can be used to calculate the potential impacts using the equation:

$$IP(j) = \sum_{i=1}^{n} Q_i CF(j)_i \tag{16.1}$$

where $IP(j)$ is the impact potential for impact category j, Q_i is the emission of substance i to a certain compartment (e.g. soil, air, surface water, groundwater) and $CF(j)_i$ is the characterisation factor for emission of substance i to the compartment to category j. Application of this equation is exemplified in Examples 16.4 and 16.5.

Example 16.4

According to the Intergovernmental Panel on Climate Change (IPCC) (Forster *et al.*, 2007), in a 100-year perspective, emissions of 1 kg methane (CH_4) will cause the same degree of global warming as 25 kg CO_2, corresponding to 25 kg CO_2-eq. kg^{-1} CH_4, while 1 kg nitrous oxide (N_2O) will cause the same degree of global warming as 298 kg CO_2 or 298 kg CO_2-equivalents per 1 kg N_2O. If treatment of manure corresponding to 1 functional unit resulted in emissions of 4.2 kg CO_2, 0.046 kg N_2O and 0.12 kg CH_4 then the total impact potential is:

$$IP(\text{global warming}) = 4.2 \text{ kg} \times 1 \text{ kg } CO_2\text{-eq. kg}^{-1} + 0.046 \text{ kg} \times 298 \text{ kg } CO_2\text{-eq. kg}^{-1}$$
$$+ 0.12 \text{ kg} \times 25 \text{ kg } CO_2\text{-eq. kg}^{-1} = 20.9 \text{ kg } CO_2\text{-eq.}$$

Example 16.5

Returning to the slurry separation example in Figure 16.1, the global warming potential for the conventional and separation scenario are calculated in Table 16.2. The impact potential within the impact category "global warming potential" is clearly larger for the conventional manure management scenario.

Table 16.2   *Global warming potential of the conventional scenario and the separation scenario.*

i	Conventional			Separation		
	Q_i (kg)	CF_i (CO_2-eq.)	IP (kg CO_2-eq.)	Q_i (kg)	CF_i (CO_2-eq.)	IP (kg CO_2-eq.)
CO_2	38.5	1	38.5	40.9	1	40.9
CH_4	2.1	25	51.7	1.6	25	40.6
N_2O	0.01	298	2.7	−0.002	298	−0.6
Total			92.9			80.9

Q_i is the emission of substance i, CF_i is the characterisation factor for substance i to global warming and IP is the impact potential.

Determining characterisation factors is obviously a difficult and computational intensive task, and a predefined impact assessment method with the most important impact categories and predetermined characterisation factors for common/frequently occurring emissions is generally used. This means that the LCA analyst does not have to personally estimate characterisation factors for the most common emissions. The most frequently applied impact assessment methods include EDIP 2003 (Hauschild and Potting, 2005), IMPACT 2002+ (Jolliet *et al.*, 2003), Stepwise (Weidema *et al.*, 2008) and ReCiPe (www.lcia-recipe.net). The most common impact methods are included in LCA software so the calculations are usually performed automatically upon request.

16.6.2   Normalisation

After the characterisation step there are some optional steps that can be used to help make the results easier to understand. Normalisation is the first optional step, which in general is considered an improvement of the communicability of the assessment results. In normalisation, the impacts are normalised with respect to a reference point, which in general is considered reliable and objective (e.g. by EUROstat, UN Environment Programme, IPCC, World Health Organization, UN Food and Agriculture Organization, etc.). One way to deal with normalisation is by dividing the characterised impact potentials by the impact of an average person in a chosen target group or area. Normalisation enables the LCA analyst to evaluate whether the contributions to one impact category associated with the production of one functional unit are large compared with other contributions to the impact category. This is achieved simply by normalising impacts in each category with the contribution for which an "average person" is responsible in that category in a relevant target area. The normalisation procedure can be described by the equation:

$$NIP(j) = \frac{IP(j)}{NR(j)} \tag{16.2}$$

where $NIP(j)$ is the normalised impact potential in impact category j and $NR(j)$ is the normalisation reference. The normalization is exemplified in Example 16.6.

Example 16.6

Table 16.3 lists the impact potentials, followed by the normalised impact potentials, the weighted impact potentials and the overall weighted impact potentials for the example with conventional and separation technologies for handling manure illustrated in Figure 16.1. The three impact categories that are analysed here are global warming potential, acidification potential and eutrophication potential. For the three impact categories that are taken into consideration, the separation scenario shows lower impact potentials. The differences range from 8.4% for acidification potential to 38.2% for eutrophication potential. Even though more energy is needed for separation and transportation in the separation scenario, its environmental impact potentials are still lower than in the conventional scenario. This is mainly due to the avoided use of mineral N and P fertiliser. An average person has a yearly contribution to global warming of 8700 kg CO_2-eq. The contribution of handling 1 ton of slurry in the conventional scenario is 92.9 kg CO_2-eq. The normalised global warming impact is 0.011 person equivalents (PE). It can also be seen that acidification is actually the category with the largest impacts, compared with global warming and eutrophication, relative to impacts otherwise produced in the area. For weighting, the normalised results need to be multiplied by a weighting factor. This factor is 1.12 for global warming, 1.27 for acidification and 1.22 for eutrophication. The data show that acidification has the highest weighted impact potential. The total weighted impact potentials for the conventional and separation scenarios indicate that the separation technology is associated with smaller environmental impacts. This means that separation technology could be an option worth considering for handling manure, seen from an environmental point of view.

Table 16.3   *Impact potentials for the conventional scenario and the separation scenario.*

	Global warming potential	Acidification potential	Eutrophication potential
Impact potential (kg [CO_2, SO_2 or NO_3^-]-eq.)			
conventional	92.9	1.6	2.2
separation	80.9	1.4	1.3
Normalised impact potential (PE)			
conventional	0.011	0.021	0.018
separation	0.009	0.020	0.011
Weighted impact potential (PET)			
conventional	0.012	0.027	0.022
separation	0.010	0.025	0.014
Total weighted impact potential (PET)			
conventional	0.061		
separation	0.049		

PE, person equivalents; PET, targeted person equivalents.

16.6.3   Weighting

Although the normalised impact potentials provide some information to the LCA analyst about how large the impact contributions are to the impact categories compared with the impact for which an average person is responsible, the normalisation procedure still does not say anything about the importance of an impact contribution in terms of the seriousness of the potential environmental impacts. This means that a product that primarily contributes to one impact category, (e.g. global warming) cannot be compared with a product that primarily contributes to another impact category (e.g. eutrophication). One question that may arise from comparing global warming and eutrophication is how many CO_2-equivalents of global warming correspond to one nitrate equivalent of eutrophication? In order to solve this problem, weighting is used. If $W(j)$ is the weighting factor for category j, then the weighted impact potential can be calculated as:

$$WIP(j) = NIP(j)W(j) \tag{16.3}$$

and subsequently the weighted impact potentials in the different impact categories can be calculated by summing the impacts in different categories:

$$WIP = \sum_{j=1}^{m} WIP(j) \tag{16.4}$$

The weighting of results is exemplified in Example 16.6.

Each impact assessment method has different ways of calculating the weights and the weighting references can be based on political targets, assuming that a category for which there are ambitious political targets is also considered a serious environmental problem. In the same way, expert judgement or monetary value can be used to calculate weighting factors. No matter which method is used to determine weighting factors, it will be highly debatable. However, the alternative is not being able to reach a conclusion about which product to choose or leaving it up to the decision maker to decide which impacts are the most important. In this sense, the chosen weighting method should only be considered as guidance on how important the impacts in the different categories could be considered.

The gradual aggregation of the, most often, complex data coming out of the inventory analysis into impact categories and weighted results is illustrated in Figure 16.4.

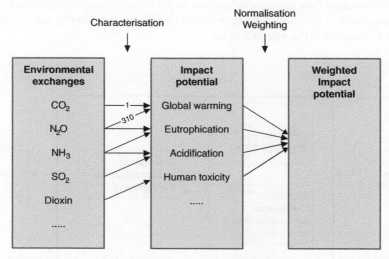

Figure 16.4 *Gradual aggregation of the large number of environmental exchanges (emissions and resource consumptions) into a smaller number of impact potentials and their subsequent aggregation into a weighted index measuring the overall severity of the impacts associated with the production of a functional unit of a product. (© University of Southern Denmark.)*

16.7   Interpretation

In the final phase of the LCA, an interpretation of the results has to be carried out. In the interpretation the findings of the inventory analysis and impact assessment are combined in order to draw conclusions and recommendations in accordance with the goal and scope. This is illustrated in Figure 16.5.

An important part of the interpretation is to test the robustness of the results. For example, if the fraction of the methane produced that is leaking from a biogas plant is estimated to be 2%, but the value could instead be 1% or 5%, the effects of decreasing the value to 1% and increasing it to 5% can be tested in the assessment. If the lower or higher estimate does not yield a change in the ranking of the scenarios being compared, the conclusions can be considered to be robust. However, if the ranking changes, any conclusion relying on this emission estimate will reflect the inherent uncertainty. There are several different, more complex, ways to test the robustness of the results, including variation analysis, sensitivity analyses, break-even point analysis and Monte Carlo simulations.

16.8   Summary

LCA is a methodology that can be used to compare the environmental impacts of products or services providing the same function. LCA has been applied in a number of studies to compare alternative manure management technologies with existing technologies. According to the ISO standards, there are four phases in an LCA: goal and scope definition, inventory analysis, impact assessment, and interpretation. The LCA practitioner has to go through these four phases. In the goal and scope phase the goal of the study is defined, and all major decisions regarding scope and modelling are made. The functional unit defines the function delivered by the assessed product or technology and has to be identical for all products or technologies that are compared. In studies of manure management technologies a functional unit could be treatment of 1 ton of manure. In the inventory analysis, all emissions and resource consumptions associated with one functional unit are calculated by making a flow model of the system. In the impact assessment, the environmental impacts of the emissions and resource consumption are assessed and compared. Finally, the results are interpreted and the uncertainty is assessed in the interpretation.

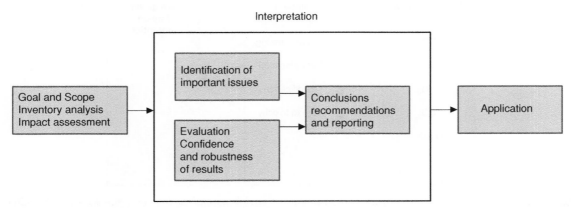

Figure 16.5 *Illustration of the way in which the information from the goal and scope, inventory analysis and impact assessment stages is combined and used in the interpretation to derive conclusions. Adapted from International Standardization Organization (2006a). (© University of Southern Denmark.)*

References

Basset-Mens, C. and van der Werf, H.M.G. (2005) Scenario-based environmental assessment of farming systems: the case of pig production in France. *Agric. Ecosyst. Environ.*, **105**, 127–144.

de Vries, M. and de Boer, I.J.M. (2010) Comparing environmental impacts for livestock products: a review of life cycle assessments. *Livest. Sci.*, **128**, 1–11.

Dijkman, T.J., Birkved, M. and Hauschild, M.Z. (2012) PestLCI 2.0: a second generation model for estimating emissions of pesticides from arable land in LCA. *Int. J. Life Cycle Assess.*, **17**, 973–986.

European Commission – Joint Research Centre (2010) *International Reference Life Cycle Data System (ILCD) Handbook – General guide for Life Cycle Assessment – Detailed Guidance*, Publications Office of the European Union, Luxembourg.

Forster, P., Ramaswamy, V., Artaxo, P., Berntsen, T., Betts, R., Fahey, D.W., Haywood, J., Lean, J., Lowe, D.C., Myhre, G., Nganga, J., Prinn, R., Raga, G., Schulz, M. and Van Dorland, R. (2007) Changes in atmospheric constituents and in radiative forcing, in *Climate Change 2007: The Physical Science Basis. Contribution of Working Group I to the Fourth Assessment Report of the Intergovernmental Panel on Climate Change* (eds S. Solomon, D. Qin, M. Manning, Z. Chen, M. Marquis, K.B. Averyt, M. Tignor and H.L. Miller), Cambridge University Press, Cambridge, pp. 129–234; http://www.ipcc.ch/pdf/assessment-report/ar4/wg1/ar4-wg1-chapter2.pdf.

Hamelin, L., Wesnæs, M., Wenzel, H. and Petersen, B.M. (2011) Environmental consequences of future biogas technologies based on separated slurry. *Environ. Sci. Technol.*, **45**, 5869–5877.

Hauschild, M.Z. and Potting, J. (2005) *Spatial Differentiation in Life Cycle Impact Assessment – The EDIP 2003 Methodology*, Danish Ministry of the Environment, Environmental Protection Agency, Copenhagen.

International Standardization Organization (2006a) *ISO 14040: Environmental Management – Life Cycle Assessment – Principles and Framework. European Standard*, ISO, Geneva.

International Standardization Organization (2006b) *ISO 14044: Environmental Management – Life Cycle Assessment – Requirements and Guidelines. European Standard*, ISO, Geneva.

Jolliet, O., Margni, M., Charles, R.L., Humbert, S., Payet, J., Rebitzer, G. and Rosenbaum, R. (2003) IMPACT 2002+: a new life cycle impact assessment methodology. *Int. J. Life Cycle Assess.*, **8**, 324–330.

Prapaspongsa, T., Christensen, P., Schmidt, J.H. and Thrane, M. (2010) LCA of comprehensive pig manure management incorporating integrated technology systems. *J. Clean. Prod.*, **18**, 1413–1422.

Weidema, B. (2003) *Market Information in Life Cycle Assessment*, Danish Environmental Protection Agency, Copenhagen.

Weidema, B.P., Hauschild, M.Z. and Jolliet, O. (2008) Preparing characterisation methods for endpoint impact assessment, in *Environmental Improvement Potentials of Meat and Dairy Products (EUR 23491 EN)* (eds B.P. Weidema, M. Wesnæs, J. Hermansen, T. Kristensen, N. Halberg, P. Eder and L. Delgado), Institute for Prospective Technological Studies, Sevilla, annex II.

17

Innovation in Animal Manure Management and Recycling

Thomas Schmidt

Technology Transfer Office, Aarhus University, Denmark

17.1   Introduction – Why is Innovation Important?

New innovations in animal waste management and recycling are brought to market each year. For each successful market introduction, however, numerous new technologies and business models fail to reach the end-user. The difference between successful commercialisation of a new innovation and falling short comes down to preparation, correct use of resources, execution and, of course, timing. This chapter addresses these issues and also introduces a number of key concepts and shows their influence on successful innovation in the field of animal waste recycling.

Innovation can take many forms and is certainly a term used liberally in different industries. One early definition sees innovation as the new combination of existing opportunities (Schumpeter, 1934). Any interpretation of innovation should take into account that it is not only about generating new ideas, concepts or technology, but also about turning those ideas into practical usages, be it commercial products, better production processes or smarter organisation of the company. For the purposes of this book, innovation is defined as the development of new technologies, methods and business models – innovations – with which companies seek to build new competitive advantages in order to gain value. In this sense, innovation is seen as a process of turning opportunities into new ideas and of putting these ideas into widely used practice (Tidd and Bessant, 2009).

Companies, in essence, exist to create value for their shareholders and this value comes from offering a product or service to a group of customers in a way that delivers revenue to the company from which a profit can be derived. Porter (1985) noted that companies build and use their core competences and competitive capabilities to deliver such products or services. A superior value offering to customers comes from the company having one or more competitive advantages over their competitors. Competitive advantage is so

Animal Manure Recycling: Treatment and Management, First Edition. Edited by Sven G. Sommer, Morten L. Christensen, Thomas Schmidt and Lars S. Jensen.

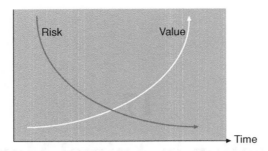

Figure 17.1   *Illustration of the effect of risk reduction on value over time. (© University of Southern Denmark.)*

crucial to a company's long-term survival that companies expend vast resources on building their competitive advantage in an attempt to create a lasting and self-enforcing effect that is termed a sustainable competitive advantage (Porter, 1985). Therefore, innovation is important to companies, because companies achieve competitive advantage through acts of innovation and they do so by approaching innovation in its broadest sense, including both new technologies and new ways of doing things (Porter, 1990). This is important, because sustainable competitive advantage allows companies to successfully navigate changes in their macro and micro environment and thus form the long-term foundation for value creation (Porter, 1985).

The major steps in the innovation process can be described as a pyramidal process by which innovations are initially searched for in a structured manner before moving to an assessment of those new innovations. This in turn results in formulation of a business case or commercialisation plan, by which the new innovation is commercialised for the benefit of the organisation. For each step in this process, from search to market launch the overall goal is to reduce risk. By reducing risk, the innovation increases value (Figure 17.1).

See Text Box – Basic 17.1 for definitions used in this chapter.

Text Box – Basic 17.1   Definitions

Incremental innovation: The development of existing methods and practices that extend products or services already on the market. It is evolutionary in nature. The incremental innovation can come about as a demand from customers – a process defined as customer pull. Examples within agriculture are a new version of a piece of software, a new add-on to a piece of machinery, or the introduction of a consultancy service in a new geographical area.

Radical innovation: The conceptualisation of ideas that are so different that they cannot be compared to any existing practices or perceptions. A radical innovation is a disruptive breakthrough that creates conceptual shifts in the marketplace. It is revolutionary in its nature. If the innovation is commercialised by a company where there is no pre-existing market – this process is typically defined as a company push. Examples within agriculture are automated farming robots, use of pharmaceutical antibodies in piglet production or the development of biofuel from manure.

Internal rate of return: The discount rate that a project's net income must exceed to be included in a company portfolio of projects. The rate is calculated as the value at which an investment has zero net present value.

Net present value (NPV): The project's total future net contribution to wealth at the time of calculation. NPV is sometimes calculated with adjustment for risk, in which case it is labelled risk-adjusted NPV (rNPV).

17.2   Innovation Typology

New innovations can come in many shapes and sizes. Therefore, it is relevant to understand the context of new innovation generation and develop an innovation typology. In this respect, it is relevant to use life cycles of products or markets as a backdrop by which different types of innovation can be described. Several versions of product life cycles and market life cycles exist. Some are relevant to describe different points in time and others are relevant to describe certain market trends or product categories. One such early and widely used product life cycle comes from Levitt (1965), who developed the model shown in Figure 17.2.

A product passes through different stages over time. In the market development stage the size of the market, sales volumes and sales growth are small. The product is also normally subject to little or no competition. The primary goal is to establish a market and build customer demand for the product. If customers gain awareness of the product and buy in to its perceived value, the product experiences increased sales and the company can expect a period of rapid sales growth described as the growth stage. In the growth stage, the company builds brand loyalty amongst the customers and focuses on increasing its market share. When a product reaches maturity, the sales growth slows or stops, and the sales volume peaks and stabilises. In the maturity stage, the market as a whole makes the most profit. The company's focus is then to defend its market share and maximise profit. When sales and associated profits consistently start to fall, the product enters the decline stage. The market for the product shrinks, thereby reducing the amount of obtainable profit. A decline may occur, because the market has become saturated, the product has become obsolete or customer preferences have changed.

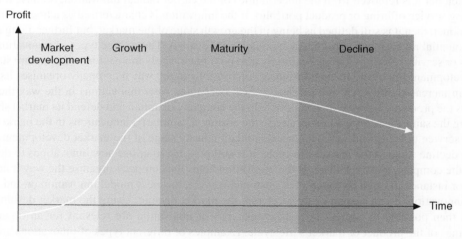

Figure 17.2   *The product life cycle. (Reprinted with permission from Levitt (1965). © 1965 Harvard Business School Publishing.)*

As markets comprise the selling of multiple products, by having a basic understanding and a framework by which to describe a product's life cycle, it is also possible to describe a market by a curve that rises and falls with revenue growth over time. Utilising the market life cycle as the descriptive context, Moore (2004) showed how different types of innovations are relevant at different times in the market life cycle, as visualised in Figure 17.3.

From Figure 17.3, it can be seen how disruptive innovations by their very nature and definition have such characteristics that they lead to new markets being developed. Therefore, such innovations occur at the origin of a market's life cycle and deliver little or no revenue in the beginning. However, if the innovation involves

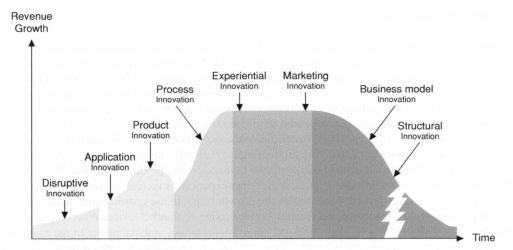

Figure 17.3   *Aligning innovation with the market development life cycle. (Based on Moore (2004). © 2004 Harvard Business School Publishing.)*

development of a product or service offering to fit a specific application, it still takes place in the growth stage of the market, but it is removed from the uncertainties of the earlier radical innovation, because it is building on a growing service offering or product portfolio. If the innovation is characterised as a line-extension of an existing product, then it is still defined as being in the growth stage of the market, but further along in terms of time and potential revenue generation than application innovation. These three types of innovation all relate to products or service offerings, but it is certainly also possible to apply innovation to the mature stages of the market development life cycle. In such instances, innovations in the way a company organises its processes can result in increased efficiency and revenue growth over time. New innovations in the way the customer experiences the product or service offering can help the company maintain and defend its market share, while still offering the same basic product or service to the customer. Similarly, innovations in the marketing of the product or service offering will maintain or extend the maturity stage of the market development life cycle. During the decline stage of the market life cycle, it is still possible to utilise new innovations to decrease the decline of the company's market share. If the innovation helps the company organise the way it conducts its business, for instance through distribution or customer service, business model innovation or, indeed, whole structural innovation, it can sustain or de-accelerate the market decline. Using the market development life cycle, it is then possible to address how different types of innovation are relevant for an organisation at different times of the product or market's life cycle. Examples of different types of innovations are provided in Example 17.1.

Example 17.1   Examples of innovations

Biosynthetic alternatives to biofuels: Researchers at the US Department of Energy's Joint BioEnergy Institute in California have used tools from synthetic biology to develop a precursor to bisabolane – a member of the terpene class of chemical compounds found in plants. Preliminary tests by the team showed that bisabolane's properties make it a promising biosynthetic alternative to Number 2 (D2) diesel fuel. The end-product would be a drop-in, renewable biofuel that could potentially replace petroleum-derived fuels or at least provide an alternative biofuel for vehicles that cannot readily be electrified

(Peralta-Yahya *et al.*, 2011). As such, this disruptive innovation would create an entirely new alternative to diesel fuel. Bioenergy production methods are described in Chapter 13, along with examples of how to assess the energy outcome.

Livestock feed supplement reduces phosphate pollution: Phytex LLC is a joint venture by Protein Scientific, an American nutraceutical biotechnology company, and JBS United, a major service-feed manufacturer in the United States. Phytex LLC offers enzymes for use in swine and poultry feeds. One phytase, called OptiPhos, enhances swine and poultry diets through increased digestibility of phytin-bound phosphorus and thereby creates substantially less phosphorus in the animal's waste. OptiPhos is based on a phytase gene developed at Cornell University by Professor Xingen Lei (Association of University Technology Managers, 2009). The genetic manipulation itself was, if not trivial, then commonly known practice and therefore could be described as being an application innovation, where the knowledge of genetic engineering was used in an application for phosphorus reduction in animal feed. The role of phosphorus in animal waste management is described in Chapter 2 and further in Chapter 15.

Opti-Chaux allows for homogeneous mixture of lime and sludge: This project is working on building an improved sludge processor for manure treatment that uses the principle of co-rotating twin-screw extrusion in order to optimise the dispersion of lime as a chemical amendment to the sludge. The machine allows for homogeneous mixing while also reducing the amount of lime used. The innovation is being developed by the French research organisation Irstea and is an example of a product innovation in Figure 17.3. Lime is used as a treatment alternative for slurry as described in Chapter 6 and shown in Table 6.1 (Irstea, 2012).

Odour reduction by use of microorganisms in feed: The international chemicals company DSM has developed an innovative direct-feed microbial called Microsource S that helps control odour emissions and ammonia (NH_3) levels from swine manure. This is important due to the risk NH_3 poses to animal and human health and the nuisance caused by odour emissions from intensified swine production described in Chapter 9. The innovation is an example of a product innovation, but also shows how changes in the regulatory environment as described in Chapter 3 can lead to new market opportunities and the development of new innovations (Frost & Sullivan, 2007).

Mobile phone app makes farm management software more accessible: The rapid development of smart phone applications available to farmers using farm management software can in itself be seen as product innovation. However, apps are also relevant examples of experiential innovation, because they represent new ways for the farmer to use and experience the farm management software. An app allows companies such as John Deere to give their customers a better slurry management capability as described in Chapter 12 without actually changing the underlying product (John Deere, 2012). This is an example of how existing technology can be modified and developed to increase the usability of the same underlying technology.

17.3   Identifying New Innovations

New innovations can come from numerous directions. They can evolve from existing technologies, be developed as answers to existing problems, appear by applying existing knowledge to new applications, or come about as a result of interaction between people or organisations from different domains. In particular, industry–science links, such as the interaction between the R&D department in an energy company and a university research group, or a machine producer and an engineering consultancy, are often thought to lead to radical innovations and are often the focal point of innovation literature, in which they are seen by some as the primary source of new innovations. Some organisations are turning to open innovation models of collaboration both to source new innovations and to commercialise innovations that fall outside their business scope. See Text Box – Advanced 17.1 for a description of open innovation. However, it is important to note that new innovations are most likely to come about in incremental steps as a result of a company's product development activities. By identifying problems or unmet needs in the market, a company can develop an understanding of its customers and can direct its research and development activities towards developing new product or service offerings to meet those needs or solve those problems.

Text Box – Advanced 17.1 Open innovation

As innovation has moved from the R&D department to the boardroom, the sources of new innovations have also crept up the to-do list of CEOs around the world. In the search for applicable models, the concept of open innovation has received increased attention during the last decade. Open innovation is a paradigm that assumes that firms can and should use external ideas as well as internal ideas, and internal and external paths to market, as they look to advance their innovations. Open innovation also assumes that the partners in the innovation process share risks and rewards. Open innovation as an academic framework originates to a large extent from the work of Chesbrough (2003). It is important to note, however, that the ideas and discussions about innovation relationships between an organisation and the stakeholders in its surrounding environment have been discussed at least since the 1960s. Open innovation is today used interchangeably with the terms user-driven innovation, distributed innovation, open source, shareware and, in some cases, as an argument against the patent system. However, none of these terms are relevant for open innovation as defined by Chesbrough (2003).

The open innovation model is used to describe how a company can expand its search strategy for new innovations beyond the boundaries of the organisation, and use the identified innovations to address existing and new markets and to find channels of diversification for projects already in the company's development portfolio. The model depicts a R&D process in a company moving from left to right (Figure 17.4). Each circle represents a research project that has the potential to turn into a marketable innovation. The dotted lines represent the organisational boundaries of the company. Research projects

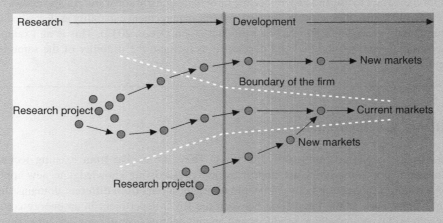

Figure 17.4 *The open innovation model. (Reprinted with permission from Chesbrough (2003). © 2003 Harvard Business School of Publishing.)*

can originate inside and outside the company. As the projects progress, they either turn into marketable innovations addressing the company's own existing or new markets, or are sold off to an outside company. The rapid development of the world economy and the pan-global and network-derived nature of many new products and service offerings means that the open innovation model is currently applicable to most companies' search strategy and, indeed, their entire innovation process. As such, Chesbrough's work should be seen as an academic model to describe the alternative to the closed product development processes of large American, Japanese and European corporations throughout the majority of the twentieth century.

Identification of new innovations involves developing search strategies. Search strategies are pre-defined and systemised methods used to organise a company's identification of new innovations (Table 17.1). They involve, firstly, a description of potential sources of new innovations. These sources can be customer groups, suppliers, academic partners, trade shows or technical papers. Secondly, a search strategy should describe who is responsible for conducting the search. This can be the marketing department, the R&D department, an outsourced consultancy firm or a mixed group of professionals from different departments in the organisation. Thirdly, the frequency by which searches for new innovations are conducted must be defined. This can be a continuous process or a periodic event such as yearly, biannual or even less frequent cycles. Finally, the outcome of the search process must be defined. The definition must be aligned with the decision-making process of the organisation that will utilise the outcome of the search – a process that can range from presenting

Table 17.1 *Examples of search strategies and how these search strategies can be operationalised in the organisation (adapted from Tidd and Bessant, 2009).*

Search strategy	Mode of operation
Sending out scouts	nominate individuals as idea hunters to track down new insights, tendencies or triggers in the surrounding environment
Scenario development	use scenario development to explore alternative possible futures and describe the products and services needed
Using the internet	detect new trends by searching and monitoring the internet through online communities, user-built content and virtual worlds
Working with active users	team with product and service users to see the ways in which they change and develop existing offerings
Deep diving	study what people actually do, rather than what they say they do
Probe and learn	use prototyping as a method to explore emergent phenomena and act as a boundary object to bring key stakeholders into the innovation process
Mobilise the mainstream	bring the mainstream user groups, trade union representatives and suppliers into the innovation process
Corporate venturing	create and deploy company-owned venture groups
Corporate entrepreneurship	stimulate and nurture the entrepreneurial talent inside the organisation
Use of brokers and bridges	search for innovations through external consultants and bridge into different industries and technical domains
Deliberate diversity	change employment criteria to hire a diverse workforce and create diverse teams
Idea generators	use creativity tools within the organisation

the outcome to decision makers without further action to using resources on developing and adapting the search results to the organisation before presenting it to decision markers.

17.4   Assessing the Potential of New Innovations

Having identified a number of new innovations, the challenge is to assess the relevance and commercial potential of these innovations. The crucial part of this process becomes mapping and addressing the potential risks of the new innovation in question, because risk and uncertainty correlate to potential value, as shown in Figure 17.1. The criteria and process by which new innovations are developed will depend on the resources and experiences available to the organisation, but the overall theme should be that in the innovation process the risk is decreased and value is built. Since risk is derived from uncertainty, one sure way to decrease the risk of a potential innovation is to increase knowledge about the innovation. Using this approach, assessment then becomes a matter of gathering and structuring information so that knowledge of the new innovation is built and presented in such a way that decision makers and stakeholders can understand the risks and make an informed decision.

According to Berneman and Denis (2002), even the most experienced innovation professional will admit that it is very difficult, if not near impossible, to pick the "winner" projects when they are initially assessed. However, by using the right processes and clearly defined assessment criteria, an organisation can avoid the "loser" projects that will drain resources without ever developing value. The boundaries for the assessment process are the time and resources available to the people responsible for the process. The decision process can be tied into the assessment process or, conversely, the assessment of new innovations is the first step in the product development decision-making process.

One example of this is the widely-adopted concept of a stage-gate model developed by Cooper (1988, 2008). In the stage-gate model, a series of decision-making gates are introduced in the product development process. The gates have pre-defined criteria on which yes/no decisions are made. Each project is developed for a given period of time with a given amount of resources in a stage until it is ready to be presented before a gate. If the project meets the criteria of the gate it moves to the next stage. If the project does not meet the criteria of the gate, it is either discontinued or repeats the previous stage. Each passage through a gate also means that the project is allocated new resources and deadlines and success criteria for the next gate are agreed. The search process for new innovations can very well be the first stage and the assessment criteria can be the basis on which yes/no decisions are made in the first gate (Cooper, 1988, 2008).

The overall themes in the assessment of new innovations are determined in the context of resources and experiences available to the organisation, but also in relation to the mission and purpose of the organisation. If the assessing entity is a research university, the goal will be to identify innovations that can be commercialised in collaboration with an industry partner. If the assessing entity is a small- or medium-sized entity (SME), the mission of the company will be closely tied to its core competencies and therefore assessment of new innovations will tie to whether the innovation complements, builds on or increases the core competencies of the SME. Conversely, if the assessing entity is a large corporation, the assessment criteria might contain elements to capture both product extension opportunities and opportunities to leverage the resources of a large corporation to develop completely new ventures or technology platforms. See Table 17.2.

In many instances, the assessment of new innovations contains categories of assessment criteria. (i) Some assessment of the level of innovation should be made (e.g. regarding the degree to which the new innovation differs from known products or services and the possibilities of obtaining intellectual property (IP) rights to protect the innovation). See Text Box – Basic 17.2 for more information about IP rights. (ii) An assessment should be made on the scalability of the new innovation in terms of production needs, geographical or durability limitations and required sales force, customer service and distribution network. (iii) Information for a proper assessment must be available on the market size in terms of number of, for instance, end-users,

Table 17.2   *Assessment criteria for new innovations at a Danish public university: the criteria are used to select which disclosed inventions will be included in the university's portfolio of new innovations that are developed into marketable innovations.*

Category	Hypothesis	Score (–2; 1; 2; 4)
Rights	Existing legal agreement or collaborations	
	Ownership situation	
	Publication need of researchers	
IP	Patentability	
	Enforcement of IP rights possible	
	Existence of dominating IP rights	
	Filing cost	
Context	Inventor group experience with innovation	
	Strategic backing of research group from management	
	Commercial product definable	
	Result of key opinion leader interview	
	Data strength	
	End-user need confirmed by industry contact	
	Resources needed to commercialize	
Commercial	Market description	
	Unique selling point of innovation	
	Value of IP in targeted industry	
	Our experience in targeted industry	
	Option to bundle with existing projects	
	Number and identity of potential licensees	
Total resource score		0.0
Total commercial score		0.0

pricing, market maturity, growth rates and market barriers, competitors, and sales channels. (iv) An assessment should also look at the human resources required in the organisation to drive the development of the new innovation. Examples are team and individual competencies, geographical location, leadership empowerment, and network description. (v) The assessment should address the potential value of the innovation to the organisation. This calculation can be made up purely of financial return on the invested resources, such as an internal rate of return percentage or NPV (Text Box – Basic 17.1), but it can also be contingent on access to new markets or customer groups, diversification of product portfolio, utilisation of un-used resources, partnering opportunities, public relations and image considerations or, indeed, a combination of some or all of the above. In all instances, it is vital to any organisation, regardless of size or mission, that the information gathered to address the assessment criteria is to the greatest extent possible based on information validated by third parties. Whether this comes from desktop research, key opinion leader interviews, focus groups or even more advanced ethnographic studies is a matter determined by the resources and time frame available.

Text Box – Basic 17.2   IP rights

IP rights refer to a set of intangible assets that are protected by law. IP rights give the owner exclusive right and privilege to a creative work. IP can be protected by both national laws and international treaties. IP can be traded as a tangible asset. The spread and use of IP rights vary based on geography and level of national economic development. The most widespread types of IP rights are patents and copyright. Utility models, industrial design rights, plant breeder's rights and trademarks are also used in many developed countries to protect IP. An important but non-protected asset within IP is confidential know-how.

Patents: Patent laws are, in effect, a contract between society and the inventor by which the owner is granted a monopoly by the state in exchange for the inventor sharing how the invention works. The monopoly granted by the state is also meant as a financial incentive for developing new inventions and promoting innovation. In general, products, methods or application can be patented, while business models or mental acts cannot. For a patent application to be granted, the invention must be universally novel, non-obvious to a person skilled in the art and industrially applicable. Patent terms vary, but generally last 20 years from the application date. Patents are prosecuted and issued nationally, although some regional systems exist to streamline the application process such as the Patent Cooperation Treaty and the European Patent Organisation. Patents are expensive to prosecute and uphold. It can take anywhere from a few years to a decade from application to issuance depending on the technical field and the workload in the patent office. Patents allow the owner of the rights to the patent to prevent others from using the claimed invention. Since parts of the invention could be patented by someone else, the patent holder cannot necessarily use the invention without rights to the dominating patents.

Copyright: Copyright is a set of exclusive rights granted through national laws to the author of an original work to use, copy, distribute, sell and adapt the work. It is required that the work be substantive and fixed in a medium. In general, copyright lasts 70 years from the death of the last author. Copyright is especially interesting in the context of innovation because it is used to protect the source code for computer software, thereby giving the owner of source code access to a competitive advantage. Note, however, that if a new source code is written to get a software program to perform the same function as a competing program, the novel software's source code is protected by its own copyright and the competing program could lose its competitive advantage on the market.

Confidential know-how: Although not protected by any kind of legislation, confidential know-how forms an important asset type of IP rights. This is because it has indefinite duration and no direct cost. Unlike patents or other types of IP rights, the only requirement is that it is kept confidential. Business assets such as trade secrets, business plans, production schedules, marketing lists are all considered confidential know-how, but they only provide the owner with a competitive advantage as long as they are kept just that – confidential.

17.5   Commercialisation of New Innovations

Regardless of the innovation typology, the goal of the innovation process must be that the identified new innovations are commercialised as a new product or service offered to the market or, alternatively, as an improvement or addition to the organisation's sustainable competitive advantage. The commercialisation process is typically described in steps or stages often referred to as value inflection points (Figure 17.5). These are key events at which point stakeholders can objectively assess that a new innovation has reached a new level of maturity, thereby decreasing the risk profile of the innovation.

In each step of the commercialisation process, the development of the new innovation requires additional capital. After each new point in the commercialisation process is reached, the value of the commercialisation project decreases slightly until the benefits of the additional development are reflected in the project's successful achievement of the next milestone. That is, following each investment there is a dip in value, because new goals have been set that the project is far from reaching, but in turn if those goals are met the value will grow significantly more, because the bar for success was set higher. Similarly, the capital

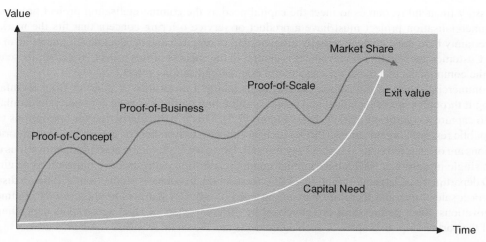

Figure 17.5 *Commercialisation process as it relates to development stage and capital need. (© University of Southern Denmark.)*

need increases as the commercialisation process advances. As the innovation is introduced on the market, an exponentially larger amount of capital is needed to support the continued commercialisation of the innovation.

In order to move forward in the commercialisation process and reach new value inflection points the organisation must execute a plan, which, for instance, can be formulated as a technology development plan, a business case or an innovation roadmap. Regardless of the terminology, some common denominators should be present. To this end, Lynn and Reilly (2002) studied some 700 teams and 50 detailed cases over a span of 10 years to identify the key success factors in the commercialisation of what they termed blockbuster products. The five factors identified as characteristic for successful commercialisation projects were:

- Commitment of senior management
- Clear and stable vision
- Improvisation
- Information exchange
- Collaboration under pressure.

As mentioned, commercialisation projects are organisation-specific, but overall address the following elements. (i) The project organisation should be clearly defined. Most organisations are not inherently built for the commercialisation of new innovations, so competencies must be sourced from various parts of the organisation and paired with external consultants where relevant. Arrangements must be made with line management to ensure that the team members have sufficient freedom and resources to contribute to the commercialisation project. (ii) The necessary internal communication and documentation must be in place. Regardless of whether the project team is hierarchical or functionally organised, internal communication is key, and the documentation of both the actual development and the commercialisation process must be in place in every step. (iii) The project leadership must have a clear vision of the commercialisation process and the necessary skills to both manage a cross-disciplinary team and navigate the larger organisation in terms of the requirements of senior management and the dependency on other departments. Equally, senior management must show actual commitment and support for the project while also providing the necessary control and guidance to satisfy their stakeholders as the fifth element. Senior management must also provide the necessary financial resources to meet the capital need as the commercialisation project progresses. (iv) The commercialisation project must have a product or service offering concept that fits the market need. Any uncertainty here must be dealt with by diligent market research and product development in an iterative process. Customer and supplier involvement is central to the success of this aspect. Together, these elements provide the components that make up a successful commercialisation project.

The commercialisation of innovations often means taking a novel idea or concept from its infancy and advancing it through steps of technical and business validation until it has been commercially available long enough to capture a significant market share (Figure 17.5). This is particularly true for innovations that originate in public research organisations (PROs) such as colleges, universities or government laboratories. Such transactions are often referred to as technology transfer. However, it should be noted that innovations originating from single inventors or SMEs have similar commercialisation paths, as have innovations originating in the R&D departments of larger corporations. PROs themselves do not have production facilities, a distribution network or a sales force and therefore they are in need of industry partners to successfully commercialise their innovations. Such partnerships are governed by different types of legal agreements depending on the nature of the partnership. Examples of such agreements are shown in Text Box – Advanced 17.2.

Text Box – Advanced 17.2   Legal Agreements used in commercialisation of innovations

New innovations are only very rarely developed exclusively from idea to fully commercialised products within one legal entity. Instead, the interconnected and globalised world of modern business demands collaboration and partnering to a very large extent. In order to successfully navigate these collaborations while also capturing and protecting value for the company, various types of legal agreements are put in place. The most common agreements with some defining features are described below.

Non-disclosure agreement (NDA; aslo know as a confidentiality agreement (CDA)): A NDA can be one-way or two-way. It should always contain a description of the subject matter being disclosed and whether it can be shown to third parties. A NDA is usually 3–5 years in length and should always contain a provision that the receiving party may not use the confidential material for purposes other than evaluation.

Material transfer agreement (MTA): These agreements cover the use of one party's physical materials by another party. A MTA should describe the receiver's intended use and the conditions under which the materials are provided. It should always very clearly define what the outcome or results of using the material can be used for, such as publication, further research or commercial activity. It is beneficial to define how new inventions are governed, if these come about as a result of the recipient's use.

Collaboration agreements: Collaboration agreements are general legal agreements that cover the interaction of two parties. Collaboration agreements can take the form of advanced consortia agreements for large infrastructure projects sponsored by international bodies with dozens of participants and they can be very limited memoranda of understanding for short interactions. Regardless of form, collaboration agreements define the purpose of the collaboration and what the parties are bringing into the collaboration in terms of manpower and financing. Collaboration agreements should also govern how the collaboration starts and ends and ideally what both parties can use the results of the collaboration for, both together and as separate parties.

Licence agreements: Licence agreements are agreements that govern the use of one party's assets by another party. Licence agreements are usually restricted to only cover the licensing of IP, but they can also cover the licensing rights to, for instance, production machinery or inventory. Licence agreements

> can be exclusive or non-exclusive or variations thereof, and can cover different geographical areas and different fields of use. Licence agreements usually define how the licensed asset can be used, by whom and at what cost. The governance and control of the licensed asset is also covered, as are terms for termination of the agreement and consequences thereof.

Since most innovations that come from PROs are at an early stage, they often need substantial development and testing until finished products are ready for the market (Thursby *et al.*, 2001). Such work can include years of laboratory development, documentation and production scale-up. Potential industry partners therefore need technical and market know-how, as well as sufficient resources to bring such innovations from early stage research results to marketable product or service offering. In cases where the technology is the foundation for a start-up company, the investors – often in the form of venture capitalists – need sufficient knowledge of both technology and business development, combined with an in-depth understanding of the local legislation pertaining to financing and company governance. These preconditions drastically narrow the number of companies for whom such innovations are relevant compared with the number of companies that could be interested were the innovation ready for market in the form of a finished product. Owing to the limited number of potential industry partners for innovations from PROs, the technical complexity of the innovations and the substantial investments needed to access them, it is essential that innovations are developed in collaboration with the right industry partners that possess the expertise and resources needed (Thursby and Thursby, 2003). This usually means that the technology transfer takes place when there is proof of concept. This is also due to the fact that PROs do not have commercial development as part of their core mission, so it is difficult for them to access the capital needed to develop the innovation through to proof of business or beyond (Figure 17.5). The benefit of this is that the industry partner who has the market insight and technical expertise is involved in the development and validation work early on, which increases the probability of success, because the right competences and resources are allocated to the commercialisation project early in the process.

17.6   Summary

Innovation is the development of new products and services offered to the market from which an organisation can derive value. It is also the way in which an organisation can leverage or increase its existing competitive advantage to gain value. As such, an innovation can take different forms and be applicable in different stages, but all types of innovation can be relevant for an organisation. New innovations arise in many different settings, and organisations must employ specific search strategies and subsequent assessment criteria to fully capture the value of the many potential ideas and new developments that exist in their environment. In order to gain value from new innovations, organisations must select and commercialise new innovations that play to their strengths and that are based on some uniform elements so that the value of the innovation is increased and documented over time as the project reaches value inflection points.

References

Association of University Technology Managers (2009) *Better Worlds Report*, AUTM, Deerfield, IL, pp. 27–31.
Berneman, L.P. and Denis, K.A. (2002) University licensing trends and intellectual capital, in *Licensing Best Practices: The LESI Guide to Strategic Issues and Contemporary Realities* (ed. R. Goldscheider), John Wiley & Sons, Inc., Hoboken, NJ, pp. 227–247.

Chesbrough, H. (2003) *Open Innovation: The New Imperative for Creating and Profiting from Technology*, Harvard Business Press, Boston, MA, p. xxv.

Cooper, R.G. (1988) The new product process: a decision guide for management. *J. Market. Manag.*, **3**, 238–255.

Cooper, R.G. (2008) Perspective: the Stage-Gate® idea-to-launch process – update, what's new, and nexgen systems. *J. Prod. Innov. Manag.*, **25**, 213–232.

Frost & Sullivan (2007) *Company Profiles: Animal Feed Additive Manufacturers and Suppliers*, Frost & Sullivan Research Services, San Antonio, TX, pp. 12-1–12-15.

Irstea (2012) The Opti-Chaux Project; http://www.irstea.fr/en/innovation/commercial-application-and-technology-transfer/technological-offers; retrieved 26 December 2012.

John Deere (2012) News release: John Deere introduces mobile farm manager application; https://www.deere.com/wps/dcom/en_US/corporate/our_company/news_and_media/press_releases/2012/agriculture/2012nov19_mobile_farm_manager.page; retrieved 26 December 2012.

Levitt, T. (1965) Exploit the product life cycle. *Harv. Bus. Rev.*, **43**, 81–92.

Lynn, G.S. and Reilly, R.R. (2002) *Blockbusters: The Five Keys to Developing Great New Products*, HarperBusiness, New York, pp. 33–174.

Moore, G.A. (2004) Innovation within established enterprises. *Harv. Bus. Rev.*, **82**, 86–92.

Peralta-Yahya, P.P., Ouellet, M., Chan, R., Mukhopadhyay, A., Keasling, J.D. and Lee, T.S. (2011) Identification and microbial production of a terpene-based advanced biofuel. *Nat. Commun.*, **2**, 483; doi: 10.1038/ncomms1494.

Porter, M.E. (1985) *Competitive Advantage*, Free Press, New York.

Porter, M.E. (1990) *The Competitive Advantage of Nations*, Macmillan, London.

Schumpeter, J.A. (1934) *The Theory of Economic Development*, Harvard University Press, Cambridge, MA.

Tidd, J. and Bessant, J. (2009) *Managing Innovation: Integrating Technological, Market and Organizational Change*, 4th edn, John Wiley & Sons, Inc., Hoboken, NJ.

Thursby, J.G. and Thursby, M.C. (2003) Industry/university licensing: characteristics, concerns and issues from the perspective of the buyer. *J. Technol. Transf.*, **28**, 207–213.

Thursby, J.G., Jensen, R. and Thursby, M.C. (2001) Objectives, characteristics and outcomes of university licensing: a survey of major, U.S. universities. *J. Technol. Transf.*, **26**, 59–72.

Index

Animal Manure Recycling: Treatment and Management, First Edition. Edited by Sven G. Sommer, Morten L. Christensen, Thomas Schmidt and Lars S. Jensen.
© 2013 John Wiley & Sons, Ltd. Published 2013 by John Wiley & Sons, Ltd.

Printed and bound by CPI Group (UK) Ltd, Croydon, CR0 4YY

16/04/2025

14658558-0005